MACHINE DESIGN

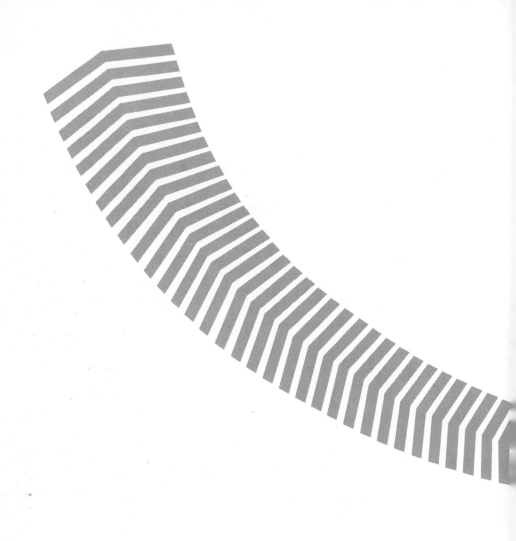

ADDISON–WESLEY PUBLISHING COMPANY

*READING, MASSACHUSETTS / MENLO PARK, CALIFORNIA /
LONDON / AMSTERDAM / DON MILLS, ONTARIO / SYDNEY*

MACHINE DESIGN

THIRD EDITION

ROBERT H. CREAMER

PROFESSOR EMERITUS
TEMPLE UNIVERSITY
COLLEGE OF ENGINEERING
AND ARCHITECTURE

**This book is in the Addison-Wesley Series
in Mechanical Engineering Technology.**

Library of Congress Cataloging in Publication Data

Creamer, Robert H.
    Machine design.

    Bibliography: p.
    Includes index.
    1. Machinery—Design. I. Title.
TJ230.C83 1983   621.8'15   83-2567
ISBN 0-201-11280-9

Reprinted with corrections, June 1984

ISBN   0-201-11280-9
HIJ-MA-93210

PREFACE

This book is a practical text covering the design of basic machine components. The level is appropriate for courses in machine design in engineering technology and industrial technology at the associate or baccalaureate degree level. The practical approach is useful to practicing designers and drafters as well as to nonmechanical majors in elementary engineering courses.

Problems are provided at the end of each chapter to give the student experience in designing frequently used components, ranging from heavy-duty to miniaturized sizes. A good background in engineering materials, shop processes, mechanics, and strength of materials is helpful in solving the problems. Extraneous information is given in many of the problems, as technical personnel should be trained to seek out only the needed information from an assortment of available data.

This text differs from the conventional text in that emphasis is placed on the total design and not merely on problem solving. Force analysis is emphasized as well as areas of judgment and good practice.

Selected references are included in Appendix C to assist the student in finding supplementary information in a library. Because textbooks and handbooks cannot possibly keep up with today's expanding mechanical design field, the student must read current technical magazines to learn of new developments.

For the most part, derivations have not been included—only the final equation in usable form. Information on the selection of electric motor drives and fluid power drives is included for additional background. Standards and governmental regulations related to design are covered at appropriate locations throughout the text. Topics on professional practice have been added.

The number of problems using metric units has been substantially increased. A balance, however, has been maintained between traditional U.S. and metric units because certain industries have not switched to the metric system. The total number of problems, both traditional U.S. and metric, has been greatly expanded.

The sections on V-belts and timing belts have been updated and enlarged. Calculations for Belleville springs have been included, material on lubrication and viscosity has been added, and various topics throughout the text have been expanded.

Cams are an important class of mechanical components and, as such, are presented in this text. Most of the curves and layouts are traditionally covered in courses in kinematics; therefore, much of this material can be omitted from machine design courses, if desired.

As the digital computer is used extensively as a problem-solving tool, a few programs are included in Appendix L in the hope that they will stimulate student interest.

The author wishes to express appreciation to the many manufacturing firms, associations, and societies who furnished illustrations and helpful design data. Special thanks are due to the educators who reviewed the manuscript and to the Addison-Wesley staff for their many helpful suggestions in developing this third edition.

Tuscaloosa, Alabama   **R.H.C.**
September, 1983

CONTENTS

1
INTRODUCTION

2
REVIEW OF MECHANICS AND STRENGTH OF MATERIALS

3
FRICTION AND LUBRICATION

4
BEARINGS

5
SHAFT DESIGN AND SEALS

6
FASTENERS, COUPLINGS, KEYS, RETAINING RINGS, WELDING AND WELD DESIGN

7
BELTING

8

CHAIN DRIVES, HOISTS AND CONVEYORS, ROPES

9

BRAKES

10

CLUTCHES

11

POWER SCREWS

12

GEARS

13
CAMS

14
SPRING DESIGN

15
FLYWHEELS

16
MISCELLANEOUS MACHINE ELEMENTS AND SPECIAL TOPICS

17
POWER UNITS

18
PROFESSIONAL PRACTICE

APPENDIXES

ANSWERS TO SELECTED PROBLEMS

INDEX

REGISTERED TRADEMARKS

The following registered trademarks, the property of the firms listed below, are used throughout the book.

AAI Corp.	Lo-Hed
AMETEK/Hunter Spring Div.	Neg'ator
The Bead Chain Manufacturing Co.	Bead Chain
The Bendix Corp.	Duo-Servo
Bethlehem Steel Co.	ANCO
Browning Manufacturing, Div. Emerson Electric Co.	Griplink, Griptwist
Cone Drive, A Unit of EX-CELL-O Corp.	Cone-Drive
Continental Screw Co., A Div. of Amtel Inc.	Taptite
Dow Chemical Co.	Dowmetal
E. I. duPont de Nemours and Co., Inc.	Teflon, Kevlar
Durametallic Corp.	Dura-Seal
Dzus Fastener Co., Inc.	Dzus
Eaton Corp.	Airflex
Eaton Corp., Engineered Fasteners Div.	Speed Clips, Speed Nuts
Federal Products Corp.	Surf-Indicator
Garlock, Inc.	Mechanipak
The Gates Rubber Co.	PowerBand
The B. F. Goodrich Co.	Rivnut
Groov-Pin Corp.	Groov-Pin
Heim, Incom International Corp.	Unibal

Illinois Tool Works, Inc., Shakeproof Division	Keps, Teks
International Nickel Co.	Monel
Lord Corp.	Flex-Bolt
McGill Manufacturing Co., Inc.	Cagerol, Camrol
MacLean-Fogg Lock Nut Co.	Whiz-lock
Nylok Fastener Corp.	Nylok
Pennwalt Corp.	S. S. White
PT Components, Inc.	P.I.V.
SPS Technologies	Flexloc, Unbrako
Thomson Industries, Inc.	Ball Bushings, Roundway
TRW, Fasteners Div. of	Palnut, Pushnut
Waldes Kohinoor, Inc.	Truarc
T. B. Wood's Sons Co.	Ultra-V, HTD, Sure-Grip
Zero-Max Co.	Zero-Max

INTRODUCTION

1

The complete design of a machine is a complex process. The designer must have a good background in such fields as statics, dynamics, and strength of materials, and in addition, must be familiar with the fabricating materials and processes. The designer must be able to assemble all the relevant facts and make calculations, sketches, and drawings to convey manufacturing information to the shop. Laboratory tests, models, and prototypes help considerably in machine design. A good designer reads technical periodicals constantly to keep abreast of new developments.

 General Considerations

Following is a list of general factors that a designer often considers before initiating a new design. Since these are general in scope, not all will apply to any particular design. Some of the questions should be thought-provoking.

1. *Cost of manufacturing.* Will the selling price be competitive? Are there cheaper ways of manufacturing the machine? Could other materials be used? Are any special tools, dies, jigs, or fixtures needed? Can it be easily inspected? Can the shop produce it? Is heat treatment necessary? Can parts be easily welded?

2. *Cost of operation.* Are power requirements too large? Will polyphase current be needed? What type of fuel will be used?

3. *Cost of maintenance.* Are all parts easily accessible? Are access panels needed? Can it be easily lubricated? Can common tools be used? Can replacement parts be obtained "off the shelf"?

4. *Safety features.* Was a suitable factor of safety used? Does the safety factor meet existing codes? Are fuses, guards, or safety valves used? Are shear pins needed? Is there any radiation hazard? Are there any overlooked "stress raisers"? Are there dangerous fumes?

5. *Packaging and transportation.* Can the machine be readily packaged for shipping without breakage? Is its size compatible to parcel-post regulations, freight-car dimensions, or trailer-truck size? Are shipping bolts necessary? Can it be pallet loaded? Is its center of gravity in a desirable location?

6. *Lubrication.* Does the system need periodic checking? Is it automatic? Is it a sealed system?

7. *Materials.* Are the chemical, physical, and mechanical properties of the machine's materials suited to its use? Is corrosion a factor? Will the materials withstand impact? Is thermal or electrical conductivity

important? Will high or low temperatures present any problems? Will design stress keep parts reasonable in size?

8. *Strength of components.* Have its dimensions been carefully calculated?
9. *Kinematics.* Does it provide the necessary motion for moving parts? Are rotational speeds reasonable? Could linkages replace cams? Which will best suit the design—belts, chains, or gears? Is intermittent motion needed?
10. *Styling.** Is the color appealing? Is the shape desirable? Is the machine well proportioned?
11. *Drawings.* Are standardized parts used? Can the drawings be simplified? Are the tolerances realistic? Is the surface finish overspecified? Must the design conform to any standards?
12. *Human engineering.* Has the operator of the equipment been considered? Are the controls conveniently located to avoid operator fatigue? Are knobs, grab bars, hand wheels, levers, and dial calibrations the right size for the average operator? Are the calibrations on the dials easily read? Are the controls easy to operate?

It should be kept in mind that no two products are alike. A day's production of miniature precision bearings might be carried by one person; yet a year's production of heavy machinery might have to be dismantled just to fit on freight cars for shipment.

 ## Types of Design

Rational design might be considered as purely mathematical, based on the laws of mechanics. For example, we can compute the size of a rivet if we know the material and the forces acting on it. *Empirical* design can be considered as conforming to existing practices with no real justification except past performance. For example, we could compute empirically the diameter of a setscrew from the size of the shaft that it will engage. Experience enters into this type of calculation mainly because of the human element in tightening this type of fastener. *Industrial* design is the term generally applied to designing for appearance as well as function. The use of streamlining, color combinations, and finishes quite often popularizes one product over another. These are important selling features. In many instances, all three of the above types of design are used.

---

*Two authoritative sources of information on styling are Van Doran, *Industrial Design*, McGraw-Hill, 1954; and Dreyfuss, *Designing for People*, Viking Press, 1974.

**Materials**

Numerous materials are available to today's designers. The function of the product, its appearance, the cost of the material, and the cost of fabrication are important in making a selection. Properties of some of the ferrous, nonferrous, and plastic materials are listed in Appendix A.

Physical and Mechanical Properties

The following definitions review some of the physical and mechanical properties that may enter into a design problem.

Stress is the internal resistance of an elastic material caused by external loading that tends to change the shape of the part. Unit stress is simply the applied load divided by the cross-sectional area subjected to the load. Stresses are classified as *tension*, *compression*, *flexure* or *bending* (combination of tension and compression), and *shear*. A body is often subjected to combinations of these primary stresses. When sufficient load is applied to a body, deformation takes place. Unit deformation can be defined as *strain*. If stress values for a test specimen are plotted against strain values, a stress-strain curve similar to Fig. 1-1 results. The stress corresponding to the highest value on the curve is known as the *ultimate stress*. The *yield stress* is located where the curve flattens out temporarily. At this point, the material is permanently deformed. It is not desirable to design a part near the point

Figure 1-1
Stress-strain diagram for mild steel.

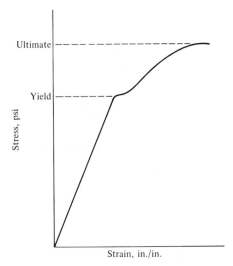

Figure 1-2
Super "L" Universal testing machine with stress-strain recorder.
Load values are given in kilograms, newtons, and pounds.
(Courtesy of Tinius Olsen Testing Machine Co.)

of failure—that is, near the stress where permanent deformation would occur.

Figure 1-2 shows a Universal testing machine equipped with a recorder for automatically plotting a stress-strain diagram. This type of machine is used for tension, compression, flexure, and shear testing.

Elasticity is the ability of a material to regain its original dimensions after a load is removed; this is contingent on keeping the applied load small enough that the elastic limit is not exceeded. *Plasticity* is the ability of a material to retain a shape after loading produces a change in form. *Stiffness* is the ability to withstand high stress without being greatly deformed. *Ductility* is the ability of a material to undergo plastic deformation when a tensile load is applied.

If sheet-metal parts are to be properly formed using a brake or drawbench, the material should be ductile. Malleability is similar to ductility, except that a compressive load is used to flatten and enlarge the part being fabricated. *Brittleness* is the opposite of ductility and malleability. A brittle material has no well-defined yield point. If a brittle test specimen is loaded with a tensile or compressive force, there is no stress value where the strain will increase without additional load; failure will occur without any appreciable deformation.

Toughness or *resilience* is the ability to withstand shock loading. The toughness of two materials can be compared by measuring (with a planimeter) the area under the stress-strain curve for each. The one with the larger area is the tougher material, regardless of ultimate strength. *Hardness* is the ability to resist penetration. Several types of hardness testers are in common use. Figure 1-3 shows a Brinell hardness tester. In this tester, the depth of penetration of a hardened steel ball is measured and converted into a Brinell hardness number. Figure 1-4 illustrates a Rockwell tester; this test also depends on the depth of penetration with a given load. Penetrators can be of the steel ball or spheroconical variety. Loading varies according to the type of penetrator used. A scleroscope is shown in Fig. 1-5. This device indicates surface hardness by measuring the height to which a diamond-tipped hammer rebounds when dropped from a given height to the metal

Figure 1-3
Brinell hardness tester.
(Courtesy of Tinius Olsen Testing
Machine Co., Inc.)

Figure 1-4
Rockwell hardness tester with digital readout.
(Courtesy of Babcock International Co., Industrial Products Group.)

Figure 1-5
Model D dial-recording scleroscope on clamping stand.
(Courtesy of Shore Instrument and Manufacturing Co., Inc.)

being tested. Tables can be found in many metallurgical texts and handbooks for use in converting hardness values from one scale to another.

Creep is the deformation of a material under steady load at elevated temperatures. This must be carefully considered when designing equipment for use at high temperatures. Some materials (for instance, lead) tend to creep at room temperatures. *Fatigue* is the failure of a part due to repeated loading or reversal of normal loading. Reversal of stresses often causes fatigue failure.

The *endurance limit* is important in any design where repeated loading and unloading occurs. When a part is subjected to this type of loading, failure occurs below the ultimate stress value. The number of cycles of stress reversal that causes failure depends on the range of imposed stresses. The endurance limit can be defined as the stress value where an infinite number of stress reversals will not cause failure. A factor of safety must be applied to the endurance stress limit of parts that are subjected to repeated loading and unloading. This limit is particularly important for rotating shafts that are subjected to bending loads. The endurance limit of a steel is often considered to be 50% of the ultimate strength. For some materials, the value is much less than this. If the stress reversal is only partial, the endurance limit may be closer to the ultimate value. It is advisable to check the *S-N* (stress versus number of cycles) diagram for a material if repeated loading or stress reversal is likely to occur.

Thermal conductivity is the ability of a material to transmit heat. Insulating materials are poor conductors of heat. One can feel quite a difference between the handle temperature of a silver spoon and that of a stainless steel spoon when each is inserted in a cup of hot coffee. The silver spoon conducts heat better than the stainless steel one. *Electrical conductivity* is the ability to conduct electricity; silver and copper are outstanding conductors of electricity and hence are commonly used for that purpose. Because of cost, however, silver is usually used only in delicate instruments. All materials expand when subjected to elevated temperatures and contract when exposed to lower temperatures. The *coefficient of expansion* is used to compute the change in linear dimensions caused by a change in temperature. Thermal stresses can also be induced and must be checked frequently. For instance, railroad tracks must be laid to provide for expansion because of the length of continuous rail and the extreme temperature variations. Long lengths of exposed piping are usually provided with expansion loops. If extreme temperatures are likely to be encountered, the melting points of materials may have to be considered.

Corrosion resistance is sometimes an important property. Oxidation problems can often be overcome by using proper protective coatings;

however, for marine applications and certain chemical plant operations, no material can be used that is not completely corrosion resistant.

Selection of Materials*

One of the first steps in designing a product is to select the material from which each part is to be made. A careful evaluation of the properties of a material must be made prior to any calculations. Appendix A lists a few properties of some of the more common engineering materials.

Steels. Various types of steel are available in such forms as rods, strips, bars, sheets, tubes, and structural shapes. One should consult manufacturers' catalogs to obtain sizes.

Cast iron. Cast iron has always been a popular engineering material. Much of this popularity lies in the fact that intricate shapes can be cast without expensive machining operations. In general, cast iron contains 2–4% carbon. The carbon is in the free state in gray cast iron and is chemically combined in white cast iron. Gray iron is easy to machine. White cast iron is difficult to machine—usually, it has to be ground.

Malleable cast iron is produced by an annealing process involving graphitization of white cast iron. Malleable iron is reasonably easy to machine and has fairly high tensile strength. A material known as semisteel is made by adding wrought iron or mild steel scrap to cast iron, thus reducing the carbon content. Semisteel is cast in the same manner as other castings. Large parts, particularly gears, can be produced with this material. Ductile cast iron is also a popular engineering material. It contains magnesium and combines the advantages of the casting process with properties that approach those of steel. Several alloy cast irons are available commercially. Many of these contain a substantial amount of nickel; some contain other alloying elements, such as copper, chromium, and silicon. In general, these alloy cast irons are corrosion resistant. Cast steel often takes the place of cast iron where higher strength is required.

Wrought iron is produced by lowering the carbon content then rolling the iron. Wrought iron is extremely ductile. It is frequently used for ornamental iron work, pipes, and some rivets. This material should not be confused with wrought steel, which is the form of steel produced from

*The student should refer to such handbooks as Brady, *Materials Handbook*, McGraw-Hill, 1971, and Mantell, *Engineering Materials Handbook*, McGraw-Hill, 1958. A helpful periodical is *Materials Engineering*, Reinhold Publishing Co.

rolling, drawing, or hammering. Plain carbon or alloy steel can be used in its production.

Bearing metals. Brass, bronze, and babbitt are three materials commonly used in bearings, as well as in other places where corrosion resistance is important. Brass is an alloy of copper and zinc; bronze is an alloy of copper and tin. Bronze is higher in cost than brass, but it is more desirable as a bearing material. Several compositions are available for these materials. Both brass and bronze can be cast, and both are easy to machine. The widely used bearing material known as babbitt is an alloy of copper, tin, and antimony. Various compositions are available that include small percentages of other elements, such as lead, arsenic, and bismuth. Babbitt is useful as a bearing material because its coefficient of friction can be made very low. None of the babbitts or white-metal alloys can withstand extreme pressures; bronze is more desirable under heavy loading conditions. All these bearing metals possess relatively good wearing characteristics, even though they are soft metals.

Aluminum. Aluminum is a corrosion-resistant material that is malleable and ductile. As it is light in weight, it is popular in certain aircraft and marine applications. It is used to provide a corrosion-resistant surface to duralumin (Alclad) sheets. Aluminum alloys are frequently used in machine parts. Aluminum-bronze alloy (aluminum, copper, iron, tin) is corrosion resistant and fairly high in strength and hardness, and can be cast.

Dowmetal® alloys. Several magnesium alloys are commercially available. These are lightweight and high in strength and can be used in sand-casting and die-casting processes. Dowmetal alloys are also used for forgings and are available in extruded shapes. All of these alloys contain a high percentage of magnesium. Sometimes heat treatment is used to increase the tensile strength and toughness of Dowmetal alloys.

Low-melting-temperature alloys. Several types of bismuth alloys have been developed by the Cerro Corporation. They have two unique characteristics. One outstanding feature is a low melting temperature; some of these melt at 117 °F, 136 °F, and 158 °F, which is far lower than the boiling point of water. Low melting temperatures make handling easier. Careful control of the other alloying elements can produce an alloy that expands upon solidification, thus reproducing intricate mold detail. Many industrial applications are possible besides the early use of these alloys as fusible elements in automatic fire alarms and sprinkler systems.

Plastics. Plastics may be divided into two chief types: thermoplastic and thermosetting. The *thermoplastic* type can be softened by heating and hardened again by cooling. *Thermosetting* plastics undergo a chemical reaction when acted upon by heat. Plastics are important engineering materials that have many desirable properties, although they are not as strong as metallic materials. Plastics are light in weight and low in cost; they can be extruded in many cases, and they provide quiet-operating mechanical parts that require no lubrication. Widely used plastics include polyethylene and acetal, acrylic, and nylon resins. Nylon has replaced metal gears, cams, bushings, and so on in some of the lighter applications. Kevlar® aramid is a high strength material with good high-temperature properties and resistance to stretch. In fiber form, it is used as a tensile member in toothed belt drives. In pulp form, it can be used as a brake material to replace asbestos —a material that has potential health and environmental hazards. Plastics producers provide a great deal of design application data; a designer should avail himself of this information when specifying plastic parts.

Copper. Copper is an important engineering material. As mentioned previously, it is valuable as an alloying element to produce various types of brasses and bronzes. Copper ranks high in electrical and thermal conductivity. It is ductile and malleable; it is also very easy to machine. Its resistance to corrosion and its color make it desirable for quite a few applications. It is readily available in sheet, wire, and rod form, and is often used in electrical wires, conductors, motors, and controls.

Beryllium copper. This is an alloy made with small amounts of beryllium and nickel. It is high in tensile and fatigue strength and is corrosion resistant, nonmagnetic, and nonsparking. Like copper, it has high thermal and electrical conductivity. It is difficult to machine and is more expensive than other copper alloys. It is frequently used for surgical instruments, springs, and dies.

Monel®. Monel is an alloy of copper and nickel. It possesses high strength, corrosion resistance, and toughness. It is often used in the marine and chemical fields for such applications as fluid meters, propellers, pumps, tanks, and heaters.

Titanium. This is an expensive, lightweight, and high-strength material developed primarily for use in aircraft. Fasteners made of titanium are available. Many parts subjected to extreme temperatures are made of titanium. It possesses high ultimate and yield strengths.

Fibers. Natural fibers from plant, animal, or mineral sources have been used for centuries in such products as rope, hoses, and containers. Synthetic fibers have been developed to replace many of the natural fibers. A fiber is usually defined as any continuous material in which the length is at least 200 times the diameter (or width, in the case of rectangular cross-sections).

Metal fibers include such materials as iron, nickel, copper, steel, and beryllium. Fibers are often combined in a matrix, such as epoxy resin, to combine desirable characteristics such as high strength and electrical conductivity or high strength and ductility. Combinations depend upon the specific applications.

Glass fibers are popular in such applications as insulation (heat, sound, electrical), fiber optics (medical and communications technology), and tires (as a reinforcement for the rubber).

Fluorocarbon fibers, which have a high resistance to heat and chemicals, are often woven or braided into materials for packings, bearings, gasket tape, and conveyor belts. Other miscellaneous materials incorporated in reinforcing fibers include alumina, graphite, boron carbide, and silicon carbide. When using or combining fibers, one should carefully check the properties of a given material—particularly its tensile strength, its modulus of elasticity, and its density—as well as any properties that relate directly to the application, such as its thermal and electrical conductivity.

 Degrees of Freedom

When a machine is designed, it is essential that the designer clearly understand the various possible movements of the object or assembly. Too often, we think only in terms of positive and negative values in the x- and y-planes. The z-plane, however, also exists. In the manufacturing field, work-holding devices cannot be designed unless all the possible degrees of freedom are explored. Figure 1-6 shows the various possibilities. Note that there are 12 degrees of freedom. These movements are important from the kinematic standpoint. Certain mechanisms must be designed to provide or restrain some of these movements. In the marine and aerospace fields the terms roll, pitch, and yaw are in frequent use, as well as the common thrust and drag. In Fig. 1-6, all axes pass through the center of gravity. *Roll* is a movement about the longitudinal centerline. *Pitch* is the alternate raising and lowering of the bow and stern of a ship with reference to a horizontal centerline through the center of gravity. *Yaw* is a clockwise or counterclockwise rotation about the vertical centerline through the center of gravity.

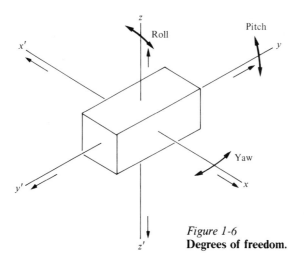

Figure 1-6
Degrees of freedom.

Some of these degrees of freedom play a prominent part in the design and application of numerical-control equipment and robots.

 **Calculations**

Mathematical calculations for components should be carefully made and preserved. Most companies provide a standard format for this information, many containing light, nonreproducible grid lines for graphs and sketches. Neatness and completeness are necessary, since others may wish to check the figures. *Each calculation should be dated!* Enough information should be given on the sheet that the designer (or anyone else) can understand what was meant, even after an extended period of time. Sketches are often useful in clarifying problems.

The following illustrates a typical calculation. Note that the basic arrangement is: (1) name the calculation; (2) give data and assumptions; (3) show formula; (4) substitute numbers for letters in the exact order; (5) show mathematical answer; and (6) state the conclusion.

1. Diameter of shaft (pure torsion)
2. Data: Torque = 480 in-lb
   Assume SAE 1045 steel with S_s = 72,000 psi
   Factor of safety = 12; working S_s = 6000 psi

3. $D = \sqrt[3]{16T/\pi S_s}$

4. $D = \sqrt[3]{(16 \times 480)/(\pi \times 6000)}$

5. $D = 0.741$ in.

6. Use $\frac{3}{4}$-in. diameter.

This decision to use a $\frac{3}{4}$-in.-diameter shaft may be modified as the design progresses. For example, this diameter might be increased to accommodate a certain desirable bearing. If several calculations are placed on the same sheet, a horizontal line should separate each one. Sheets should be numbered "Sheet__of__" so that the reader always knows whether the set is complete.

Such mathematical operations as addition, subtraction, multiplication, division, powers, and trigonometric functions can be performed with great accuracy using sophisticated desk-top and hand-held electronic calculators. However, one must exercise care in rounding off answers to appropriate values. Answers should be recorded to the same degree of accuracy as the original data. The following example illustrates solving a simple problem with an electronic calculator.

◆ *Example* ────────────────────────────────

Find the induced stress when a $\frac{7}{8}$-in. diameter rod is subjected to simple shear with an applied load of 7500 lb.

Solution.

$$S_s = \frac{P}{A} = \frac{4P}{\pi D^2} = \frac{4(7500)}{\pi(0.875)^2} = 12{,}500 \text{ psi.}$$

If the π constant (3.1415926) is used, the eight-digit display reads 12,472.549. The accuracy for the shaft diameter is three significant figures; therefore, the answer should be rounded to a value of 12,500 psi.

───

It should be noted that certain calculator models display results electronically, whereas others provide a tape. The electronic display, of course, has the disadvantage that results cannot be checked without repeating the entire process and checking the final result. Computers are also valuable tools for solving problems. Computer operation time is expensive; therefore computers should be used only for complicated, repetitive, or data-heavy problems. The so-called personal or home computer is actually a microcomputer. Certain models, with peripheral equipment, provide a means for solving typical mechanical design problems.

Laboratory Support; Preproduction Models; Mock-ups

A new product is expensive to market; competition is keen. For this reason, design support is found outside the engineering drafting room. Laboratories furnish much of the information needed to establish basic concepts; they can also be used to gauge how a product will perform in the field. Endurance testing of components and assemblies through thousands (or possibly millions) of cycles can be performed in a laboratory to simulate a customer's treatment of a product. For example, a refrigerator door is opened and slammed many times each day throughout many years of operation. Laboratory devices can be used to open and close such a door every few seconds, 24 hours a day. The results of this testing can be studied and perhaps bring new ideas to the designer's mind. Experimental shops under the supervision of the engineering department are often used to make up parts for testing or for a preproduction model. Such a model can then be studied by the interested departments within the plant. In this way, manufacturing difficulties can be found and corrected (and certain parts redesigned) before any drawings reach the manufacturing areas. Full-scale mock-ups are also helpful, particularly when piping and electrical layouts are complicated and space limitations are critical.

 ## Design Stress; Factor of Safety

It would be foolish to design any part at the ultimate stress or point of failure. It would also be foolish to design a part near the yield stress or the point where permanent deformation can take place. A factor of safety is therefore divided into the ultimate stress or the yield stress to find the design stress to be used in calculations. If cyclic loading is required, the factor is applied to the endurance limit.

The factor is in reality a margin of safety. Its value can range from 1.5 to 10, depending on such items as the type of service intended, human safety if failure should occur, the cost of repairs, unreliable information on the properties of the specific materials being used, or inferior quality of materials.

For static loads and ductile materials, the factor should be divided into the yield stress. For static loads and brittle materials (such as white cast iron) that have no well-defined yield point, the factor should be divided into the ultimate stress and the necessary stress-concentration factors should be applied, as is shown in Chapter 5 for changes in shaft diameters. In the case of cyclic loading (where there is potential fatigue failure), the factor of safety should be divided into the endurance limit. The endurance limit should include corrections for such specific characteristics as size, surface finish, stress concentration, and reliability.

For severe shock loading, a separate factor should be used. This must be based on such items as the energy of the impact and the energy absorption characteristics of the part being stressed.

The selection of any factor of safety is an important decision. It is generally decided upon by the senior members in any organization. If its value is too high, oversized parts will result; if it is too low, the ever-present risk of failure plagues the manufacturer. In the structural areas, prevailing codes specify the working stress to be used in calculations. There are also codes that provide design-stress values, such as the ASME (American Society of Mechanical Engineers) Code on Transmission Shafting. Such codes must be used where applicable.

◆ *Example*
AISI steel has an ultimate strength of 80,000 psi and a yield strength of 50,000 psi. This material is used in a machine part where the load is static. If failure should occur, there would be no danger to humans and the part could be easily replaced. Find the design stress.

Solution.

$$\text{Design stress} = \frac{50,000}{5} = 10,000 \text{ psi.}$$

◆ *Example*
White cast iron is used for a part that is subjected to a steady compressive force with no reversal. Replacing the part is cumbersome, but human safety need not be considered since there is adequate mechanical guarding of the area. Find the design stress if the ultimate compressive stress is 667 MN/m².

Solution.

$$\text{Design stress} = \frac{667}{8} = 83.3 \text{ MN/m}^2.$$

It should be noted that the number 8 was selected as a matter of judgment.

 **Metrication**

The metric system of measurement is widespread, and the term *SI units*—derived from *Système International d'Unités*—can be found in various engineering standards produced by United States manufacturers. Firms with worldwide manufacturing facilities and those actively engaged in changing over to the metric system find that giving measurements in both inches and millimeters is advantageous. The most outstanding feature of the metric system is the fact that it has a decimal base; thus, calculations are simplified.

SI units are classified in three categories: base units, derived units, and supplementary units. Table 1-1 shows the seven base units; these are well-defined units that are considered dimensionally independent. It is interesting to note that the unit for mass is the only one that contains a prefix; a *kilogram* represents 1 000 grams. Another characteristic of the International System is that commas are not used to indicate thousands; instead, a space is left open between groups of three digits on each side of the decimal point. For example, 12,350,421 would be written as 12 350 421 when using metric units.

The term *meter* is frequently spelled *metre*—the French spelling. Advocates of the French form point out that the metric unit for length measurement might otherwise be confused with a measuring instrument. Although pronounced differently, the term *micrometer* can represent a precision measuring tool or a linear measurement. The measure of volume known as a *litre* or *liter* can also be spelled in two different manners. This unit is currently in use with the International System and may eventually be replaced with *decimetre*. In this text, *metre* and *litre* will be spelled as *meter*

Table 1-1
SI Base Units

Quantity	Name	Symbol
length	meter	m
mass	kilogram	kg
time	second	s
electric current	ampere	A
thermodynamic temperature*	kelvin	K
luminous intensity	candela	cd
amount of substance	mole	mol

*°K = 273.15 + °C, where C = Celsius, formerly centigrade.

and *liter*. Practice varies in regard to the use of the terms *metrification* or *metrication*. In this text, the latter will be used.

The second class of SI units is known as *derived units*. These units can be formed by combining base units in accordance with algebraic relationships to define typical scientific and engineering quantities such as area, volume, density, velocity, and others. Table 1-2 shows a partial listing of derived SI units that are typically used in mechanical design work. Supplementary SI units deal with angles only and are listed in Table 1-3. Prefixes with SI units indicate magnitude. Table 1-4 lists the multipliers for various prefixes used with metric units. Whenever possible, avoid the prefixes hecto, deka, deci, and centi. The remaining prefixes represent steps of 1 000.

The International Organization for Standardization (ISO) makes the following recommendations for using prefixes with SI units:

1. Prefix symbols should be printed with upright roman type without spacing between the prefix and the unit symbol. For example, *milligram* is correct; *milli gram* is incorrect.
2. An exponent affixed to a symbol containing a prefix indicates that the multiple or submultiple of the unit is raised to the power expressed by

Table 1-2
SI Derived Units (Partial Listing)

Quantity	Name	Symbol
area	square meter	m^2
volume	cubic meter	m^3
frequency	hertz	Hz
mass density, density	kilogram per cubic meter	kg/m^3
speed, velocity	meter per second	m/s
angular velocity	radian per second	rad/s
acceleration	meter per second squared	m/s^2
angular acceleration	radian per second squared	rad/s^2
force	newton	N or $kg \cdot m/s^2$
pressure	pascal	Pa
kinematic viscosity	square meter per second	m^2/s
dynamic viscosity	newton-second per square meter	$N \cdot s/m^2$
work, energy, quantity of heat	joule	J or $N \cdot m$
power	watt	W or J/s
thermal conductivity	watt per meter kelvin	$W/(m \cdot K)$
stress (mechanical)*	newton per square meter	N/m^2

*Certain organizations prefer to use the term *pascal* (Pa) for mechanical stress. A pascal is equal to one N/m^2.

Table 1-3
SI Supplementary Units

Quantity	Name	Symbol
plane angle	radian	rad
solid angle	steradian	sr

the exponent. The following examples illustrate this recommendation:

$$1 \text{ mm}^3 = 10^{-9} \text{ m}^3 \quad \text{and} \quad 1 \text{ mm}^{-1} = 10^3 \text{ m}^{-1}.$$

3. Compound prefixes (formed by the juxtaposition of two or more SI prefixes) should not be used. For example, 1 nm is acceptable, but 1 mμm is not correct.

There are a number of commonplace units that are outside of the International System but are frequently applied in conjunction with the SI base or derived units. These are listed in Table 1-5, together with appropriate symbols. Also listed are typical values in terms of SI units. One must remember that the term *minute* can have two meanings. One pertains to time, and the other relates to angular measurement. Note that the symbols are different.

Table 1-4
Prefixes of SI Units

Prefix	Symbol	Multiplier for Unit
tera	T	10^{12}
giga	G	10^{9}
mega	M	10^{6}
kilo	k	10^{3}
hecto	h	10^{2}
deka	da	10^{1}
——	——	$10^{0} = 1$
deci	d	10^{-1}
centi	c	10^{-2}
milli	m	10^{-3}
micro	μ	10^{-6}
nano	n	10^{-9}
pico	p	10^{-12}
femto	f	10^{-15}
atto	a	10^{-18}

Table 1-5
Widespread Units Used with SI Units

Name	Symbol	Value in SI Unit
minute	min	$1 \text{ min} = 60 \text{ s}$
hour	h	$1 \text{ h} = 60 \text{ min} = 3\,600 \text{ s}$
day	d	$1 \text{ d} = 24 \text{ h} = 86\,400 \text{ s}$
degree	°	$1° = (\pi/180) \text{ rad}$
minute	′	$1' = (1/60)°$
second	″	$1'' = (1/60)'$
liter	l	$1 \text{ l} = 1 \text{ dm}^3 = 10^{-3} \text{ m}^3$
tonne	t	$1 \text{ t} = 10^3 \text{ kg}$

The International Organization for Standardization (ISO) recommends the following to assure uniformity in the use of units:

1. The product of two or more units is preferably indicated by a centered dot immediately between the two units, unless there is no risk of confusion with another unit symbol. For example, the product of newtons and meters could be expressed as N • m or N m, but not mN.
2. When a derived unit is formed from two others by division, a solidus (/), a horizontal fraction line, or negative powers should be employed. Thus, velocity expressed in meters per second can be shown as

$$m/s, \quad \frac{m}{s}, \quad \text{or} \quad m \cdot s^{-1}.$$

3. The solidus should not be repeated on the same line unless ambiguity is resolved by parentheses. For example, an acceleration can be shown as m/s^2 or $m \cdot s^{-2}$, but not as m/s/s.

To assist in the transition from the U.S. to the metric system, a number of manufacturing firms prefer to use "dual dimensioning." Each dimension on a drawing is listed in inches and millimeters (or meters, in the case of large equipment). There are advantages and disadvantages to such a system. On the positive side, this procedure forces employees to think in terms of metric units. However, inspectors will note only the dimension expressed in the same units as their measuring equipment. Use of the dual system also places an extraordinary number of dimensions on a given drawing; this can lead to errors in interpreting the views. Also, when design changes are made, two sets of numbers must be changed.

The impact of metrication within an organization can be reflected in the revision of the firm's engineering standards. If U.S. units are replaced with metric ones, care must be exercised to provide the needed accuracy for each dimension. The usual practice is to use an electronic calculator rather than a chart or slide rule. Calculators are available for such conversions. In addition to the usual functions, built-in constants are provided for making such conversions as: feet to meters, yards to meters, inches to centimeters, miles to kilometers, pounds to kilograms, ounces to grams, Fahrenheit degrees to degrees Celsius (formerly called centigrade), quarts to liters, gallons to liters, and ounces to cubic centimeters.

Engineering standards vary extensively. However, there are a number of key points in which various firms use similar metrication standards. Typical items include:

1. an indication on the drawing that information is provided in metric units; this may take the form of displaying the term *metric* prominently on the drawing format;
2. an indication of the usual form of linear units, such as millimeters, centimeters, or meters;
3. a system for rounding off numbers when converting existing U.S. dimensions to appropriate metric units;
4. customary tolerances for the metric dimensions;
5. preferred drill, punch, and thread sizes;
6. recommended surface roughness values in micrometers, such as 0.050, 0.100, 0.20, 0.40, 0.80, 1.6, 3.2, 6.3, and 12.5 to replace the popular microinch values of 2, 4, 8, 16, 32, 63, 125, 250, and 500; also appropriate values for waviness height, waviness width, and roughness width; and
7. manufacturing data such as bend-allowance charts and similar information.

Appendix J shows selected U.S./metric conversions. Although a few of these are *exact*, most conversion factors have been rounded for simplification. If extreme accuracy is desired, a student should refer to conversions supplied through recognized authoritative standards such as those of the American Society for Testing and Materials, National Aeronautics and Space Administration, and others.

 **Robotics**

An expanding technology centers around the use of robots for performing various industrial processes. These applications—and the design of the units—are known as *robotics*. Originally, robots were made to somewhat resemble humans. The usual configuration consisted of a torso, head, arms, and legs. A few movements by arms and legs, coupled with flashing lights for eyes, attracted attention at exhibits and museums. These movements were mechanically or electrically actuated through the use of controls.

The modern industrial robot is a much more complex device that usually simulates only a few of the many human movements. The automatic pilot used on aircraft is actually a robot with limited but very important capabilities, such as the ability to automatically maintain altitude, keep the plane steady, and follow a preset course. The industrial robot is a multifunctional device that is completely programmable and can thus perform a number of tasks. It can be programmed by means of a computer and also by mechanical methods. Often, the mechanical control is a drum not too different from the device used in player pianos. The moving parts can be moved electrically, pneumatically, or hydraulically.

Many of the industrial models simulate arm and finger movements. Such actions make the robot useful in foundries, metal-working shops, and assembly shops. Typical applications are spray painting, spot welding, die casting, sand casting, and loading/unloading operations.

Although there are always societal problems associated with any type of automation, it should be noted that robots can perform dangerous operations under extremely hazardous working conditions that would not be desirable for humans. Environmental conditions in the workplace are not too essential for robots. If (sightless) robots work in a separate area from humans, lighting can be reduced to that normally found in warehouses. Temperature and humidity variations, radiation hazards, and poor air quality do not inhibit robot operation if the device has been properly designed. Also, these devices can be worked continuously without the normal breaks, vacations, and limited working hours mandated for human workers.

Through sophisticated instrumentation, robots can simulate several of the human senses. By using optical fibers, they can distinguish shapes and sizes—similar to the sense of sight. With proper programming and instrumentation, a robot can respond to voice commands and use its manipulators and "sense of touch" to actuate switching devices and carry out varied, sometimes delicate, operations.

The design of industrial robots is a challenging field that incorporates wide knowledge of electrical and mechanical controls.

Figure 1-7 illustrates a welding application using an industrial robot. The robot welds a base assembly for a computer mainframe. Forty-four seam welds (each two inches in length) are made in a period of less than 12 minutes. The welds are made from a variety of angles. The equivalent manual operation took 45 minutes.

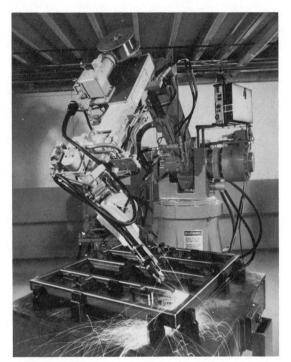

Figure 1-7
Welding application using industrial robot.
(Courtesy of Cincinnati Milacron.)

 Summary

The design of any machine is a complex problem. Much judgment must be exercised; even so, failures sometimes happen. Experience should always be helpful in any design situation. The student should recognize that many of the topics presented in this chapter represent specialties to certain individuals. Indeed, experts spend a lifetime in some of these rather narrow areas.

Careful calculations are necessary to ensure the validity of a design. Calculations never appear on drawings, but are filed away. This is done for several reasons. If a part fails, it is useful to know how the defective components were originally designed. Also, calculations from past projects can form an "experience file." When a similar design is needed, past records are a great help.

Checking calculations (and dimensions on drawings) is of utmost importance. The misplacement of one decimal point can ruin an otherwise acceptable project. For example, if one designed a bracket to support 100 lb when it should have been figured for 1000 lb, it would surely fail. All aspects of design work should be checked and rechecked.

Finally, a successful designer does all he or she can to keep up to date. New materials and production methods appear daily. Drafting and design personnel can lose their usefulness by not being versed in modern methods and materials. An individual can develop his or her professional skills through numerous channels. Appendix E lists some of the technical societies and trade associations. Many of these sponsor expositions, journals, seminars, and other means of communication to assist interested individuals in keeping abreast of current trends. Also, schools and company training departments organize continuing education courses for this specific purpose. The design field offers unlimited opportunities, but only to those who are qualified. Many of the products that will be designed ten years from now (or even sooner) have not even been considered yet.

Questions for Review

1. Differentiate between *industrial design* and *empirical design*.
2. What effect will a large factor of safety have on the size of components?
3. Why should engineering calculations be dated?
4. What should be considered before designing an automatic pencil sharpener for a pencil manufacturer who plans to use the device continuously for an eight-hour day?
5. What is the purpose of making a mock-up?
6. What factors should be considered when choosing a factor of safety?
7. Why is human engineering important in most design problems?
8. Carefully check the lubricating systems used in a current-model automobile. List each assembly that requires lubrication, and indicate whether it is continuously or periodically lubricated.
9. List the common types of fasteners (bolts, nuts, rivets, self-tapping screws, and so on) that are used in current-model automobiles. Give probable reasons for the use of each type, such as (1) easy assembly, (2) appearance, (3) low cost, and so on.
10. Compare a 10-year-old window-unit air conditioner with one currently on the market. What industrial design improvements have been made?

11. Referring to current technical periodicals, list ten new developments in the field of mechanical product design.
12. Why is yield stress important to a designer?
13. How does elasticity differ from toughness?
14. How does hardness differ from toughness?
15. List two applications where fatigue could be a serious consideration.
16. List three examples of die-cast parts. Indicate why this process was desirable in the examples.
17. What advantages do plastics have over metals as materials for machine parts? What are some disadvantages?
18. What is meant by the term *SI units?*
19. What advantages does the metric system have over the decimal-inch system?
20. What are the advantages of using dual dimensioning on production drawings? What disadvantages result from this practice?
21. Name six basic SI units and indicate the quantity that each measures.
22. Differentiate between the terms *kilogram* and *newton* with respect to the quantity each represents.
23. Check current periodicals for five new robot applications. For each, indicate the operation(s) being performed, the method of control, the actuation of the moving parts, and the environmental conditions.

REVIEW OF MECHANICS AND STRENGTH OF MATERIALS

2

It is impossible to design any of the traditional mechanical components without a complete understanding of the forces that act upon such components. A thorough understanding of mechanics and strength of materials is essential before undertaking the study of any phase of machine design. The material presented in this chapter reviews *some* of the highlights in the general areas of statics, kinematics, dynamics, stresses, and beams.

 ## Statics Definitions

A few of the fundamental terms used in statics are as follows:

Force is defined as an action between two bodies that tends to cause some change in the receiving body. The force may cause a change in position or a change in appearance. The term *load* is often used in conjunction with force. However, load is usually defined as a weight distributed over a given area. To be completely defined, a force must have an established magnitude, direction, point of application, and line of action.

A *vector* is a graphical representation of force, velocity, acceleration, or any other vector quantity. Figure 2-1 shows a vector representing a force of 100 lb acting on a body. In graphical solutions, the vector must be accurately drawn to a suitable scale. In Fig. 2-1, the vector definitely represents a pushing action on the body. If it were moved to the dotted position, it would represent a pulling action on the body. *Scalar* quantities possess magnitude values only.

A *couple* is a pair of equal and opposite parallel forces that tend to cause rotation of a body. Couples cannot be collinear.

A *rigid body* is one that is not appreciably deformed by the action of any external force or load.

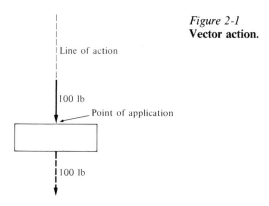

Figure 2-1
Vector action.

Line of action

100 lb

Point of application

100 lb

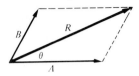

Figure 2-2
Resultant by a parallelogram method.

The term *resultant* refers to the single force that, acting alone, can replace two or more forces acting together. Figure 2-2 shows the resultant *R* of forces *A* and *B*. In this particular example, the construction is made by completing a parallelogram using vector values for *A* and *B*. In this "parallelogram" method, the vectors must be constructed so that either the tail ends or the arrow ends are concurrent. In Fig. 2-3, the resultant is found from a force polygon constructed by joining one vector to another in a tail-end to arrow-end process. The resultant is the closing line; the arrowhead is placed on the end of the resultant vector that "opposes" the others.

An *equilibrant* is a single force that can place a system of forces in equilibrium. It is equal in magnitude, but opposite in direction, to the resultant of a force system.

Components of forces can be found by reversing the procedure shown in Fig. 2-2. It might be said that *A* and *B* are components of force *R*. It is often convenient in practical problems to find the horizontal and vertical components of a force or series of forces. Finding the resultant of a force system is called *composition;* finding components of a force is called *resolution*.

A force system is in *equilibrium* when the algebraic sum of all the vertical and horizontal components of the external forces and reactions is equal to zero. Thus, $\Sigma V = 0$ and $\Sigma H = 0$.

Reactions are external effects caused by some type of loading on a body. Figure 2-4 shows a simple truss with a 1000-lb force applied. Since the load is applied midway between the supports, the reactions are 500 lb each.

A *principle of transmissibility* should be evident; so far as the reactions (external effects) are concerned, it makes no difference whether the 1000-lb load is applied as shown in Fig. 2-4 or is shifted along its line of action and suspended from the lower chord (shown dotted). It would, however, greatly affect the internal conditions of all the structural members.

Concurrent forces are forces whose lines of action meet in a single point. Concurrency is important in analyzing forces in structures and machines.

Figure 2-3
Force polygon.

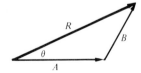

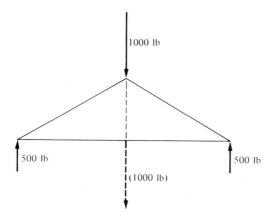

Figure 2-4
Principle of transmissibility of forces.

A *free-body* analysis is an important tool that can be used to solve force problems. The free body is a section that is "removed" from the rest of the structure or machine for analysis. Figure 2-5 shows the free-body diagram for the left end of a simple truss. This indicates that three forces act at this one point; equilibrium exists. If the inclination of all vectors is known and there are no more than two unknown magnitudes, the system can be solved graphically or mathematically.

 Force Systems

Systems of forces can be classified in the following ways:

1. coplanar
   a) collinear
   b) concurrent
   c) nonconcurrent (including parallel)
2. noncoplanar
   a) concurrent
   b) nonconcurrent (including parallel)

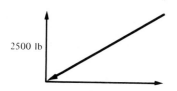

Figure 2-5
Free-body diagram of truss joint.

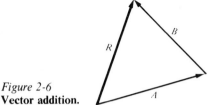

Figure 2-6
Vector addition.

Vector Addition

Scalar quantities can be added or subtracted by means of simple arithmetic, since direction is not involved. In vector addition (or subtraction), direction is considered. In Fig. 2-6, forces A and B are added vectorially to yield the resultant. This force polygon is constructed by accurately laying out each vector to scale and at the proper inclination. Note that, in addition, the tail end of each vector joins the arrow end of the preceding vector. If equilibrium exists for the force system, the closing line is the resultant or the equilibrant. If it is the resultant, the arrow on the closing line opposes the others. If it is the equilibrant, the tail and arrow conform to those of the other vectors. Figure 2-7 shows that several vectors can be added in one simple figure. The order in which the vectors are added has no effect on the answer. A vector equation for the forces shown in Fig. 2-7 can be written without the use of any illustration:

$$R = A \longmapsto B \longmapsto C \longmapsto D.$$

However, unless inclinations are defined, the algebraic style of equation is useless except for indicating that vector quantities are involved. In the force system of Fig. 2-7, the dotted lines indicate intermediate resultants that need not be drawn. The dotted lines indicate that A and B are added; this resultant is then combined with force C to yield another intermediate resultant. Force D is then added to yield the total resultant of the entire force system.

Figure 2-7
Vector addition: force polygon.

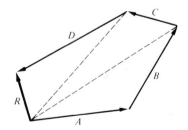

Mathematical Solutions

Force systems can be accurately analyzed using simple trigonometry. The following examples show some of the techniques used.

◆ *Example* ————————————————————————————

Using the mathematical approach, find the resultant of the forces shown in Fig. 2-8. Determine the inclination in addition to the magnitude.

Solution. By the law of cosines,

$$R = \sqrt{(60)^2 + (40)^2 - 2(40)(60)\cos 80°};$$

$$R = \sqrt{3600 + 1600 - 4800(0.174)} = 66.1 \text{ lb.}$$

By the law of sines,

$$\frac{40}{\sin \theta} = \frac{66.1}{\sin 80°}.$$

Thus

$$\sin \theta = \frac{40 \sin 80°}{66.1} = \frac{40(0.985)}{66.1} = 0.596$$

and

$$\theta = 36.6°.$$

This is sometimes notated as θ_x, since it is measured from the horizontal.

Figure 2-8
Resultant of two forces.

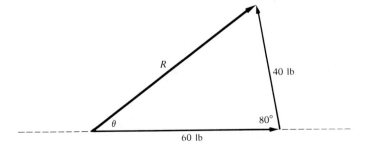

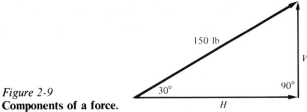

Figure 2-9
Components of a force.

◆ *Example* ───────────────────────────────

Find the vertical and horizontal components of the 150-lb force shown in Fig. 2-9.

Solution. Solve the right triangle by applying the following trigonometric functions:

$V = 150 \sin 30° = 150(0.5) = 75$ lb;

$H = 150 \cos 30° = 150(0.866) = 130$ lb.

◆ *Example* ───────────────────────────────

Find the resultant (magnitude and inclination) of the force system shown in Fig. 2-10(a). Prepare a table for the vertical and horizontal components of each force; be careful to apply positive and negative signs for direction. Positive signs can be used for components pointing upward or to the right. Then construct the force polygon with the resulting Σ-values.

Figure 2-10
Resultant of a concurrent, coplanar force system.

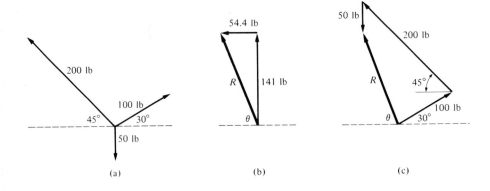

Solution. The vertical and horizontal components are as follows:

Force	V	H
100	$100 \sin 30° = 50$	$100 \cos 30° = 86.6$
200	$200 \sin 45° = 141$	$-200 \cos 45° = -141$
50	-50	
	$\Sigma V = +141$	$\Sigma H = -54.4$

The force polygon is shown in Fig. 2-10(b). By the Pythagorean theorem,

$$R = \sqrt{(\Sigma V)^2 + (\Sigma H)^2}\,;$$

$$R = \sqrt{(141)^2 + (54.4)^2} = 151 \text{ lb.}$$

Then

$$\theta = \arctan \frac{\Sigma V}{\Sigma H}\,;$$

$$\theta = \arctan \frac{141}{54.4} = \arctan 2.59 = 68.9°.$$

Figure 2-10(c) shows one graphical method for finding the resultant. The three vectors are laid out at the proper inclination and to a proper force scale. The closing line of the polygon represents the resultant. Note the direction of the resultant vector.

Static Equations of Equilibrium

The term *moment* is often used in mechanics and strength of materials. Moment is defined as the product of a force and the perpendicular distance of its line of action from a reference axis. Figure 2-11 shows a simple moment.

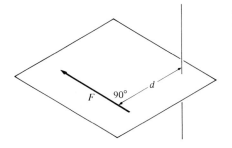

Figure 2-11
Moment.

For equilibrium conditions, the sum of the moments about a point must be equal to zero. Thus, we have these important equations of equilibrium:

$$\Sigma F_x = 0; \quad \Sigma F_y = 0; \quad \Sigma F_z = 0; \quad \Sigma M = 0. \tag{2-1}$$

Simple examples of concurrent forces in equilibrium are presented in the following two problems. The first example is one where the forces are all in one plane; the second uses noncoplanar forces.

◆ *Example* _____

Find the "stresses" in *AB* and *BC* of Fig. 2-12(a) using both graphical and mathematical methods. (*Note:* The term *stress* is often used to mean *force* in this type of problem.)

Graphical solution. Draw a free-body diagram of the forces at point *B* as shown in Fig. 2-12(b). (This merely isolates this joint from the structure.) The directions of *AB* and *BC* are as yet unknown. Next, construct a force polygon (Fig. 2-12c) by laying out the known vector (100 lb) vertically to a suitable force scale. Through either end of this known vector, draw lines *AB* and *BC* parallel to the members in the space diagram. These lines will thus intersect at point *X*, establishing the remainder of the force polygon. Arrowheads can now be placed in the force diagram for vectors *BC* and *AB*. Since the three forces are in equilibrium, the tail end of each vector attaches to the arrow end of the preceding one. Thus, directions are established. Now, compare the vector directions found in the force polygon with the appropriate unknowns in the free-body diagram. We find that *AB* pulls away from the joint (indicating tension) and *BC* points toward the joint (indicating compression). Magnitudes of the vectors are found by scaling from the force diagram.

Figure 2-12
Free-body analysis of a joint.

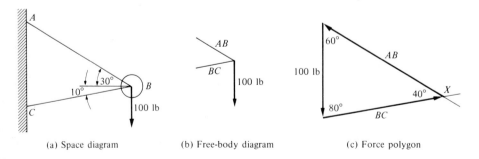

(a) Space diagram   (b) Free-body diagram   (c) Force polygon

Mathematical solution. Referring to the force polygon of Fig. 2-12(c), and applying the law of sines, we obtain the following equation:

$$\frac{AB}{\sin 80°} = \frac{100}{\sin 40°}.$$

Thus

$$AB = \frac{100 \sin 80°}{\sin 40°} = \frac{100(0.985)}{0.643} = 153 \text{ lb.}$$

Also,

$$\frac{BC}{\sin 60°} = \frac{100}{\sin 40°}.$$

Thus

$$BC = \frac{100(0.866)}{0.643} = 135 \text{ lb.}$$

Another mathematical approach—carried out without constructing a force polygon—uses two of the equilibrium equations, as follows:

$$\Sigma F_x = 0;$$
$$(BC)\cos 10° - (AB)\cos 30° = 0;$$
$$0.985(BC) - 0.866(AB) = 0.$$

Therefore,

$$(BC) = 0.879(AB).$$
$$\Sigma F_y = 0;$$
$$(BC)\sin 10° + (AB)\sin 30° - 100 = 0;$$

substituting,

$$0.879(0.174)(AB) + 0.5(AB) = 100.$$

Therefore,

$$(AB) = 153 \text{ lb}$$

and

$$(BC) = 0.879(AB),$$

or

$$(BC) = 0.879(153) = 135 \text{ lb.}$$

In this example, vectors pointing to the right or upward were considered positive.

◆ *Example* ————————————————————————

Figure 2-13(a) shows a tripod with equal legs spaced equally on level ground. The legs are 52 in. long. A 20 lb instrument rests 48 in. above the ground. Find the force in each leg.

Figure 2-13
Noncoplanar, concurrent equilibrium example.

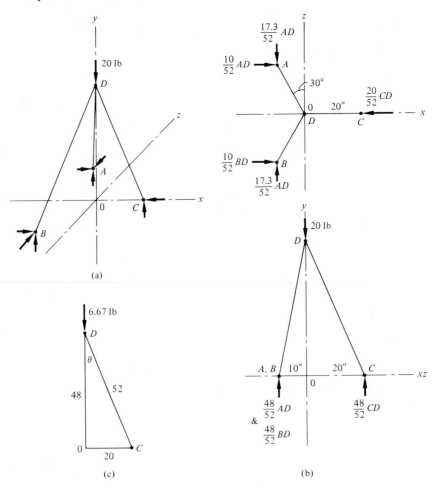

(a)

(b)

(c)

Solution. This problem can be solved in a number of ways. The first method presents a general technique for this type of problem that can be simplified because the legs are equal in length and spacing. Figure 2-13(a) shows the tripod oriented so that one leg is in the x-axis, which also bisects the other two legs. The legs then form a triangle with sides of 5, 12, and 13. Therefore, distance CO is 20 in., and the distance for both AB and AD (measured along the x-axis) is $20 \sin 60°$, or 17.3 in., as shown in the xz plane of Fig. 2-13(b). Three simultaneous equations of equilibrium can then be written as follows:

$$\Sigma F_x = 0;$$

$$\frac{10}{52} AD + \frac{10}{52} BD - \frac{20}{52} CD = 0.$$

$$\Sigma F_y = 0;$$

$$\frac{48}{52} AD + \frac{48}{52} BD + \frac{48}{52} CD - 20 = 0.$$

$$\Sigma F_z = 0;$$

$$\frac{17.3}{52} AD - \frac{17.3}{52} BD = 0.$$

If the equilibrium equation for F_x is multiplied by 2.4, CD can be eliminated. Since AD must equal BD as shown in the equation for F_z, the numerical value for AD is readily shown to be 7.23 lb. This is also the value for BD and CD.

In drawing the free-body diagram, it is usually helpful to make plan and front views (x-z and x-y planes) and to sketch the various components at the proper locations. These are shown in Fig. 2-13(b).

The use of the moment equation can simplify this problem. Referring to the front view in Fig. 2-13(b), moments can be taken about the two coincident points (A, B) as follows:

$$\Sigma M_{A, B} = 0.$$

Then,

$$\frac{48}{52} CD(30) - 20(10) = 0,$$

$$CD = \frac{200(52)}{48(30)} = 7.23 \text{ lb},$$

and

$$CD = AD = BD = 7.23 \text{ lb}.$$

Another fast solution to this simple problem can be found by referring to Fig. 2-13(c) and noting the triangle formed by the line of action of the 20 lb load, the ground, and one of the legs. The load must be distributed evenly to the three legs; therefore, the force at the top of this triangle is 20/3, or 6.67 lb. Then,

$$\cos \theta = \frac{48}{52} = \frac{6.67}{CD}$$

or

$$CD = \frac{52(6.67)}{48} = 7.23 \text{ lb},$$

which also equals the force in the other two legs.

Resultant of Noncoplanar, Concurrent Forces

The resultant of three mutually perpendicular forces can be found by obtaining the resultant of two of them and combining that value with that of the third, bearing in mind both magnitude and direction. Or, the total resultant for the three forces can be found in one step, and then the angles can be obtained in the x, y, and z planes. The following example shows this simple procedure.

◆ *Example* ─────────────────────────

Find the resultant and the angles for three mutually perpendicular concurrent forces with the following values: $F_x = 5$ N, $F_y = 4$ N, $F_z = 3$ N.

Solution.

$$R = \sqrt{F_x^2 + F_y^2 + F_z^2}\,;$$

$$R = \sqrt{5^2 + 4^2 + 3^2} = 7.07 \text{ N}.$$

$$\cos \theta_x = \frac{5}{7.07} = 0.707, \quad \theta_x = 45°.$$

$$\cos \theta_y = \frac{4}{7.07} = 0.566, \quad \theta_y = 55.5°.$$

$$\cos \theta_z = \frac{3}{7.07} = 0.424, \quad \theta_z = 64.9°.$$

If the forces do not all lie in the x, y, and z planes, x, y, and z components can be obtained and added algebraically to find the resultant.

Parallel Forces

The resultant of parallel forces can also be found by graphical or mathematical methods. In Fig. 2-14, the resultant of the 100-lb and 50-lb forces is equal to the sum of the two forces, or 150 lb. It lies between the two forces, but closer to the 100-lb force. The location can be found by inverse proportion. First, lay out a vector representing 50 lb on the line of action of the 100-lb force on the same side of the reference line. Then construct a vector equal to the 100-lb force on the line of action of the 50-lb force, but on the opposite side of the reference line. A line scribed from the extremities of these displaced vectors establishes the pivot point O, which is on the line of action of the resultant. This construction applies if the two forces act in the same direction. If the two forces act in opposite directions,

Figure 2-14
Resultant of parallel forces.

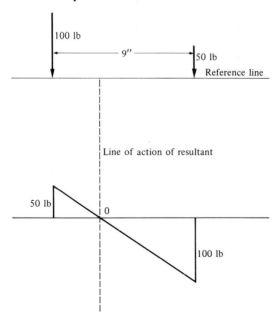

the resultant is equal to the difference of the numerical values, and lies outside the two on the side of the larger. Its direction is the same as the larger of the two forces. For the graphical solution in this case, the 100-lb and 50-lb forces are transposed, but placed on the *same* side of the reference line. Then the line connecting the extremities of the transposed vectors locates point *O* for the line of action of the resultant. The following example shows the analytical approach to parallel-force problems.

◆ *Example*

Find the location of the resultant of the parallel forces in Fig. 2-14.

Solution. Let x = distance from the 100-lb force to the line of action of the resultant. Let $9 - x$ = distance from the 50-lb force to the line of action of the resultant. Then, take moments about point *O*. Consider clockwise moments as positive and counterclockwise moments as negative. Thus, $50(9 - x) - 100x = 0$. Solving for x, we find that $x = 3$ in.

Parallel Reactions

Inverse-proportion methods can also be used to determine reactions. In most practical problems, the reactions become important design figures. Let us solve a reaction problem by both graphical and analytical methods.

◆ *Example*

In Fig. 2-15, find the reactions at *A* and *B* caused by the 400-lb load.

Graphical solution. Project the 400-lb vector horizontally to the line of action of either of the unknown reactions. Then a line drawn from this

Figure 2-15
Reactions: parallel lines of action.

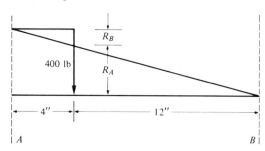

point to the reference line at the other reaction divides the original 400-lb vector into two components, R_A and R_B. These can then be scaled.

Mathematical solution. Take moments about either of the unknown reactions. Let us assume that we take moments about R_A. Then

$$4(400) - 16R_B = 0 \qquad \text{or} \qquad R_B = 100 \text{ lb.}$$

R_A is then found by subtracting R_B from 400 lb or by taking moments about R_B. (Here, it is equal to 300 lb.) Both reactions act upwards and satisfy the static equation of equilibrium $\Sigma F_y = 0$.

Nonparallel Reactions

When a machine or structural member is subjected to a load acting at an angle—such as P in Fig. 2-16(a)—one of the reactions must support the horizontal component of P. In this case, let us assume that P is known in magnitude and direction. Let us also assume that R_A takes all the horizontal component of P. This can be done by hinging the left end and providing a smooth surface at the right end; a roller or smooth surface dictates a vertical reaction for R_B. If the part is supported by bearings, the bearing on the left end must be able to handle a thrust load.

We can solve R_A and R_B graphically by noting that equilibrium exists. This means that all the lines of action of the external forces (including reactions) must meet in a point. This concurrency point O can be found by extending the lines of action of force P and R_B. Then a line scribed from the point of application of R_A to point O shows the inclination of R_A. Refer to Fig. 2-16(a). Now that the line of action of R_A is established, a force polygon can be constructed as shown in Fig. 2-16(b). The known force P is drawn to a suitable scale, and R_A and R_B are established by drawing vectors parallel to the aforementioned lines in Fig. 2-16(a) and joining the extremities of the known vector P.

To solve mathematically, resolve P into horizontal and vertical components as shown in Fig. 2-16(c). Since R_A takes all the horizontal reaction, $P_H = R_{A_H}$. Moments can be taken about R_A to solve for R_B. Thus

$$bP_V - aR_B = 0,$$

since

$$\Sigma M = 0,$$

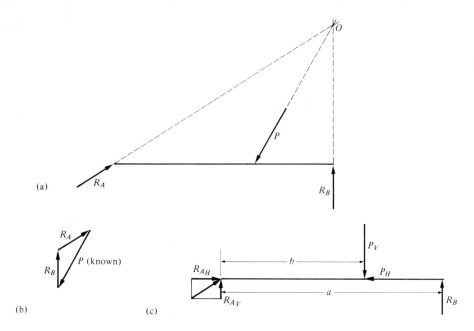

Figure 2-16
Nonparallel reactions.

or

$$R_B = \frac{bP_V}{a}.$$

Then R_{A_V} can be solved using the fact that $\Sigma M_B = 0$ or $\Sigma F_y = 0$. After both R_{A_H} and R_{A_V} have been solved, the resultant of the two (which is the reaction at the left end) can be found in magnitude and direction.

Overhung Mounting

Straddle mounting is an arrangement in which the load is placed between the two supports (or bearings). *Overhung* mounting is one in which a substantial part of the load is placed outside one of the reaction positions. The following example shows the mathematical approach to finding reactions with an overhung load.

◆ *Example* _____

In the configuration of Fig. 2-17, a beam is supported at points A and B. If the total weight of the beam (uniformly distributed) is 100 lb, find the reactions at A and B.

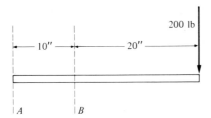

Figure 2-17
Overhung mounting.

Solution. The weight of the beam can be considered as concentrated at its center of gravity, which is 15 in. from R_A. Taking moments about A, we obtain

$$15(100) + 30(200) - 10R_B = 0.$$

Thus

$$R_B = 750 \text{ lb.}$$

This reaction is upward. The reaction at A can be found by taking moments about B or adding vertical forces algebraically. The latter method produces the following:

$$\Sigma F_y = 0;$$

$$200 + 100 - 750 + R_A = 0.$$

Thus

$$R_A = 450 \text{ lb.}$$

Its direction is downward.

In all beam or shaft calculations, one should consider the weight of the member in addition to the external loads. The weight of a beam is often neglected in textbook problems to simplify a calculation.

 Motion

By definition, *kinematics* is the study of motion and its application by means of such components as gears, cams, belts, chains, and other devices that transmit motion. The important considerations are:

1. *Displacement:* the distance between a beginning and ending point, without considering the path followed.

2. *Velocity:* the distance covered per unit of time. Velocity is a vector quantity.
3. *Acceleration:* the rate of change of velocity with respect to time. Like velocity, it is also a vector quantity. Both velocity and acceleration are considered in linear and angular terms.

A few important definitions and equations are presented as review material.

Linear-Angular Velocity Relationship

In designing certain mechanical components, it is often necessary to convert linear-velocity values to angular-velocity values or vice versa.

Angular velocity may be expressed by the following general equation:

$$\omega = \frac{\theta}{t}, \tag{2-2}$$

where

ω = angular velocity,

θ = angular displacement,

t = time.

Often θ is expressed in radians and t in seconds. Thus, ω would be given in radians per second. One radian is equal to 57.3 deg; 1 deg is equal to 0.01745 rad. Another useful relationship is

$$v = r\omega, \tag{2-3a}$$

where

v = linear velocity,

r = radius from center of rotation.

The value of v is frequently expressed in feet per second (fps), in which case r is expressed in feet. In terms of revolutions per minute, the following relationships are useful:

$$\omega = 2\pi N, \qquad \text{since 1 rev} = 2\pi \text{ rad}; \tag{2-3b}$$

$$v = \pi DN = 2\pi rN, \qquad \text{where } N = \text{revolutions per minute.} \tag{2-3c}$$

Angular velocity is popularly expressed in rpm because this value is easy to obtain with (a) counters used in conjunction with timers, (b) tachometers that measure rpm values directly, or (c) stroboscopes that can be calibrated to give rpm values directly. Velocity is often expressed in miles per hour (mph), feet per minute (fpm), or feet per second (fps). A dimensional analysis of all units is a necessity.

Displacement-Velocity-Acceleration Equations

The following listing of conventional formulas is useful in setting up many types of projects involving kinematics.

Linear	*Angular*	
$s = v_1 t + \dfrac{at^2}{2},$	$\theta = \omega_1 t + \dfrac{\alpha t^2}{2},$	**(2-4a)**
$v_2 = v_1 + at,$	$\omega_2 = \omega_1 + \alpha t,$	**(2-4b)**
$v_2{}^2 = v_1{}^2 + 2as,$	$\omega_2{}^2 = \omega_1{}^2 + 2\alpha\theta,$	**(2-4c)**

where the subscripts 1 and 2 refer to initial and final values after a period of time. Popularly used units in these equations are as follows:

v = linear velocity (fps),

s = linear displacement (ft),

t = time (sec),

ω = angular velocity (rad / sec),

α = angular acceleration (rad / sec^2),

θ = angular displacement (rad).

Normal and Tangential Accelerations

Normal components of acceleration for a point on a rotating or rolling wheel are found from the following equation:

$$a_n = v\omega = \frac{v^2}{r} = \omega^2 r. \qquad\qquad \textbf{(2-5)}$$

Tangential accelerations are given by the equation

$$a_t = \alpha r. \qquad\qquad \textbf{(2-6)}$$

The normal (centripetal) component of acceleration of a point on a rotating body is directed toward the center of rotation. The tangential component of acceleration acts at 90 deg to the normal in the direction of the acceleration (or deceleration). The four cases shown in Fig. 2-18 can be explained as follows. (In all four cases, the normal component of acceleration is directed toward the center of the wheel.)

a) In Fig. 2-18(a), the wheel is rotating clockwise about a fixed center and is being accelerated. Thus, the a_t vector points clockwise. The vector equation is

$$a_A = a_{n_A} \longmapsto a_{t_A}.$$

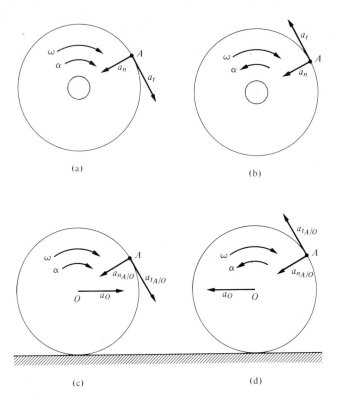

Figure 2-18
Accelerations at a point on rotating and rolling wheels.

b) In Fig. 2-18(b), the wheel is also rotating clockwise, but it is being decelerated. Thus a_t is directed counterclockwise. The total acceleration is given by the above equation.

c) In Fig. 2-18(c), the wheel is rolling to the right and is being accelerated. The normal and tangential components of the acceleration at point A with respect to O can be calculated; however, these must be added vectorially to the acceleration at point O. The acceleration at point O on a rolling wheel will always be numerically equal to the tangential acceleration at a point on the periphery of the wheel with respect to the center point of the wheel. The vector equation is

$$a_A = a_O \overset{+}{\longrightarrow} a_{n_{A/O}} \overset{+}{\longrightarrow} a_{t_{A/O}}.$$

d) The wheel of Fig. 2-18(d) is also rolling toward the right, but it is being decelerated; $a_{t_{A/O}}$ and $a_{n_{A/O}}$ can be calculated.

Note the direction of $a_{t_{A/O}}$ due to deceleration; also note that a_O points to the left because of deceleration. The vector equation will be exactly the same as in (c); however, the value and direction of the total acceleration at point A will be different.

 ## Work, Energy, and Power

Power and efficiency are two important considerations in comparing various machines. In simple terms, *work* is the result of multiplying a force exerted on a body by the distance that the body is displaced. Thus, units for work are often such units as in-lb and ft-lb. Note that time is not involved in any definition of work. When time enters the problem, the term *power* results. Thus power can be defined as work per unit of time. Typical power units include ft-lb/min, ft-lb/sec, and in-lb/min. In engineering work, the term *horsepower* is usually used. By definition, 1 hp is equal to 33,000 ft-lb of work per minute, or 550 ft-lb/sec. The following relationships are useful in calculations:

$$\text{hp} = \frac{FV}{33,000} = \frac{F}{33,000} \frac{2\pi rN}{12},$$

where

F = applied force (lb),

V = linear velocity (fpm),

N = angular velocity (rpm),

r = radius (in.).

Torque can be defined as a product of applied force and the distance from the point of application to the radius to produce center of rotation or torsion. In formula form,

$$T = Fr. \tag{2-7}$$

The previous horsepower equation can be simplified by replacing the product of Fr with the torque T and combining the numerical values into one constant. Thus

$$\text{hp} = \frac{NT}{63\,000},$$

where

$$T = \text{torque (in-lb).*} \tag{2-8a}$$

Many practical applications can be found for this relationship.

---

*Computer program in Appendix L.

In the metric system, the unit of power is the kilowatt (or watt, in the case of small values). Torque is expressed in newton-meters (or newton-millimeters). One horsepower is equal to 0.746 kilowatts. Torque is found by dividing power by angular velocity; however, angular velocity must be converted to radians per second. The following relationships should be helpful in solving problems:

$$\text{torque} = \frac{\text{power}}{\text{angular velocity}},$$

or

$$T = \frac{\text{kW} \, (1\,000)(60)}{2\pi N} \quad \text{N} \cdot \text{m},$$

or

$$\text{kW} = \frac{2\pi NT}{60\,000}. \tag{2-8b}$$

Dynamic equations of motion depend on Newton's second law. This law states that if an unbalanced force is applied to any body or particle, the body or particle must be accelerated. In equation form,

$$F = ma = \frac{Wa}{g}, \tag{2-9}$$

where

F = force (lb),

m = mass (slugs),

W = weight (lb),

a = acceleration $(\text{ft} / \sec^2)$,

g = acceleration of gravity (approximately 32.2 ft / sec^2).

The following illustrative example shows how the use of this equation can affect design loads.

◆ *Example* ——————————————————————————

If a 150-lb man stands on a platform scale in an elevator that is accelerated upward at 10 ft/sec^2, what will the platform scale read?

Solution. Since only vertical forces are involved, the following relationship can be established (P is the scale reading):

$$P - W = \frac{Wa}{32.2},$$

$$P - 150 = \frac{150(10)}{32.2},$$

$$P = 196.6 \text{ lb}.$$

Clearly, design loads must be greatly increased if extensive changes in velocity are anticipated.

Energy is the capacity for performing work. Energy is available in many forms: chemical, mechanical, and electrical. In certain types of machinery, energy is received in one form and converted into another to produce the desired work. Two general categories for classifying energy are:

1. *Potential energy:* the energy that a body has because of its position or configuration. Typical examples of potential energy are (1) water stored at an elevation, (2) a compressed spring, and (3) an upper blanking die before being released.

2. *Kinetic energy:* energy due to motion. Kinetic energy can be expressed by the following formula:

$$K = \frac{mv^2}{2},$$

(2-10)

where

K = kinetic energy (ft-lb),

m = mass (slugs),

v = velocity (fps).

One must remember in using this equation that mass (expressed in slugs) is equal to the weight of the body divided by the acceleration of gravity (approximately 32.2 ft/sec^2). The acceleration of gravity in SI units is 9.807 m/s^2.

In evaluating any equipment, efficiency is of the utmost importance. Efficiency is usually expressed in percentage form. Thus,

$$\text{efficiency} = \frac{\text{output}}{\text{input}} \times 100$$

or

$$e = \frac{\text{output}\,(100)}{\text{input}},$$

(2-11)

where e = efficiency (in percent). In general, *work* output is used in this equation. Efficiency is always less than 100%; thus, output is smaller in numerical value than input. The work lost is usually converted to frictional heat.

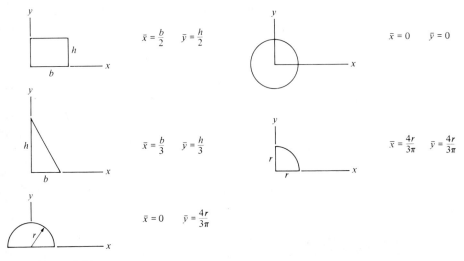

Figure 2-19
Centroids of common areas.

Centroids

A *centroid* can be defined as the center of gravity of a plane figure (assuming the material is the same throughout). Figure 2-19 shows the centroids of a few common geometrical shapes. It is possible to establish $\bar{x}$, $\bar{y}$, and $\bar{z}$ for solid geometrical shapes. Thus, three coordinates establish the center of gravity for the three-dimensional figure.

The centroids for irregularly shaped plane figures can be found by subdividing the figure into common shapes with known centroids. A general relationship for this can be written by taking moments about the *x-*, *y-*, or *z*-axis. In the case of $\bar{x}$, this can be expressed as

$$A\bar{x} = A_1 x_1 + A_2 x_2 + \cdots, \qquad (2\text{-}12)$$

where

A = area of composite figure (sq in.),

$\bar{x}$ = centroidal distance from *y*-axis (in.).

For this relationship, areas are represented by vectors with positive signs. Openings such as holes are shown by vectors with negative signs. The following problem illustrates setting up such a relationship.

◆ *Example* _____

Find the centroidal distance $\bar{x}$ from the y-axis for the composite figure shown in Fig. 2-20.

Solution

$$\bar{x} = \frac{A_1 x_1 + A_2 x_2 - A_3 x_3}{A_1 + A_2 - A_3},$$

$$\bar{x} = \frac{12(2) + 36(7) - 2(8)}{12 + 36 - 2} = 5.65 \text{ in.}$$

Figure 2-20
Centroid of a composite figure.

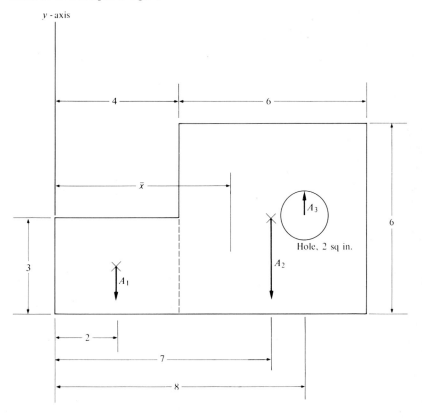

Moment of Inertia

In many types of engineering problems—particularly the study of beams and columns—the *moment of inertia* of the section is needed. It is defined as the sum of the products obtained by multiplying each elemental area of a plane surface by the square of its distance from a given axis. The usual units for moment of inertia are inches to the fourth power (in^4). A useful index for evaluating strength in bending is the moment of inertia of a section divided by the distance from the neutral axis (through the centroid) to the outermost fiber. This ratio is known as the *section modulus*. Values for the section modulus are expressed in inches to the third power (in^3).

In Fig. 2-21, if a plane area is considered in the plane *x-y*, then

$$I_x = \int y^2 \, dA \quad \text{and} \quad I_y = \int x^2 \, dA, \tag{2-13}$$

where I = moment of inertia (in^4).

Figure 2-22 shows moment-of-inertia values for familiar cross sections (section-modulus values are also included for convenience in working many types of problems). Values for I and I/c for structural shapes (I-beams, channels, angles, and so on) are readily found in various handbooks.

Figure 2-21
Moment of inertia.

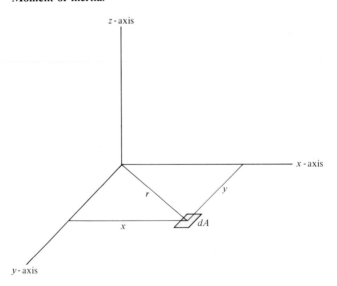

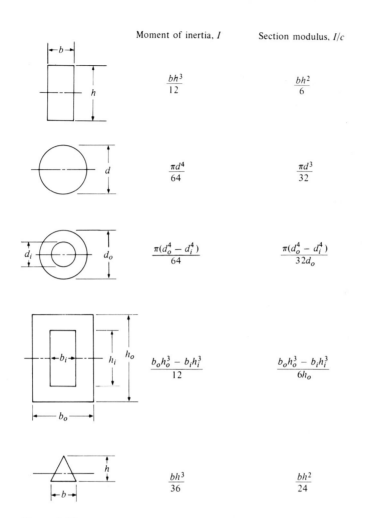

	Moment of inertia, I	Section modulus, I/c
	$\dfrac{bh^3}{12}$	$\dfrac{bh^2}{6}$
	$\dfrac{\pi d^4}{64}$	$\dfrac{\pi d^3}{32}$
	$\dfrac{\pi(d_o^4 - d_i^4)}{64}$	$\dfrac{\pi(d_o^4 - d_i^4)}{32d_o}$
	$\dfrac{b_o h_o^3 - b_i h_i^3}{12}$	$\dfrac{b_o h_o^3 - b_i h_i^3}{6h_o}$
	$\dfrac{bh^3}{36}$	$\dfrac{bh^2}{24}$

Figure 2-22
Centroidal moment-of-inertia formulas for common shapes.

Polar Moment of Inertia

Refer again to Fig. 2-21. The *polar moment of inertia* is the moment of inertia of a plane area with respect to a perpendicular to the plane area. This value is useful in handling torsion problems, such as shafting design. Let J represent the polar moment of inertia. Then

$$J_z = \int r^2 \, dA = \int (x^2 + y^2) \, dA = \int x^2 \, dA + \int y^2 \, dA \qquad \textbf{(2-14)}$$

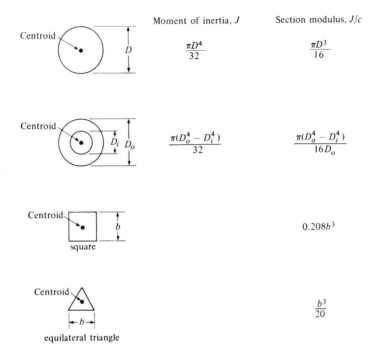

| | Moment of inertia, J | Section modulus, J/c |

$$\frac{\pi D^4}{32}$$

$$\frac{\pi D^3}{16}$$

$$\frac{\pi (D_o^4 - D_i^4)}{32}$$

$$\frac{\pi (D_o^4 - D_i^4)}{16 D_o}$$

$$0.208 b^3$$

$$\frac{b^3}{20}$$

Figure 2-23
Polar moment of inertia formulas for common shapes.

or

$$J_z = I_y + I_x.$$

Figure 2-23 shows the values of polar moment of inertia and (J/c) for common shapes. The polar moment of inertia is used in torsion calculations.

Parallel-Axis Theorem

A useful relationship for finding a moment of inertia based on any axis parallel to the one through the centroid is as follows:

$$I = \bar{I} + Ad^2, \tag{2-15a}$$

where

I = moment of inertia based on new axis (in^4),

$\bar{I}$ = moment of inertia through centroid (in^4),

d = distance between parallel axes (in.),

A = area of section (in^2).

This theorem is useful in finding the moment of inertia of a composite figure made up of common shapes. The parallel-axis theorem also applies to the polar moment of inertia. Thus,

$$J = \bar{J} + Ad^2. \hspace{4cm} \textbf{(2-15b)}$$

The following example shows an application of the parallel-axis theorem.

◆ *Example* —————————————————————

Find the axial moment of inertia for the angle shown in Fig. 2-24 based on the axis x-x.

Solution. First, divide the angle into areas A_1 and A_2. Then find the location of the centroid for the angle (composite figure):

$$\bar{y} = \frac{A_1\bar{y}_1 + A_2\bar{y}_2}{A_1 + A_2} = \frac{12(3) + 6(1)}{12 + 6} = 2.33 \text{ in.}$$

Figure 2-24
Moment of inertia: parallel-axis theorem.

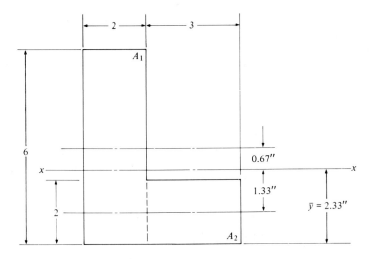

Moment-of-inertia values for each of these areas are

$$I_1 = \frac{b_1 h_1{}^3}{12} = \frac{2(6)^3}{12} = 36 \text{ in}^4 \qquad \text{for area 1,}$$

$$I_2 = \frac{b_2 h_2{}^3}{12} = \frac{3(2)^3}{12} = 2 \text{ in}^4 \qquad \text{for area 2.}$$

Applying the parallel-axis theorem for the composite figure and basing the moment of inertia on the centroid of the composite figure, we obtain

$$\bar{I} = \left(I_1 + A_1 d_1{}^2\right) + \left(I_2 + A_2 d_2{}^2\right),$$

$$\bar{I} = 36 + 12(0.67)^2 + 2 + 6(1.33)^2 = 54 \text{ in}^4.$$

Radius of Gyration

The radius of gyration is shown by the following formula:

$$k = \sqrt{I/A}, \tag{2-16}$$

where k = radius of gyration (in.). The parallel-axis theorem can be applied to the radius of gyration in the following manner:

$$k^2 = \bar{k}^2 + d^2.$$

 Shear and Moment Diagrams

A thorough understanding of the construction and application of shear and moment diagrams is essential in solving problems in mechanics, strength of materials, and machine design. Such diagrams are useful in solving beam and shaft problems. The following problem shows a simple beam with a uniformly distributed load, a slight overhang, and a concentrated load.

◆ *Example* ──────────────────────────────

Construct shear and moment diagrams for the beam shown in Fig. 2-25.

Solution. As a first step, find the reactions R_1 and R_2:

$$\Sigma MR_1 = 0$$

or

$$7(14)(100) + 8(200) - 10R_2 = 0.$$

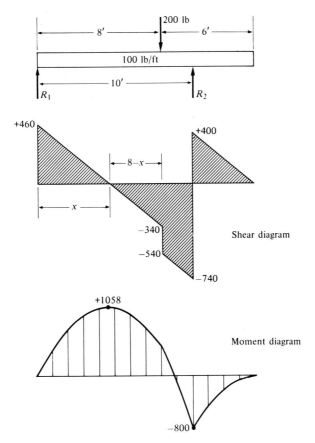

Figure 2-25
Shear and moment diagrams.

Thus, $R_2 = 1140$ lb. Then $R_1 = 460$ lb. This can be found by subtracting 1140 from the total load of 1600, since $\Sigma F_y = 0$. If preferred, moments can be taken about R_2 to solve for this value.

Construct the shear diagram by plotting vertical shear values along the entire length of the beam. Upward external forces can be considered positive and downward loads negative. It is essential to establish certain critical points. These are:

1. at the location of each concentrated load;
2. at the beginning and end of each uniformly distributed load;
3. at the beginning and end of each nonuniform load, as well as enough intermediate points to establish the slope of the curve.

Positive values are placed above the reference line and negative values below. The left end of the beam has a vertical shear load of $+460$ lb. One foot to the right, the value is 100 lb less, since the uniformly distributed load is 100 lb/ft. This procedure continues until the line of action of a concentrated load is reached. At such a point, the shear diagram goes vertically down (or up, in some cases). In most cases, the diagram can be drawn by locating the key points and then scribing the connecting lines.

Distance x (a point of zero shear) can be found from the similar right triangles. Thus,

$$\frac{460}{x} = \frac{340}{8 - x}$$

or

$$x = 4.6 \text{ ft.}$$

Moment values can be computed and a moment diagram constructed with positive values above the reference line and negative values below the line. In each case, we consider the beam and its loading to the left of the point about which we are taking moments. At 4.6 ft, $M = 460(4.6) - 100(4.6)(2.3) = 1058$ lb-ft (max. positive). At 10 ft, $M = 460(10) - 100(10)(5) - 200(2) = -800$ lb-ft (max. negative).

The usual practice is to compute bending moments at each external load or reaction, at the beginning and end of a distributed load, and at points where the shear value is zero. For use in most calculations, it is desirable to find the maximum moments in lb-in.

The maximum positive value is located 4.6 ft from the left end of the beam; the maximum negative value is located 10 ft from the left end. These are the locations where the shear value is zero.

Beam Deflections

The allowed deflection of a beam is a critical part of design. In structural design, too much deflection can cause damage to floor and ceiling materials. In shaft design, too much deflection can lead to bearing damage and also cause transmission elements to give poor performance. Thus, deflections must be known, and the design must hold deflections to reasonable limits. Structural deflection is often limited to a distance equal to $\frac{1}{360}$ of the span.

Several methods can be used to compute deflections. In this review, the moment-area method will be considered. Two theorems are involved in this method. These can be stated as follows:

1. The angle between the tangents to the elastic curve of a beam at any two points is equal to the area of the portion of the moment diagram between these two points divided by the product of the modulus of elasticity and the moment of inertia.
2. The vertical displacement of a point A on the elastic curve of a beam from the tangent to the elastic curve at another point B is equal to the moment of the area of the bending moment diagram (between the two points) with respect to A divided by the product of the modulus of elasticity and the moment of inertia.

The following examples apply this method to two elementary beam problems.

◆ *Example* —————————————————————————

Figure 2-26 shows a cantilever beam with a concentrated load at the free end. Find the maximum deflection at the free end and the angle θ.

Solution. Figure 2-26(a) shows the beam arrangement with letter designations. Figure 2-26(b) shows the elastic curve greatly exaggerated (the length of a beam is large compared to the linear deflection). Figure 2-26(c) illustrates the moment diagram with the centroidal distance from the free end. Let y represent the deflection. From the second theorem,

$$y = \frac{(\text{area})\bar{x}}{EI} = \frac{(L/2)(-PL)(2L/3)}{EI} = -\frac{PL^3}{3EI}\text{ in.}$$

The negative sign merely indicates that the deflection is below the original, or unloaded, position of point A. From the first theorem,

$$\theta = \frac{(\text{area})}{EI} = \frac{(L/2)(-PL)}{EI} = -\frac{PL^2}{2EI}\text{ rad.}$$

◆ *Example* —————————————————————————

Find the maximum deflection for the simple beam shown in Fig. 2-27(a). Assume that the load is 1000 lb, the span is 10 ft, $I = 15.4$ in⁴, and $E = 30 \times 10^6$ psi.

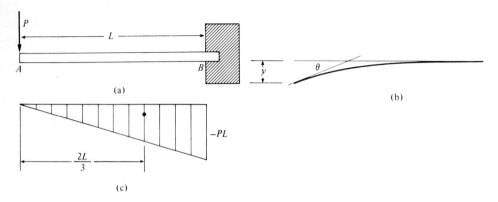

(a)

(b)

(c)

Figure 2-26
Deflection at the free end of a cantilever beam.

Solution. The maximum deflection occurs at the midpoint. A tangent to the elastic curve at this point is a horizontal line:

$$y = \frac{(\text{area})\bar{x}}{EI} = \frac{(1/2)(PL/4)(L/2)(2/3)(L/2)}{EI} = \frac{PL^3}{48EI}$$

$$= \frac{1000(10 \times 12)^3}{48(30 \times 10^6)(15.4)} = 0.078 \text{ in.}$$

Figure 2-27
Deflection of a simple beam.

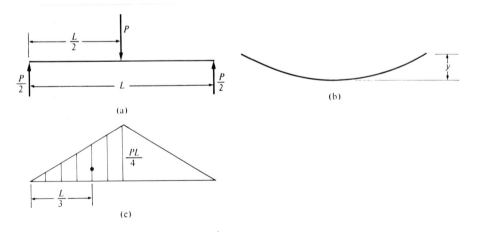

(a)

(b)

(c)

This is well within the value of $\frac{1}{360}$ of the span, which would be $\frac{1}{360}(120) = 0.333$ in.

Figure 5-4 in the chapter on shafting shows a few formulas for maximum deflections. Deflections for many beam arrangements are given in various handbooks.* If a beam is subjected to more than one single type of loading, the principle of superposition can be applied by adding the deflection formulas.

 Simple Stresses

Stress is defined as the internal resistance of a material to the action of external forces. From the engineering standpoint, unit stress has greater significance. *Unit stress* is usually expressed in pounds per square inch (psi); it is applied force per unit of area. The three fundamental stresses are *tension*, *compression*, and *shear*. These result from loading as shown in Fig. 2-28. The body in Fig. 2-28(a) is subjected to tensile stress; the body in Fig. 2-28(b) is subjected to compressive forces that induce compressive stress in the cross section. The rivet in Fig. 2-28(c) is subjected to shearing forces; thus, shearing stress is induced in the cross-sectional area of the rivet. For tension and compression, the following formula applies:

$$S = \frac{P}{A},$$

(2-17)

Figure 2-28
Simple stresses.

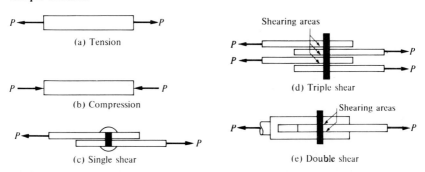

(a) Tension

(b) Compression

(c) Single shear

(d) Triple shear

(e) Double shear

*Refer to Roark, *Formulas for Stress and Strain*, 4th ed. McGraw-Hill, 1965.

where

S = induced stress (psi),

P = applied external load (lb),

A = area of cross section (in^2).

For shearing stress, the following equation is used:

$$S_s = \frac{P}{A},$$

(2-18)

where

S_s = induced shearing stress (psi).

Other units are the same as for tension and compression. If a rivet is subjected to double shear, there are two areas, $2a$, subjected to the shearing action. Thus, $2a$ is substituted for A. It is possible to have multiple shearing areas; the pin in Fig. 2-28(d) has three shearing areas. For this figure, the formula would be $S_s = P/3a$.

Figure 2-28(e) shows a pin with two shearing areas. For this arrangement, the formula becomes $S_s = P/2a$. Note that there is no load eccentricity with an *even* number of shearing areas. The applied forces are collinear; this is the ideal situation. Clevis mountings for pneumatic and hydraulic cylinders resemble this configuration.

Strain

From the engineering viewpoint, strain is defined as deformation. Unit strain is the ratio of a change in length to the original length. Thus, the equation for unit strain or deformation is

$$\delta = \frac{\Delta L}{L},$$

(2-19)

where

δ = unit deformation (in./in.),

ΔL = change in length (in.),

L = original length (in.).

Modulus of Elasticity

Hooke's law states that stress is proportional to strain within the elastic limit. Beyond the elastic limit, permanent deformation occurs. The yield point on a stress-strain diagram is the point where the curve flattens out before resuming its upward path. Hooke's law applies to the elastic limit, which is slightly below the yield point. For most materials, the modulus of elasticity is the same for tension or compression. The shear modulus for elasticity is the ratio of shearing stress to shearing deformation. This value is somewhat less than the modulus for tension and compression. The shear modulus for elasticity depends on Poisson's ratio. Poisson's ratio is the ratio of lateral unit deformation to axial unit deformation. Figure 2-29 shows diagrammatically the effect of applying a compressive force to a body. This ratio varies for different materials. For steel, Poisson's ratio is often taken as 1/4; for brass and bronze, a typical value is often 1/3. The following equations show the modulus of elasticity for tension and compression (E) and the shear modulus (G):

$$E = \frac{\text{stress}}{\text{strain}} = \frac{S}{\delta} = \frac{PL}{A(\Delta L)} \qquad \textbf{(2-20)}$$

and

$$G = \frac{E}{2(1 + \mu)}, \qquad \textbf{(2-21)}$$

where

E = modulus of elasticity (psi),

G = shear modulus of elasticity (psi),

S = stress (psi),

δ = strain (in./in.),

P = load (lb),

L = original length (in.),

ΔL = change in length (in.),

A = cross-sectional area (in.^{2}),

μ = Poisson's ratio.

Figure 2-29
Ratio of lateral to axial strain.

Thermal-Expansion Stresses

Any material expands when its temperature is elevated and contracts when its temperature is lowered. If the material is restrained from expanding or contracting, stresses are induced. The resulting equation can be written:

$$S = EC(t_2 - t_1), \tag{2-22}$$

where

C = coefficient of linear expansion (in./in./°F temperature change),

$t_2 - t_1$ = temperature change (°F).

◆ *Example* ────────────────────────────────────

A 10-in.-long, $1\frac{1}{8}$ in. steel bar is welded between two supports. If the bar is installed at a temperature of 70 °F, how much stress is induced if the temperature is increased to 130 °F? The coefficient of expansion for steel is 6.7×10^{-6}.

Solution

$$S = EC(t_2 - t_1) = 30 \times 10^6(6.7 \times 10^{-6})(130 - 70) = 12{,}060 \text{ psi.}$$

──

Note that the length drops out of the derived equation and that area is not needed if stress is being calculated. If axial force is needed, then the area is required for the equation $P = SA$. Temperature stresses can be utilized to good advantage in certain areas of design. For example, bands can be expanded by heating, then placed over solid cylinders and permanently affixed by shrink fitting. Also, segments of flywheels can be held together by links designed for expanding and shrinking into position.

Beam Stresses

Beams are subjected to various stresses caused by the type of loading. Figure 2-30 shows a simple beam carrying a single concentrated load in the midsection. The uppermost fibers in the beam are compressed by the applied load. The fibers at the bottom of the beam are lengthened because of the tensile load along the bottom. The shortening of the upper surface of the beam is equal to the lengthening of the bottom part. The change from tension to compression occurs at the neutral axis and is shown in Fig. 2-31. Thus, the distance from the neutral axis to the outmost fiber of a beam

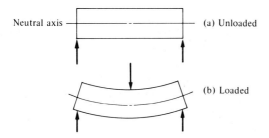

Neutral axis —   — (a) Unloaded

                                  — (b) Loaded

Figure 2-30
Beam subjected to bending.

plays a significant part in designing a beam for strength in bending or flexure. As previously mentioned, the moment of inertia of the cross section based on the neutral axis provides a measure for the strength in bending. Flexural stress may be computed from the following relationship:

$$S = \frac{Mc}{I} = \frac{M}{Z}. \qquad \text{(2-23)}$$

The preceding flexure formula is one of the standard relationships used in designing beams. It is often written in the form

$$\frac{M}{S} = \frac{I}{c}.$$

If the maximum moment and the allowable flexural stress are known, one can solve for I/c. Then, knowing the section modulus (I/c or Z), one can establish suitable dimensions for the required cross section. Handbook tables provide moment-of-inertia and section-modulus values for the common structural shapes. Where shapes such as those shown in Fig. 2-22 are desired, the appropriate section-modulus values can be used. If the chosen shape has two unknown dimensions, one must be assumed; then the other unknown can be derived.

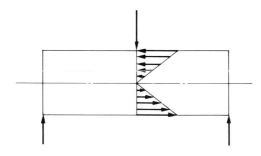

Figure 2-31
Stresses in a loaded beam.

◆ *Example* —————————————————————————

For the arrangement in Fig. 2-25, a rectangular section with a width of 1.25 in. is to be used. The ultimate tensile strength is 80,000 psi, and the ultimate shearing strength is 60,000 psi. For a factor of safety of 10, find the height of the beam.

> *Solution.* First, the maximum allowable moment is worked out to be 1058(12) = 12,700 lb-in. The allowable flexural stress is found by dividing the ultimate tensile stress by the factor of safety. Then
>
> $$\frac{M}{S} = \frac{I}{c} = \frac{bh^2}{6}$$
>
> for a rectangular section; or
>
> $$h = \sqrt{\frac{6M}{bS}} = \sqrt{\frac{6(12,700)}{1.25(8000)}} = 2.76 \text{ in.}$$

Note that this calculation does not include the weight of the rectangular beam. To be technically correct, one should recalculate, including the weight of the beam.

————————————————————————————————————

If a different width were used, the height of the beam would also be different. Sometimes the width has to be a certain value; then the appropriate height value is computed. In other cases, the height has to be a certain value; then the width becomes the unknown. Available stock sizes often dictate the choice of dimensions.

In the design of beams, horizontal and vertical shear should be checked. If strips of metal are stacked as shown in Fig. 2-32 to form a beam, and a sufficient load were applied to produce a large deflection, there would be a sliding (horizontal) action between strips, illustrating horizontal shear.

Figure 2-32
Horizontal shear stress in a beam.

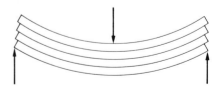

Horizontal shearing stresses at the centroidal axes are determined by the following equations:

$$S_H = \frac{3V}{2A} \qquad \text{for rectangular sections,} \qquad \textbf{(2-24a)}$$

$$S_H = \frac{4V}{3A} \qquad \text{for circular sections,} \qquad \textbf{(2-24b)}$$

where

V = maximum vertical value from shear diagram (lb),

A = cross-sectional area (in^2).

Vertical shear in I-shaped beams is given by the following:

$$S_V = \frac{V}{td}, \qquad \textbf{(2-25)}$$

where

t = web thickness (in.),

d = depth of beam (in.).

Torsional shear stress is found using a method similar to that used to determine bending stress. The following equation is used:

$$S_s = \frac{Tc}{J} \qquad \text{or} \qquad \frac{T}{S_s} = \frac{J}{c}, \qquad \textbf{(2-26)}$$

where

T = twisting moment (lb-in.).

Figure 2-23 shows (J/c)-values for some of the common shafting shapes.

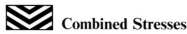 **Combined Stresses**

In two-force members, only direct tensile or compressive stresses are involved. In isolated torsion applications, shearing stress alone may be induced. However, in many engineering applications, parts are subjected to multiple loads (or components) that produce combined stresses. Most shafts are subjected to shearing stresses caused by torsion as well as flexural stress caused by the action of such components as gears, pulleys, cams, and similar devices. The frame of a "C" clamp must be designed for a combination of bending and tensile stresses. A twist drill can simultaneously undergo a compressive force, a twisting moment, and a bending moment, thus producing three stresses in combination. Although there are many stress combinations, we shall consider only two here.

Figure 2-33
Direct stress combined with flexure.

Direct Stress Combined with Flexure

In Fig. 2-33, a beam is subjected to a force P acting at an angle other than horizontal or vertical. Force P can be resolved into two components, P_V and P_H. The horizontal component produces a tensile stress throughout the beam; P_V produces a bending moment on the cantilever beam. Thus, at point A in the diagram, tensile stress is induced because of the direct stress and the bending stress. At point B, tensile stress is induced because of the direct load; also, compressive stress results from the flexure produced by P_V. The following formula applies to this type of situation:

$$S = \pm \frac{P}{A} \pm \frac{Mc}{I}. \tag{2-27}$$

To distinguish between the two types of stress, one can assign a plus sign for tension and a minus sign for compression.

◆ *Example* _____

For the rectangular beam in Fig. 2-33, assume that the force $P = 100$ lb acts 60 deg below the horizontal and at a distance of 5 in. horizontally from the fixed support. Assume that the beam is 0.5 in. in width and 1 in. in depth. Find the stress at points A and B.

Solution. First, resolve the force P into horizontal and vertical components:

$$P_y = 100 \sin 60° = 86.6 \text{ lb};$$

$$P_x = 100 \cos 60° = 50.0 \text{ lb}.$$

Then

$$S = \frac{P}{A} \pm \frac{Mc}{I} = \frac{P}{A} \pm \frac{6M}{bh^2} = \frac{50}{(0.5)(1)} \pm \frac{6(5)(86.7)}{0.5(1)^2};$$

$$S = 100 \pm 5200.$$

Thus

$$S_A = 100 + 5200 = 5300 \text{ psi} \quad \text{(tension)},$$
$$S_B = 100 - 5200 = -5100 \text{ psi} \quad \text{(compression)}.$$

Flexure Combined with Torsion

The maximum resultant shearing stress can be found from the following equation:

$$S_{s_{max}} = \sqrt{(S/2)^2 + S_s^2} = \frac{\sqrt{S^2 + (2S_s)^2}}{2}. \tag{2-28}$$

The maximum resultant tensile or compressive stress can be found from the following relationship:

$$S_{max} = \frac{S}{2} + S_{s_{max}}. \tag{2-29}$$

It should be noted that these equations are limited to combinations of normal and shear stresses only. Failure theories that deal with triaxial stresses are more complex. Examples of such theories can be found in many handbooks and textbooks on strength of materials.

 **Metric Units**

The following metric terms are applicable to mechanics and strength of materials. Although conversion factors are included in Appendix J, a few multipliers are presented here along with typical physical quantity values to give the student insight into the relative magnitudes of units in the U.S. and metric systems.

1. *Force.* The term *force* should not be confused with the term *mass*. By definition, a newton is the force required to accelerate a mass of one kilogram one meter per second per second. In statics problems, the term newton, *not* the term kilogram, should be used to indicate force. It is technically incorrect to use the term kilogram-force (kgf), although this is commonly used in manufacturers' catalogs in conjunction with torque and stress units. Weighing scales calibrated in kilograms are really indicating mass, not the force actually exerted. Capacity of hoisting

devices may indicate capacity in kilograms (kg) or tonnes (t). As a basis for comparison, one kilogram is equal to 2.2 lbs and a tonne is equal to 1 000 kilograms. Popular units for expressing forces are the newton (N), the kilonewton (kN), the meganewton (MN), and the giganewton (GN).

2. *Displacement, Velocity, Acceleration.* The base unit for distance in the metric system is the meter. Large distances are traditionally expressed in kilometers and short ones in millimeters. *The term centimeter is generally not used in engineering practice.* For comparison, one inch is equal to 25.4 millimeters and one mile is roughly 1.6 kilometers.

The larger values of velocity are generally expressed as kilometers per hour (km/h); the smaller ones are normally stated as meters per second (m/s). Accordingly, linear acceleration values are often expressed as meters per second squared (m/s^2).

The radian is a supplementary unit in the SI system and is almost universally used in calculating angular velocity and angular acceleration. Thus, rad/s and rad/s^2 are well-established relationships in the kinematics branch of mechanics.

3. *Torque, Work, Energy, Power.* The U.S. foot-pound designation for work or torque is properly replaced by newton-meter in the metric system. However, it is not unusual to find torque wrenches calibrated in kg · cm and kg · m units. Also, the torque applied to small torsion springs is often given in kg · mm.

Energy is usually specified in joules (J). By definition, one joule is equal to one newton-meter. To convert foot-pounds to joules, merely multiply the foot-pound figure by 1.355. The joule is also used to express the thermodynamic quantity of heat. To convert from British thermal units (Btu) to joules, multiply by 1 055.

The appropriate metric unit for power is the watt (W), which is a derived SI unit based on joules per second. One horsepower is equal to 746 watts, or 0.746 kilowatts.

4. *Moment of Inertia, Section Modulus of Area.* Appropriate units for moment of inertia and section modulus in the metric system are 10^6 mm^4 and 10^3 mm^3, in place of the inches^4 and inches^3 traditionally used in the United States.

It is interesting to note that practice varies in the literature distributed by various steel mills regarding the properties of sections rolled to metric dimensions. Structural cross sections such as I-beams, H-beams, T-sections, and angles are given consistently in millimeters. However, moment of inertia is often given as cm^4 (not 10^6 mm^4) and the section modulus as cm^3 (not 10^3 mm^3). Conversion is simple; however, the numerical figures will differ substantially because of the differences in powers of ten between the two designations.

A beam length can be expressed in meters or millimeters. It is necessary to convert from meters to millimeters when using the flexure formula to be compatible with other units used in the formula. Also, if the allowable flexural stress is specified in MN/m², appropriate changes must be made to conform to the other units.

Although loads on beams should be expressed as newtons, the term kilogram-force (kgf) is frequently used. The conversion from kilogram-force to newton can be made by multiplying the kgf value by 9.81.

5. *Stress.* The preferred designation for stress is the meganewton per square meter (MN/m²). Again, certain commercial catalogs use the term kilogram-force per square millimeter (kgf/mm²), decanewton per square millimeter (daN/mm²), or pascal (Pa) to express this quantity. For this reason, a student should be knowledgeable in the various units that might be encountered in seeking design information. The following approximate relationships are presented in order to become acquainted with magnitudes used in practice:

$$60,000 \text{ psi} = 413 \text{ MN/m}^2,$$

$$60,000 \text{ psi} = 42.3 \text{ kgf/mm}^2,$$

$$60,000 \text{ psi} = 41.3 \text{ daN/mm}^2,$$

or

$$60,000 \text{ psi} = 413 \times 10^6 \text{ Pa}.$$

6. *Modulus of Elasticity.* The most desirable unit for calculations involving the modulus of elasticity is the giganewton per square meter (GN/m²). A typical modulus of elasticity value (tension) for steel is generally 30,000,000 psi; in metric units, this value is roughly 207 GN/m².

◆ *Example* ————————————————————————

Find the force required to punch a 25-mm hole in a 20-gage (0.9525 mm) steel plate if the shearing strength of the material is 400 MN/m².

Solution.

$$P = S_s A = S_s \pi D t,$$

$$P = 400 \,(10^3)(\pi)(25)(10^{-3})(0.9525)(10^{-3}),$$

$$P = 29.9 \text{ kN/m}^2.$$

◆ *Example* ————————————————————————

A simply supported wood beam has a rectangular cross section of 50-mm wide and 100-mm deep. A concentrated load rests at the midpoint of its

4-meter span. Find the allowable load if the allowable tensile stress is 8.3 MN/m².

Solution.

$$\frac{M}{S} = \frac{I}{c} = Z = \frac{BH^2}{6} \text{ (for rectangular section)};$$

Maximum moment $= \dfrac{PL}{4}$.

Therefore

$$P = \frac{4BH^2S}{6L} = \frac{4(0.05)(0.1)^2(8.3)(10^6)}{6(4)};$$

$$P = 692 \text{ N.}$$

 Summary

The intent of this chapter is to provide a brief review of some concepts in the general area of mechanics and strength of materials. Every area covered is extremely complex in detail; thus, constant reference must be made to texts that deal in more depth with certain phases of design.

Texts and handbooks in the structural field present formulas for beams under various conditions of loading. Many mechanical components, such as shafts, are often loaded in a similar manner. The use of handbook data can save much time in finding maximum moments or deflections. For situations not covered in such listings, the student should exercise care in constructing and interpreting shear and moment diagrams.

Other mathematical tables (in addition to the well-known tables of logarithms and trigonometric functions) can be very useful by promoting accuracy and saving time. Typical tables found in handbooks and commercial catalogs include:

1. weights, areas, and circumferences of round steel bars;
2. functions of numbers, such as squares, cubes, square roots, cube roots, reciprocals, and so on;
3. root areas of screw threads;
4. linear-rotational speed relationships;
5. torque, horsepower, and speed relationships;
6. circular segments;
7. exponential values of e;
8. radians-degrees tables;
9. linear nut advancement for screw-thread rotation.

Hundreds of nomographs have been published for the common engineering equations. These are valuable, but serve to best advantage as a check on calculations made by the designer.

Statics, kinematics, and kinetics play important roles in machine-design work. Thus, a designer should have much background and reference material for each of these areas.

Whether a problem should be solved graphically or mathematically depends on the accuracy needed. In the analytical solution, the vector-analysis approach is usually helpful in establishing the mathematical solution. Carefully drawn sketches, with vectors, are usually prerequisites for arriving at correct mathematical solutions to problems.

Questions for Review

1. Explain why shifting of weight in a freight elevator could present serious problems.

2. Why is the collision of two vehicles traveling 60 mph more than twice as serious as a collision at 30 mph? Use sample figures to justify the conclusion.

3. Differentiate between displacement, velocity, and acceleration.

4. What is meant by brake horsepower? Why is efficiency important in the design of any equipment?

5. Differentiate between potential energy and kinetic energy. Give an example of each.

6. Indicate appropriate metric units for each of the following: (a) stress, (b) work, (c) power, (d) torque, (e) bending moment, and (f) moment of inertia.

Problems

1. A 1000-lb force acts at the end of a 10-ft cantilever beam. The section modulus for the steel beam is 8.0 in^3. How much stress is induced, and where does the maximum stress occur?

2. It is necessary to drill a 1-in. hole in the web of a 10-in. I-beam. Which of the following locations is preferable from the flexural standpoint? (a) 3 in. from the top. (b) 5 in. from the top. (c) 7 in. from the top.

3. For a given material, the modulus of elasticity is 100 GN/m^2 in tension and 40 GN/m^2 in shear. Find Poisson's ratio.

4. A 25-mm shaft is keyed to a 300-mm diameter pulley and transmits 3 kW of power. The keyed assembly rotates at 1725 rpm. What is the tangential force at the key and at the edge of the pulley? What is the peripheral speed of the pulley in m/s?

5. A 6.5-mm steel wire is 4-m long and stretches 10-mm under a given load. If the modulus of elasticity is 200 GN/m^2, what is the applied load?

6. What force is necessary to punch a 1-in. hole in a $\frac{1}{8}$-in. steel plate if the ultimate shearing stress is 60,000 psi and the ultimate compressive stress is 80,000 psi?

7. A grinding wheel rotates at 1750 rpm and has a surface speed of 2290 ft/min. Find the diameter of the wheel in inches.

8. A piston reciprocates 100 times per minute (single strokes). The length of the stroke is 6 in. What is the average wristpin velocity in ft/min? If the center of the crankshaft is on the same centerline as the piston, find the crankpin velocity in ft/min and the angular velocity of the crank in rpm.

9. A $\frac{1}{8}$-in.-diameter wire is used in a chain link. The allowable stress (tensile) is 10,000 psi for the link material. What load can the chain safely support?

10. Figure 2-34 is a diagram of a jib crane. The load can be moved the entire length of the horizontal beam. What position would be considered when designing (a) the beam, (b) the cable and fittings, (c) the wall supports? If the cable were longer for the same length of beam, would the "stress" of the cable be increased or decreased?

11. How much will a 25-ft cable be lengthened if the unit strain is 0.00067 in./in.?

12. A 1.5-m vertical post is subjected to a force of 9 000 newtons acting at the top at an angle of 45 deg. The round post has an outside diameter

Figure 2-34
Problem 10.

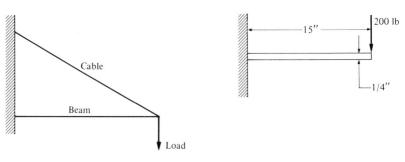

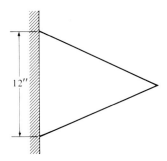

Figure 2-35
Problem 13.

of 100 mm and an inside diameter of 75 mm. Find the maximum induced stress and its location.

13. In the cantilever beam of Fig. 2-35, the modulus of elasticity is 200 GN/m². Find the maximum induced beam stress at the wall and at the midpoint.

14. A C-clamp has a basic T-section with a cross-sectional area of 1.81 sq in. The section modulus is 0.55 in³. The maximum allowable stress for the frame is 8000 psi. If the distance from the centerline to the centroid of the frame is 5 in., how much load can be placed between the jaws?

15. A U-bolt supports a load of 6000 lb. The cross section of the bolt has a diameter of $\frac{1}{2}$ in. How much stress is induced in the sides of the bolt?

16. A 1-in.-diameter shaft has a 2-in.-diameter collar resting on a support. The axial load on the shaft is 10,000 lb and the thickness of the collar is $\frac{1}{2}$ in. How much shear stress is induced?

17. Find the diameter of a wheel rotating at 180 rpm with a rim velocity of 1000 ft/min.

18. What is the kinetic energy of a 1000-lb weight moving at 15 mph?

19. A 2-ton weight is lowered at a constant acceleration of 2 ft/sec². What is the cable "stress"? If the weight is raised at the same rate of acceleration, what is the cable "stress"?

20. In Fig. 2-36, dimension $a = 12$ in., $b = 9$ in., and $c = 21$ in. Find the maximum bearing reaction for a load of 1800 lb.

The following problems are provided for practice in the use of metric units.

21. The management of a plant wishes to provide a speed limit sign on the entrance road, giving both U.S. and metric units. The speed limit is 25 mph. What limit should be shown in metric? Round off the figure to an appropriate value and indicate the appropriate metric units.

22. Find forces in each member of the truss shown in Fig. 2-37 and label *T* or *C* for tension or compression. Solve mathematically and graphically.

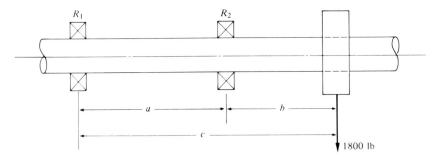

Figure 2-36
Problem 20.

23. The speedometer of a vehicle changes from 20 km/h to 80 km/h in 10 seconds. Find the acceleration for this period and express the answer in m/s^2.

24. Find the force needed to punch a 10-mm hole in a 1.5-mm steel plate. Assume that the ultimate shearing stress is 400 MN/m^2.

25. The modulus of elasticity (tension) of a material is $207 \times 10^9 \, N/m^2$ and the transverse modulus of elasticity (shear) is $82.7 \times 10^9 \, N/m^2$. Find Poisson's ratio.

26. A hollow rivet has an outside diameter of 5 mm and an inside diameter of 3 mm. If the allowable shearing stress is $400 \times 10^6 \, N/m^2$, what maximum shearing force can the rivet sustain if it is subjected to double shear?

27. A circular rod is 12.8 mm in diameter and carries gauge marks placed 50.8 mm apart. When the rod is subjected to an axial force of 31 150 newtons, the distance between the gauge marks is 50.86 mm. Find the modulus of elasticity of this material.

28. A picture frame weighs 1.2 kg and is suspended by a wire 1 m long when slack, attached at the two ends. When the frame is hung, the center of the wire is 50 mm higher than a horizontal line through the

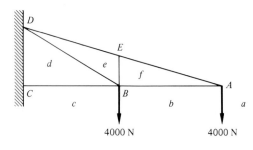

Figure 2-37
Problem 22.

two ends. Find the "stress" in the wire. (*Note:* Express the answer in kilograms, although force values should be expressed in newtons when using the metric system.)

29. A vertical load of 400 newtons acts at the end of a horizontal rectangular cantilever beam 2 m long and 25 mm wide. If the allowable bending stress is 130 MN/m², find the depth of the beam in mm.

30. Find the radius of gyration for a triangular section with 30-mm sides.

31. A vehicle with 0.75-m diameter wheels is traveling at a speed of 25 km/h. What is the rotational speed of a wheel in rpm? If the rear-axle ratio is 4/1, what is the engine speed?

32. A simply supported timber beam is 50 mm by 200 mm in cross section and 4 m long. If the fiber stress is not to exceed 8.3 MN/m² and the beam weight is neglected, find the maximum mid-span concentrated load that the beam can support if the 200-mm dimension is vertically oriented.

33. An I-beam rolled to the following metric dimensions extends 3 m from a building:

$$\text{depth} = 460 \text{ mm},$$
$$\text{width} = 190 \text{ mm},$$
$$\text{web thickness} = 8.5 \text{ mm},$$
$$I = 294 \times 10^6 \text{ mm}^4,$$
$$I/c = 1300 \times 10^3 \text{ mm}^3.$$

The allowable flexural stress must not exceed 138 MN/m². Neglecting the beam weight, find the maximum load that can be supported at the end of this cantilever beam. Express the load in newtons.

34. Find the vertical shearing stress induced if the metric I-beam described in Problem 33 is subjected to a load of 60 000 newtons.

35. An 80-mm DIN (*Deutsche Industrie Normen*—The West German Industrial Standard Specification) I-beam is simply supported at its ends with a span of 4 m and carries a uniformly distributed load of 100 kg per meter. The following properties are listed for this beam:

$$\text{cross-sectional area} = 7.57 \text{ cm}^2,$$
$$\text{weight} = 5.94 \text{ kg/m},$$
$$\text{moment of inertia} = 77.8 \text{ cm}^4,$$
$$\text{section modulus} = 19.5 \text{ cm}^3.$$

Considering the applied load and the weight of the beam, calculate the induced flexural stress. Express your answer in MN/m².

36. A 50-mm diameter steel shaft transmits 10 kW to a pulley rotating 100 rpm. Find (a) the shearing stress induced in the key; and (b) the

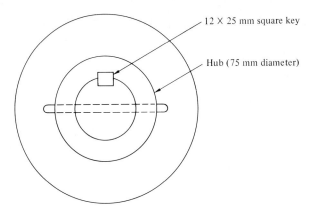

Figure 2-38
Problem 36.

diameter for a steel pin that could replace the key, assuming that the pin will be made of the same material and will be stressed to the same value as the shaft. See Fig. 2-38.

37. The crankshaft of an engine rotates at 1200 rpm. The connecting rod is 6 in. long and the effective radius of the crank circle is 2 in. When the crank makes an angle of 30° with respect to the horizontal plane, what is the piston velocity? Express your answer in fpm.

38. For the beam loading shown in Fig. 2-39, calculate the reactions at A and B and draw the shear and moment diagrams, labeling key points. Neglect the weight of the beam.

39. Calculate $\bar{x}$ and $\bar{y}$ for the shape shown in Fig. 2-40.

40. Find the forces in each leg of a tripod positioned as shown in Fig. 2-41.

Figure 2-39
Problem 38.

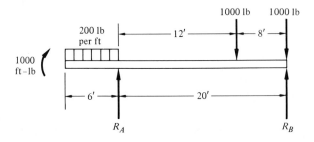

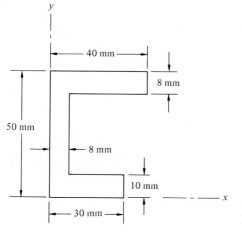

Figure 2-40
Problem 39

Figure 2-41
Problem 40

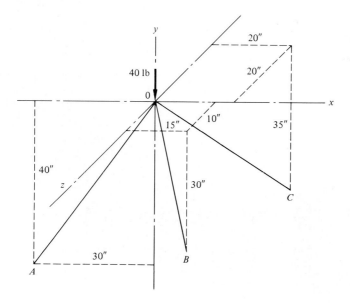

FRICTION AND LUBRICATION

3

Friction plays an important part in mechanical design. In such components as clutches, brakes, and belts, a large amount of friction is needed so that the parts will function properly. For moving parts such as rotating shafts in bearings and sliders or pistons in guides or cylinders, it is important to keep friction at a minimum. In the case of moving parts, friction plays a huge part in the overall efficiency of the machine.

 ## Sliding Friction

In pure sliding friction, one part moves relative to another part. The coefficient of friction can be determined experimentally by noting the amount of pull required to overcome friction:

$$f = \frac{F}{N},$$ (3-1)

where

f = coefficient of friction,

F = pull required to overcome friction (lb),

N = normal force or reaction (lb).

The normal force is due to the weight acting on the surface. The value of f depends on several variables. These include:

1. the surface roughness (microinches, root mean square), which depends on the materials used and the process by which the surfaces were finished;
2. the rubbing velocity of the two surfaces; and
3. the condition of the surfaces (dry, completely lubricated, partially lubricated).

Table 3-1 gives a few typical coefficient-of-friction values.

Table 3-1
Typical Coefficient of Static Friction Values (Dry Conditions)

Contacting Surfaces	Coefficient of Friction
steel–steel	0.15–0.30
steel–carbon graphite	0.25
steel–woven asbestos	0.30–0.60
steel–brake lining	0.40–0.50
cast iron–cast iron	0.15–0.20
cast iron–leather	0.30–0.60
cast iron–bronze	0.10–0.20
bronze–bronze	0.20

Friction and the Inclined Plane

By definition, the coefficient of friction is equal to F/N. The friction angle is then defined as $\phi = \arctan F/N$. If the angle a plane makes with the horizontal is less than the friction angle for an object on the plane, the body will remain stationary unless a force is applied to move it up or down the incline. If the plane angle is larger than the friction angle, a force *must* be applied to the object to move it up the plane or to prevent it from sliding down the incline. When the plane angle becomes large enough, the object tips. (Tipping will be neglected in this discussion.) In all inclined-plane problems, the equations of static equilibrium apply.

In Fig. 3-1(a), the plane angle is larger than the friction angle and impending movement is *up* the plane. Vectors F and N are components of

Figure 3-1
Friction and the inclined plane.

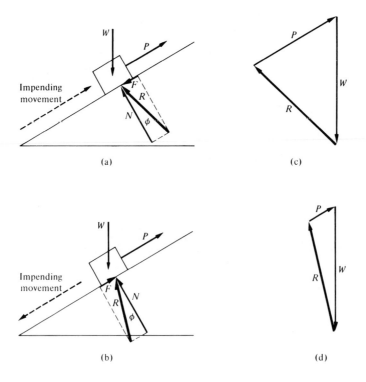

R, parallel and perpendicular, respectively, to the plane. The direction of F is always opposite to the direction of impending movement. If the block is being moved up the plane, the value of P is the sum of F and the component of W that is parallel to the plane.

In Fig. 3-1(b), the plane angle is also larger than the friction angle, but impending movement is *down* the incline. Note that components of R are (1) F parallel to the plane and (2) N perpendicular to the plane. Again F is directed opposite to the direction of impending motion. Therefore, the force P required to lower the load is the difference between the component of W parallel to the plane and the value of F.

Thus with a certain plane angle and a given coefficient of friction between the plane and a body, it requires a greater pull to move the load up the plane than it does to lower the load down it. The following example shows the mathematical and graphical approaches to this type of problem.

◆ *Example* ———————————————————————

For Fig. 3-1(a) and 3-1(b), assume that the block weighs 200 lb, the plane angle is 30 deg, and the coefficient of friction is 0.3. (a) Find the pull required to start the block up the plane; solve mathematically. (b) Find the pull needed at P if the weight is being lowered down the incline; solve analytically. (c) Check both solutions graphically.

Solution. (a) Using the 30-deg incline as the reference axis, write an equation of equilibrium for the "horizontal" components of P, W, and R (see Fig. 3-1a):

$P - W \sin 30 - fW \cos 30 = 0$;

$P - (200)(0.5) - (0.3)(200)(0.866) = 0$;

$P = 152$ lb.

(b) Follow a similar procedure for downward movement, referring to Fig. 3-1(b):

$P - W \sin 30 + fW \cos 30 = 0$;

$P - (200)(0.5) + (0.3)(200)(0.866) = 0$;

$P = 48$ lb.

(c) Establish friction angle ϕ and the inclination of R for impending upward movement (Fig. 3-1a) and impending downward movement (Fig. 3-1b) by laying out 10 units on N and 3 units perpendicular to N. Construct force polygons as shown in Fig. 3-1(c) and 3-1(d) to a suitable force scale, then scale the value of P for each case. Check these figures with the mathematical solutions.

Rolling Resistance

Rolling resistance is often referred to as rolling friction, which is not quite correct. Figure 3-2 shows graphically the problem of rolling resistance. When a wheel or ball rests on a flat surface, the wheel or ball may be flattened somewhat, the surface over which it rolls may be indented, or some combination of the two. In any event, enough horizontal force must be applied at the axle to "climb over the hump." With steel wheels running on steel tracks, this indentation is very minute; thus it is easy to push such a device. With an automobile wheel running on a concrete highway, most of the flattening is in the tire. If a loaded wheelbarrow (with a rubber tire) is pushed through soft ground, flattening occurs in both the ground and wheel. Referring again to Fig. 3-2, note that the point of application of the reaction between the wheel and the flat surface lies somewhat ahead (distance a) of the vertical line through the center of the wheel. If this flattening does not produce too large a value for a, then $\sin \theta$ is equal to $\tan \theta$, and the following equation applies:

$$a = \frac{Fr}{W},$$ (3-2)

where F and W are expressed in pounds or weight units, and a and r are expressed in inches or linear units.

It should be obvious from the previous discussion that, for a given weight to be pushed, the surfaces (wheel and flat surface) should be hard and the wheel diameter should be as large as possible. Therefore, hand trucks usually have large wheels. In the case of automobiles, however, the

Figure 3-2
Rolling resistance.

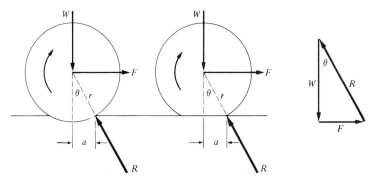

Table 3-2
Typical *a* Values for Rolling Resistance

Contacting Surfaces	*a* Values (inches)
steel–steel (hardened)	0.003–0.004
steel–asphalt	0.05–0.20
steel–concrete	0.05
ball bearings	0.0001
roller bearings	0.0012

drive is through the wheels, and traction (with a loss of efficiency) becomes important.

Table 3-2 shows a few typical *a* values for rolling resistance. It should be noted that values vary widely because of surface conditions. Best results can be determined experimentally.

◆ *Example* ―――――――――――――――――――――――――――

A $1\frac{3}{4}$-in.-diameter shaft is supported by two sleeve bearings. The total load on the two bearings is 2800 lb. Find the horsepower lost in friction if the coefficient of friction between shaft and bearing is 0.1 and the shaft rotates 200 rpm.

Solution. The load per bearing is 1400 lb. The frictional force is

$$F = fN = (0.1)(1400) = 140 \text{ lb.}$$

Rubbing velocity is

$$V = \frac{\pi DN}{12} = \frac{(\pi)(1.75)(200)}{12} = 91.6 \text{ ft/min.}$$

The horsepower loss per bearing is thus

$$\frac{FV}{33,000} = \frac{(140)(91.6)}{33,000} = 0.389 \text{ hp,}$$

and total horsepower loss is 2(0.389) = 0.778 hp.

◆ *Example* ―――――――――――――――――――――――――――

A wheelbarrow with load weighs 20 kg. On level ground, the 0.3-m steel wheel makes an indentation 30-mm deep. Find the horizontal force (in newtons) that must be exerted at the axle to start movement, neglecting friction at the axle bearing.

Solution. Referring to Fig. 3-2, find distance a as follows:

$$\theta = \text{arc cos } \frac{150 - 30}{150} = 36.9°;$$

$$a = r \sin \theta = 150 \sin 36.9° = 90.1 \text{ mm};$$

then,

$$F = \frac{Wa}{r} = \frac{(20)(9.81)(90.1)}{150} = 117.9 \text{ N.}$$

 ## Journal Friction and the Friction Circle

When a complicated series of links is connected by oversized pins (shafts), the analysis can be somewhat simplified by using a *friction circle*; this is a specialized and valuable technique for certain segments of the industrial community. When a shaft rotates in a sliding or sleeve-type bearing, friction results—one surface is rubbing against another surface. In its simplest form, the calculation for this type of friction is similar to that for sliding friction, and uses the same equation. Figure 3-3 illustrates a shaft rotating in a bearing with an exaggerated amount of clearance. With no friction (and no movement), the reaction would be R_n. The reaction with friction would be R_f. The slope of the shaft is dictated by the coefficient of friction. In other words, θ is the angle of friction and $\tan \theta = F/R_n$. If r is equal to the radius of the shaft, then $r \sin \theta$ is the radius of a friction circle tangent to the line of action of the reaction R_f. For small angles, $\sin \theta$ is equal to $\tan \theta$; therefore, the radius of the friction circle is equal to $f \times r$. To simplify further, *the*

Figure 3-3
Journal friction.

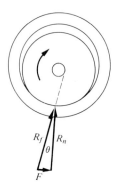

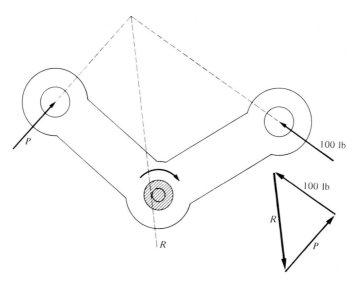

Figure 3-4
Rocker-arm application.

diameter of the friction circle is equal to the diameter of the shaft times the coefficient of sliding friction.

To use the friction circle as a method of analysis, reaction lines are first drawn tangent to the friction circle. In the case of a shaft rotating in a sleeve-type bearing, the reaction is drawn tangent to the friction circle on the side that opposes the rotation of the shaft. In many linkages, both shaft and bearing turn; in any case, however, there is relative motion of one part with respect to the other. The line of action is drawn on the side of the "close-in" between the parts or the side that reduces the mechanical advantage. This latter statement can best be understood by studying the complete force polygon—friction will always impede motion, thus causing a greater operating force.

Figure 3-4 shows a rocker-arm application. Force *P* must overcome the 100-lb force and friction in the bearing. In this figure, the line of action of the 100-lb resistance and of the operating force are defined. The friction-circle diameter is found by multiplying the diameter of the shaft by the given coefficient of friction. Since the device is in equilibrium under the action of the 100-lb force, the operating force *P*, and the reaction at the bearing, all three lines of action must intersect. A line is drawn from the intersection of the 100-lb resistance and the operating force *P* tangent to the friction circle on the side that would reduce the mechanical advantage. From the completed force diagram, it can be seen that *P* would be smaller if this line were drawn on the right side rather than the left. If friction were not considered, this line would be drawn through the shaft center.

It is more often difficult to apply this analysis to two-force (tension or compression) members. Without friction, the centerline would be the line of action of the reaction. When the friction circle is used, the centerline is displaced to the appropriate side of the friction circle and the force diagram is drawn. Figure 3-5 illustrates some rules to follow in designing tension and compression members. For clarity, the side of the friction circle to which the displaced centerline should be drawn is indicated by only a short line. Arrows on the two-force links show the direction of impending motion relative to the reference link. In applying these rules, it is often helpful to orient the figure to approximate the position of the linkage.

Figure 3-6(a) illustrates some of this analysis. Free-body diagrams are drawn adjacent to the links to aid in understanding the action of forces and reactions. Figure 3-6(b) shows the completed problem using sliding friction graphically, and the friction circle applied to all shaft-bearing situations.

Figure 3-5
Friction circle: two-force members.

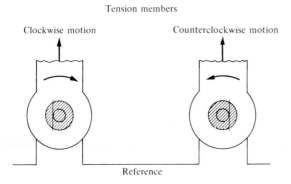

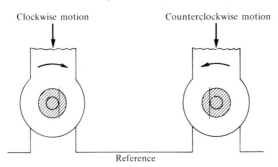

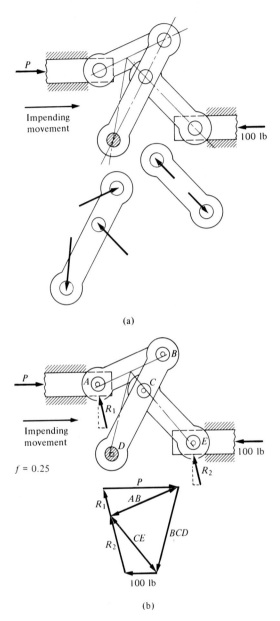

(a)

(b)

Figure 3-6
An example of a friction-circle analysis.

◆ *Example* ――――――――――――――――――――――――――

Figure 3-7(a) shows an arm with oversized pins 1 in. in diameter. If the coefficient of friction is 0.2 and clockwise motion is impending, find the operating force P required to overcome friction and start lifting the 200-lb weight.

Solution. Without friction, the summation of the moments about the centerline of the fixed pivot gives the result. Thus

$$2(200) = 4P \quad \text{or} \quad P = 100 \text{ lb.}$$

Figure 3-7
An example of a friction-circle analysis.

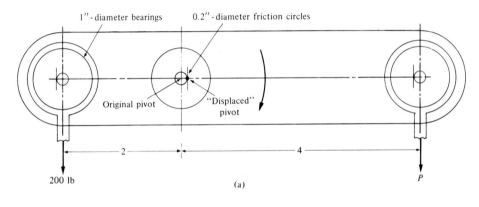

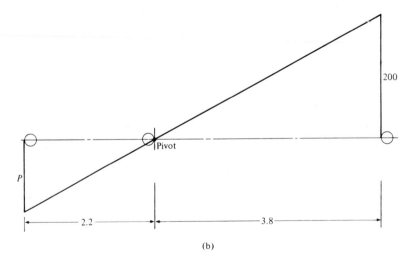

With friction, friction circles must be constructed at the center of each pin; the diameter of the friction circle is equal to 0.2 times 1, or 0.2 in. The vertical centerlines are then displaced (shown in the figure by a dash only) and drawn tangent to the friction circle on the side of the "close-in" between the shaft and bearing. Moments can then be taken using the newly established "centerlines." Thus

$$200(0.1 + 0.1 + 2.0) = (4.0 - 0.1 - 0.1)P;$$

$$2.2(200) = 3.8P;$$

$$P = 116 \text{ lb.}$$

By inverse proportion, the same result can be obtained graphically using the displaced "centerlines" for constructing the vectors and the horizontal centerline as the baseline. This is shown in Fig. 3-7(b). A force scale must be used in laying out the magnitude of the 200-lb vector; P is read on the left displaced "centerline."

The frictional torque for a shaft/sleeve arrangement similar to that shown in Fig. 3-3 can be found mathematically from the following equation:

$$T_f = rfW$$

where

T_f = torque required to overcome friction (in-lb or N · m),

r = radius of shaft (in. or m),

f = coefficient of sliding friction,

and

W = applied load (lb or N).

In most metric shaft applications, millimeters are used for the shaft radius, yielding a torque value in N · mm.

 **Pivot Friction**

Quite often, the frictional moments at pivots must be determined. Actually, a pivot is a simple type of bearing. Power is lost in any pivot; thus it is essential to determine how much power is lost through friction. Equations (3-3a) and (3-3b) give the torque required to overcome friction: $T_f =$

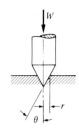

Figure 3-8
Conical pivot.

frictional torque (in-lb), f = coefficient of friction, W = load (lb), and r = radius (in.).

Conical pivot. See Fig. 3-8:

$$T_f = \frac{2}{3}\left(\frac{r}{\sin\theta}\right)fW. \qquad\qquad (3\text{-}3a)$$

Flat pivot. See Fig. 3-9:

$$T_f = \frac{2}{3}rfW. \qquad\qquad (3\text{-}3b)$$

◆ *Example* ────────────────────────────────────

Referring to Fig. 3-8, compare the frictional torque for conical points of 90° and 60°, assuming the following conditions: shaft diameter = 1.25 in.; r = 3/8 in.; W = 50 lb; and coefficient of friction = 0.15.

Solution.

$$T_f - \frac{2}{3}\left(\frac{r}{\sin\theta}\right)fW;$$

$$T_f = \frac{2}{3}\left(\frac{0.375}{\sin 45°}\right)(0.15)(50) = 2.65 \text{ in-lb (for 90° point)};$$

$$T_f = \frac{2}{3}\left(\frac{0.375}{\sin 30°}\right)(0.15)(50) = 3.75 \text{ in-lb (for 60° point)}.$$

Figure 3-9
Flat pivot.

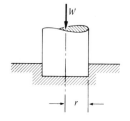

◆ *Example* ――――――――――――――――――――――

Find the horsepower lost to friction in a flat-pivot application, assuming the following conditions: coefficient of friction = 0.15; shaft diameter = 1.25 in.; axial force = 50 lb; and rotational speed of the shaft = 450 rpm.

Solution.

$$T_f = \tfrac{2}{3}rfW = \tfrac{2}{3}(0.625)(0.15)(50) = 3.125 \text{ in-lb};$$

$$\text{hp} = \frac{NT}{63{,}000} = \frac{450(3.125)}{63{,}000} = 0.0223 \text{ hp lost}.$$

 **Collar Friction**

See Fig. 3-10. Frictional torque must be taken into account in designing collars and disc clutches. The frictional radius for a collar can be calculated from the following equation:

$$r_f = \frac{2}{3}\left(\frac{r_o^{\,3} - r_i^{\,3}}{r_o^{\,2} - r_i^{\,2}}\right),\qquad\qquad\text{(3-4)}*$$

where

r_o = outside radius (in.),

r_i = inside radius (in.).

The frictional torque (in-lb) is then

$$T_f = \frac{2}{3}\left(\frac{r_o^{\,3} - r_i^{\,3}}{r_o^{\,2} - r_i^{\,2}}\right)fW.\qquad\qquad\text{(3-5)}*$$

 **Lubrication**

When two surfaces rub together, power is lost to friction, heat is developed, and wear takes place. The purpose of a lubricant is to separate the two rubbing surfaces somewhat and to reduce the abrasion. Oils and greases were developed to fill this need. Full lubrication tends also to reduce the temperature at the bearing surfaces. Figure 3-11 shows roughly what takes place when a lubricant separates the rubbing surfaces.

*Computer program in Appendix L.

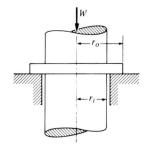

Figure 3-10
Collar friction.

Most moving mechanical parts require some form of lubrication. The type of component dictates some of the lubrication requirements, but the location of the machine (environmental conditions) also has an effect on the type of lubricant (and lubricating system) needed. Some components that require attention are rotating shafts in sleeve bearings, antifriction bearings (ball, roller, needle), slides, guides, ways, and cylinders. Dissimilar materials usually cause less surface damage than similar materials. Some of the more popular bearing materials are discussed in Chapter 4.

Properties of Lubricants

Producers of petroleum products perform many tests on their products. A few of these are listed below:

1. *Flash point:* the temperature at which the vapor above the surface will ignite when a flame is passed over the surface.
2. *Fire point:* the temperature at which the oil will release sufficient vapor to support combustion.
3. *Gravity, API:* a method of testing developed by the American Petroleum Institute that relates to specific gravity as follows:

$$\text{gravity, API(deg)} = \frac{141.5}{\text{specific gravity}} - 131.5. \tag{3-6}$$

Figure 3-11
Lubricant action.

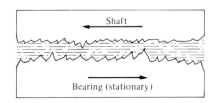

4. *Pour point:* the lowest temperature at which an oil will pour.
5. *Viscosity:* a measure of an oil's ability to flow.

 Viscosity and Viscosity Index*

From the designer's standpoint, these two properties are probably more important than any of the others. Various tests have been developed for determining viscosity, including, among others, the Saybolt, Redwood, Engler, and kinematic viscosity tests. Conversion charts are available for converting from one system to another. A field test for approximating Saybolt viscosity can be made with a viscosimeter that utilizes the falling sphere. This viscosimeter consists of two tubes, each containing a small steel ball. One ball is in a tube with a standardized (calibrated) oil; the other is in a tube with an oil sample. To carry out a test, the device is tilted. When the ball in the standard oil reaches a prescribed point, the position of the other ball is noted along a scale calibrated in Saybolt Seconds. Thus an approximation of the viscosity can be made.

Figure 3-12 shows a schematic sketch of the Saybolt viscosimeter. The sample of oil is heated to a prescribed temperature (usually 100, 130, or 210 °F). When the oil has reached the proper temperature, a plug is withdrawn sharply from the standard orifice, and the time (in seconds) required to fill a 60-ml container is found. This viscosity is known as Saybolt Seconds Universal (SSU). For high-viscosity oils, Saybolt Furol viscosity is obtained by using the same equipment with a larger orifice. The Furol times are approximately one-tenth those of the Universal for a given oil.

To simplify viscosity units, the Society of Automotive Engineers has standardized certain ranges of viscosities. Since they are interested in cold-weather and warm-weather applications, they grade the oil at two temperatures: 0 and 210 °F. For example, SAE 30 oil would have an SSU range of 58 to 69 at 210 °F, corresponding to an SSU of 538 at 100 °F.

Table 3-3 shows how some SAE lubricants compare at 100 °F. Saybolt and kinematic viscosity ranges are listed for typical SAE values.

Absolute viscosity can be defined as the tangential force per unit area required to move one parallel plane with respect to another at unit velocity. The planes are separated for a unit distance by the fluid being analyzed for viscosity. *Kinematic viscosity* is usually expressed in stokes or centistokes. The stoke is considered as square centimeters/second; the centistoke is $\frac{1}{100}$

*For a complete discussion of lubricant testing and conversion from kinematic viscosity to Saybolt viscosity, refer to ASTM Standard (1972, part 17).

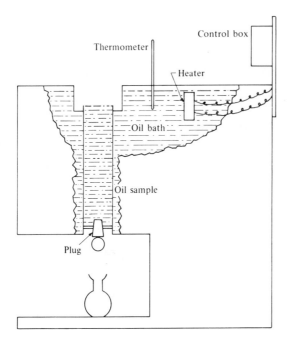

Figure 3-12
Saybolt viscosimeter.

Table 3-3.
Comparison of SAE Lubricants at 100 °F

Lubricant	Saybolt viscosity, SSU	Kinematic viscosity, centistokes
SAE 10	165 to 240	35.4 to 51.9
SAE 20	240 to 400	51.9 to 86.6
SAE 30	400 to 580	86.6 to 125.5
SAE 40	580 to 950	125.5 to 205.6
SAE 50	950 to 1600	205.6 to 352
SAE 60	1600 to 2300	352 to 507
SAE 70	2300 to 3100	507 to 682

stoke. The following relationship can be used for converting kinematic to absolute values, or vice versa:

$$Z = Z_k \rho, \tag{3-7}$$

where

Z = absolute viscosity (cP),

Z_k = kinematic viscosity (centistokes),

ρ = density (gm/cu $\cdot$ cm).

Appendix G lists some viscosity conversion factors.

Kinematic viscosity can be measured by using calibrated viscosimeter tubes (glass) containing a sample of oil immersed in a constant-temperature water bath at the usual standardizing temperatures. This viscosimeter is essentially a unit containing two glass bulbs connected by a capillary tube. The oil sample is raised (usually with a vacuum pump) to a position above an upper timing mark. A timer or stopwatch is used for measuring the time required for the meniscus of the oil sample to pass from this timing mark to a lower one. A calibration constant is applied to the efflux time to obtain the kinematic viscosity in centistokes. Kinematic viscosities from 0.4 to 16 000 centistokes can be measured by this method.

Viscosity index (V.I.) is a measure of an oil's tendency to change in viscosity with changes in temperature. The viscosity index is merely an empirical number that relates these changes. The value 0 V.I. was assigned to the behavior of relatively sensitive oils; 100 V.I. was assigned to relatively insensitive oils. Values can go outside the reference range of 0 to 100. Viscosity-index improvers are sometimes added to lubricating oils so that severe temperature changes will not affect the viscosity too greatly. The viscosity index is an important property of an oil when equipment is used outdoors.

Viscosity in Lubrication*

Plain journal bearings provide a good illustration of the role of viscosity in hydrodynamic lubrication. A bearing journal running under load will assume an eccentric position such as shown in [Fig. 3-13]. The rotating journal drags oil

*Courtesy of Texaco's magazine *Lubrication*, Vol. 64:1 (1978).

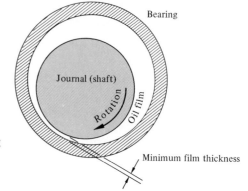

Figure 3-13
**Schematic of journal bearing
cross section.**
(Courtesy of Texaco's
magazine *Lubrication*.)

into the wedge-shaped clearance by virtue of the oil's viscosity. As the gap
between the bearing and journal surfaces becomes progressively narrower,
pressure builds up in the fluid film. The pressure developed in this way is
known as *hydrodynamic* pressure and maintains the separation of the journal
and bearing, thus supporting the load with the fluid film.

Minimum film thickness in hydrodynamically lubricated bearings decreases as the surface speed and the lubricant viscosity decrease and as the load
increases. The relationships are frequently expressed by means of a convenient
grouping of these quantities called the *duty parameter*, $\eta V/W$, where η is the
viscosity of the lubricant in the film, V is the surface velocity of the rotating

Figure 3-14
**Minimum film thickness as a
function of duty parameter.**
(Courtesy of Texaco's
magazine *Lubrication*.)

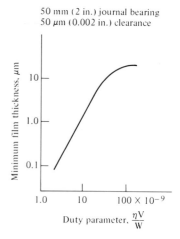

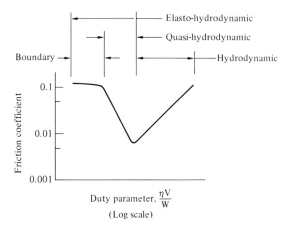

Figure 3-15
Coefficient of friction as a function of duty parameter.
(Courtesy of Texaco's magazine *Lubrication*.)

journal, and W is the bearing load. [Figure 3-14] shows the calculated variation of minimum film thickness with the duty parameter for a plain journal bearing. As the value of the duty parameter decreases, a lower limit of the minimum film thickness is reached. Below this limit, classical hydrodynamic lubrication no longer prevails. Minimum film thicknesses in hydrodynamically lubricated journal bearings typically range between 0.0025 mm (0.0001 in.) and 0.076 mm (0.003 in.).

Even though the surfaces in a journal bearing are separated by a fluid film, there is nevertheless friction between them. The friction results from the shearing of the lubricant as it is being "dragged" into the sector of the minimum film thickness. Since viscosity is a fluid's resistance to shearing force, the friction decreases with a decrease in viscosity. Journal bearing friction also depends on the surface speed and the load and can be plotted as a function of the duty parameter, as shown in [Fig. 3-15]. The minimum friction coefficient corresponds roughly to the lower limit of film thicknesses. Further decrease of the duty parameter results in increasing friction because of intermittent contact of the minute high points (asperities) of the surface textures. This situation is referred to as *mixed* or *quasi-hydrodynamic* lubrication because the load is only partially supported hydrodynamically. Continued reduction of the duty parameter leads to a region of *boundary lubrication*. In this regime, the load is no longer supported by the hydrodynamic pressure, and the conventional concept of viscosity is no longer the predominant influence in the lubrication of the surfaces.

Lubrication Systems

Lubrication of moving parts is important for proper functioning. Inadequate lubrication can cause additional friction, resulting in overheated parts and possible failure. Various systems are used for providing oil where needed. Some are listed below.

1. *Bath lubrication.* In bath lubrication, the bearings are completely submerged in a bath of oil. Although this is effective, it creates turbulence and prohibits proper filtering of the lubricant to remove impurities.
2. *Splash lubrication.* In a splash lubrication system, rings or dippers are attached to the shaft (or are made an integral part of it). As the shaft rotates, oil is scooped up from the reservoir and splashed out over the moving parts.
3. *Pressure lubrication.* In a pressure lubrication system, oil is pumped to the point where lubrication is needed. Afterwards, it is returned to a reservoir, filtered, and recirculated. In the process, it has a good opportunity to dissipate some of the heat it has acquired.
4. *Gravity lubrication.* Gravity lubrication can be provided in a number of ways. One possibility is to fill a reservoir that feeds by gravity to the various key points. This can be a central oiling system leading to many bearings. One reservoir can serve several lines, but if one line is clogged,

Figure 3-16
Multiple-feed oiler, electric solenoid control.
(Courtesy of Trico Manufacturing Company.)

one bearing receives no oil. Several commercial units are available that have sight drip fittings; this eliminates the danger of clogging. Figure 3-16 illustrates such a device; this model has a solenoid-operated control that stops oil flow when the machine being lubricated becomes idle.

5. *Oil-hole lubrication.* A number of different types of oil-hole covers and oil cups are available. Figure 3-17 shows some of those commonly used. They are available in threaded form and in shoulder drive form (the latter are pushed into a drilled hole, providing a snug fit). They can be obtained with hinged covers, threaded covers (which must be removed if oil is to be applied), and spring-loaded ball valves. Oil cups are available for side feed, 45-deg-angle feed, and underfeed, as well as conventional top feed.

6. *Hand feed.* This system is largely outdated; it requires a person to oil by hand intermittently. Any slip-up in a schedule could cause lubrication failure and damage to mating parts.

Grooving of bushings. Oil grooves are provided in bushings to aid in supplying oil to the proper places. This is necessary because a shaft resting in a bearing squeezes out the oil. Furthermore, as the shaft starts to turn, the oil is wedged against one side by viscous drag, thus increasing the pressure. Oil grooves serve as reservoirs and redistribution centers for the lubricant. Grooving is more important at low rotational speeds than at high speeds. Figure 3-18 shows a typical bushing with an oil groove. The edges of the groove are usually chamfered to prevent the oil "shearing" and to provide easier distribution.

Extreme-Pressure Lubricants

In such applications as antifriction bearings (ball, roller, needle) and gear trains, friction is low because there is only point or line contact between the mating surfaces. However, excessively high unit pressures can result because of the small contact area. At the point of contact between the ball and raceway of a ball-bearing assembly, unit pressures can be as high as 200,000 psi. Likewise, the line contact between meshing gears can produce high unit pressures—nearly as high as ball bearings. Each line (or small area) of contact changes as the meshing gear rotates; thus, the pressures are instantaneous only. Extreme-pressure lubricants include additives that keep friction losses and wear to a minimum.

Greases

Basically, a grease is a lubricating oil to which a thickener such as metallic soap has been added (although there are other methods for producing greases). Additional additives, such as viscosity-index improvers, ex-

Figure 3-17
Miscellaneous oil cups, oil-hole covers, and sight glasses.
(Courtesy of Gits Bros. Manufacturing Co.)

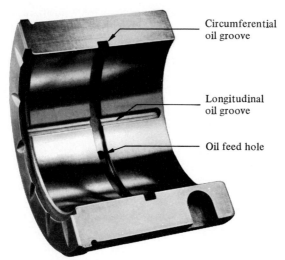

Circumferential
oil groove

Longitudinal
oil groove

Oil feed hole

Figure 3-18
Bearing half showing oil grooves.
(Courtesy of Federal-Mogul Corp.)

treme-pressure agents, rust preventatives, antioxidants, and so on, go into modern greases. Greases are usually used where sealing is impossible or inadequate. Sometimes greases replace oils in lubricating inaccessible parts. A grease has an apparent viscosity, since it does not flow in a manner similar to oil. In most cases, it has to be forced.

Dry Lubricants

Oils and greases come to mind first when one thinks of lubricants. However, a lubricant can be in the form of a solid, a liquid, or a gas. In the next chapter, air as a lubricant will be discussed in the section describing air bearings.

Of the dry (or solid) lubricants, graphite is probably the commonest and least expensive. In powdered form, it is a household lubricant, frequently used in locks. In flake form, it is often used in greases and oils. In its finely divided colloidal form, it is used in penetrating oils. If it is used in oils and the oil dries away, a layer of graphite will remain on the surface to provide some further lubrication.

Molybdenum disulfide is another common dry lubricant. It can be used as the powdered mineral (dry), or suspended in oils or greases. Like graphite, it has a tendency to "plate" onto surfaces. Boron nitride is a very expensive, white powdered mineral used as a dry lubricant. Niobium diselenide is a new dry lubricant with a high electrical conductivity. In powdered form, it can be applied to the rubbing surfaces, or it can be used with sintered metals. It is useful in extremely high-temperature applications. Its coefficient of friction is exceptionally low.

In general, dry lubricants are used where combustible oils and greases cannot be used. Graphite and molybdenum disulfide are stable to about 650 °F; at this point, they begin to oxidize.

 **Summary**

Friction can be an asset or a liability. In certain applications, it is desirable to use materials with a high coefficient of friction—clutches, brakes, and belts are notable examples. Where one part moves relative to another component (either by rotation or translation), friction is not desirable. The resulting abrasive action promotes wear, generates heat, and lowers the overall efficiency of the machine.

The two main classes of friction are sliding friction and rolling resistance (usually erroneously called rolling friction). The coefficient of sliding friction depends on the surface finishes, the materials used, the rubbing velocity, and the type of lubricant (if any) used between the mating surfaces. The friction circle can be a valuable, time saving tool for analyzing shaft friction; however, its use must be thoroughly understood, or errors will result. Rolling-resistance values depend on the hardness of the surfaces and on the diameter of the wheel or ball.

The selection of a proper lubricant is essential to the effective operation of most types of machinery. The viscosity of an oil is one of its important properties; if extreme temperature variations are encountered, an oil with a high viscosity index is preferable. The design of the lubrication system is important and should be considered in the beginning stages of machine design. Bearing pressures and the rubbing velocities of the shaft-bearing surfaces must be considered when selecting a lubricant. Dry lubricants are often used where high temperatures are encountered. Lubrication design is a highly specialized field, but much information is readily available in handbook and chart form.

Questions for Review

1. Explain why tests for viscosity are conducted at more than one temperature.
2. Differentiate between the terms *viscosity* and *viscosity index*. Explain why each of these properties is important in mechanical design.

3. Explain the significance of surface hardness and wheel (or roller) diameter in rolling resistance applications.

4. What is meant by the term *duty parameter?*

5. What advantages do dry lubricants have over liquid ones? What disadvantages?

6. Explain the features of five different types of lubrication systems. List an outstanding advantage for each type.

7. What is the purpose of oil grooves in large bearings?

8. What factors influence the value for a coefficient of friction?

9. Give five applications where a wheel diameter influences rolling resistance.

Problems

1. A 30-in.-diameter wheel with a frictionless axle is embedded 1 in. deep in the ground. It supports a load of 100 lb.
a) What horizontal force (applied at the axle) is needed to start the wheel rolling? Solve graphically, and explain why the formula $F = Wa/r$ might not produce the same answer.
b) What mathematical solution would give the same result as the graphical analysis?

2. A crate resting on a concrete floor has a mass of 90 kg.
a) If a horizontal force of 350 N is required to move it, what is the coefficient of friction?
b) If the crate were resting on a ramp inclined 30 deg to the horizontal, would it slide down? Prove your answer.

3. A 45 kg crate rests on two rollers 100 mm in diameter. If the coefficient of rolling resistance is 2.5 mm (both top and bottom), how much horizontal force must be applied to the crate to put it in motion?

4. A $\frac{1}{2}$-in. vertical shaft rotates 100 rpm in a 90-deg conical pivot bearing similar to that shown in Fig. 3-8. It sustains a load of 50 lb. Find the horsepower lost in friction if the coefficient of friction is 0.1.

5. A $\frac{3}{4}$-in. vertical shaft rotates 250 rpm in a flat pivot bearing (see Fig. 3-9). The coefficient of friction is 0.1 and the load on the shaft is 300 lb. Find the frictional torque.

6. Find the horsepower lost when a collar is loaded with 1000 lb, rotates at 25 rpm, and has a coefficient of friction 0.15. The outside diameter of the collar is 4 in. and the inside diameter is 2 in.

7. A 1-in. horizontal shaft rotates at 500 rpm in a sleeve-type bearing. The coefficient of friction is 0.15. Calculate the horsepower lost in the bearing if the reaction between the shaft and the bearing is 800 lb.

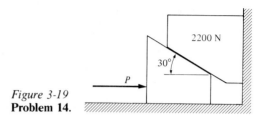

Figure 3-19
Problem 14.

8. A 2-in. horizontal shaft rotates in a sleeve-type bearing. The coefficient of friction is 0.1 and the shaft applies a load of 500 lb. Find the frictional resistance graphically and mathematically.

9. An oil has a specific gravity of 0.82. Find the gravity, API.

10. A 100-lb box rests on a plane inclined 15 deg with the horizontal. The coefficient of friction is 0.3. Find the force (parallel to the plane) to move the box (a) if upward movement is impending and (b) if downward movement is impending.

11. Find the frictional radius for a single disc clutch; the disc has an outside diameter of 10 in. and an inside diameter of 6 in.

12. A freight car weighs 60 tons.
   a) If the coefficient of rolling resistance is 0.015 in. and the wheel diameters are 33 in., what draw-bar pull is required to move the car horizontally?
   b) What pull would be required to move the car up a 1% grade? (A 1% grade is an elevation of 1 ft per 100 ft of level distance.)

13. A slide rests on 5 steel balls. The coefficient of rolling resistance is 0.025 in. between the slide and the balls, and 0.020 in. between the balls and the base. The weight of the slide is 400 lb. Find the horizontal force needed to set the slide in motion if the balls are 1 in. in diameter.

14. Find the horizontal push required if the wedge shown in Fig. 3-19 is to raise the block. The coefficient of friction at all sliding surfaces is

Figure 3-20
Problem 15.

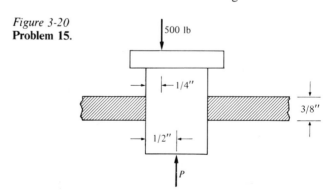

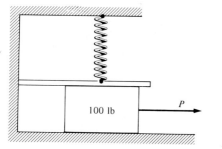

Figure 3-21
Problem 18.

0.4. Calculate force P mathematically and check your answer graphically.

15. In Fig. 3-20, find the force required at the bottom centerline of the valve stem if the coefficient of friction is 0.2. Solve graphically for (a) upward movement impending and (b) downward movement impending.

16. A bearing sustains a load of 4450 N. The shaft diameter is 100 mm, the coefficient of sliding friction is 0.01, and the shaft speed is 400 rpm. Find the horsepower lost in the bearing.

17. A 760 mm wheel carries a load of 4450 N. The coefficient of rolling resistance is 0.5 mm. What force is necessary to roll the wheel horizontally?

18. In Fig. 3-21, the spring exerts a force of 50 lb against the 100-lb block. If the coefficient of sliding friction is 0.3, find the force P necessary to overcome friction.

19. In Fig. 3-22, use the friction-circle analysis to determine the force P required to cause the impending movement of the drum to be clockwise.

Figure 3-22
Problem 19.

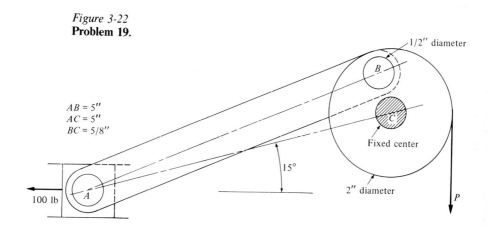

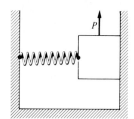

Figure 3-23
Problem 20.

Link *A*–*B* is 5 in. from the center of one bearing to the center of the other bearing. The coefficient of sliding friction is 0.3. All pins are $\frac{1}{2}$ in. in diameter. If impending movement of *P* were upward, would the value be the same?

20. In Fig. 3-23, find the force *P* required to move the 100-lb block upward if the spring force is 40 lb and the coefficient of friction is 0.3.

21. In Fig. 3-24, find the force *P* required to start the block up the plane if the coefficient of friction is 0.2 and the diameter of the wheel bearing is 1 in. The diameter of the wheel is $2\frac{1}{2}$ in. Neglect friction between the cable and the wheel (pulley). The weight of the block is 100 lb. If the block were being lowered, would the value of *P* be different? Solve graphically.

22. A 50-mm-diameter shaft supported by two sleeve bearings carries a load of 13.3 MN. The shaft rotates at 150 rpm. If the coefficient of sliding friction between the shaft and bearings is 0.1, how much power is lost in friction?

23. A dolly has four wheels, each 75 mm in diameter. Find the force needed to move the dolly horizontally if the total weight supported by the four

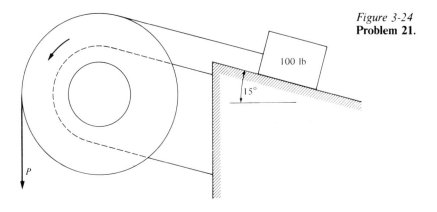

Figure 3-24
Problem 21.

100 lb

15°

P

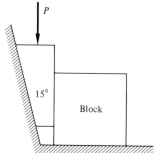

Figure 3-25
Problem 27.

wheels is 4450 N, the coefficient of rolling resistance is 0.25 mm, and bearing-journal friction is not considered.

24. Find the horsepower lost in moving a four-wheeled, $\frac{1}{2}$-ton car horizontally at 15 mph if it is equipped with 20-in. wheels and 2-in. bearings. The coefficient of sliding friction is 0.1 and the coefficient of rolling resistance is 0.01 in.

25. A 12-in.-diameter hoisting drum lifts a load of 500 lb at a speed of 60 ft/min. It is driven by a 2-in.-diameter shaft; the coefficient of sliding friction in the shaft bearings is 0.1. Find the horsepower required to meet these conditions.

26. A hydraulic oil has a viscosity of 60 Saybolt universal seconds at 100 °F. Convert to centistokes (cSt) and to meters^2/second units.

27. Find the force P required to counteract the 10 N downward force exerted by the block if the coefficient of sliding friction is 0.3 at all surfaces. Assume that the wedge is weightless. Refer to Fig. 3-25.

28. In Fig. 3-26, the coefficient of sliding friction is 0.3 between the 100-lb block and the vertical surface. Assume that the pin connections are frictionless. Find the value of P needed to start the block moving.

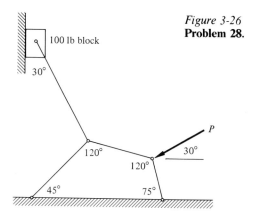

Figure 3-26
Problem 28.

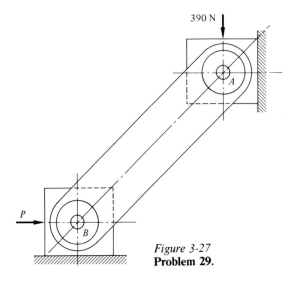

390 N

A

P

B

Figure 3-27
Problem 29.

29. Referring to Fig. 3-27, find the value of *P* required to move the lower slider if the bearings are each 25 mm in diameter, their frictional coefficients are both 0.3, the force on the upper slider is 390 N, distance *AB* is 118 mm, and the rod is positioned 45° above the horizontal. Solve graphically using a force scale of 1 mm = 10 N.

BEARINGS

4

Sliding Bearings (Bushings)

The simplest type of bearing consists of a hole in a frame or housing that contains some type of bushing. The usual procedure is to provide a tight fit between the bushing and the housing and then allow a running fit between the bushing and the shaft or rod. Bushings are used for rotating shafts; they are also used to guide such components as piston rods, regular guide rods, and any other type of guide or slide that translates. In design work, it is frequently necessary to decide whether to use plain sliding bearings (bushings) or antifriction bearings (ball, roller, needle). The type of application will often dictate the preferable type. It should be noted that the antifriction bearings do entail friction, but less friction than bushings. The major advantage of the sliding bearing is cost—it is much cheaper to produce. In the total design, however, the life expectancy of the bearing and the cost (plus inconvenience) of replacement must be considered. Friction losses are much greater for the bushing. However, radial space can be saved by using bushings, which take up much less radial space than ball and roller bearings. In designing bearings, one must consider the rotational speed, the heat-dissipation problem, and the intended use of the machine—continuous, intermittent, reversing, and so on.

L / D Ratio; Allowable Pressures

The length-to-diameter ratio of a bushing is important. Allowable bearing pressures are based on pressure per square inch of projected area. The projected area is defined as the product of the length of the bearing and the nominal inside diameter. Figure 4-1 shows the L- and D-dimensions of a simple bearing of this type. The L/D ratio is seldom less than 1. If it is too

Figure 4-1
Plain bearing (bushing).

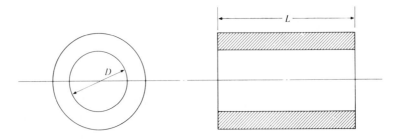

Table 4-1
Suggested Values for *L / D* Ratio and Allowable Bearing Pressures

Type of application	*L / D*	Allowable pressure, psi
Electric motors	2	100 to 200
Pumps	2	80 to 100
Machine tools	2–4	80 to 100
Main bearings (marine)	1–1.5	400 to 600
Crankpins (marine)	1–1.5	1000 to 1200
Main bearings (automotive)	0.5–0.8	500 to 600
Crankpins (automotive)	0.5–0.8	1500 to 1800

small, lubrication becomes a real problem. If the ratio is greater than 4, there may be alignment problems and additional lubrication problems. In general, the most practical values for L/D ratio fall between $1\frac{1}{2}$ and 2. Many "off-the-shelf" bushings are available in this range. Ratios of less than 1 are sometimes necessary to maintain compactness in the design. This is particularly true in the case of automotive crankpins and wristpins; however, the system of lubrication in this case allows their effective use. Laboratory testing of bearings with the proposed lubrication system is always helpful and desirable. Table 4-1 lists a few suggested values for L/D ratios and allowable bearing pressures based on the projected area of the bushing. The formula for L/D ratio is as follows:

$$A = LD \quad \text{and} \quad p = \frac{F}{A};$$

therefore

$$p = \frac{F}{LD};\qquad\qquad\text{(4-1)}$$

where

A = projected area (in^2),

F = load on bearing (lb),

p = allowable pressure (psi),

L = length of bearing (in.),

D = nominal shaft diameter or inside diameter of bushing (in.).

◆ *Example* ────────────────────────────────────

An electric motor has a $\frac{7}{8}$-in. shaft supporting a load of 150 lb. The bearing pressure is to be kept at 100 psi. Find a suitable length for the bearing and the L/D ratio.

Solution.

$$L = \frac{F}{pD} = \frac{150}{100(\frac{7}{8})} = 1.71 \text{ in.;}$$

$$\frac{L}{D} = \frac{1.71}{0.875} = 1.95.$$

────────────────────────────────────

Oil Grooves

Proper selection and placement of oil grooves in a bushing is important. Figure 4-2 shows a typical bushing with a longitudinal groove that extends nearly to the end of the bearing. It is important that it not extend all the way, for if it did, oil would immediately be lost from the bearing.

The oil hole that feeds oil to a bushing is drilled through the housing and the bushing. The lubricant is then supplied through the oil hole—by an oil cup or a central lubricator, in most applications. Where bath or splash methods are employed, this is not necessary, since sufficient oil reaches all areas.

Figure 4-3 shows the action of a lubricant when a shaft rotates in a bearing. In Fig. 4-3(a), the shaft is stationary; metal-to-metal contact is made at the bottom of the bushing. Theoretically, this is line contact; actually, a grooving action also extends a few degrees on either side of the

Figure 4-2
Plain bearing with axial groove and oil hole.

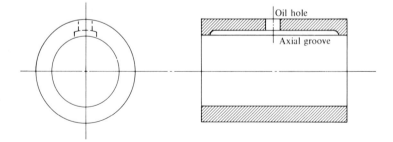

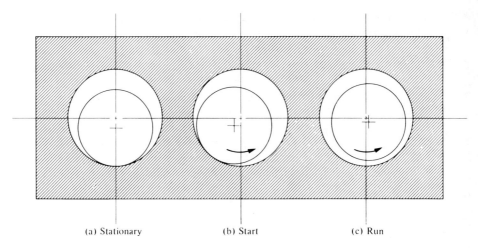

(a) Stationary   (b) Start   (c) Run

Figure 4-3
Shaft–bearing relationships.

vertical centerline. As the shaft starts to turn counterclockwise (Fig. 4-3b), the shaft tends to climb up the left side of the bearing. This forces the lubricant in a counterclockwise direction, producing a higher pressure on the lower left side of the journal. After the shaft attains operating speed (Fig. 4-3c), the lubricant under high pressure is forced under the bottom of the shaft, which "floats" on the lubricant a little below and to the right of the center of the bearing. The exact operating position of the journal depends on many variables, such as (1) the radial clearance between the two components, (2) the rotational speed of the shaft, (3) the surface finishes of each of the parts, (4) the type of oil used, and (5) the method by which the lubricant is introduced to the bushing. Needless to say, any change in the shaft load could upset provisions made for the preceding items. In general, the load and speed are specified in the design problem; the designer is responsible for adjusting the other variables.

Types of Lubrication

Perfect lubrication (thick film) is lubrication that maintains a complete film of lubricant between the surfaces of the shaft and bearing. *Boundary lubrication* (thin film) is the type where metal-to-metal contact can sometimes occur. This should not be confused with *dry operation*. With no lubricant, there is a continuous abrasive action between the two materials; this causes additional wear and generation of heat. Under light operating conditions, some devices do not require any lubrication. If conditions do not warrant lubrication, it should not be included in the design.

Diametral Clearance; Sliding Bearings

The *diametral clearance* is the difference between the diameter of the shaft and the inside diameter of the bearing in which it rotates. The *radial clearance* is one-half the diametral clearance. The purpose of a clearance is to provide space for an oil film to form. If the clearance is too small, the film cannot fit into the available space (for perfect lubrication); if it is too large, the oil pressure will not rise enough to lift the shaft above the bottom of the bearing. A rule of thumb is to make the diametral clearance equal to 0.001 to 0.0015 in. per inch of shaft diameter. However, this must be modified for various bearing materials. Laboratory verification and past experience with similar designs are important in selecting suitable values. The ASME standards for cylindrical fits also serve as a helpful guide.

 Porous Bearings

Bearings made of powdered metal are popular in light applications. They are made by compressing metal powder in a die cavity with heavy tonnage presses. Brittle briquettes are formed, which are then sintered. The sintering process is one of heating the powder to a temperature slightly below the critical point and then carefully controlling the cooling process. The finished bearing is porous; approximately 25% (by volume) of the bearing consists of pores or air spaces. These air spaces can then be filled with a nongumming lubricating oil. When a shaft rotates in such a bearing, it is lubricated by drawing oil out of the pores. When the shaft ceases to turn, the oil is drawn back into the voids by a capillary action. The strength of the bearing is not as great as one that is 100% metal; however, by carefully designing the bearing and keeping its load within limits, satisfactory results can be obtained. With porous lubrication, oil reaches all parts of the shaft. Many metal powders may be used. For bearing applications, phosphor bronze is the most popular, although a ferrous-based powder is sometimes used to provide greater strength. Figure 4-4 shows a method for replenishing the oil supply. Note that no oil hole or groove is necessary in the bushing; the oil hole in the housing merely supplies oil to the surface of the bearing. It then travels to all sections of the bushing by capillary action. Tooling charges are high initially, but if a large number of bearings of a given size are to be produced, unit costs are very reasonable. This same powder-metallurgy process can also be used for making mechanical components in large quantities. Reentrant angles cannot be handled unless additional work is

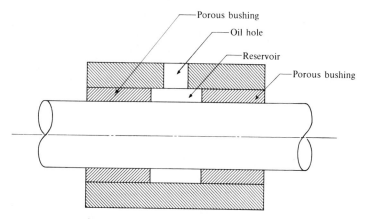

Figure 4-4
Replenishing oil in a porous bearing.

done after the part is formed. The porous metal can be machined, although special tools are usually required.

Porous bearings are commonly available ("off the shelf") in the sleeve type or the flange type. Powdered metal bearings are used primarily for radial applications, although thrust bearings made of such material are sometimes used. Dimensional tolerances of 0.001 in. are obtainable without additional operations. This type of bearing is particularly useful where oil drips could contaminate a product (food, cloth, etc.) or where oil drips could create a safety hazard.

Oilless Bearings

Porous metal bearings that contain their own oil supply are sometimes classified as oilless bearings; this is a misnomer. True oilless bearings use no oil. Some oilless bearings are sleeves impregnated with slugs of graphite or some other solid lubricant. Nylon bearings (sleeve and collar types) are oilless; they need no lubricant because they have excellent wear characteristics and are used only with light design loads. Nylon bearings are often used in light assemblies where noise levels must be kept low.

 ## Materials for Sleeve Bearings

The demands on a bearing material are severe; hence, it must have unusual properties. It should be strong enough to endure pressures caused by the load on the shaft. It must be malleable enough to assume the true position of the journal—this is necessary because the usual manufacturing tolerances allow some leeway in mounting components. After a new machine is "broken in," the inside bore of a bushing is often not concentric with the outside diameter. A bearing material should also be wear resistant, have a low coefficient of friction, and be able to withstand the temperatures encountered. (Even though a bearing is designed for thick-film or perfect lubrication, thin-film or boundary lubrication will occur as the machine is being started and brought up to operating conditions.) The material should also be corrosion resistant and have a coefficient of expansion similar to that of the housing that contains it.

Bearings can be either of solid construction or made in two parts. If a bearing is made in two parts, suitable lugs must be provided to prevent relative movement in the axial and radial directions. Some of the lined bushings are of the split type. Babbitt metals are frequently used for lining bearings. Babbitt is approximately 90% tin; the remainder is antimony and copper. The tin is quite often replaced with a lead base. The modulus of elasticity of lead-based babbitt is lower and thus it is more malleable; it also has a lower coefficient of friction. Babbitt metal cannot be used for high pressures; where higher pressures are needed, bronze linings are more suitable. Bronze is basically an alloy of copper and tin, but lead is usually added when it is used as a "bearing bronze." In general, a high tin content will produce greater strength. As with the babbitted bearings, a high lead content will provide a lower coefficient of friction and will help the bearing assume the position of the shaft. As a rule of thumb, the wall thickness of a bronze bushing is equal to $\frac{1}{8}$ in. per inch of shaft diameter. In some bearings, a layer of bearing bronze is used as a first lining; this is then covered with a thin layer of babbitt. In this way the corrosion-resistance of the babbitt and its low coefficient of friction are realized. If the babbitt does wear away under unusual and unanticipated loading, the high strength of the bronze can be used to good advantage.

Other bearing materials include the porous bronze and porous iron mentioned previously. The porous iron is not corrosion resistant; this could lead to trouble in certain applications. Nylon and Teflon® are popular among the plastic materials. Rubber is often used as a bearing material in marine applications and in constructing many types of pumps. The rubber permits contaminants such as sand and grit to pass through the bearing without damaging the shaft. It should be noted that in marine applications, water is often used as the lubricant. Aluminum alloys have found numerous uses, particularly in the automotive field. These relatively inexpensive alloys are high in strength and resistant to corrosion. Copper alloyed with cadmium or silver permits higher operating temperatures. A cast-iron bearing used in combination with a hardened steel shaft is probably the least expensive

Figure 4-5
Heavy-wall half bearing.
(Courtesy of Federal-Mogul Corp.)

type, but it is not corrosion resistant and it will score the shaft if contaminants are present. Scores of materials and combinations are available to the designer; all the properties and characteristics must be evaluated before a selection is made.

Figure 4-5 shows an automotive sleeve-type half bearing. Such bearings were originally developed for main bearings and connecting-rod bearings of reciprocating engines. A heavy series uses a greater wall thickness. The light series is available in sizes from $\frac{3}{4}$-in. shaft diameter up to 5 in. Heavy series start at $1\frac{1}{2}$-in. shaft diameter and have the same maximum size as the light series. Note the provision for locking the two halves together, and also the annular groove with oil hole for lubrication. Distributing grooves are provided on sizes where the length-to-diameter ratio (L/D) is greater than 0.80, so that axial lubrication will be more complete. Distributing grooves do not extend to the edge of the bearing. Although this type of bearing is classified as automotive, it is commonly used in heavy-machinery applications, such as large pumps, turbines, and compressors.

 Heat-radiating Capacity of Sliding Bearings

No matter how complete the lubrication may be in a sliding bearing, friction is always present. As the shaft rotates inside the bearing, heat is generated. This heat must be dissipated as much as possible. This can be done in several ways:

1. Provide surplus circulating oil.
2. Provide adequate air cooling.
3. Provide water cooling system.

The operating temperature of a bearing should be kept below 140 °F if possible. When an oil is heated to above 150 °F, its viscosity changes abruptly, and oxidation follows.

The heat-radiating capacity of a sliding bearing can be computed from the following empirical formula, originally developed by Axel K. Pederson:

$$Q = \frac{(t_o - t + 33)^2}{k}, \qquad\qquad\qquad \text{(4-2a)}$$

where

Q = heat-radiating capacity expressed in ft-lb/min-sq in. of projected area;

t_o = probable operating temperature of bearing (°F),

t = ambient temperature of surrounding air (°F),

k = const = 31 (for bearings of heavy construction and good ventilation) or 55 (for bearings of light construction in still air).

In many instances, the important calculation is to find the probable operating temperature for the bearing. This must be done to determine whether some type of cooling system must be incorporated into the design. The value of t_o can be found by rearranging the preceding equation:

$$t_o = \sqrt{Qk} + t - 33.$$

The value of Q is found from the following relationship:

$$Q = \frac{FV}{LD}, \qquad\qquad\qquad \text{(4-2b)}$$

where

F = frictional resistance, or the product of the coefficient of friction and the bearing load (lb),

V = rubbing velocity (fpm),

LD = projected area (sq in.).

The following example shows how this information can be applied to a bearing problem.

◆ *Example*

A heavy-duty sleeve bearing is used with good ventilation. The bearing has a diameter of 2 in. and a length of 4 in. The load on the bearing is 1100 lb. A shaft rotates 200 rpm in the bearing and transmits no heat to the bearing other than that due to shaft-bearing friction. The coefficient of friction is 0.02 and the ambient temperature is 70 °F. Find the probable operating temperature of the bearing.

Solution.

$$Q = \frac{FV}{LD} = \frac{0.02(1100)(2/12)(200)\,\pi}{4(2)} = 288 \text{ ft-lb/min-sq in.;}$$

$$t_o = \sqrt{Qk} + t - 33 = \sqrt{(288)(31)} + 70 - 33 = 131.5\,°F.$$

From this calculation, it appears that cooling water would not be necessary, since the ventilation of the bearing will keep the probable operating temperature at a reasonable level. For the given bearing dimensions, a slightly larger value for coefficient of friction, load, or speed will increase the operating temperature to a point where additional cooling will be needed.

 **Air Bearings**

Air is sometimes used as a lubricant. However, several problems must be carefully considered. Air must be of the proper pressure to maintain the desirable part clearances. Also, the air must be completely free of contaminants; otherwise, serious surface damage will result. Failure of an air supply can be disastrous. Air bearings can be of the radial type or the thrust type. Figure 4-6 shows a typical arrangement for a thrust application. An air line is connected and air is introduced at the bottom and center of the spindle. The bottom section of the shaft is often recessed directly above the air connection. Air is compressed and escapes around the shaft (air is a compressible fluid and must be considered as such). This type of bearing cannot withstand heavy loads, since a suitable air film cannot be maintained. Other gases besides air can be used. Air bearings warrant consideration in applications demanding extremely high rotational speeds or uniformly low frictional drag.

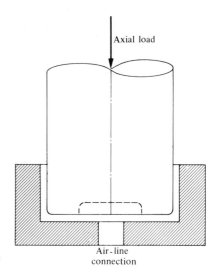

Axial load

Air-line
connection

Figure 4-6
Air bearing, thrust type.

Magnetic Bearings

In a magnetic bearing, the shaft position is controlled through the adjustment of an electromagnet energizing current; the most common arrangement has the electromagnets in the housing. Since the shaft and the stationary housing make no contact, there are no friction surfaces; thus there is neither wear nor need for lubrication, and the assembly is almost noiseless. However, there are magnetic losses due to eddy currents and hysteresis. Often a magnetic bearing system is comprised of both radial and thrust bearings.

Rotating Sleeve Bushings

Another approach to bushing design is to provide a tight fit between a shaft and a bronze bushing and then a running fit between the bushing and the housing. In such an application, the bushing runs in the bore of the housing (or steel sleeve or liner). The size of the wearing surface is increased, since the working surface is based on the outside diameter of the bushing and not on the inside diameter. For example, if we assume that a 1-in.-diameter shaft uses a bushing with a $\frac{1}{8}$-in. wall thickness, the effective circumference would increase from $\pi(1) = 3.14$ in. to $\pi(1.25) = 3.93$ in.; this is an increase of 25%. A larger contacting surface means that the bearing can hold more lubricant; this tends to increase the life of the bearing.

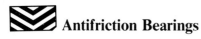 **Antifriction Bearings**

Antifriction bearings can be classified in a number of different ways. For instance, they can be self-aligning or nonself-aligning. They can be used for radial applications, thrust applications, or a combination of radial and axial use. Typical types of antifriction bearings are (1) ball, (2) roller, and (3) needle. Antifriction bearings are used in various design situations and are subjected to all kinds of abuse. Manufacturers give rated loads under more or less standard conditions; if a designer uses them under any other set of conditions, he or she must modify the design loads accordingly. Most manufacturers give the bearing ratings in pounds, based on a certain number of expected hours of operating life at a certain number of revolutions per minute.

Advantages of Antifriction Bearings

Although the initial cost of antifriction bearings is high, they have some advantages over sliding bearings. The outstanding feature is the reduced frictional loss. Imagine, for instance, pushing a 100-lb crate over a concrete floor. If we placed two rollers under the crate, it would be much easier to handle—although guiding it could present a problem! If we used four balls under the crate, it would roll easily and could be moved in any direction with little effort. If we consider the actual contact under these three sets of circumstances, we find that we have (1) surface contact, (2) line contact, and (3) point contact, respectively. The greatest frictional losses occur with sliding contact; the least, with point contact. From the pressure standpoint, the reverse is true; thus, unit pressure under a surface is small compared with unit pressure concentrated in one point. If balls are used in any application, the surfaces on which they operate must be exceedingly hard. With any type of rolling contact, there is less abrasion; thus, wear is much less of a problem. Ball bearings are usually designed with separators to prevent balls from rubbing each other; this lessens wear. Along with this absence of wear, there is less need for elaborate lubrication design. Some antifriction bearings contain a sealed-in lubricant that in most instances lasts for the life of the machine. In other cases, the lubricant is replaced only after a prescribed number of hours of operation, as part of a preventive maintenance program. Antifriction bearings can be operated at much higher speeds than sliding bearings.

From the standpoint of space utilization, ball bearings use a large amount of diametral space, but very little axial space. If diametral space is at a premium, needle bearings are more desirable than ball bearings, although needle bearings need more axial space. Antifriction bearings are

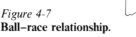

Figure 4-7
Ball–race relationship.   (a)   (b)

good from the standpoint of loading. Ball bearings of the double type can be used to sustain heavy loads; roller bearings are ideal for heavy loads. While antifriction bearings require considerable accuracy in mounting, precise action from the machine will result in better performance. Maintenance is trivial for antifriction bearings as compared with sliding bearings. When machine parts do require servicing, interchanging parts with antifriction bearing mountings is easy, since no break-in is required. A user of antifriction bearings should evaluate all of the various types available before making a selection.

Friction in Antifriction Bearings

Ball, roller, and needle bearings are certainly not frictionless; however, compared to sliding bearings, friction and friction losses are very low. The average coefficient of friction in a ball bearing is approximately 0.001; some bearings have a still lower value. Contact between a ball and a race is known as *point contact*. As the transmitted load passes through a small point, pressure values are excessively high. On the other hand, friction loses are low. In the case of roller and needle bearings, the load is transmitted along a thin line; this is known as *line contact*. From the kinematic standpoint, point or line contact between two components is referred to as *higher pairing;* surface contact is referred to as *lower pairing.*

Figure 4-7 shows the relationship that can exist between the ball and the race of an antifriction bearing. (Both figures in this sketch are exaggerated.) In Fig. 4-7(a), the curvature of the race is larger than the curvature of the ball. This is true point contact and aids in reducing friction; however, pressure at the point of contact is relatively large. In Fig. 4-7(b), the radius of the curvature is the same for both ball and race. This configuration reduces the unit pressure between the two parts, but increases friction. A proper compromise between the two situations is the ideal arrangement.

Ratings of Bearings

Although boundary dimensions are fairly well standardized among various manufacturers, load ratings are not. Similarity in dimensions makes interchangeability possible; however, a designer must check the specific manufacturer's ratings and rating methods before substituting one make of bearing for another.

Since antifriction bearings are used in various areas of manufactured products, each application must be carefully analyzed. Most manufacturers specify a basic *dynamic capacity*. This is a radial rating (or thrust rating, in the case of thrust bearings) expressed in pounds under certain conditions. Four factors that often enter into such ratings are *life*, *speed*, *ring rotation*, and *type of loading*. Each item usually becomes a factor that modifies the rated load.

Life. In general, the lives of bearings are inversely proportional to the cubes of the respective loads. Certain manufacturers use the so-called B-10 life. The B-10 life, expressed in hours, is the number of hours that 90% of the bearings in a given group will exceed before the first evidence of fatigue develops. The average life is approximately five times the minimum B-10 life. Some manufacturers may base their rated loads on a B-10 life of 1000 hr; others may base their ratings on 500, 1500, or 3800 hr, or any other convenient figure. The rated loads are then adjusted by applying the appropriate factor (usually from a chart, formula, or nomograph) to correct the given load under "standard" conditions to the user's desires. In certain types of equipment, the amount of use varies from hour to hour and day to day; in other types, a machine operates steadily for an eight-hour shift. For this reason, many engineers prefer a rating in millions of revolutions (or cycles). This is particularly true in automotive applications; an automobile runs at different rates of speed and for various intervals of time. If one knows the total miles that a car will travel in its useful life, a few simple calculations reveal the total number of revolutions that a particular shaft may turn. Thus, the total number of revolutions is a more useful value than total hours of operating life.

Speed. A speed factor is frequently employed in rating bearings. Bearings can be rated at 500, 700, or 33.3 rpm, or at any other suitable figure. The dynamic load values are then corrected by a suitable factor when any speed other than the "standard" one is used. If a firm rates its bearings at 500 rpm, the factor will be 1 for this speed; anything above or below this speed will vary the factor on either side of unity.

Ring rotation. The rating for a bearing depends on whether the inner ring or outer ring rotates. With some exceptions, the inner race of a bearing turns with the shaft. The exceptions dictate an adjustment in the rated values for a bearing. These values can be found from the manufacturer's catalog.

Type of loading. Equivalent loads are frequently used to simplify bearing calculations. If a radial bearing has radial loads only, the radial rating from the catalog is used; however, in most cases it is corrected for the three preceding conditions. If a thrust bearing is subjected to axial loads only, the bearing rating is used subject to correction of the three preceding items. In the majority of cases, however, the loading of a bearing is a combination of radial and thrust loads. In such cases, the use of an equivalent load is desirable. If the load is predominantly radial, a certain additional amount is added to the radial load to allow for the thrust loading; the total amount is the equivalent load, and radial rating tables are used. If the predominant load is one of thrust, the reverse procedure is used. The amount added to the predominant value for the minor value depends on the type of bearing used.

 Types of Ball Bearings

Ball bearings are classified in several different ways. At this point, a few of the commonly used general types will be discussed. These types are classified according to general construction; they are not grouped in terms of

Figure 4-8
Nomenclature of typical single-row ball bearing.
(Courtesy of New Departure-Hyatt Bearings, a division of General Motors Corp.)

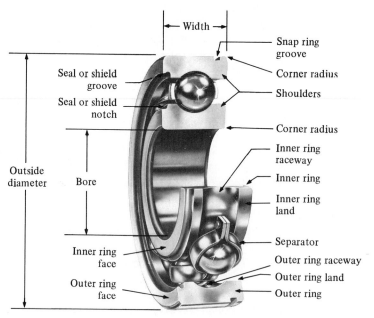

(a)   (b)

Figure 4-9
(a) Single-row Conrad bearing. (b) Maximum-capacity (loading groove) bearing.
(Courtesy of New Departure-Hyatt Bearings, a division of General Motors Corp.)

shields, seals, and snap rings. Figure 4-8 shows the nomenclature of a typical single-row ball bearing.

1. The first type is known as the Conrad type of bearing; it is a deep-groove type. It is assembled by placing the inner ring in an eccentric position with respect to the outer ring. The appropriate number of balls are then introduced into the larger space, after which the rings are returned to a concentric position. Then the balls are evenly spaced, and the two-part separator is installed from either side. This provides equal spacing on a permanent basis. Since the shoulders of the grooves are uniform, a fairly large axial load can be applied in either direction. The Conrad is primarily a radial-type bearing. Because of the type of construction, only a limited number of balls can be used; this in turn limits the radial capacity of the bearing. (See Fig. 4-9a.)

2. The *maximum-capacity* ball bearing is easily recognized by the loading grooves. Figure 4-9(b) shows this type of bearing. Some of the balls are introduced by spreading the rings eccentrically; the remaining balls are inserted through the loading grooves. The gap at the loading grooves can be widened slightly by an assembly fixture. A greater number of balls can be inserted in this type of bearing than in the Conrad type; thus, its radial-load rating is higher. The loading grooves do not extend

to the bottom of the grooves in the races. This type can carry some thrust load; the assembly can be supplied with shields, seals, or snap rings.

3. Figure 4-10 illustrates an *angular-contact* type of bearing. Note the relative sizes of the shoulders. On the outer ring, one shoulder is heavy and one is counterbored. This type of bearing is assembled by expanding the outer ring and then inserting the inner ring and balls. Once assembled, the balls cannot be removed. The ratings for this type are higher than for the Conrad type. In addition, the angular-contact type can handle larger axial loads than the Conrad type. Most configurations of the angular-contact bearing are designed to take the thrust load in one direction only. Some manufacturers produce a duplex bearing designed to take thrust loads in either direction.

4. Figure 4-11 illustrates a *double-row* ball bearing. This is generally considered a heavy-duty ball bearing. Larger radial ratings are possible because the number of balls is doubled. Loading grooves are used to provide the maximum number of balls. Since the contact-angle lines diverge outward, a small amount of misalignment is possible. It must be remembered that exact alignment is usually essential in bearing applications. Double-row bearings permit reasonable values for thrust loads.

5. *Ball thrust* bearings handle axial loads only. Figure 4-12 shows one configuration of this type. The use of this bearing is somewhat restricted, since no radial load is permitted. This type is usually used when two radial bearings need to be used in addition to the thrust bearing. Naturally, it would be more desirable to find ball bearings that

Figure 4-10
Angular-contact bearing.
(Courtesy of New Departure-Hyatt Bearings, a division of General Motors Corp.)

Figure 4-11
Double-row bearing.
(Courtesy of New Departure-Hyatt
Bearings, a division of General
Motors Corp.)

could handle the axial load, and eliminate the thrust bearing. A careful force analysis often dictates whether or not a strictly thrust type is absolutely necessary.

Ball Bearing Selection

Antifriction bearings are preengineered by the manufacturers; thus, a designer *selects* bearings from the catalog information. The following de-

Figure 4-12
Ball thrust bearing.
(Courtesy of SKF Industries, Inc.)

scribes a typical process for making such a selection—but a restricted one, since only partial SKF catalog information is presented. The following equation is used to show the relation between load and life:

$$L_{10} = \left(\frac{C}{P}\right)^3,$$ (4-3)

where

L_{10} = rating life in millions of revolutions,

C = load that will allow one million revolutions (the basic load rating in pounds),

P = applied load to the bearing (in pounds).

The rating life can be calculated from the equation

$$L_{10} = \frac{16,667}{N}\left(\frac{C}{P}\right)^3,$$ (4-4)

where

N = rpm.

If the bearing has been chosen and thus C is known, the allowable load can be found from the equation

$$P = \frac{C}{C/P},$$ (4-5)

or by using the nomogram shown in Fig. 4-13. In this modified equation, L_{10} represents the life in hours. In cases where there is a significant variation in speed, such as in vehicular applications, the rating life may be expressed in miles. The following equation can be used for such cases:

$$L_{10} = 49.6\ W\left(\frac{C}{P}\right)^3,$$ (4-6)

where

L_{10} = life in miles and

W = wheel diameter in inches.

In using these equations, it must be assumed that proper application conditions are met, such as proper lubrication, alignment, and temperature limits. Fatigue life can be improved by providing better than adequate lubrication or by using bearings made from superior steels obtainable through specialized metallurgy, heat treatment, or metal processing techniques. Also, one can apply a life adjustment factor for reliability varying from the usual 90% indicated in the footnote on the nomogram. In this case,

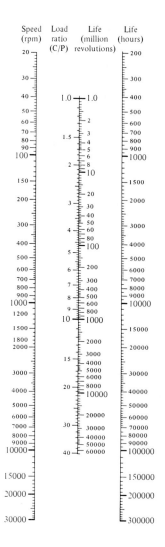

Figure 4-13
Basic load ratio–life nomogram for ball bearings. The life is expected to be exceeded by 90% of the bearings.
(Courtesy of SKF Industries, Inc.)

the designated life is expected to be exceeded by 90% of the bearings; the median life is approximately five times as long.

In rating tables, there is a listing for static load used for bearings that do not rotate or oscillate. Such bearings are not subject to fatigue, but permanent deformations developing on the load-carrying surfaces of the bearing establish a limit on their load-carrying capacity.

Ball bearings are often subjected to a thrust or axial load as well as to a radial load. In such cases, it is necessary to convert these forces into a so-called equivalent load. This is done by use of the following equation:

$$P = XF_r + YF_a, \tag{4-7}$$

where

    P = equivalent load,

    F_r = actual radial load,

    F_a = actual thrust load,

    X — radial factor, and

    Y = thrust factor.

The values X and Y are listed on the rating tables.

Table 4-2 illustrates a typical page of ball bearing ratings—in this case, the Series 62 single-row, deep-groove bearings.

The following examples illustrate the procedure for using a nomogram and a rating table. It should be noted that complete dimensional data are included for each bearing.

◆ *Example* _____

Find the thrust load that a 6215 ball bearing can carry at 800 rpm if the required rating life is 10,000 hours.

Solution. The basic load rating for a 6215 bearing is 11,450 lb (see Table 4-2). Using the equation

$$L_{10} = \frac{16,667}{N} \left(\frac{C}{P} \right)^3,$$

$$P = \frac{C}{\left(\dfrac{NL_{10}}{16,667} \right)^{1/3}} = \frac{11,400}{\left(\dfrac{800(10,000)}{16,667} \right)^{1/3}} = 1456 \text{ lb},$$

or, using the nomogram in Fig. 4-13,

$$C/P = 7.8,$$

and therefore

$$P = \frac{C}{C/P} = \frac{11,400}{7.8} = 1462 \text{ lb}.$$

◆ *Example* _____

Bearing 6210 is used to carry a radial load of 2 224 newtons and a thrust or axial load of 1 779 newtons at a constant inner-ring speed of 720 rpm. What bearing life can be expected?

Table 4-2
Ratings of Ball Bearings

Series 62 single row, deep groove ball bearings

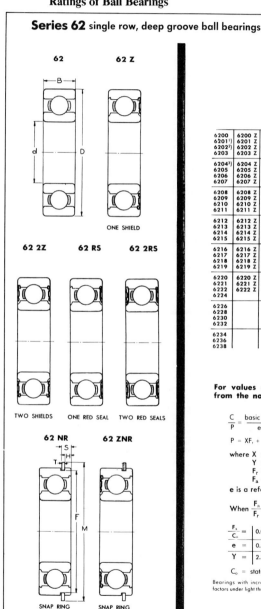

62   62 Z

ONE SHIELD

62 2Z   **62 RS**   **62 2RS**

TWO SHIELDS   ONE RED SEAL   TWO RED SEALS

62 NR   **62 ZNR**

SNAP RING   SNAP RING
ONE SHIELD

BEARING NUMBER							Nominal	
							d	
							mm	in.
6200	6200 Z	6200 2Z	6200 RS	6200 2RS	6200 NR	6200 ZNR	10	.3937
6201[1]	6201 Z	6201 2Z	6201 RS	6201 2RS			12	[1] .4724
6202[2]	6202 Z	6202 2Z	6202 RS	6202 2RS			15	[2] .5906
6203	6203 Z	6203 2Z	6203 RS	6203 2RS	6203 NR	6203 ZNR	17	.6693
6204[3]	6204 Z	6204 2Z	6204 RS	6204 2RS	6204 NR	6204 ZNR	20	[3] .7874
6205	6205 Z	6205 2Z	6205 RS	6205 2RS	6205 NR	6205 ZNR	25	.9843
6206	6206 Z	6206 2Z	6206 RS	6206 2RS	6206 NR	6206 ZNR	30	1.1811
6207	6207 Z	6207 2Z	6207 RS	6207 2RS	6207 NR	6207 ZNR	35	1.3780
6208	6208 Z	6208 2Z	6208 RS	6208 2RS	6208 NR	6208 ZNR	40	1.5748
6209	6209 Z	6209 2Z	6209 RS	6209 2RS	6209 NR	6209 ZNR	45	1.7717
6210	6210 Z	6210 2Z	6210 RS	6210 2RS	6210 NR	6210 ZNR	50	1.9685
6211	6211 Z	6211 2Z	6211 RS	6211 2RS	6211 NR	6211 ZNR	55	2.1654
6212	6212 Z	6212 2Z	6212 RS	6212 2RS	6212 NR	6212 ZNR	60	2.3622
6213	6213 Z	6213 2Z	6213 RS	6213 2RS	6213 NR	6213 ZNR	65	2.5591
6214	6214 Z	6214 2Z	6214 RS	6214 2RS			70	2.7559
6215	6215 Z	6215 2Z	6215 RS	6215 2RS	6215 NR	6215 ZNR	75	2.9528
6216	6216 Z	6216 2Z	6216 RS	6216 2RS	6216 NR	6216 ZNR	80	3.1496
6217	6217 Z	6217 2Z	6217 RS	6217 2RS	6217 NR	6217 ZNR	85	3.3465
6218	6218 Z	6218 2Z					90	3.5433
6219	6219 Z	6219 2Z					95	3.7402
6220	6220 Z	6220 2Z	6220 RS	6220 2RS			100	3.9370
6221	6221 Z	6221 2Z					105	4.1339
6222	6222 Z	6222 2Z			6222 NR		110	4.3307
6224							120	4.7244
6226							130	5.1181
6228							140	5.5118
6230							150	5.9055
6232							160	6.2992
6234							170	6.6929
6236							180	7.0866
6238							190	7.4803

For values of C/P at different speeds, the life is found from the nomogram on the inside back cover.

$$\frac{C}{P} = \frac{\text{basic load rating from table}}{\text{equivalent load (from formula below)}}$$

$P = XF_r + YF_a$

where X = a radial factor given below
$\quad\quad\quad Y$ = a thrust factor given below
$\quad\quad\quad F_r$ = the radial load, calculated
$\quad\quad\quad F_a$ = the thrust load, calculated
e is a reference value given in the table.

When $\dfrac{F_a}{F_r}$ $\begin{cases} \text{is smaller than or equal to e use } X = 1 \text{ and } Y = 0 \\ \text{is greater than e use } X = 0.56 \text{ and } Y \text{ from table} \end{cases}$

$\dfrac{F_a}{C_o} =$	0.014	0.028	0.056	0.084	0.11	0.17	0.28	0.42	0.56
e =	0.19	0.22	0.26	0.28	0.30	0.34	0.38	0.42	0.44
Y =	2.30	1.99	1.71	1.55	1.45	1.31	1.15	1.04	1.00

C_o = static load rating from table.

Bearings with increased internal clearance will have somewhat more favorable X and Y factors under light thrust load. In this respect consult SKF.

Table 4-2
(Continued)

Bearing Dimensions D mm	D in.	B mm	B in.	F in.	M in.	S in.	H in.	T in.	Max. Fillet Radius[4]) in.	Preferred Shoulder Diameter in. Shaft	Preferred Shoulder Diameter in. Housing	Brg. Wt. lb	Balls No.	Balls Diam. in.	Static Load Rating C_0 lb	Basic Load Rating C lb	Approx. Speed Limit[5]) rpm	Basic Brg. No.
30	1.1811	9	.3543	1.109	1 23/64	.120	.078	.042	.024	.500	.984	.07	7	3/16	440	805	25000	**6200**
32	1.2598	10	.3937						.024	.578	1.063	.09	7	15/64	685	1180	23000	**6201**
35	1.3780	11	.4331						.024	.703	1.181	.11	8	15/64	790	1320	20000	**6202**
40	1.5748	12	.4724	1.500	1 3/4	.120	.078	.042	.024	.787	1.380	.15	8	17/64	1000	1650	18000	**6203**
47	1.8504	14	.5512	1.756	2 1/16	.136	.094	.042	.039	.969	1.614	.25	8	5/16	1390	2210	15000	**6204**
52	2.0472	15	.5906	1.958	2 17/64	.136	.094	.042	.039	1.172	1.811	.29	9	5/16	1560	2420	13000	**6205**
62	2.4409	16	.6299	2.347	2 21/32	.190	.125	.065	.039	1.408	2.203	.45	9	3/8	2250	3360	11000	**6206**
72	2.8346	17	.6693	2.709	3 5/64	.190	.125	.065	.039	1.614	2.559	.66	9	7/16	3070	4440	9400	**6207**
80	3.1496	18	.7087	3.024	3 13/32	.190	.125	.065	.039	1.811	2.874	.85	9	15/32	3520	5040	8400	**6208**
85	3.3465	19	.7480	3.221	3 19/32	.190	.125	.065	.039	2.008	3.071	.94	9	1/2	4010	5660	7700	**6209**
90	3.5433	20	.7874	3.417	3 51/64	.220	.125	.095	.039	2.205	3.268	1.00	10	1/2	4450	6070	7100	**6210**
100	3.9370	21	.8268	3.811	4 3/16	.220	.125	.095	.059	2.441	3.583	1.34	10	9/16	5630	7500	6500	**6211**
110	4.3307	22	.8661	4.205	4 37/64	.220	.125	.095	.059	2.717	3.976	1.75	10	5/8	6950	9070	5900	**6212**
120	4.7244	23	.9055	4.536	5 3/32	.265	.156	.109	.059	2.913	4.370	2.22	10	21/32	7670	9900	5400	**6213**
125	4.9213	24	.9449						.059	3.110	4.567	2.41	10	11/16	8410	10800	5100	**6214**
130	5.1181	25	.9843	4.930	5 1/2	.265	.156	.109	.059	3.307	4.764	2.59	11	11/16	9250	11400	4800	**6215**
140	5.5118	26	1.0236	5.324	5 57/64	.297	.188	.109	.079	3.504	5.118	3.13	10	3/4	10000	12600	4500	**6216**
150	5.9055	28	1.1024	5.718	6 9/32	.297	.188	.109	.079	3.740	5.512	3.94	11	25/32	12000	14400	4200	**6217**
160	6.2992	30	1.1811						.079	3.937	5.906	5.00	10	7/8	13600	16600	3900	**6218**
170	6.6929	32	1.2598						.079	4.213	6.220	5.90	10	15/16	15600	18800	3700	**6219**
180	7.0866	34	1.3386						.079	4.409	6.614	7.20	10	1	17800	21100	3500	**6220**
190	7.4803	36	1.4173						.079	4.606	7.008	8.47	10	1 1/16	20100	23000	3300	**6221**
200	7.8740	38	1.4961	7.624	8 3/8	.339	.219	.120	.079	4.803	7.402	10.2	10	1 1/8	22500	24900	3100	**6222**
215	8.4646	40	1.5748						.079	5.197	7.992	12	9	1 3/16	22600	25100	2900	**6224**
230	9.0551	40	1.5748						.098	5.669	8.504	13	9	1 1/4	25000	26900	2600	**6226**
250	9.8425	42	1.6535						098	6.063	9.291	17.5	10	1 1/4	27800	28800	2400	**6228**
270	10.6299	45	1.7717						.098	6.457	10.079	20	11	1 1/4	30600	30400	2200	**6230**
290	11.4173	48	1.8898						.098	6.850	10.866	26.5	12	1 1/4	33400	32000	2000	**6232**
310	12.2047	52	2.0472						.118	7.402	11.496	32	12	1 3/8	40400	36700	1900	**6234**
320	12.5984	52	2.0472						.118	7.795	11.890	34	11	1 1/2	44100	37500	1800	**6236**
340	13.3858	55	2.1654						.118	8.189	12.677	41	11	1 5/8	51700	44100	1700	**6238**

[1])This bearing is also available with ½" (.4998-.4995) bore as 465898, 465898 Z, 465898 2Z, 465898 RS and 465898 2RS.
[2])This bearing is also available with ⅝" (.6248-.6245) bore as 466041, 466041 Z, 466041 2Z, 466041 RS and 466041 2RS.
[3])This bearing is also available with ¾" (.7495-.7491) bore as 465966, 465966 Z, 465966 2Z, 465966 RS and 465966 2RS.
[4])The maximum fillet on the shaft or in the housing, which will be cleared by the bearing corner. These values apply to all corners except corner adjacent to snap ring, for which maximum in all cases is 0.020".

[5])This refers to oil lubrication and moderate load. With grease lubrication it is generally not practical to use speeds higher than ⅔ of those shown.

Note: Bearings 6200 to 6220 inclusive are also available in ABEC 5 tolerances, see page 76.
When shield is required on same side as snap ring, order as ZNBR.

For values of C/P at different speeds, the life is found from the nomogram in Fig. 4-13. (Extracted from SKF® *Ball Bearings*, January 1981, SKF Industries, Inc. Bearings Group by special permission.)

Solution. Table 4-2 gives ratings in pounds; therefore, the loads in newtons must be converted to pounds:

$$F_a = \frac{1\,779}{4.448} = 400 \text{ lb}$$

and

$$F_r = \frac{2\,224}{4.448} = 500 \text{ lb.}$$

Then,

$$\frac{F_a}{F_r} = \frac{400}{500} = 0.8.$$

C_o and C values can be obtained from the rating table:

$$C_o = 4450$$

and

$$C = 6070.$$

$$\frac{F_a}{C_o} = \frac{400}{4450} = 0.09.$$

For

$$\frac{F_a}{C_o} = 0.09,$$

the value of e is approximately 0.29. Since

$$\frac{F_a}{C_o} = 0.8,$$

which is larger than

$$e = 0.29,$$

the application factors are $X = 0.56$ and $Y = 1.53$ (by interpolation). Then, the equivalent load

$$P = 0.56\,(500) + 1.53\,(400) = 892\text{ lb},$$

and

$$\frac{C}{P} = \frac{6070}{892} = 6.80.$$

For a C/P value of 6.80 and a rotational speed of 720 rpm, the nomogram shows a bearing life of 7300 hours. This value could also be calculated by the equation

$$L_{10} = \frac{16{,}667}{N}\left(\frac{C}{P}\right)^3 = \frac{16{,}667}{720}\left(\frac{6070}{892}\right)^3 = 7278\text{ hours}.$$

Special Types

There are numerous types of special bearings; some of these are in the ball-bearing family, some are not. Figure 4-14 shows a sectional view of a superquiet "O"-type bearing. Note the construction of the outer ring with a

Figure 4-14
Super-quiet "O" series bearing.
(Courtesy of Fafnir Bearing, a
division of Textron, Inc.)

nonmetallic section. In applications where noise is an important factor, bearings that are so isolated have great value.

Ball Bushings® are illustrated in Fig. 4-15. These provide low friction and excellent wear characteristics for linear motions; lubrication and maintenance problems are minimized. Three or more oblong circuits of balls are enclosed in the bushing. The linear shaft contacts a straight side of each circuit; the balls are returned in a clearance provided in the sleeve.

A ball-bearing (b/b) spline is shown in Fig. 4-16. This coupling device has mating inner and outer members containing axial races, usually of the Gothic-arch form. Circulating bearing balls minimize friction and lash. The stick-slip common to usual splines is eliminated. A b/b screw is shown in Chapter 11 with the translation screws. Such a screw minimizes friction and produces high efficiency. This recirculating ball device is an actuator; the balls operate between the helical race in the nut and the screw. Patented deflectors divert the ball path from the race into a tubular guide that carries

Figure 4-15
Ball-Bushing® bearing.
(Courtesy of Thomson
Industries, Inc.)

Figure 4-16
Ball-bearing spline.
(Courtesy of Saginaw Steering Gear,
a division of General Motors Corp.)

them diagonally across the outside of the nut and returns them to the active raceway. Although it is most prominent in the steering mechanisms of automobiles, other uses have been found for this versatile device. Precision control and long life have popularized this type of actuator.

Figure 4-17 is a cutaway drawing of the Unibal® spherical bearing. Note that provision for lubrication has been made. This type of bearing is completely self-aligning. This self-aligning characteristic allows its use in situations where precise linear alignment is difficult to obtain.

Miniature Ball Bearings

The necessity of miniaturization in the mechanical field has contributed to the development of exceptionally small antifriction bearings. The Anti-Friction Bearing Manufacturers Association considers any ball bearing with

Figure 4-17
Unibal® bearing.
(Courtesy of the Heim, Incom
International, Inc.)

an outside diameter less than $\frac{3}{8}$ in. a miniature bearing. Bearings as small as 0.1250 in. (outside diameter) and 0.0400 in. (inside bore) are commercially available. Several types that can be readily obtained include the plain type with or without shields, the flanged type with or without shields, and a pivot type containing three small balls. Friction values are exceptionally low for miniaturized bearings. They are expensive because of the extreme precision and great care needed in producing and handling them. "White-room" conditions are necessary in their production; special high-grade lubricants are necessary for proper functioning. Packaging is an important phase of production. Plastic seals are used to cap special vials for preservation; a proper amount of oil is enclosed with the bearings. Miniature bearings are extensively used in the instrument field.

Besides regular miniature ball bearings, others are made with "white-room" precision. Special types used in space projects include guide, gimbal, synchro, rotor, and turbine bearings. Special bearing assemblies can be found in cam followers, shaft guides, tape guides, and ball carriages.

Types of Roller Bearings

Because roller bearings have line contact rather than point contact, they have greater load-carrying capacity than ball bearings. Like ball bearings, roller bearings are available in many forms, each type designed to fill a particular need. The simplest version is the single-row type illustrated in Fig. 4-18(a). These can be obtained with a flange on one ring only to permit

Figure 4-18
(a) Single-row cylindrical roller bearing.
(b) Double-row cylindrical roller bearing.
(Courtesy of SKF Industries, Inc.)

Figure 4-19
Loadstar roller bearings.
(Courtesy of New Departure-Hyatt
Bearings, a division of General
Motors Corp.)

a small amount of axial shaft movement. Dismounting is easy in this case, even if both rings are mounted with a tight fit. Figure 4-18(b) shows a double-row type. This type is popular where extreme accuracy is desired. Figure 4-19 shows a Loadstar roller bearing. This bearing uses larger rollers with closer spacing resulting from the use of an X-bar for roller retainment rather than the conventional spacer bar. The X-bar provides better race-to-roller proportions; lubrication is more effective in this type of construction.

A spherical roller bearing is shown in Fig. 4-20. This type is available with a straight cylindrical bore or with a tapered bore. These bearings are self-aligning; a small amount of misalignment between the housing and the shaft has no ill effects on their performance. Spherical roller thrust bearings can also be obtained. These are valuable in applications where the load is

Figure 4-20
Spherical roller bearing.
(Courtesy of SKF Industries, Inc.)

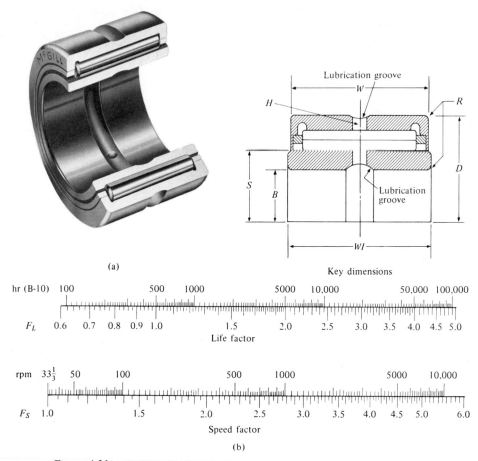

(a)

Key dimensions

Life factor

Speed factor

(b)

Figure 4-21
(a) Cagerol bearing. (b) Catalog information about Cagerol bearing.
(Courtesy of McGill Manufacturing Company, Inc.)

primarily one of thrust, or of thrust combined with smaller radial loads. The bearing is self-aligning. These can be used where the thrust load is introduced from either the horizontal or vertical direction. Figure 4-21(a) shows a Cagerol® roller bearing with an inner ring. On some types, the rollers act directly against the shaft. Since it uses little diametral space, this type is useful where space is limited. A tapered roller bearing is used where thrust and radial loads are present. Various angles are used; if the thrust load is greater than the radial load, the angle is greater. Figure 4-22 shows a typical tapered roller bearing. Wheel bearings in automobiles are of the tapered roller type.

Figure 4-22
Tapered roller bearing.
(Courtesy of The Timken
Company.)

Roller-Bearing Ratings

Every bearing manufacturer has its own method of rating bearings. For this reason, a designer has to consult a specific manufacturer's catalog to get appropriate rating information. Table 4-3 and Fig. 4-21(b) give examples of some of the information contained in catalogs.* It should be noted that these tables are abridged—the primary purpose of presenting the material here is to familiarize the student with the procedure to be followed. For this particular antifriction bearing, the basic dynamic load rating is based on a life of one million revolutions or the bearing capacity of $33\frac{1}{3}$ rpm for 500 hr of B-10 life. This could also be considered as 2500 hr of average-fatigue life. The following example shows how the charts can be applied to a design problem.

◆ *Example* _____

Select a bearing with a minimum B-10 life of 10,000 hr. The operating speed is 600 rpm and the radial load (design load) is 1200 lb.

> *Solution.* For a B-10 life of 10,000 hr, the life factor is 2.46; for a speed of 600 rpm, the speed factor is 2.38. These values are obtained from the charts of Fig. 4-21. The design load must then be multiplied by the life and speed factors. Thus the required bearing capacity is
>
> 1200(2.46)(2.38) = 7020 lb.
>
> From the abridged listing, the MR-18 is the smallest-diameter bearing (1.6250 in. outside diameter) that would meet these require-

*Extracted from Bearing Catalog No. 80a, 1973, McGill Manufacturing Company, Inc., by special permission.

Table 4-3
Ratings of MR Series Cagerol® Bearings (Partial Listing)

McGill Number	McGill Number	B	D	W	WI	H
Outer ring & roller assy.	Inner ring only	+.0000 Tol.	+.0000 Tol.	+.000 −.005	+.000 −.005	Hole Dia.
MR-10-N	MI-6-N	.3750	1.1250	.750	.760	1/8
MR-10-N	MI-7-N	.4375	1.1250	.750	.760	1/8
MR-10	MI-6	.3750	1.1250	1.000	1.010	1/8
MR-12-N	MI-8-N	.5000	1.2500	.750	.760	1/8
MR-12-N	MI-9-N	.5625	1.2500	.750	.760	1/8
MR-12	MI-8	.5000	1.2500	1.000	1.010	1/8
MR-14-N	MI-10-N	.6250	1.3750	.750	.760	1/8
MR-14-N	MI-11-N	.6875	1.3750	.750	.760	1/8
MR-14	MI-10	.6250	1.3750	1.000	1.010	1/8
MR-16-N	MI-12-N	.7500	1.5000	.750	.760	1/8
MR-16	MI-12	.7500	1.5000	1.000	1.010	1/8
MR-16	MI-13	.8125	1.5000	1.000	1.010	1/8
MR-18-N	MI-14-N	.8750	1.6250	1.000	1.010	1/8
MR-18-N	MI-15-N	.9375	1.6250	1.000	1.010	1/8
MR-18	MI-14	.8750	1.6250	1.250	1.260	1/8
MR-18	MI-15	.9375	1.6250	1.250	1.260	1/8
MR-20-N	MI-16-N	1.0000	1.7500	1.000	1.010	1/8
MR-20	MI-16	1.0000	1.7500	1.250	1.260	1/8
MR-22-N	MI-18-N	1.1250	1.8750	1.000	1.010	1/8
MR-22	MI-17	1.0625	1.8750	1.250	1.260	1/8
MR-22	MI-18	1.1250	1.8750	1.250	1.260	1/8

Courtesy of McGill Manufacturing Co., Inc.

ments. The basic dynamic table value of 8070 is larger than the 7020 lb required. It should be noted that many of the larger sizes would also meet these requirements. In some instances, it might be convenient to use a larger size to simplify other aspects of the design.

 **Needle Bearings**

Needle bearings are somewhat similar to roller bearings. The needle bearing has a large number of rollers that are somewhat smaller in diameter; since the rollers are smaller in size, a full complement can be used. Thus, greater load-carrying capacity is possible. The coefficient of friction is larger for

Table 4-3
(Continued)

R	Shaft Dia.				HSG. Bore Dia.					
Max. fillet for shaft & hsg.	Rotat. shaft	Tol. +.0000	Stat. shaft	Tol. +.0000	Rotat. HSG.	Tol. +.0000	Stat. HSG.	Tol. +.0000	Aircraft static capacity	Basic dynamic capacity
.025	.3755		.3747		1.1247		1.1257		6600	3350
.025	.4380		.4372		1.1247		1.1257		6600	3350
.025	.3755		.3747		1.1247		1.1257		9800	4580
.040	.5005		.4997		1.2497		1.2507		8800	4010
.040	.5630		.5623		1.2497		1.2507		8800	4010
.040	.5005		.4997		1.2497		1.2507		13000	5480
.040	.6255		.6247		1.3747		1.3757		9500	4110
.040	.6880		.6872		1.3747		1.3757		9500	4110
.040	.6255		.6247		1.3747		1.3757		14100	5620
.040	.7505		.7497		1.4997		1.5007		11700	4670
.040	.7505		.7497		1.4997		1.5007		17400	6380
.040	.8129		.8121		1.4997		1.5007		17400	6380
.040	.8754		.8746		1.6247		1.6257		18400	6480
.040	.9379		.9371		1.6247		1.6257		18400	6480
.040	.8754		.8746		1.6247		1.6257		24400	8070
.040	.9379		.9371		1.6247		1.6257		24400	8070
.040	1.0004		.9996		1.7497		1.7507		21700	7140
.040	1.0004	−.0005	.9996	−.0005	1.7497	−.0007	1.7507	−.0007	28700	8900
.040	1.1255		1.1246		1.8747		1.8757		22700	7180
.040	1.0630		1.0621		1.8747		1.8757		30000	8950
.040	1.1255		1.1246		1.8747		1.8757		30000	8950

needle bearings than for roller bearings. A typical value for needle bearings might be 0.0025, compared to 0.0015 for the roller bearing. The prime advantage of needle bearings is the small amount of radial space taken up by the small rollers or needles. Several types are available. Figure 4-23 shows a drawn-cup needle bearing. Because of its construction, it cannot take a sudden overload as well as a heavy-duty type; the wall section of the outer race is small compared to other types. When needle bearings are used without an inner ring (direct contact with shaft), the shaft must be properly hardened if it is to function properly. The drawn-cup type of bearing should not be used with a split housing unless it is protected by a sleeve.

Needle bearings are used extensively in cam roller followers. Such a follower is illustrated in Fig. 13-7 under "Cam Followers." Costs can be reduced by using the drawn-cup type of needle bearing rather than the heavy-duty type. Aircraft needle bearings are made with an inner ring; some

Figure 4-23
Drawn-cup needle bearing.
(Courtesy of The Torrington
Company.)

of these are made with a double row of needle thrust bearings. Needle thrust bearings are relatively low in cost but are high in capacity. They can be used for loads that are completely axial.

 ## Lubrication in Antifriction Bearings

Although friction values are low for the ball, roller, and needle bearings, lubrication is still necessary. Point or line contact reduces friction greatly. However, unit pressures are extremely high. For this reason, extreme-pressure (EP) lubricants are often needed. Many antifriction bearing types include a special provision for introducing the lubricant. One should consult the manufacturer for the best possible lubricant, and specify it in the design.

 ## Special Bearings

Special types of bearings are available for performing specific functions. Figure 4-24 illustrates some applications of Roundway® bearings and ways. From the construction, it should be noted that the rollers are chain mounted and operate on round ways. This reduces friction and cuts wear. Mounting is simplified; the height is adjustable, and self-alignment further reduces many of the usual problems. The load capacity is greater than for ball bushings. The coefficient of friction is approximately 0.005 for a single bearing and 0.007 for the standard V-mounted units.

Rod ends with male or female threads on one end and the antifriction bearing on the other are used extensively in the aircraft field. Special care and extreme precision are incorporated in aircraft bearings. Certain bear-

Figure 4-24
Roundway® bearings.
(Courtesy of Thomson Industries, Inc.)

ings are designed for reciprocating devices, where the action on the bearing is one of oscillation.

 Transmission Accessories

Many types of bearings are mounted in a housing custom-designed for a particular machine. In other cases, the designer uses commercial power-transmission units. These are rated in the same manner as bearings. Some types are available for sleeve bearings; many are available with the commonly used antifriction bearings. Figure 4-25 shows a light-duty pillow block. In using such a device, the designer must check carefully the ratings and the corrections for speed and life, as well as mounting dimensions. Provisions for shaft position must be made, and mounting bolts in the proper locations must be provided.

Figure 4-26 illustrates a flange cartridge. This type of unit is often used where a shaft continues through a wall opening or where there is a vertical

Figure 4-25
Light-duty pillow block.
(Courtesy of Fafnir Bearing, a
division of Textron, Inc.)

Figure 4-26
Flange cartridge.
(Courtesy of Fafnir Bearing, a
division of Textron, Inc.)

plate for convenient mounting. Positioning this unit is relatively easy if proper mounting holes are provided. Ratings must be checked in the same way as with a similar bearing. Seals and shields are selected and specified when needed; lubrication is checked and incorporated in the design where necessary. Transmission units are found in light-duty, medium-duty, and heavy-duty classes.

Rubber units are frequently used in air-conditioning and ventilating applications where it is desirable to reduce noise and vibration. Take-up units are often used where the position of the transmission unit must be adjusted. This is particularly true in belt applications.

 **Thrust Washers**

Thrust washers are members of the bearing family. For analysis, see the section on collar friction in Chapter 3. The action of a thrust washer is similar to that of a collar. Thrust washers must be lubricated to minimize friction. Because of the nature of the component, boundary or partial lubrication is all that can be attained. Usually, thrust washers are made of steel or solid bronze. Oil grooves are often cut into the washer to assist in the lubrication process. Various oil-groove configurations are used. Some thrust washers are spherical. Many are fabricated with holes, tabs, scallops, or lugs. The final shape of the thrust washer depends on the use to which it is placed and the loads that act upon it.

 ## Wheels and Casters

Mechanical and electrical equipment is often mounted on wheels or casters to provide mobility. Hundreds of possible wheel and caster assemblies are commercially available. Usual wheel sizes range from 6 in. to 16 in. in diameter. Tires on wheels are often made of solid rubber. Pneumatic tires are also available.

Casters are available in hundreds of styles for thousands of purposes. There are light, medium, and heavy-duty types; some are equipped with wheel brakes, some are not. An important classification of casters is based on whether or not they are equipped for swiveling. Several types of bearings are used; some are made of porous metal, others employ antifriction bearings. Three types of antifriction bearings are commonly used: the ball bearing, the standard roller bearing, and the tapered roller bearing. Casters can be equipped with spring action to absorb shock.

The usual range of caster-wheel diameters is from 2 in. to 8 in., although smaller ones are available. Wheel materials include soft rubber, hard rubber, phenolic resin, drawn steel, semisteel, and hard composition. The wheel material is selected according to the surface on which the caster will roll and the amount of rolling resistance that may be permitted. Certain casters have plates on the top; also, various types of stems are available. If plates are used, the designer must carefully check the plate size for compatability with the design; also, the designer must check the bolt pattern with respect to the size of bolts needed and the spacing between bolts.

Manufacturers frequently give load ratings. These are usually not the loads that produce failure, but rather the maximum desirable load to ensure relatively easy movement. The diameter is an important consideration. If the wheel is large, the horizontal operating force is smaller than it would be for a small-diameter wheel. The width of the wheel is also important. For heavier loads, extra width will help distribute the load and prevent floor damage. The larger wheel width, however, requires a greater operating force. On swivel-type casters, the axial (vertical) load is usually carried on either ball bearings or on tapered roller bearings.

Figure 4-27 shows one typical type of caster. A designer must consult manufacturers' catalog information to justify the proper selection for a particular product.

Figure 4-27
Regular-duty caster.
(Courtesy of The Fairbanks
Company.)

 Calculations of Bearing Loads

Regardless of the type of bearing under consideration, the actual bearing cannot be selected or designed until a complete force analysis is made. Before making a force analysis, tentative locations for bearings must be selected; sometimes, these positions may have to be changed when the numerical values for the loads are excessive. The general scheme of the assembly will often dictate where the bearings are likely to be placed. For example, a gear reduction box will be sized when the gears are selected; the bearings will probably then be mounted on the sides of the housing for convenience and compactness. Calculations for bearing loads are simple problems in mechanics; placement of the bearings often involves judgment on the part of the designer.

The following sketches show some common applications. In each sketch, the bearing is represented by the symbol ⊠ on either side of the shaft. This symbol is used regardless of the type of bearing. Many bearings support radial loads only; this implies that the forces from the other components act perpendicularly to the centerline of the shaft. Some bearings primarily support thrust loads; that is, they support loads parallel to the axis of the shaft. Other bearings must be designed to lend radial and axial support to a shaft.

Mechanical Components Causing Bearing Loads

A bearing is somewhat similar to a wall (or foundation) supporting a truss. In the case of a truss, vertical loads must include the weights of the various members and roof, plus the weight of overhead cranes, conveyors, and so on. Some of the truss loads will not be vertical. Mechanical bearings, must support the weights of the shafts and components, plus additional loading transmitted from such components as belts, chains, gears, cams, and connecting rods. In other cases, all loading will be vertical; in many cases, combined loading will be involved. Such components as propellers and helical gears induce thrust loads; drilling operations cause thrust loads that, in some cases, are greater in value than the radial loads in the equipment. When a shaft is used vertically, the weight of the shaft and supported parts must be included in the thrust-bearing reaction.

Following are some typical loading situations that influence the bearing load. For pulley (belt) or sprocket (chain) drives, the maximum load is equal to the pull on the tight side plus the pull on the slack side, assuming these two pulls are parallel as shown in Fig. 4-28. Figure 4-29 shows a case where the pulls are not parallel. The equilibrant (the single force needed to place the system in equilibrium) can be found graphically or mathematically. From the force diagram, the vertical component of E is $F_2 \sin 45$; the horizontal component of E is $F_1 + F_2 \cos 45$. Then,

$$E = \sqrt{(F_2 \sin 45)^2 + (F_1 + F_2 \cos 45)^2}.$$

This equation holds for an angle of 45 deg; any other situation is handled in a similar manner.

In Fig. 4-30, the weight on the drum cable would be in equilibrium with the reaction at the center of the drum and the operating force P, unless the weight was being accelerated. If the load were accelerated upward, the total load on the hoisting cable would be $W + Wa / g$. The bearing would then have to support this total load plus the operating force P. In most cases, the load is raised by applying sufficient torque to the shaft.

Figure 4-31 shows a pitch circle for a spur-tooth gear. If this is the driven gear in mesh with a driving gear, the load is applied by this vector

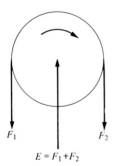

Figure 4-28
Pulley loads.

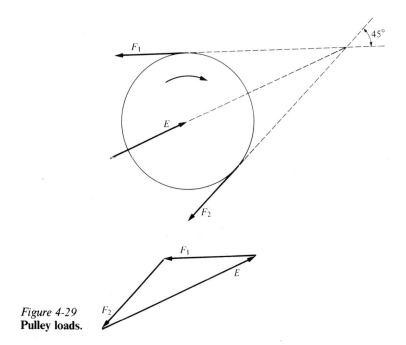

Figure 4-29
Pulley loads.

acting at the pressure angle of the gears ($14\frac{1}{2}$ or 20 deg). This action is discussed in detail in Chapter 12.

When connecting rods engage crankshafts, the force is applied through the centerline of the connecting rod to the offset of the crankshaft. The amount of the offset affects loading on the crankshaft bearings. The transmission of forces from a cam to a follower (whether flat-faced or circular) is always along the common normal at the point of contact. This

Figure 4-30
Drum and cable loads.

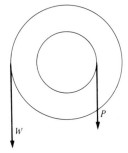

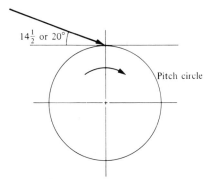

Figure 4-31
Spur-gear loads.

common normal changes as the cam rotates. It is therefore necessary to determine the maximum value; this can usually be done by careful inspection. Quite often, cams are heavy components and thus the cam weight must be considered.

The maximum loads for bandsaws are similar to those caused by belts and chains, except that the cutting action of the teeth influences the pull on the tight side of the belt. This value depends on the number of teeth per inch, the type of teeth, the type of material being sawed, and the thickness of the stock being cut.

The thrust load on twist drills is sometimes quite large and greatly affects the bearing design. For some alloy steels and larger drill sizes, thrust loads may be as high as 1800 lb. Radial loads must also be considered.

Figure 4-32 shows a situation where the load from a component to a shaft is applied between two shaft bearings. Before the bearings can be selected, the reaction values must be determined. The usual practice is to make both bearings identical for simplicity, so only the maximum value would be found. In this case, it would be R_2, since the load is closer to R_2 than to R_1. The value of R_2 can be found by taking moments about R_1. Thus

$$R_2 = \frac{Pa}{c}.$$

If dimensions a and b are made equal, R_1 and R_2 will be equal. This would be ideal; however, mounting problems may sometimes prevent this. Reactions are considered to act through the centerline of the bearings in all cases.

Figure 4-33 shows the bearing reactions with an overhung load. This method of mounting should be avoided wherever possible, since it causes a large reaction at one of the bearings (R_2, in this case). This means that a bearing with a greater capacity must be used at this point. The nature of the

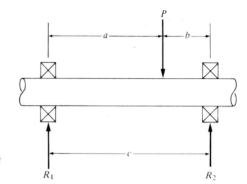

Figure 4-32
Bearing reactions: straddle mounting.

equipment often forces this type of mounting to be used. The output shaft of such equipment as gasoline engines, electric motors, and speed reducers must protrude to be usable. Quite often, the length of the shaft is kept to a minimum so that the overhung load will not be placed too far away from the bearing. From the sketch of Fig. 4-33, it is obvious that R_2 would equal the sum of P and R_1 if dimensions a and b were made equal. In using this type of mounting, it is desirable to keep the load P as close to R_2 as possible. In Fig. 4-33, R_2 is the larger reaction and can be found by taking moments about R_1. Thus

$$R_2 = \frac{Pc}{a}.$$

Note that R_1 is downward to place the system in equilibrium.

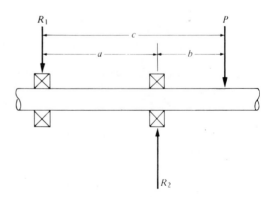

Figure 4-33
Bearing loads: overhung mounting.

Sometimes a mounting is used in which some of the loads fall inside the supporting bearings and some are overhung. The same procedure is used; that is, moments are taken about one of the unknown reactions. The other reaction can then be found by taking moments or by summing the vertical forces and reactions.

 ## Summary

Any equipment that uses rotating shafts or reciprocating rods must be equipped with suitable bearings. Basically, a designer has a choice between sliding (sleeve) bearings and antifriction bearings. Such a choice must be made by considering the total design in terms of function and preciseness; preciseness affects the total cost. In considering costs, one must think in terms of initial cost, the consequences of failure, and the cost of routine maintenance. Some of the features (both good and undesirable) of each type of bearing are listed below.

The Sliding Bearing

1. small radial space requirement
2. large axial space requirement
3. low pressures
4. high frictional losses
5. difficult lubrication (grooves needed)
6. misalignment permitted
7. breaking-in required
8. difficult replacement
9. quiet

The Antifriction Bearing

1. high initial cost
2. low frictional losses and show less wear
3. precise mounting needed
4. no breaking-in required
5. predictable life
6. large radial space requirement (except for needle bearings)
7. small axial space requirement
8. high unit pressures
9. higher speeds possible
10. fewer maintenance problems
11. greater precision possible

Bearing accessories such as pillow blocks and shaft hangers are readily available as off-the-shelf items. Ratings on such equipment are readily available to a designer; thus, their performance can be predicted.

Casters and wheels are easily purchased and are useful for providing mobility. Acceptable ratings are obtainable and can be helpful when incorporated into a design with other components.

Proper bearing design (coupled with good shaft design) can do much to increase the efficiency and reliability of a machine. Preloading on antifriction bearings reduces deflection within the bearing; this increases the rigidity of the shaft.

Questions for Review

1. Sketch various lubrication groove patterns for sliding bearings. Indicate the advantages and limitations of each type.

2. Discuss the undesirable features of a large L/D ratio. Why is a low L/D ratio not practical?

3. Why are some thrust washers made in spherical form?

4. List some limitations of gas (or air) bearings. What desirable features do they have?

5. What advantages do antifriction bearings have over sleeve (sliding) bearings? What disadvantages?

6. Why is the rated capacity of a Conrad-type bearing lower than that of a comparable size of bearing equipped with loading grooves?

7. What is meant by the B-10 life of an antifriction bearing? How does it compare with the average fatigue life?

8. In an antifriction bearing, is the load rating when the outer ring is rotating different from what it is when the inner ring is rotating? Why?

9. In antifriction-bearing terminology, what is meant by the term "boundary dimensions"? Why are they important?

10. What lubrication problems are encountered with antifriction bearings? How do these differ from sliding-bearing lubrication problems?

11. What is the purpose of shields on antifriction bearings?

12. List some typical methods of mounting ball bearings.

13. Why are miniature precision bearings assembled under "white-room" conditions?

14. Discuss the interchangeability of antifriction bearings as compared with sliding bearings.

15. What types of antifriction bearings are available for linear motions?

16. What is a spherical bearing? What distinguishes it from other types?

17. How does a shaft hanger differ from a pillow block?

18. What advantage does a split-pillow block have over the regular type?

19. What type of application would use a flange cartridge?

20. What advantages do ball or roller thrust bearings have over thrust washers? What disadvantages?

21. Why are some roller bearings tapered? Give two applications for tapered roller bearings.

22. How do swiveling casters differ in construction from the nonswiveling type?

23. Why are some antifriction bearings preloaded?

24. How does the Anti-Friction Bearing Manufacturers Association help the bearing industry? How does the AFBMA help the bearing user?

25. Why should a designer check the availability (and delivery) of antifriction bearings before specifying them?

Problems

1. A 36-mm shaft uses a sleeve bearing that sustains a load of 4000 N. If the allowable bearing pressure is 1.3 MN/m², find the length of the bearing and the L/D ratio.

2. A 20-mm shaft uses sleeve bearings. The total load per bearing is 2000 N. An L/D ratio of 2.5 is desired. What is the bearing pressure?

3. A 22-mm shaft is supported by sleeve bearings in an arrangement similar to Fig. 4-34. The load is 2200 N. The sleeve bearings have an L/D ratio of 1.5. Find the maximum bearing pressure.

Figure 4-34
Problem 3.

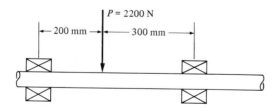

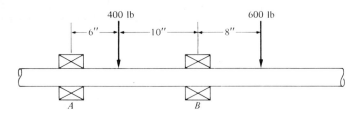

Figure 4-35
Problem 7.

4. A sleeve bearing is to have an L/D ratio of 1.0 and an allowable bearing pressure of 0.5 MN/m². Find the inside diameter and the length of the bearing if it is to sustain a load of 2550 N.

5. A sleeve bearing has an outside diameter of 1.50 in. and a length of 2 in. The wall thickness is $\frac{3}{16}$ in. The bearing is subjected to a radial load of 450 lb. Find the bearing pressure and the L/D ratio.

6. A thrust washer has an inside diameter of 12 mm and an outside diameter of 75 mm. If the allowable bearing pressure is 0.6 MPa, how much axial load can it sustain?

7. A 1.25-in-diameter shaft is mounted in sleeve bearings as shown in Fig. 4-35 and subjected to two radial loads. Each of the sleeve bearings has a length of 2.25 in. What is the maximum bearing pressure? Which bearing has the maximum?

8. The sleeve bearings in Fig. 4-36 are to be identical for the sake of simplicity. They are each to have an L/D ratio of 1.6 and an allowable bearing pressure not to exceed 100 psi of projected area. Find the value for the inside diameter and the length. Specify which bearing the design is based on, then find the pressure on the other bearing.

9. A light-duty sleeve bearing operates in still air at an air temperature of 70 °F and sustains a load of 350 lb. A shaft rotates inside the bearing

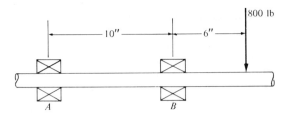

Figure 4-36
Problem 8.

at 600 rpm. The coefficient of friction is 0.015 between shaft and bearing. The shaft diameter is 1.75 in. and the bearing length is 2.50 in. Find the probable operating temperature of the bearing.

10. A heavy-duty, well-constructed bearing has good ventilation and an operating temperature of 140 °F. The ambient temperature is 85 °F. Find the total heat-radiating capacity in Btu/min. (*Note:* 1 Btu is equal to 778 ft-lb.) The inside diameter of the bearing is 1.5 in. and the length is 2.5 in.

11. Using the ratings in Table 4-3 and the charts of Fig. 4-21(b), select a roller bearing for a fatigue life of 80,000 hr and a speed of 400 rpm. The radial design load on the bearing is 950 lb.

12. Using the ratings in Table 4-3 and the charts of Fig. 4-21(b), find how much radial load can be placed on an MR-20 bearing if it is to operate at 600 rpm and have a B-10 life of 12,000 hr.

13. A plain sintered bronze bearing has an inside diameter of 10 mm, an outside diameter of 14 mm, and a length of 16 mm. A radial load of 88 newtons acts on the bearing. Find the L/D ratio and the bearing pressure in pascals.

14. A flanged bearing has the following dimensions:

    inside diameter = 12 mm,

    outside diameter = 18 mm,

    length = 12 mm,

    flange diameter = 22 mm,

    flange width = 3 mm.

    How much thrust load can this bearing sustain if the allowable bearing pressure is 345×10^3 pascals? Express your answer in newtons.

15. In Fig. 4-29 find the bearing load (E) if F_1 = 1200 N, F_2 = 400 N, angle = 60° (instead of 45°), and pulley diameter = 200 mm. Then find the torque for the pulley. Do not consider friction at any point.

16. A small vehicle uses 6206 ball bearings. The radial load on each wheel is 800 lb, and the wheel diameter is 24 in. Find the expected life in miles under these conditions.

17. A 6208 ball bearing has an inner-ring rotational speed of 840 rpm and carries an axial load of 1335 newtons and a radial load of 2670 newtons. Find the expected life of this bearing mathematically. Check your results using the nomogram of Fig. 4-13.

18. A sleeve bearing has an outer diameter of 75 mm, an inner diameter of 55 mm, and a length of 50 mm. The radial load is 2000 newtons. Find the bearing pressure (in MPa) and the L/D ratio.

SHAFT DESIGN
AND SEALS

5

Shafts are usually of circular cross section; either solid or hollow sections can be used. A hollow shaft weighs considerably less than a solid shaft of comparable strength, but is somewhat more expensive. Shafts are subjected to torsion, bending, or a combination of these two; in unusual cases, other stresses might also become involved. Careful location of bearings can do much to control the size of shafts, as the loading is affected by the position of mountings. After a shaft size is computed, its diameter is often modified (upward only) to fit a standard bearing. Calculations merely indicate the minimum size.

 ## Common Shaft Sizes

Table 5-1 lists some of the common available sizes for steel (round, solid) shafting. These are nominal sizes only. A designer must accurately compute the exact size so that it will fit properly into bearings. Since any machining is costly, a minimal amount of metal should be removed from stock sizes. Any metal removed in certain locations changes the shaft diameter in various axial positions. Therefore, proper radii must be provided to minimize stress concentrations. Information on stress-concentration factors is provided in Figs. 5-7 and 5-8. Abrupt changes in diameter without sufficient radii produce so-called "stress raisers." Thus it is desirable—from the standpoint of stress—to provide large radii. However, large radii also make

Table 5-1
Common Shaft Sizes (Solid Circular)

$\frac{1}{2}$	$1\frac{7}{16}$	$2\frac{3}{8}$	$3\frac{7}{8}$
$\frac{9}{16}$	$1\frac{1}{2}$	$2\frac{7}{16}$	$3\frac{15}{16}$
$\frac{5}{8}$	$1\frac{9}{16}$	$2\frac{1}{2}$	4
$\frac{11}{16}$	$1\frac{5}{8}$	$2\frac{5}{8}$	$4\frac{1}{4}$
$\frac{3}{4}$	$1\frac{11}{16}$	$2\frac{3}{4}$	$4\frac{7}{16}$
$\frac{13}{16}$	$1\frac{3}{4}$	$2\frac{7}{8}$	$4\frac{1}{2}$
$\frac{7}{8}$	$1\frac{13}{16}$	$2\frac{15}{16}$	$4\frac{3}{4}$
$\frac{15}{16}$	$1\frac{7}{8}$	3	$4\frac{15}{16}$
1	$1\frac{15}{16}$	$3\frac{1}{8}$	5
$1\frac{1}{16}$	2	$3\frac{1}{4}$	$5\frac{1}{4}$
$1\frac{1}{8}$	$2\frac{1}{16}$	$3\frac{3}{8}$	$5\frac{7}{16}$
$1\frac{3}{16}$	$2\frac{1}{8}$	$3\frac{7}{16}$	$5\frac{1}{2}$
$1\frac{1}{4}$	$2\frac{3}{16}$	$3\frac{1}{2}$	$5\frac{3}{4}$
$1\frac{5}{16}$	$2\frac{1}{4}$	$3\frac{5}{8}$	$5\frac{15}{16}$
$1\frac{3}{8}$	$2\frac{5}{16}$	$3\frac{3}{4}$	6

Table 5-2
Typical Metric Shaft Diameters (mm)

4	40	110
5	45	120
6	50 (1.969″)	130
7	55	140
8	60	150 (5.901″)
9	65	160
10	70	170
12 (0.472″)	75 (2.953″)	180
15	80	190
17	85	200 (7.874″)
20	90	220
25 (0.984″)	95	240
30	100 (3.937″)	260
35	105	280 (11.02″)

it difficult to mount such other components as pulleys, cams, gears, and so on because of radius interference. Often, the bore of such other components has to be chamfered to clear radii at the point where a shaft changes diameter. In the final analysis, a compromise has to be made between ideal shaft radii and the undercutting of other components.

 Metric Shaft Sizes

The diameters of shafts made compatible with metric-sized bores of mechanical components (such as antifriction bearings) are specified in millimeters. Although any shaft size can be turned to provide extremely accurate fits, Table 5-2 shows popular nominal sizes.

 Torsion

Equations for a shaft in pure torsion are listed below; these equations are for round solid and round hollow sections only:

$$T = S_s\frac{J}{c} = \frac{S_s\pi D^3}{16} = \frac{63,000\ (\text{hp})}{N} \qquad \text{for solid shafts;} \qquad \textbf{(5-1)}*$$

$$T = S_s\frac{J}{c} = \frac{S_s\pi\left(D_o^{\,4} - D_i^{\,4}\right)}{16D_o} = \frac{63,000\ (\text{hp})}{N} \qquad \text{for hollow shafts;} \qquad \textbf{(5-2)}$$

*Computer program in Appendix L.

where

T = torque (in-lb),

S_s = design stress in shear (psi),

D = diameter of solid shaft (in.),

D_o = outside diameter of hollow shaft (in.),

D_i = inside diameter of hollow shaft (in.),

hp = horsepower,

N = revolutions per minute,

J = polar moment of inertia (in^4),

c = distance from neutral axis to outermost fiber (in.).

◆ *Example* _____

Compute the diameter of a solid shaft that rotates 100 rpm and transmits 1.2 hp. The design stress for shear is to be 6000 psi and the shaft is subjected to torsion only.

Solution.

$$T = \frac{63,000 \; (\text{hp})}{N} = \frac{(63,000)(1.2)}{100} = 756 \text{ in-lb};$$

$$D = \sqrt[3]{\frac{16T}{\pi S_s}} = \sqrt[3]{\frac{(16)(756)}{(3.14)(6000)}} = 0.863 \text{ in. (minimum).}$$

Use $\frac{7}{8}$-in. diameter.

◆ *Example* _____

A solid steel torsion shaft is needed with a capacity of 2.5 kW when rotating at 1200 rpm. Find the minimum diameter if the allowable shear stress is 40 MN/m^2.

Solution.

$$T = \frac{\text{kW} \; (1000)(60)}{2\pi N} = \frac{2.5(1000)(60)}{2\pi(1\,200)} = 19.89 \text{ N} \cdot \text{m}.$$

$$D = \sqrt[3]{\frac{16T}{\pi S_s}} = \sqrt[3]{\frac{16(19.89)}{\pi(40)(10)^6}} = 0.013 \; 6 \text{ m, or 13.6 mm.}$$

Torsional Deflection (Solid Shaft)

The amount of twist in a shaft is important. One rule of thumb is to restrict the torsional deflection to one degree in a length equal to 20 diameters. For example, if the active part of a shaft is 40 in. and the shaft diameter is 2 in., 1 deg of torsional deflection would be permitted. In some applications, the angle of twist must be smaller than this. The following equation applies to torsional deflection:

$$D = \sqrt[4]{\frac{32TL}{\pi G\theta}}, \tag{5-3}$$

where

L = length of shaft subjected to twist (in.),

G = shear modulus of elasticity (psi),

θ = angle of twist (rad).

Figure 5-1 shows the angle of twist (greatly exaggerated) that appears when torque is applied to a shaft.

◆ *Example* ─────────────────────────

A 2-in. lineshaft transmits 20 hp and rotates at 200 rpm. Two pulleys spaced 30 in. apart cause a torsional deflection. If the shear modulus of elasticity is 12,000,000 psi, find the angle of twist in degrees.

Figure 5-1
Torsional deflection of shaft.

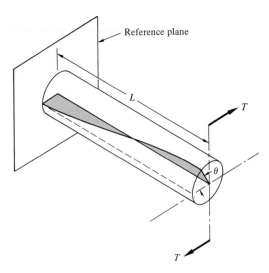

Solution.

$$T = \frac{63{,}000 \ (\text{hp})}{N} = \frac{(63{,}000)(20)}{200} = 6300 \text{ in-lb};$$

$$\theta = \frac{32TL}{\pi GD^4} = \frac{(32)(6300)(30)}{\pi(12{,}000{,}000)(2)^4} = 0.0100 \text{ rad};$$

$$0.0100 \times 57.3 = 0.573 \text{ deg.}$$

◆ *Example* ─────────────────────────────

Find the diameter of a shaft for which torsional deflection will not exceed one degree in a length of 20 diameters. The shaft is to transmit 40 hp at 200 rpm. Assume that the shear modulus of elasticity is 12,000,000 psi.

Solution.

$$T = \frac{63{,}000 \ (\text{hp})}{N} = \frac{(63{,}000)(40)}{200} = 12{,}600 \text{ in-lb};$$

$$D = \sqrt[4]{\frac{32TL}{\pi G\theta}} = \sqrt[4]{\frac{(32)(12{,}600)(20D)}{\pi(12{,}000{,}000)(0.01745)}} = \sqrt[3]{\frac{(32)(12{,}600)(20)}{\pi(12{,}000{,}000)(0.01745)}};$$

$$D = \sqrt[3]{12.3} = 2.31 \text{ in. (minimum).}$$

Note. The change in the order of the radical is justified because D was removed from under the radical. If this is done by substituting $L = 20D$ in the equation

$$\theta = \frac{32T(20D)}{\pi GD^4},$$

it is evident that

$$\theta = \frac{32T(20)}{\pi GD^3}.$$

Strength vs. Stiffness (Round Solid Torsion Shaft)

For any given set of shaft conditions, there is a diameter below which the angle of twist and not the maximum allowable stress becomes the controlling factor in the design of a shaft. The following example shows how to determine this value.

◆ *Example* _____

For a round solid torsion shaft, find the diameter below which the angle of twist instead of the maximum allowable shear stress becomes the controlling factor. Assume that the allowable shear stress is 6000 psi, the transverse modulus of elasticity is 12×10^6 psi, and the allowable twist is 0.1 degree per foot of length.

Solution. Equating T,

$$\frac{S_s \pi D^3}{16} = \frac{\pi G \theta D^4}{32 L} \; ;$$

$$\frac{6000(\pi) D^3}{16} = \frac{\pi(12 \times 10^6)(0.1)(\pi/180) D^4}{32(12)} \; ;$$

$$D = \frac{6000(32)(12)(180)}{16(12 \times 10^6)(0.1)\pi} = 6.875 \text{ in.}$$

 Hollow vs. Solid Shafting

The following problem illustrates how to compare solid and hollow shafts. (Refer to Fig. 5-2.)

◆ *Example* _____

(a) Find the inside and outside diameters of a hollow shaft that will replace a 3-in. solid shaft made of the same material. The hollow shaft should be

Figure 5-2
Solid versus hollow shafting.

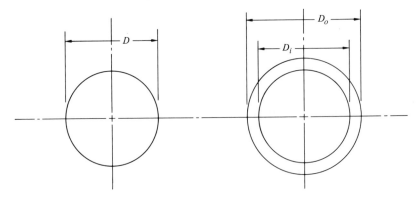

equally strong in torsion, yet weigh half as much per foot. (b) If the hollow shafting material costs 20% more per pound per foot, what percent saving in cost is effected by using the hollow?

Solution. (a) Two simultaneous equations must be set up, one equating J/c for hollow and solid shaft sections, the other equating cross-sectional areas such that one hollow-shaft cross-sectional area equals one-half solid-shaft cross-sectional area:

Solid   *Hollow*

$$\frac{\pi D^3}{16} = \frac{\pi\left(D_o^{\,4} - D_i^{\,4}\right)}{16 D_o};$$

$$\frac{1}{2}\frac{\pi D^2}{4} = \frac{\pi}{4}\left(D_o^{\,2} - D_i^{\,2}\right).$$

Simplifying, we obtain

$$D^3 = \frac{D_o^{\,4} - D_i^{\,4}}{D_o} \quad\quad \text{or} \quad\quad D_i^{\,4} = D_o^{\,4} - D^3 D_o,$$

$$\frac{D^2}{2} = D_o^{\,2} - D_i^{\,2} \quad\quad \text{or} \quad\quad D_i^{\,2} = D_o^{\,2} - \frac{D^2}{2}.$$

Thus

$$D_i^{\,4} = D_o^{\,4} - D^2 D_o^{\,2} + \frac{D^4}{4}.$$

Equating $D_i^{\,4}$, we find

$$D_o^{\,4} - D^3 D_o = D_o^{\,4} - D^2 D_o^{\,2} + \frac{D^4}{4},$$

$$D^2 D_o^{\,2} - D^3 D_o - \frac{D^4}{4} = 0.$$

Applying the quadratic equation, we obtain

$$D_o = \frac{D^3 \pm \sqrt{D^6 + 4 D^2 (D^4/4)}}{2 D^2} = \frac{D^3 + D^3\sqrt{2}}{2 D^2} = \frac{D}{2} + \frac{D\sqrt{2}}{2}. \quad\quad \textbf{(5-4a)}$$

Substituting $D = 3$, we obtain

$$D_o = \tfrac{3}{2} + \tfrac{3}{2}(1.414) = 3.621 \text{ in.}$$

Substituting in either of the simplified simultaneous equations, solve for D_i:

$$D_i^2 = D_o^2 - \frac{D^2}{2} \quad \text{or} \quad D_i = \sqrt{D_o^2 - \frac{D^2}{2}} \; ; \qquad \textbf{(5-4b)}$$

$$D_i = \sqrt{(3.621)^2 - \frac{(3)^2}{2}} = \sqrt{13.1 - 4.5} = 2.93 \text{ in.}$$

Area of solid-shaft cross section $= (\pi/4)D^2 = 0.785(3)^2 = 7.07 \text{ in}^2$. (b)
Area of hollow-shaft cross section $= (\pi/4)(D_o^2 - D_i^2) = 0.785[(3.621)^2 - (2.93)^2] = 3.56 \text{ in}^2$.

$$\text{Percentage saving} = \frac{7.07 - 1.20(3.56)}{7.07}(100) = 40\%.$$

◆ *Example* ————————————————————————————

A 1-in. solid shaft is to be replaced with a hollow shaft of equal torsional strength having an outside diameter of 2 in. Find the inside diameter and the percentage of weight saved.

Solution. For the solid shaft,

$$T = S_s \frac{\pi D^3}{16}.$$

For the hollow shaft,

$$T = S_s \frac{\pi (D_o^4 - D_i^4)}{16D_o}.$$

Next, equate T for each section, since they are to have equal torsional strength. Thus

$$S_s \frac{\pi D^3}{16} = S_s \frac{\pi (D_o^4 - D_i^4)}{16D_o}.$$

This simplifies to $D^3 D_o = D_o^4 - D_i^4$. Then

$$D_i^4 = D_o^4 - D^3 D_o,$$

$$D_i^4 = (2)^4 - (1)^3(2),$$

$$D_i^4 = 16 - 2 = 14,$$

$$D_i = \sqrt[4]{14} = 1.93 \text{ in.}$$

The area of the solid shaft is

$$\frac{\pi D^2}{4} = \frac{\pi (1)^2}{4} = 0.785 \text{ in}^2.$$

The area of the hollow shaft is

$$\frac{\pi (D_o^2 - D_i^2)}{4} = \frac{\pi \left[(2)^2 - (1.93)^2\right]}{4} = 0.216 \text{ in}^2.$$

The percentage saving in weight depends on the cross-sectional areas. Thus

$$\text{percentage saving} = \frac{0.785 - 0.216}{0.785}(100) = 72.5\%.$$

Combined Torsion and Bending (Solid Shafting)

A shaft is often subjected to combined torsion and flexure. There are numerous ways of computing a shaft diameter under these conditions. The simplest is to compute equivalent bending and twisting moments for the shaft and then substitute these values into the regular equations for torsion and bending. Equations for equivalent moments are as follows:

$$T_E = \sqrt{M^2 + T^2}, \tag{5-5}$$

$$M_E = \frac{M + T_E}{2}, \tag{5-6}$$

where

T_E = equivalent twisting moment (in-lb),

M_E = equivalent bending moment (in-lb).

The torsion equation then becomes

$$D = \sqrt[3]{\frac{16 T_E}{\pi S_s}} \tag{5-7}$$

and the bending (or flexure) equation becomes

$$D = \sqrt[3]{\frac{32 M_E}{\pi S}}. \tag{5-8}$$

Both equations must be solved; the larger of the two diameters is then used for the calculated size. It is important to remember that the allowable shearing stress is used in the *torsion* formula; the allowable stress in tension is used in the *flexure* (or bending) formula. The following example shows how this method is applied. Assume that the design dictates the bearing locations; if they were located closer to the vertical load, the effects of bending might not be too serious. Note that, in all cases, distances are figured from the centerlines of such components as bearings, pulleys, sprockets, gears, and so on. The design stress values for flexure should be based on a factor of safety that includes some provision for endurance (this type of loading on a shaft induces reversed bending, which could lead to fatigue). The endurance limit is often assumed to be 50% of the ultimate strength before applying a factor of safety.

◆ *Example* ─────────────────────────────

In Fig. 5-3, the shaft transmits 10 hp at 500 rpm. Assume that the design stresses are 6000 psi (shear) and 8000 psi (tension). Compute the diameter. (Neglect shaft weight.)

Solution. First, compute the two reactions. Here, these are 600 lb on the left bearing and 200 lb on the right bearing. The maximum moment is then $(600)(5) = 3000$ in-lb. Thus

$$T = \frac{63,000 \, (\text{hp})}{N} = \frac{(63,000)(10)}{500} = 1260 \text{ in-lb};$$

$$T_E = \sqrt{M^2 + T^2} = \sqrt{(3000)^2 + (1260)^2} = 3260 \text{ in-lb};$$

$$M_E = \frac{M + T_E}{2} = \frac{3000 + 3260}{2} = 3130 \text{ in-lb};$$

$$D = \sqrt[3]{\frac{16 T_E}{\pi S_s}} = \sqrt[3]{\frac{(16)(3260)}{\pi(6000)}} = \sqrt[3]{2.77} = 1.40 \text{ in.};$$

$$D = \sqrt[3]{\frac{32 M_E}{\pi S}} = \sqrt[3]{\frac{(32)(3130)}{\pi(8000)}} = \sqrt[3]{3.99} = 1.59 \text{ in.}$$

The safe shaft size for the given conditions is the larger of the two, or 1.59 in.

─────────────────────────────────────

Figure 5-4 lists some typical formulas for maximum moments and deflections for various shaft conditions.

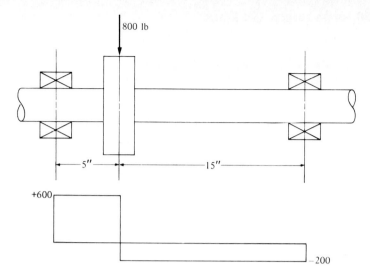

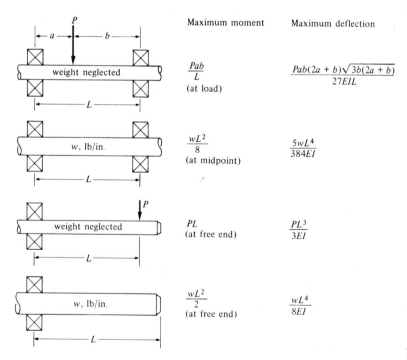

Figure 5-3
Combined bending and torsion.

Figure 5-4
Maximum moments and deflections for typical shaft conditions. For various loadings of continuous shafting, one should completely evaluate shear and moment diagrams to obtain the maximum moment. Deflection can be found by the area-moment method.

	Maximum moment	Maximum deflection
weight neglected	$\dfrac{Pab}{L}$ (at load)	$\dfrac{Pab(2a+b)\sqrt{3b(2a+b)}}{27EIL}$
w, lb/in.	$\dfrac{wL^2}{8}$ (at midpoint)	$\dfrac{5wL^4}{384EI}$
weight neglected	PL (at free end)	$\dfrac{PL^3}{3EI}$
w, lb/in.	$\dfrac{wL^2}{2}$ (at free end)	$\dfrac{wL^4}{8EI}$

169

Tension or Compression Combined with Torsion

In some applications, a thrust load is combined with torque, resulting in a combination of tension (or compression) and shear. In such cases, the maximum resultant stresses exceed the original stresses. A twist drill is subjected to torsion and compression as it is drilling through a piece of metal. A propeller spinning in air or water is another example of combined thrust and torque. The following equations give the maximum resultant stresses encountered under such conditions:

$$S_{s_{max}} = \frac{\sqrt{S^2 + 4S_s^2}}{2},$$

$$S_{max} = \frac{S}{2} + S_{s_{max}},$$

where

S = tensile (or compressive) stress due to axial load,

S_s = shearing stress due to torsion.

◆ *Example* ————————————————————————————

Find the maximum shearing stress and maximum tensile stress caused by an axial load of 500 lb and a torque of 600 in-lb acting on a 1-in.-diameter solid shaft.

Solution.

$$S = \frac{P}{A} = \frac{P}{(\pi/4)(1)^2} = \frac{500}{0.785} = 637 \text{ psi};$$

$$S_s = \frac{16T}{\pi D^3} = \frac{(16)(600)}{\pi(1)^3} = 3077 \text{ psi};$$

$$S_{s_{max}} = \frac{\sqrt{S^2 + 4S_s^2}}{2} = \frac{\sqrt{(637)^2 + 4(3060)^2}}{2} = 3075 \text{ psi};$$

$$S_{max} = \frac{S}{2} + S_{s_{max}} = \frac{637}{2} + 3077 = 3396 \text{ psi}.$$

Note that the maximum resultant stresses are larger than the original stresses. In this type of calculation, the usual procedure is to assume a diameter for the shaft, calculate the resultant stresses, and then see whether they fall within the prescribed limits. If the results exceed the allowable stresses or are lower than the allowable stresses, a new diameter should be assumed and new calculations made. Previous design experience is very helpful in handling this type of situation. A knowledge of approximate shaft size saves much time in arriving at a suitable final solution.

Mohr's circle provides a graphical representation of and check for the mathematical solution to the problem of shafts subjected to torsion (shear) and axial tension or compression. In Fig. 5-5, the vertical axis can be used for shear and the horizontal axis for the direct stress of tension or compression. The calculated value for direct stress is laid out from point 0 to point B. The induced torsional stress is drawn from point B vertically to point C. The radius for Mohr's circle is found by drawing a line from point C to the midpoint A, located between 0 and B. The center of the circle is A. The distance from A to C gives the value for $S_{s_{max}}$, and the horizontal distance from 0 to D represents S_{max}.

Figure 5-5
Mohr's circle solution for example.

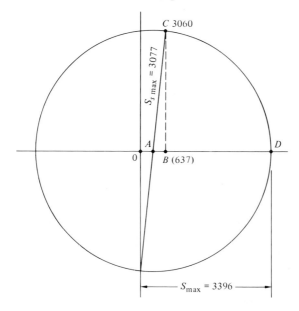

Keyseats in Shafts

Various types of keys, such as flat, square, and Woodruff keys, are used to connect shafts to other components. The resulting keyseat weakens the shaft. A rule of thumb is to reduce the design stress by 25% in any case where a shaft has a keyseat. For example, if 8000 psi were the allowable stress in torsion, 6000 psi would be used if the shaft had to have a keyseat.

Critical Speeds

Like any beam, an overhanging shaft or one supported between two bearings will deflect under its own weight. Thus, its mass center will not be collinear with the centerline of the bearings. The situation is worse if the shaft carries rotating discs such as pulleys, gears, flywheels, cams, and so on. At some critical speed, the period of vibration of the center of the mass under the opposing centripetal and elastic forces will equal the period of rotation of the shaft. Dangerous vibrations will occur at this critical speed, and the shaft will be dynamically unstable. At this speed, the angular velocity in revolutions per minute will equal the natural frequency. It is considered good practice to operate the shaft at least 20% above or below this speed. The following equations can be used to determine critical speeds under certain conditions.

Round steel shaft (of uniformly distributed weight):

$$\omega_c = \sqrt{\frac{384EI}{5L^3}\frac{12g}{W}} \quad \text{and} \quad N = \frac{60\omega_c}{2\pi}, \tag{5-9}$$

where

W = total weight (lb),

ω_c = critical speed (rad/sec),

N = rotational speed (rpm),

D = shaft diameter (in.),

L = effective length (in.),

I = moment of inertia (in^4),

E = modulus of elasticity (psi),

g = acceleration of gravity (32.2 ft/sec^2).

If E is taken to be 30,000,000 psi and the density of steel is 0.28 lb/cu in., the critical speed in rev/min is

$$N = \frac{60}{2\pi} \sqrt{\frac{(384)(30,000,000)(\pi D^4)(12)(32.2)}{(5)(L^3)(64)(\pi D^2/4)(L)(0.28)}}.$$

This simplifies to

$$N = 4,270,000 \frac{D}{L^2} \quad \text{or} \quad \omega_c = 446,000 \frac{D}{L^2}. \tag{5-10}$$

This is the lowest critical speed. Other critical speeds can be found by multiplying the above equations by 4, 9, 16, 25, and so on.

◆ *Example* ———————————————————————————

A 2-in. shaft is supported by bearings 90 in. apart. The shaft carries no rotating discs; its weight is 0.28 lb per cu in., its modulus of elasticity is 30,000,000 psi, and it has a uniform (round) cross section. Find the first two critical speeds.

Solution.

$$N = 4,270,000 \frac{D}{L^2} = \frac{4,270,000(2)}{(90)^2} = 1054 \text{ rpm.}$$

The next critical speed would then be (4)(1054) = 4216 rpm.

———

Disc mounted midway between bearings (neglecting shaft weight):

$$\omega_c = \sqrt{\frac{576EIg}{WL^3}}. \tag{5-11}*$$

Disc mounted *a* **in. from left bearing and** *b* **in. from right bearing** (neglecting shaft weight):

$$\omega_c = \sqrt{\frac{36EILg}{a^2 b^2 W}}. \tag{5-12}*$$

It should be noted that the preceding equations assume that anti-friction bearings are used; therefore, the shaft behaves like a simply supported beam. If the shaft is mounted with sleeve-type bearings, the shaft behaves like a fixed-end beam and different equations should be used.

Critical Speed for System

The critical speed of a shaft carrying a number of discs of various masses can be approximated by Dunkerley's equation, simplified form of

*I is the moment of inertia of the *shaft alone* and W is the weight of the *disc alone*.

which is:

$$\frac{1}{\omega_{c_s}^2} = \frac{1}{\omega_{c_1}^2} + \frac{1}{\omega_{c_2}^2} + \frac{1}{\omega_{c_3}^2} + \cdots \qquad (5\text{-}13)$$

Here, ω_{c_s} is the critical speed for the system (in rad/s). The other ω_c values represent the critical speed of the shaft (with no discs) and the critical speed of each disc alone, assuming that the shaft is massless.

◆ Example

A steel shaft carrying a pulley and a gear is supported by ball bearings at each end. The following critical speeds have been determined:

ω_c for shaft = 100 rad/s,

ω_c for pulley = 125 rad/s,

ω_c for gear = 125 rad/s.

Find the critical speed for the system.

Solution.

$$\frac{1}{\omega_{c_s}^2} = \frac{1}{\omega_{c_1}} + \frac{1}{\omega_{c_2}} + \frac{1}{\omega_{c_3}};$$

$$\frac{1}{\omega_{c_s}^2} = \frac{1}{(100)^2} + \frac{1}{(125)^2} + \frac{1}{(125)^2} = \frac{3.5625}{15,625}.$$

Therefore

$$\omega_{c_s} = 66.2 \text{ rad/s}.$$

Since vibrations in machinery can be very serious, it is good practice to check each design thoroughly (mathematically) and then verify results with carefully conducted laboratory tests.

Flexible Shafts

Flexible shafts can be used for remote control or for power transmission. Commercial sizes range from 0.150 in. to 0.750 in. They can be used to transmit power from an input source to an output around many obstacles that would ordinarily dictate an elaborate and expensive system of gearing. Thus, one shaft assembly can replace many parts. A typical flexible-shaft assembly consists of a cable (spring, wire rope, or cable), a flexible casing (sheath), shaft end fittings, and casing end fittings. One of the common cable end fittings is a formed square. Typical applications include speedometer drive, radio controls, spotlight controls, dental drills, and small portable tools. The cable is made of built-up sections wound around a single wire, each successive layer wrapped in the direction opposite to the preceding one. Small-size flexible cables can operate at speeds of 20,000 rpm. Larger sizes normally operate at speeds of 1750 to 3600 rpm. In power-drive applications, the direction of the outermost helix is important; for clockwise rotation, a left-hand helix should be used; for counterclockwise rotation, a right-hand helix.

The torque capacities of various flexible shafts should be obtained by referring to manufacturers' catalogs. Minimum operating radii should also be carefully checked before incorporating a given shaft in a design. Proper lubrication is important in power drive shafts.

Figure 5-6 shows how flexible shafts can be used.

Figure 5-6
Flexible-shaft application.
(Courtesy of S. S. White®.)

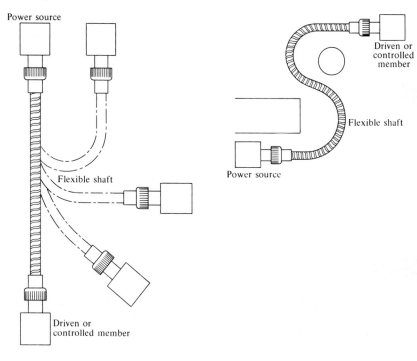

Noncircular Shafts

Shafts often have a cross section of some shape other than round. The blade of a screwdriver has a rectangular cross section; most radio and television control shafts are made round with a flattened section to avoid using a setscrew; and some shafts end with a triangular section to avoid using a key. Socket-wrench handles contain a square shaft for engaging the removable sockets. Valve stems are usually machined to a square section on the end to engage the hand wheel. An approximation for the maximum shearing stress can be found by applying the torsion formula $S_s = Tc/J$ and using the polar moment of inertia of the section under consideration. However, this equation gives only an approximation, because the shearing stress is not uniformly distributed in a noncircular section. An equation for a square section is as follows:

$$S_s = \frac{4.8T}{B^3},$$ (5-14)

where B is the side of the square section (in.).

A hexagonal section is frequently found in conveyor shafts and in certain small tools, such as setscrew wrenches. The length of such wrenches is limited to control the amount of torque that one might apply manually. Since the stress is not uniformly distributed in this type of cross section, the polar section modulus does not give an accurate value. The approximate stress can be found by dividing the torque by $0.2 F^3$, where F is the distance across the flat part of the hexagon.

Linear Deflection of Shafts

As a rule of thumb, the bending deflection of a shaft should be limited to 0.01 in. per foot of length between supports. Greater stiffness is sometimes necessary. The type of bearing in the support can dictate the amount of deflection; if the bearings are self-aligning, linear deflection does not present so great a problem. If the bearings are not self-aligning, binding can occur in the bearing itself; this could lead to serious difficulty and possibly malfunctioning of parts.

## Materials for Shafts

Shafts up to 3 in. in diameter are usually made of cold-rolled steel. The carbon content of the steel may range from 0.3 to 0.5%. Shafts from 3 in. to 5 in. are either cold-rolled or forged; shafts above 5 in. must be forged and then machined. In very light applications, plastics can be used for shafting materials, particularly in electrical applications, where their insulating properties are useful for safety reasons. In addition, plastics can help reduce production costs.

 Power Take-offs

Lineshafts can be continuous shafts with an input at one location and several outputs on either side of the input. Gears or pulleys can be used to supply power to a shaft or to take power off the lineshaft to operate other equipment. Two typical situations are explained in the following illustrative problems.

◆ *Example* _____

Figure 5-7 shows a lineshaft rotating at 150 rpm and supplying 42 hp at *A*. Power is removed at *B*, *C*, and *D* in the amounts shown. For the sake of simplicity, assume that the shaft is subjected to torsion alone. Calculate the diameter based on torsion where the allowable stress is 8000 psi.

Solution. The shaft must be designed for the section that transmits the largest horsepower value. Referring to the figure, we find the following values:

between *A* and *B*: 12 hp,

between *A* and *C*: 20 + 10 = 30 hp,

between *C* and *D*: 10 hp.

Figure 5-7
Power take-offs.

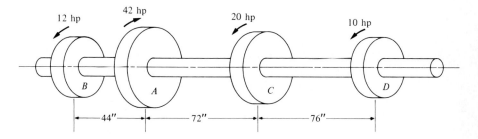

Thus the shaft must be designed on the basis of 30 hp:

$$T = \frac{63,000 \text{ (hp)}}{N} = \frac{63,000(30)}{150} = 12,600 \text{ in-lb};$$

$$D = \sqrt[3]{\frac{16T}{\pi S_s}} = \sqrt[3]{\frac{16(12,600)}{\pi(8000)}} = 2.00 \text{ in. (minimum)}.$$

This value is the minimum diameter based on strength. Torsional deflection should also be checked to make certain that the angle of twist is within reasonable limits.

◆ *Example*
In the arrangement shown in Fig. 5-8(a), power is received on a 24-in.-diameter pulley at A from a horizontal belt. Pulley A weighs 200 lb. On the same shaft as pulley A is another pulley, B, which weighs 75 lb and has a diameter of 12-in. Pulley B transmits power to another machine with belt pulls as shown. Find a suitable shaft diameter for an allowable shearing stress of 6000 psi and an allowable flexural stress (including endurance) of 5000 psi. (Neglect the shaft weight.)

Solution. First, consider the shaft as a beam with vertical components (Fig. 5-8b) and horizontal components (Fig. 5-8c). Determine reactions R_1 and R_2 for both the vertical and horizontal loading. Vertical loading at B is equal to 75 lb + (300 + 100)(0.707), or 358 lb. Then,

$$R_{2_V} = \frac{15(200) + 35(358)}{40} = 388 \text{ lb},$$

$$R_{1_V} = 358 + 200 - 388 = 170 \text{ lb}.$$

Next, construct the vertical moment diagram as shown in Fig. 5-8(d).
The horizontal loading at B is (300 + 100)(0.707) = 283 lb. Next, find R_{1_H} and R_{2_H}:

$$R_{2_H} = \frac{15(1500) - 35(283)}{40} = 315 \text{ lb},$$

$$R_{1_H} = 1500 - 283 - 315 = 902 \text{ lb}.$$

Next, construct the horizontal moment diagram as shown in Fig. 5-8(e).
The maximum moment is located at point A. Its value can be found from the Pythagorean theorem. Thus

$$M = \sqrt{(2250)^2 + (13,630)^2} = 13,800 \text{ in-lb}.$$

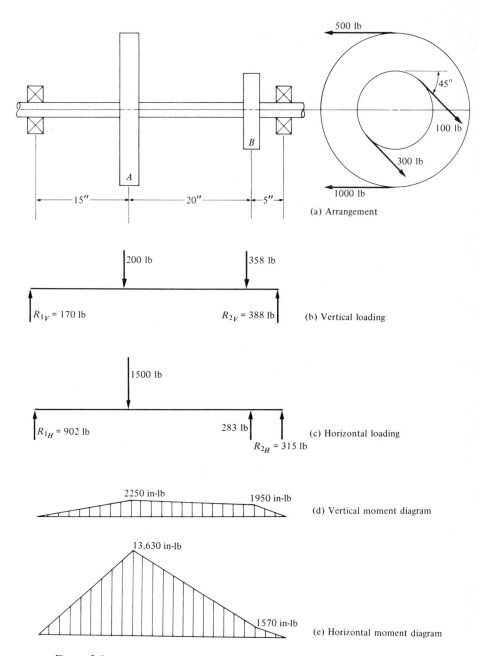

(a) Arrangement

(b) Vertical loading

(c) Horizontal loading

(d) Vertical moment diagram

(e) Horizontal moment diagram

Figure 5-8
Pulley-shaft arrangement.

Next, the torque is found at each pulley:

$$T_A = (1000 - 500)(12) = 6000 \text{ in-lb},$$

$$T_B = (300 - 100)(6) = 1200 \text{ in-lb}.$$

The larger value is T_A; thus, T_A should be used as the twisting moment in the following equations:

$$T_E = \sqrt{M^2 + T^2} = \sqrt{(13{,}800)^2 + (6000)^2} = 15{,}100 \text{ in-lb},$$

$$M_E = \frac{M + T_E}{2} = \frac{13{,}800 + 15{,}100}{2} = 14{,}500 \text{ in-lb}.$$

Now compute the shaft diameter by using the equivalent twisting moment and the equivalent bending moment:

$$D = \sqrt[3]{\frac{16 T_E}{\pi S_s}} = \sqrt[3]{\frac{16(15{,}100)}{\pi (6000)}} = 2.34 \text{ in.},$$

$$D = \sqrt[3]{\frac{32 M_E}{\pi S}} = \sqrt[3]{\frac{32(14{,}500)}{\pi (5000)}} = 3.09 \text{ in.}$$

The shaft must be at least 3.09 in. in diameter; $3\frac{1}{8}$ in. would be the logical choice for a common shaft size.

This discussion has been limited to shaft design on the basis of strength alone. Torsional deflection should be checked to make sure that its value is reasonable.

In most shafting, torsion is the primary consideration. Torsion induces shearing stresses. In many situations, tension (or compression) can also be induced. In the case of flexure (bending), both tensile and compressive stresses combine with the torsional shearing stresses. Axial loads can induce tensile or compressive stresses that combine with the torsional shearing stresses. Some shaft assemblies are subjected to torsional, axial, and flexural loads. Design becomes extremely complicated in such instances.

Laboratory study is of great help in designing shafts that will be subjected to unusual loading. It is desirable to have S-N studies (stress versus number of cycles) made where unusual reversals of stress are anticipated. Stress studies with plastic models and polarized light are also helpful.

A decision often has to be made between hollow and solid shafting. For a given torque, the use of hollow shafting can effect a great saving in weight. Hollow shafting, however, is higher in cost and uses more radial space than its solid counterpart. Shafting is usually round. However, squared-off sections can sometimes eliminate the use of setscrews or keys, with a resultant saving in cost. Valve handles ordinarily have a square hole to fit a square section of the valve stem (shaft).

 ## Stress Concentration Factors*

When loads are applied to mechanical components, induced stress values are higher than average at certain locations, such as the fillet of a keyseat in a shaft, the fillet at the base of a gear tooth, and the fillet at the change in cross-sectional area of a shaft. The design stress must be set high enough to provide safety at these localized areas, which are popularly called "stress raisers."

A stress concentration factor can be defined as the ratio of the maximum stress to the average stress:

$$K_t = \frac{S_{max}}{S}, \tag{5-15}$$

where

S_{max} = maximum stress (psi),

S = average stress (psi),

K_t = stress concentration factor.

The subscript t is used to indicate *theoretical*. If the critical area is subjected to a three-dimensional effect such as that caused by torsion on a shaft fillet, another subscript may be added to K_t to distinguish shear from tension, compression, or flexure. The factor is then written K_{ts}.

Stress concentration factors are determined experimentally, for the most part. Several methods are employed; a few of these are:

1. *Photoelastic study.*[†] A plastic model of the part being studied is made and viewed with some type of polariscope. The plastic material is double-refracting when stressed. Polarized light shows the strains; a polarizing screen or analyzing filter is needed to reveal the direction and magnitude of stresses (Fig. 5-9).

   The test sample can be machined from a block of plastic that is free of residual stresses, or the part can be cast from a liquid form of the plastic. Still another technique is to coat the actual machine part with

---

*For a complete discussion of design stress concentration factors, see R. E. Peterson, *Stress Concentration Factors*, Wiley, 1974.

†For a complete treatment of photoelasticity, refer to M. M. Frocht, *Photoelasticity*, Volumes I and II, Wiley, 1941, 1948.

Figure 5-9
Photoelastic study of a key in a keyseat.
(Courtesy of the Photolastic Division of Measurements Group, Inc.,
Raleigh, North Carolina, USA.)

a plastic coating; the latter technique, however, is not suitable for three-dimensional analysis. Regardless of the material used, the part must be subjected to loads simulating those found in actual operation. Stress patterns can be photographed. From this study, stress concentration factors can be established for use in design calculations.

2. *Use of strain gages.* This method involves detecting a change in electrical resistance that indicates a highly stressed area or location. It is effective in large structural sections, but generally not for small mechanical components. The use of the previously mentioned plastic coating reveals patterns over a larger area than those covered by strain gages.

3. *Moiré method.* In this method, two sets of closely spaced lines are superimposed and viewed with transmitted light, producing a mottled pattern. This "silk screen" effect can be noted on window screens or curtains when light passes through. The Moiré method is particularly useful in evaluating stresses in a part subjected to many types of primary stresses, such as torsion and bending combined.

Developing stress concentration factors involves much preparation and the use of special equipment. To apply the factors to design problems, one merely finds the maximum stress at a critical section of a machine part.

Two typical stress concentration factor charts are shown in Figs. 5-10 and 5-11. Figure 5-10 shows stress concentration factors for a shaft with a

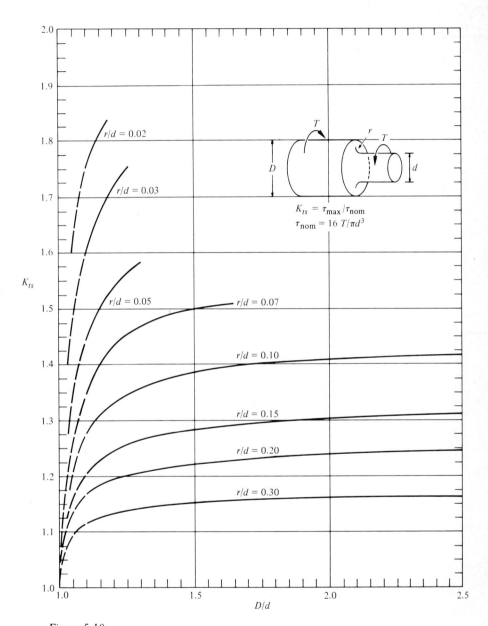

Figure 5-10
Stress concentration factor, K_{ts}, for a shaft with a shoulder fillet under torsion.
(Extracted from R. E. Peterson, *Stress Concentration Factors*, John Wiley & Sons, Inc., 1974, by permission.)

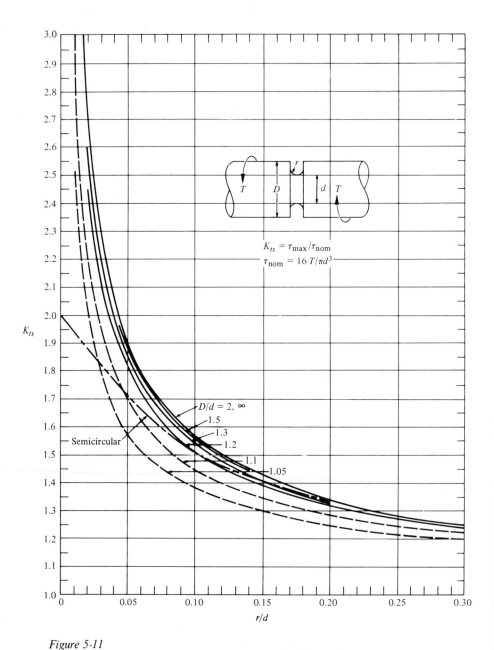

Figure 5-11
Stress concentration factor, K_{ts}, for a grooved shaft in torsion.
(Extracted from R. E. Peterson, *Stress Concentration Factors*, John Wiley & Sons, Inc., 1974, by permission.)

shoulder fillet under torsion. Figure 5-11 lists values for a grooved shaft under torsion. The following examples show how these factors can be applied.

◆ *Example* ————————————————————————

The average stress without safety factor for the material in a shaft subjected to torsion is 60,000 psi. The shaft has a change in section from a 1-in. diameter to a $\frac{5}{8}$-in. diameter with a fillet radius of $\frac{1}{8}$ in. On what value of stress should the shaft be designed before applying a factor of safety?

Solution. The ratio D/d is 1 divided by $\frac{5}{8}$, or $D/d = 1.6$. The ratio r/d is equal to $\frac{1}{8}$ divided by $\frac{5}{8}$, or $r/d = 0.2$. Referring to Fig. 5-10, we find that K_{ts} is 1.23. Then, $S_{s_{max}} = K_{ts}, S_{s_{ave}} = 1.23(60,000) = 73,800$ psi.

Increasing the radius of the fillet reduces the stress concentration at this point. Thus, fillets should be made as large as possible without interfering with adjoining surfaces of mating parts.

◆ *Example* ————————————————————————

A $1\frac{1}{2}$-in.-diameter torsion shaft has a groove radius of $\frac{1}{8}$ in. The shaft diameter at the groove is 1 in. The shaft is made of cold-drawn Monel. Find the design stress for a factor of safety of 8, assuming the shearing stress of Monel is $\frac{5}{8}$ of the ultimate tensile stress.

Solution.

$$\frac{D}{d} = \frac{1.50}{1.00} = 1.5;$$

$$\frac{r}{d} = \frac{0.125}{1.000} = 0.125.$$

Thus $K_{ts} = 1.48$ from Fig. 5-11. The ultimate shearing stress of Monel is $(\frac{5}{8})(100,000) = 62,500$ psi. The design stress is then equal to $(62,500/8)$ 1.48 = 11,560 psi.

Dynamic Seals

In general, the use of dynamic seals involves greater study than static seals. Before designing or selecting a seal, the type of movement should be studied. Rotation, translation, or a combination of the two basic types of motion may be involved. It is important to know the rubbing velocity. In the case of rotating members, rotational speed is usually known; from the diameter of the shaft, rubbing velocity (or surface speed) of the shaft has to be determined. It is also important to know what fluids are being sealed. Leakage always presents a problem; leakage depends on the viscosity of the fluid and on the clearance between the shaft member and the seal. Clearances must be kept small; the amount of leakage increases greatly as the clearance values increase.

In selecting seals, it is also necessary to design against the probable temperature conditions and pressure or vacuum values. The shaft or piston rod must have the proper surface finish. Surface finishes frequently range from 5 to 30 microinches root-mean-square. In mounting components, it is important to consider concentricity. If the center of the shaft is not concentric with the center of the seal, one side of the seal will wear away, causing leakage. Shaft run-out is also important; excessive shaft deflection or too much play in the mounting bearings may cause difficulties that, in turn, are detrimental to proper sealing. The hardness of the shaft surface sometimes dictates the material to be used in the lip of the dynamic seal.

An important part of any oil seal is the lip. Various configurations are available for the lip element. The lip shape depends on many of the previously mentioned considerations. In general, there are two basic types of shaft seals, one with a spring and one without. Several types of springs are used in shaft seals; the most popular is the "garter" spring.

Figure 5-12 illustrates a typical springless oil seal. These are available in a wide variety of configurations and sizes. A typical spring-loaded type is shown in Fig. 5-13. The use of the spring ensures uniform seal pressure against the shaft. Seals are easily obtainable for surface speeds up to 3000 ft/min. Depending on the selected materials, temperature ranges from $-65\,°F$ to $225\,°F$ are available. Certain seals using special materials can tolerate even higher temperatures.

Mechanical Seals

Commercial types of mechanical seals are available in many sizes. These have several outstanding features. They tend to reduce wear on shaft

Figure 5-12
Typical springless-type dynamic seal.

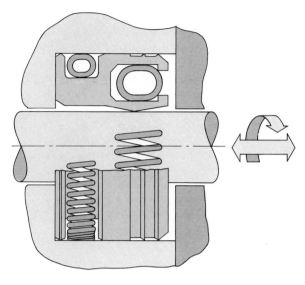

Figure 5-13
Spring-loaded seal.
(Courtesy of Bal-Seal Engineering Company.)

surfaces, and they eliminate fluid leakage. Figure 5-14 shows a Dura-Seal®;
the cutaway section reveals the internal construction of the seal.

Figure 5-15 shows a cutaway view of a PK Mechanipak® seal. Fluid
pressure enters from the left side; leakage is prevented by the seal at the
right side of the assembly. A stop collar is provided for positioning,
although the seal can also be installed against a snap ring or shaft shoulder.
Seals are commonly available for shaft sizes ranging from $\frac{3}{8}$ in. to 3 in. and
for shaft speeds up to 2000 rpm.

Seal Materials

Depending on operating conditions, a wide range of materials may be
used. The lip material can be felt, cork, rubber, synthetic rubber, or leather.
Teflon® is a synthetic material that is often used in seal design. The metallic
components of the seal can be steel, brass, stainless steel, or other noncorro-
sive metals. The application of the seal usually dictates the choice of
material for both the lip and the body of the mechanical seal.

Figure 5-14
Dura-Seal® mechanical seal.
(Courtesy of Durametallic
Corporation.)

Figure 5-15
Mechanipak® seal.
(Courtesy of Garlock, Inc.)

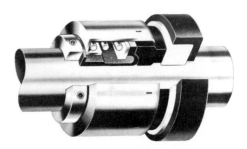

 **Summary**

Shafts are essential mechanical components and should be carefully designed for both strength and stiffness. Stiffness is particularly important
where a torsional deflection might place one keyed component out of phase
with another driven part. In certain types of machinery, such an event could
lead to a serious malfunction.

Loading on shafts must be analyzed to determine what stresses are
induced. Some possibilities are:

1. torsion alone,
2. flexure alone,
3. torsion combined with bending,
4. torsion combined with bending and axial stress.

Appropriate equations can be used after evaluating the principal stresses
encountered. With stress reversal, endurance *must* be considered.

Stress concentrations occur at keyseats, annular grooves, holes, and fillets at cross-sectional changes. Factors should be applied to average stress values to compensate for these "stress raisers."

When shaft sizes are calculated, the minimum sizes are usually increased to fit the bores of such components as pulleys, couplings, bearings, or gears. Critical shaft speeds should be calculated. Then, to prevent dangerous vibrations, these values should be avoided as operational speeds.

Flexible shafting is useful for transmitting torque around corners. The bending radius of the shaft must be held to values recommended by the manufacturer. The use of flexible shafting eliminates the gears and pulleys ordinarily needed to effect angular drives. The use of flexible shaft drives is limited to applications where torsional deflection is not a prime consideration.

Questions for Review

1. In designing the rear axle of an automobile, what items need to be considered?

2. List ten applications for noncircular shafting. Indicate why the noncircular section is used in each case.

3. List ten applications for nonmetallic shafting. Indicate why the material is used in each case.

Problems

1. A round steel shaft transmits $\frac{3}{4}$ hp at 1750 rpm. The shaft is subjected to torsion only and the design stress is 7000 psi. Find the diameter.

2. A 25-mm round steel shaft transmits 5 kW and rotates at 200 rpm. If this shaft is subjected to pure torsion, what shearing stress is induced?

3. A round steel shaft transmits 0.375 kW at 1800 rpm. The shear modulus of elasticity is 80 GN/m². The torsional deflection is not to exceed 1 deg in a length equal to 20 diameters. Find the shaft diameter.

4. A round steel shaft rotates at 200 rpm and is subjected to a torque of 225 N · m and a bending moment of 340 N · m. The allowable shearing stress is 40 MN/m²; the allowable tensile stress is 53.3 MN/m². Find the diameter.

5. A 1-in.-diameter shaft is to be replaced with a hollow shaft of the same material, weighing half as much, but equally strong in torsion. The

outside diameter of the hollow shaft is to be $1\frac{1}{2}$ in. Find the inside diameter.

6. A shaft rotates at 1000 rpm and transmits $\frac{3}{4}$ hp. The shaft is subjected to torsion only and the working stress is 6000 psi. Find the diameter.

7. Find the diameter of a shaft for which the angle of twist under torsion is not to exceed 1 deg in 20 diameters. The shaft transmits 3.75 kW and rotates at 100 rpm. The modulus of elasticity (shear) is 80 GN/m²; the modulus of elasticity (tension) is 200 GN/m².

8. A solid shaft is subjected to a bending moment of 6780 N · m and a torque of 12 300 N · m. The allowable shearing stress is 40 MN/m² and the allowable tensile stress is 53.3 MN/m². Find the diameter of the shaft.

9. A shaft supported on bearings 200 mm apart transmits 187 kW at 200 rpm. The maximum bending moment is 2712 N · m. The allowable shearing stress is 53.3 MN/m², and the allowable bending stress is 46.7 MN/m² because of unusual loading conditions. Find the shaft diameter.

10. A 100-mm solid shaft is to be replaced with a hollow shaft equal in strength under torsion and made of the same material. The outside diameter of the hollow shaft is to be 125 mm. What should be the inside diameter? The allowable shearing stress is 40 MN/m².

11. A 2-in.-diameter shaft is 10 ft long and receives 60 hp at its midpoint from a pulley. A machine at the left end uses 25 hp; one at the right end uses 35 hp; $E = 30,000,000$ psi and $G = 12,000,000$ psi. The shaft rotates at 200 rpm. Find the maximum shearing stress and the relative angle of twist between the two ends.

12. A $\frac{1}{4}$-in. shaft is subjected to torsion alone. If the design stress is 6000 psi, how much horsepower can it transmit if it is rotating at 2000 rpm?

13. A shaft is subjected to an equivalent bending moment of 30,000 in-lb and an equivalent twisting moment of 40,000 in-lb. If the working strengths are 6000 psi in torsion and 8000 psi in bending, what should the shaft size be?

14. A $\frac{1}{4}$-in. shaft is designed with a working stress of 7000 psi in shear. If it rotates at 1725 rpm, how much horsepower can it safely transmit? If the shear modulus of elasticity is 12,000,000 psi, what is the torsional deflection in degrees per foot of length?

15. A $\frac{3}{4}$-in. round steel shaft transmits $\frac{3}{4}$ hp at 1750 rpm while being subjected to an axial force of 400 lb. Find the resultant shearing and compressive stresses.

16. A $\frac{7}{8}$-in. round steel shaft rotates at 100 rpm and transmits 1.25 hp while being subjected to an axial force of 5000 lb. Find the resultant shearing and compressive stresses.

17. A 2-in. round steel shaft is mounted between bearings 60 in. apart. Considering only the weight of the shaft (0.28 lb/cu in.), decide whether it would be safe to operate this shaft at 2000 rpm. Justify your answer.

18. A $\frac{1}{2}$-in. round steel shaft is mounted on bearings 30 in. apart. Find its lowest critical speed.

19. A 50-mm round torsion shaft rotates at 500 rpm. Torsional deflection is not to exceed 0.02 radians for its 600 mm length. The shear modulus of elasticity is 85 GN/m²; the allowable shearing stress is 55 MN/m². Find its capacity in kW based on (a) torsional stiffness and (b) torsional strength.

20. A ship's propeller shaft is 5 in. in diameter. The thrust load is 12,000 lb and the torque is 150,000 in-lb. Find the resultant maximum shearing and compressive stresses.

21. A 1-in.-diameter shaft has a single disc weighing 75 lb mounted midway between two bearings 20 in. apart. Find the lowest critical speed in rpm. Neglect the weight of the shaft. Assume that the modulus of elasticity is 30,000,000 psi.

22. In Fig. 5-3, the shaft diameter is 1.58 in. Find the lowest critical speed. Neglect the shaft weight. Assume that the modulus of elasticity is 30,000,000 psi.

23. A 3-in. shaft carries a flywheel mounted on bearings spaced 12 in. on centers. The flywheel weighs 1.5 tons. Find the vertical shearing stress induced when the flywheel is stationary.

24. A socket wrench has a $\frac{1}{2}$-in. square shaft engaging the socket. If a force of 50 lb is exerted at the end of a 12-in. handle, how much shearing stress is induced in the square shaft section?

25. A $\frac{1}{8}$-in.-diameter round steel shaft was designed for a torsional stress of 8000 psi.
    a) If the shaft is subjected to pure torsion, how much torque can it transmit?
    b) If it rotates at 2000 rpm, what horsepower can it transmit?

26. A $\frac{1}{2}$-in.-valve stem is ground square at the end to accommodate a hand wheel. If a torque of 75 in-lb is applied to the 3-in. hand wheel, how much shearing stress is induced? Neglect stress concentrations at the point where the stem changes shape.

27. A 5-in. solid torsion shaft made of steel is to be replaced with a hollow one weighing 25% less, but 25% stronger. Find the inside and outside diameters of the hollow shaft.

28. A hollow shaft has an outside diameter of 4 in. and an inside diameter of 3 in. Find the diameter of a solid shaft with equal strength in torsion.

29. A $\frac{1}{4}$-in.-diameter solid shaft has an allowable stress of 6000 psi and is subjected to pure torsion. If its rotational speed is 1800 rpm, how much horsepower can it transmit safely?

30. A $\frac{1}{4}$-in. steel shaft transmits 30 in-lb of torque. The effective length of the shaft is 12 in.; the modulus of elasticity in shear is 12,000,000 psi. Find the angular deflection in degrees.

31. A solid steel shaft is subjected to a twisting moment of 10,000 in-lb and a maximum bending moment of 8000 in-lb. The allowable bending stress (including endurance) is 8500 psi and the permissible shearing stress is 8000 psi. Find a suitable diameter for the shaft.

32. A 1-in. round steel shaft rotates at 600 rpm and handles 3 hp while being subjected to an axial load of 1000 lb in compression. Find the maximum compressive and shearing stresses induced.

33. A shaft is made of AISI 1020 cold-drawn steel. A factor of safety of 10 is to be used and is to be based on the ultimate strength. The shaft is to be designed for 5 hp at 200 rpm.
    a) Calculate the minimum diameter based on torsional strength.
    b) Find the diameters for a hollow shaft that has equal torsional strength but that weighs one-half as much.

34. A $1\frac{1}{2}$-in. Monel shaft is used in a torsional application. Based on an operating speed of 100 rpm and a safety factor of 12, how much horsepower can it transmit? Assume the shearing stress is $\frac{3}{4}$ of the ultimate tensile stress.

35. A $\frac{1}{2}$-in. brass shaft rotates at 1200 rpm and transmits $\frac{1}{8}$ hp. If it is subjected to pure torsion, how many degrees of twist will it have if the effective length is 2 in., the ultimate tensile strength is 55,000 psi, the modulus of elasticity is 15×10^6 psi, and the shear modulus of elasticity is 5.6×10^6 psi?

36. Two shafts are being considered in a design project. One is a hollow shaft with an outer diameter of 4 in. and an inner diameter of 3 in.; the other shaft is solid and has a diameter of 2.75 in. Both shafts are made of AISI 1020 cold-rolled steel. On a percentage basis, how does the

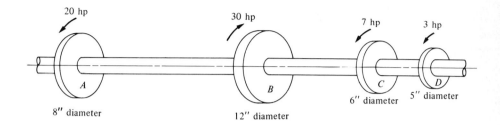

Figure 5-16
Problem 37.

hollow shaft compare with the solid shaft with respect to the following?
a) torsional strength
b) weight
c) cost (assume that the hollow shaft costs 20% more per pound)

37. Referring to Fig. 5-16, design a solid shaft subjected to pure torsion such that the shearing stress will not exceed 6000 psi. The continuous shaft rotates at 100 rpm. Assume that the shaft is made of AISI 1040 steel. Power is supplied at pulley B; take-offs are at A, C, and D. Distances between the pulleys are as follows: $AB = 30$ in., $BC = 10$ in., and $CD = 8$ in.
a) Calculate the shaft size based on strength.
b) Select a suitable common shaft size.
c) Calculate the torsional deflection (maximum) between the power input pulley and the furthest output pulley.
d) Find the factor of safety based on the selected shaft size.

38. A solid steel shaft rotates at 100 rpm and transmits a torque of 110 $N \cdot m$. The allowable shear stress is 40 MN/m^2. Based on torsion alone, calculate the minimum diameter.

39. A 12-mm steel shaft is subjected to torsion only. How much torque can be safely transmitted if the allowable shearing stress is 40 MN/m^2?

40. A stepped torsion shaft has diameters of 16 mm and 12 mm and a fillet radius of 2 mm. The shaft is subjected to a torque of 12.5 $N \cdot m$. Find the maximum induced stress caused by the fillet.

41. A 12-mm steel shaft has a U-shaped annular groove with a groove radius of 1 mm. The shaft diameter at the root of the groove is 8 mm. If the shaft is subjected to a torque of 4 $N \cdot m$, find the maximum induced shear stress.

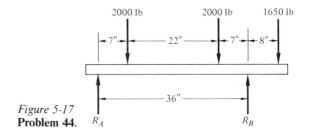

Figure 5-17
Problem 44.

42. A round steel shaft transmits 50 kW, rotates at 160 rad/s, and is 3 m long. Based on torsion, calculate the minimum shaft diameter if the allowable shearing stress is 40 MN/m² and the transverse modulus of elasticity is 68 GN/m². Also, find the angle of twist for the 3-m length, based on the calculated minimum diameter.

43. A steel shaft operates at 188 rad/s and must handle 2 kW of power. The shearing stress is not to exceed 40 MN/m² and the transverse modulus of elasticity is 80 GN/m². Calculate the minimum shaft diameter based on pure torsion and the minimum diameter based on stiffness if the angle of twist cannot exceed 1 degree in a length of 20 diameters.

44. A round steel shaft rotates at 12.8 rpm, transmits six horsepower, and is subjected to vertical loading as shown in Fig. 5-17. If the allowable stresses for flexure and shear are 10,000 psi and 8000 psi respectively, what minimum diameter is acceptable?

Figure 5-18
Problem 47.

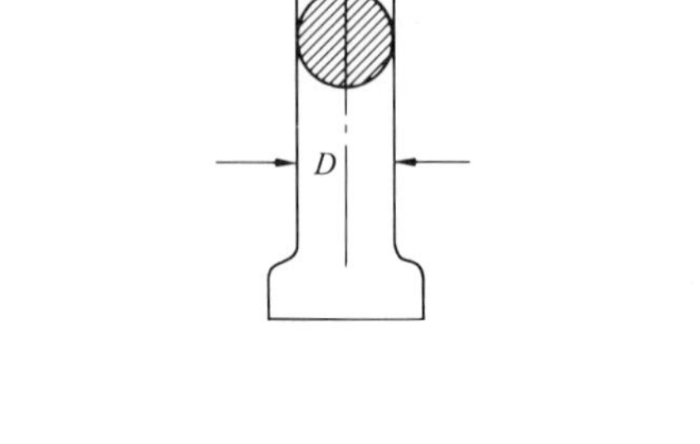

45. A 75-mm solid round shaft 3 m long is subjected to a constant torque of 11 000 N · m at each end, together with an axial tensile load of 0.31 MN also applied at each end. Find the maximum tensile and compressive stresses.

46. Solve Problem 45 using the Mohr's circle analysis.

47. In Fig. 5-18, the allowable flexural stress is 53.3 MN/m², the allowable torsional stress is 40 MN/m², and the transverse modulus of elasticity is 80 GN/m². The turning forces are applied at A and B to produce a maximum torque of 200 N · m. Find suitable diameters for D and d if the angle of twist in the stem cannot exceed 1 degree in a length of 20 diameters.

FASTENERS, COUPLINGS, KEYS, RETAINING RINGS, WELDING AND WELD DESIGN

6

 **Fasteners**

Fasteners are essential components of almost every design. They can be classified into three broad categories: removable, semipermanent, and permanent. *Removable* fasteners are those that can be easily removed with hand tools and without damaging any parts. Ordinary nuts and bolts are typical examples. *Semipermanent* fasteners can easily be removed, but with some damage to the fastener. Cotter pins, together with bolts and nuts containing a self-locking insert such as nylon, are examples of this type. *Permanent* types are those installed so as to remain permanently in place; they would not be removed for routine maintenance work. Rivets and drive pins fall into this category. Certain borderline types of fasteners are difficult to classify.

Force Analysis of Fasteners

Many riveted or bolted connections are subjected to direct shear. In such problems, the shearing area is the minimum cross-sectional area that is influenced by the active force or forces. Care must be taken to determine whether single, double, or multiple shear is present. Fasteners subjected to double shear have two shear areas that carry the load. Some fasteners are subjected to tensile loads. If screw threads are subjected to axial loads, the threads may strip; this is a shear failure. Tensile loads on a bolt produce tensile stresses within the bolt. These are computed in the same manner as with any tensile member. Tightening stresses are frequently considered in designing bolts.

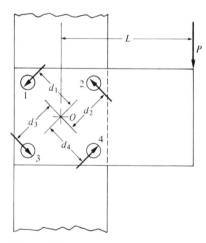
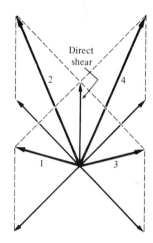

Figure 6-1
Eccentric loading: rivets or bolts.

Combined stresses occur in certain fastening problems. Often the fastener is subjected to both a primary stress and a secondary stress. The following discussions consider two fastening situations where combined stresses are present. These analyses are similar for both bolted and riveted connections.

Figure 6-1 shows an eccentric loading on fasteners. The primary force is one of direct shear and has a value of $P/4$. Thus each fastener supports its share of the vertical load. However, each rivet or bolt is subjected to a secondary shearing force induced by the twisting moment of each fastener about the centroid of the fasteners taken as a group (point O). Summing moments about a polar axis through point O gives the equation

$$PL = F_1 d_1 + F_2 d_2 + F_3 d_3 + F_4 d_4.$$

The following relationships should also be noted:

$$\frac{F_2}{F_1} = \frac{d_2}{d_1} \quad \text{or} \quad F_2 = \frac{d_2 F_1}{d_1};$$

$$\frac{F_3}{F_1} = \frac{d_3}{d_1} \quad \text{or} \quad F_3 = \frac{d_3 F_1}{d_1};$$

$$\frac{F_4}{F_1} = \frac{d_4}{d_1} \quad \text{or} \quad F_4 = \frac{d_4 F_1}{d_1}.$$

Substituting these values in the moment equation, we have

$$PL = \frac{F_1}{d_1}\left(d_1{}^2 + d_2{}^2 + d_3{}^2 + d_4{}^2\right).$$

From this equation, the secondary force for fastener number 1 can be determined. The total force on fastener number 1 is the vector sum of the direct force and the secondary force. If the allowable stress is known, appropriate cross-sectional areas can be computed. In this particular case, $d_1 = d_2 = d_3 = d_4$; thus the secondary forces are all equal, although they act at different angles. The resultant force is shown graphically for each fastener. In practice, one would design for the maximum force value and make all fasteners identical in size. The vector sum of the primary and secondary values can be determined graphically or mathematically. It is important to know the directions of the forces as well as their magnitudes.

Another typical fastening problem is the bolted (or riveted) bracket. Several configurations are possible; Fig. 6-2 shows one possibility. In this figure, assume that the bolts are located as shown. The three bolts are subjected to a vertical shearing force P carried equally by all three fasteners. Tensile pulls on the bolts may be found by taking moments about the lower

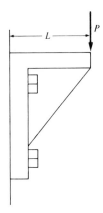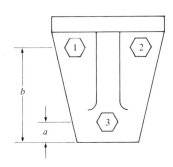

Figure 6-2
Bolted bracket.

edge. Thus, the following equations may be written:

$$P_1 = P_2 = \frac{b}{a}P_3$$

and

$$PL = (P_1 + P_2)b + P_3a.$$

Combining, we have

$$PL = \frac{2b^2P_3}{a} + P_3a.$$

The value of P_3 can be obtained; then the proper values for P_1 and P_2 can be found by substitution. The greatest values will be the pulls on fasteners 1 and 2. The shear area for stripping is the circumference at the root of the thread times the axial length of thread engagement. In connections using bolts and nuts, the nut thickness is often found from this calculation.

The preceding analysis is based on tensile pulls only. However, fasteners are also subjected to direct shear. It is usually advisable to check the fasteners for stripping and for direct shear, then design for the worst condition. This avoids the lengthy computation of combined stresses. If the value of L in Fig. 6-2 is small, direct shear is usually the basis for design. If it is large, pulling or stripping of the threads is the basis.

From the practical standpoint, all three fasteners are made the same size; this helps prevent mounting errors and allows the stocking of fewer sizes of fasteners. Bolt number 3 could be smaller in size, if desired. From the strength standpoint, it is desirable to place the bolts (or rivets, or screws) as close as possible to the top of the bracket. Direct shear can be completely eliminated from the fastener system if the bracket rests against some type of ledge at the bottom edge.

In selecting bolt positions, one must consider the appearance of the bracket, the tightening space needed for the bolts, and the available space.

Blind Fasteners

Blind fasteners are defined as fasteners that can be installed from one side only. Several types of blind rivets are available and are extensively used in sheetmetal applications, particularly in the aircraft industry. One type is an assembly (two pieces) that is placed in aligned holes in the parts being assembled. A tool then exerts an upward force on the center portion. A mandrel head forms the blind side of the rivet and remains wedged in place; the stem breaks at a predetermined section. The stem is then removed. Another type of blind rivet is a Rivnut® (see Fig. 6-3), which is a one-piece unit with internal threads. The blind side is formed by upsetting with a threaded heading tool. This tool exerts an upward force that forms a bulge in a predetermined location. When the tool is removed, the threads can be used with regular threaded fasteners to secure other components to the assembly. Still another type of blind fastener is the explosive rivet. This is a one-piece fastener with a small, carefully centered cavity running the full length of the shank. This cavity contains an explosive charge; when heat is applied to the head, the charge explodes and forms the rivet.

Assembly and Disassembly Considerations

When using the various types of threaded fasteners, such as nuts and bolts, self-tapping screws, and others, it is desirable to think carefully about

Figure 6-3
Rivnut®.
(Courtesy of the B. F. Goodrich Company.)

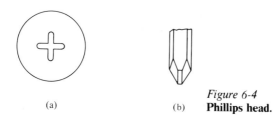

(a)

(b)

Figure 6-4

Phillips head.

the methods that may be used in assembling the components. An intelligent choice of the type of head can then be made. A slotted head is used if assembly (and disassembly) is to be done with a blade-type screwdriver. This type of head would be convenient for household use, but not for mass production. A Phillips head (cross recess) is more desirable in production work, because the tapered, cross-type blade will not easily slip out as the fastener is being secured. Window trim in most automobiles is fastened with this type; it lends itself well to mass assembly methods.

Figure 6-4(a) shows a Phillips head with the cross recess. The driving tool with cross-type blade is depicted in Fig. 6-4(b).

A clutch head can be used with power tools where the bit is shaped to fit the recess; the fastener can then be removed with an ordinary blade-type screwdriver. Hexagonal heads are also desirable for power-tool assembly. Some of these include a slot for an ordinary screwdriver, so that the average user can remove the fastener if necessary. Socket-head fasteners are also frequently used; however, special tools are needed to remove this type.

Appearance of Fasteners

The appearance of a product can be greatly improved by choosing properly finished fasteners. Plating (cadmium, zinc, chromium, and other nonferrous metals) is frequently used. Stainless steel fasteners can add to the final appearance. Regular bolts are obtainable in many finishes and many materials.

Types of Rivets

Rivets are available with several types of heads, in a wide range of diameters and lengths, and made of many materials. Typical materials are wrought iron, mild steel, aluminum alloys, and copper alloys. Figure 6-5 shows button, high-button or acorn, pan, cone, and countersunk heads. Many others are available. Hollow rivets, usually made from a copper alloy, are extensively used in light assemblies.

For diameters and lengths, it is advisable to consult an industrial hardware catalog. Rivets are available with diameters up to $1\frac{1}{4}$ in. and

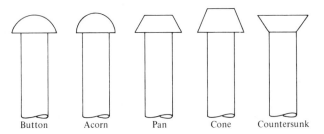

| Button | Acorn | Pan | Cone | Countersunk |

Figure 6-5
Common rivet heads.

lengths up to 7 in. Like bolt lengths, rivet lengths are found by measuring from the headless end to the maximum head diameter. It is important to understand that grip length depends on total length. The grip length is the distance from the head to the newly formed head after the rivet is installed. Grip length is important. If it is figured on the low side, improper heading may occur. If it is figured on the high side, the plates being riveted will not be adequately secured.

In using rivets, one must consider the fact that they are permanent fasteners. They cannot be removed without destroying the rivet; sometimes other parts are damaged as well. Therefore, rivets should not be used in any application where routine maintenance requires removal.

U.S. Screw Threads

Since many fasteners use screw threads, it is well to have a good understanding of the various types of threads and their terminology. Most fasteners make use of the 60-deg V-thread. Screws for power transmission are discussed in Chapter 11. One of the important thread definitions is *pitch*, defined as follows:

$$\text{pitch} = \frac{1}{\text{number of threads per inch}}.$$

The pitch is expressed in inch units. The term *pitch* is defined in the same way for all types of thread.

The Unified Standard thread is commonly used in the United States, Great Britain, and Canada; it is illustrated in Fig. 6-6.

Series and fits for Unified Standard threads. The following series are available in the Unified Standard thread; however, not all of these series are available in all sizes of thread:

1. coarse (UNC),
2. fine (UNF),
3. extra-fine (UNEF),
4. 8-pitch (8N),
5. 12-pitch (12N),
6. 16-pitch (16N).

The *coarse* thread is one of the more popular series. It is relatively easy to assemble and disassemble, and thus is used for common assembly work. The *fine* series is used where vibration problems are frequently encountered, such as in automotive applications. The *extra-fine* series is often used in aircraft-instrument applications, particularly in threaded parts used for making fine adjustments. High-grade alloy steel is usually used for fasteners in the UNEF series. The *8-pitch* series is extensively used in bolting high-pressure pipe flanges. The *12-pitch* series is usually used in the heavy-machinery field. The *16-pitch* series is available in the larger sizes only and

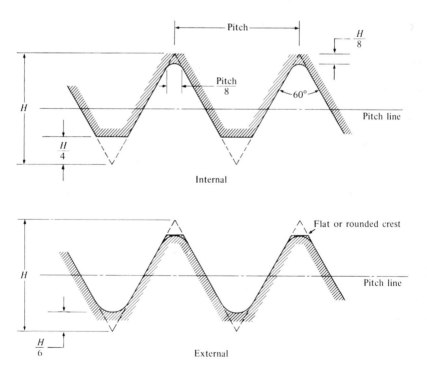

Figure 6-6
Nomenclature for Unified Standard screw threads.

hence is considered a fine thread. It is used for adjustment collars and for large bearing retainer nuts.

Basically, three fits are available in the Unified Standard thread; these are classified as 1, 2, and 3. A letter following the class is used to distinguish external threads (A) from internal threads (B). For example, a thread specified as $\frac{1}{4}$-in.–20 UNC-2B indicates an internal thread, a major diameter of $\frac{1}{4}$ in., a pitch of $\frac{1}{20}$ in. (20 threads per inch), and a Class 2 fit. Depth or length of thread is listed after the specification; this was not indicated in the preceding illustration. Class 1 threads are designed with the loosest fits; with this class, fast assembly is possible. Class 2 is the most widely used. Class 3 threads have the tightest fits; this class is used in precision work only. Variations of fits are obtained by mixing one class for the external thread with another class for the internal thread.

Metric Screw Threads

Screw threads in the metric system do not align with the U.S. system. Pitches and diameters are expressed in millimeters and are not interchangeable with the Unified Standard coarse or fine threads or with any other U.S. threads. The designation for a metric thread gives the nominal size and the pitch, expressed in millimeters and prefixed with the letter M. For example, if we would like to replace a $\frac{1}{2}''$-13 UNC thread with its nearest metric counterpart, we have two choices, one larger and one smaller. The larger thread is M14 × 2, which has a diameter of 0.551 inches. The smaller thread is the M12 × 1.75, which has a diameter of 0.472 inches. Tolerancing also differs from the U.S. system.

Let's examine a complete thread specification in the metric system, such as that of the M14 × 2 thread. Tolerance information follows the thread size, such as: M14 × 2–5h. The number 5 indicates the tolerance grade (tolerance); the letter h shows the tolerance position (allowance). Small letters are used for external threads; capital letters indicate internal threads. One must refer to metric thread tables to get information on recommended tolerance and allowance figures.

Table 6-1 gives a partial list of metric screw threads from approximately the $\frac{1}{4}$ inch to the $\frac{3}{4}$ inch sizes. None of the metric threads exactly match the American standard; therefore, none are interchangeable.

Screw Fasteners

Figure 6-7 shows the popular screw-type fasteners. The most common is the *through bolt*, which is used with a nut and lock washer. These components are used to join two parts together when occasional removal is desired.

Table 6-1
Partial List of Metric Screw Sizes

M6 × 1	M14 × 2
M6 × 0.75	M14 × 1.5
M7 × 1	M14 × 1.25
M7 × 0.75	M15 × 1.5
M8 × 1.25	M16 × 2
M8 × 1	M16 × 1.5
M9 × 1.25	M17 × 1.5
M9 × 1	M18 × 2.5
M10 × 1.5	M18 × 2
M10 × 1.25	M18 × 1.5
M10 × 1	M19 × 2.5
M11 × 1.5	M20 × 2.5
M12 × 1.75	M20 × 2
M12 × 1.50	M20 × 1.5
M12 × 1.25	

Wrenches are needed for both the nut and the bolt. Semifinished bolts and nuts have a washer face on the bearing surface; needless to say, this is not of much value unless the surfaces on which they rest are also finished. Spot facing around bolt holes is necessary to ensure proper contact and avoid unnecessary stresses on the fastener. The holes into which the bolts fit must have sufficient clearance—as a rule of thumb, approximately $\frac{1}{32}$ in. based on the diameter.

Studs are used where a plate is bolted to a rather large and cumbersome part. A good example of this application is the head on a gasoline engine. The stud is threaded on both ends; actually, these can be different threads if desired. The stud screws directly into the larger member and clears the covering plate; then a nut is secured to the free end. If many studs are used to secure a sealing gasket, a uniform tightening torque must be applied with a torque wrench to ensure an even pressure on the gasket.

Machine screws and *cap screws* are similar to studs. They are used primarily to secure a plate to another component. However, these fasteners are used for lighter applications. Cap screws are usually finished to add to their appearance. A cap screw can have a hexagon head, a fillister head (as in Fig. 6-7), or several possible types of socket heads (see Fig. 6-8). Machine screws are usually used for smaller parts. These are available with a fillister head, oval head, flat head, or round head. It is wise to check the available lengths when using screw fasteners. Standard length increments usually vary by $\frac{1}{8}$ in. up to 1 in. bolt length; the increment then increases to $\frac{1}{4}$ in. from 1 to 4 in. Sizes over 4 in. are not very common, so their availability should be

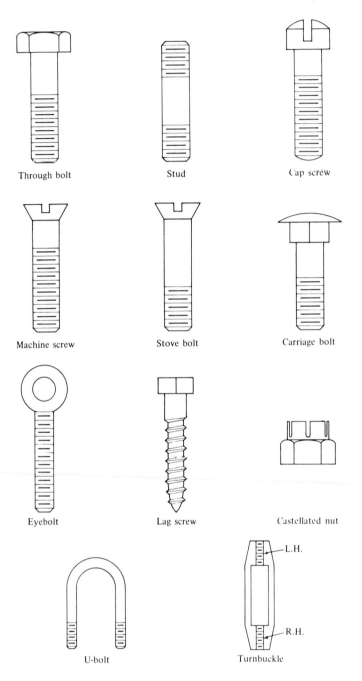

Figure 6-7
Common screw fasteners.

Figure 6-8
Unbrako³⁰ socket-head cap screw.
(Courtesy of SPS Technologies.)

carefully checked. The engagement length of a cap screw or machine bolt depends on the material that it engages. For steel in steel, an engagement length equal to one bolt diameter is often used. If brass is used, the engagement length is $1\frac{1}{2}$ times the diameter; in other nonferrous metals and in plastics, an engagement length equal to two bolt diameters is frequently used. A calculation for stripping (shearing) is desirable so that sufficient thread engagement can be provided.

Stove bolts are widely used, inexpensive fasteners that require a nut (usually square). A loose fit between the nut and bolt makes assembly easy.

A *carriage bolt* is characterized by a square section immediately under the head. The head cannot be held with a screwdriver or any type of wrench. Square nuts are usually used on the other end. As the nut is turned, the carriage bolt is held stationary by the square section. This section seats in a square hole when a carriage bolt is used in metal and wedges into the wood when it is used for fastening metal to wood.

Eyebolts are used to provide a hoisting ring in certain pieces of equipment. They are usually attached in much the same manner as through bolts; washers and nuts are used on the threaded ends.

Lag screws resemble bolts on the head end and wood screws on the threaded end. They are primarily used to secure machinery to the base of a crate during shipping. A lag screw is tightened with an ordinary wrench.

A *castellated nut* is used in conjunction with a cotter pin to prevent a nut from unthreading. The hole for the cotter pin is drilled after it is determined how far the nut will thread onto the bolt. Slotted hexagon nuts operate on the same principle, but are heavier in construction.

A U-*bolt* is used with two nuts to secure parts together. Typical applications include fastening leaf springs and clamping cable ends.

A *turnbuckle* is used with two threaded rods. The turnbuckle is constructed with internal threads, one end left-handed and the other right-handed. As the turnbuckle is rotated, the rod ends are either drawn closer together or farther apart, depending on the direction of rotation. This device is used to tighten cables or to strengthen an assembly by exerting a tensile pull on two parts that have a tendency to separate.

Other common fasteners (not shown in Fig. 6-7) include the *thumb nut*, the *thumb screw*, and the *cap* or *acorn nut*. The thumb nut or screw is used where finger tightening is desired. The cap or acorn nut is used to cap threaded rod ends. A popular application is in the area of shop furniture, where extended threaded rods could snag clothing.

Washers

Washers are necessary components that go along with threaded fasteners. *Plain washers* are used to distribute the load over a larger area. *Spring lock washers* are used to place an axial load between a bolt and nut so that the nut cannot easily unthread. Washers are frequently made of wrought steel, brass, phosphor bronze, or stainless steel.

Setscrews

Setscrews are another form of fastener. They are usually used to prevent relative circular motion between two parts, such as shafts and pulleys. They can be used only where the torque requirements are low. Their major advantage is that they are easily removed. They can be classified according to the type of point, the type of head, or a combination of the two. Although square heads are available, they are not desirable from the safety standpoint; they are easier to remove, however, with common hand tools. Safety requirements dictate headless setscrews for rotating parts. These are available in slotted, hexagonal-socket, or fluted-socket form. Points can be classified as cone, flat, cup, knurled cup, full dog, half dog, and oval. As the torque-carrying capacity of setscrews is exceptionally small, they are sometimes used in conjunction with a flat key. Another way of increasing their torque-carrying capacity is to use two setscrews (90 deg apart) or three setscrews (120 deg apart).

In using setscrews, it is important to select the proper point for a specific application. The type of materials used in the joined components influences the selection, as does the hardness of the materials. Consideration must be given to how often the parts are to be disassembled. The following discussion is intended to give guidelines for selection, together with a few of the outstanding features of each type of point. Refer to Fig. 6-9(a).

The *cone point* is usually used if the two joined parts are to have a permanent position with respect to each other. The part where the point rests should be spotted with the same angle as the point. The cone angle is made either 118 or 90 deg. This angle is determined by the ratio of the setscrew length to the nominal diameter. If the setscrew length is equal to or smaller than the nominal diameter, the angle is made 118 deg. If the setscrew length is greater than the nominal diameter, the cone angle is 90 deg. Usually, the point is spotted to one-half its length. When properly spotted, the cone type of point provides the greatest axial and torsional holding power of any setscrew. A proper design will provide sufficient shear

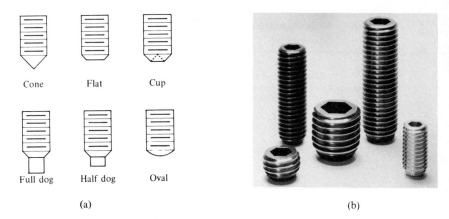

Figure 6-9
(a) Setscrew points. (b) Unbrako® socket setscrews.
(Fig. 6-9b courtesy of SPS Technologies.)

strength in the area above the point. If the cone point has adequate hardness, it can be used in pivot applications.

The *flat point* is used if frequent adjustment is necessary between the two parts being secured. It is particularly useful with hardened shafts. A section on the shaft should be flattened to receive the flat point. This type does no damage to the parts and permits a relatively large number of threads for a given length of setscrew.

The *cup point* is the most widely used type. It is generally not used against hardened shafts. Assembly is rapid and requires no appreciable preparation. It cannot be used unless the damage caused by the cup point is allowable. The knurled cup point has counterclockwise knurls that prevent the screw from backing off in poorly tapped holes. Such types are desirable where vibration problems exist.

The *full dog* and *half dog* setscrews are used if members are to be joined permanently at a given location. In either type, a hole must be drilled for receiving the point. This hole must be drilled to the same size as the setscrew dog point; otherwise, the clearance will eventually elongate the hole. Half-dog points are much more common than full-dog points. The half-dog point can be used in the same type of application as the flat setscrew. However, the wall thickness of the collar part must be large because the protruding length of the dog point reduces the number of active threads in a given length of setscrew.

The *oval* type can be used in the same circumstances as the cup point. The cup point seldom requires spotting; the oval type, however, usually

Table 6-2
**Maximum Dog-Point Diameters for
American Standard Setscrews**

Size	Maximum Dog-point Diameter	Size	Maximum Dog-point Diameter
5	0.083	$\frac{3}{8}$	0.250
6	0.092	$\frac{7}{16}$	0.297
8	0.109	$\frac{1}{2}$	0.344
10	0.127	$\frac{9}{16}$	0.391
12	0.144	$\frac{5}{8}$	0.469
$\frac{1}{4}$	0.156	$\frac{3}{4}$	0.563
$\frac{5}{16}$	0.203		

requires spotting on the part that receives the point. A groove is often cut and shaped either axially or annularly to accommodate the oval point. This type is used when frequent adjustment is necessary. Proper positioning of grooves assists the operator in making adjustments.

Figure 6-9(b) shows typical Unbrako® socket setscrews with a knurled-cup point.

Table 6-2 lists the American Standard setscrew sizes that are available with the dog point. For convenience in designing, the maximum diameters are given for the dog point. Holding ability is based on the cross-sectional area of the dog point; if failure occurs, this area fails in shear. The following example shows how holding power is computed.

◆ *Example* ————————————————————

A clamp is secured to a vertical slide by means of a half-dog setscrew with a nominal diameter of $\frac{1}{4}$ in. How much load can the clamp sustain if the allowable shearing stress for the setscrew is 18,000 psi?

Solution. From Table 6-2, the dog diameter is found to be 0.156 in. Then,

$$P = S_s A = \frac{S_s \pi D^2}{4} = \frac{18,000(3.14)(0.156)^2}{4} = 344 \text{ lb.}$$

Holding power of setscrews. In specifying a setscrew, one of the chief concerns of a designer is its static holding power. Depending on the application, this holding power can be either axial or torsional.

Table 6-3 lists the torsional and axial holding power of Unbrako®
setscrews with Class 3A screw threads seated in Class 2B tapped holes. The
holding power was defined as the minimum load required to produce 0.01
in. of relative movement between shaft and collar. A plain cup point was
used in sizes 0 to 3 and a knurled cup point in sizes 4 through 1 in. Note
that the tabulated axial and torsional holding powers are typical strengths
and should be used accordingly; specific safety factors should be used that
are appropriate to the application. Good results have been obtained with a
factor of 1.5 to 2.0 under static load conditions and of 4.0 to 8.0 for various
dynamic situations. It should also be noted that the boldface values indicate
the holding powers of setscrew sizes on the basis of a screw diameter
roughly half the shaft diameter. All holding-power values are based on the
assumption that the recommended seating torque was applied.

The following examples show how setscrews are selected.

◆ *Example*
A collar is secured to a $\frac{1}{2}$-in.-diameter vertical post by means of a $\frac{1}{4}$-in.-
diameter cup-pointed setscrew. Assume a factor of safety of 1.8. How much
vertical load can the setscrew safely support?

> *Solution.* The holding power of a $\frac{1}{4}$-in. setscrew is 1000 lb, as shown in
> Table 6-3. Applying a factor of safety of 1.8 to this figure, we obtain a
> safe load of 1000/1.8, or approximately 556 lb.

◆ *Example*
A pulley is secured to a $\frac{7}{8}$-in. shaft by two $\frac{3}{8}$-in. cup-pointed setscrews. If a
factor of safety of 4 is applied and the pulley rotates uniformly at 1200 rpm,
how much horsepower can it transmit?

> *Solution.* One $\frac{3}{8}$-in. setscrew has a torsional holding force of 875 lb.
> Therefore, two setscrews would hold 1750 lb. However, a factor of
> safety of 4 is to be applied. Thus, the 1750 lb must be divided by 4,
> yielding a value of 438 lb at the point. The torque value can be found
> by multiplying 438 lb by the shaft radius. Thus,

$$T = 438(7/8)(1/2) = 192 \text{ in-lb}$$

and

$$\text{hp} = \frac{NT}{63,000} = \frac{1200(192)}{63,000} = 3.66 \text{ hp.}$$

Table 6-3
Torsional and Axial Holding Power of Unbrako ® Socket Setscrews,
UNRC or UNRF Thread, Plain or Plated, Seated Against Steel Shaft

Nom. size	Seating torque inch-lbs.	Axial holding power (pounds)	Shaft Diameter (Shaft Hardness Rc 15 to Rc 35) Torsional holding power inch-lbs.											
			$\frac{1}{16}$	$\frac{3}{32}$	$\frac{1}{8}$	$\frac{5}{32}$	$\frac{3}{16}$	$\frac{7}{32}$	$\frac{1}{4}$	$\frac{5}{16}$	$\frac{3}{8}$	$\frac{7}{16}$	$\frac{1}{2}$	$\frac{9}{16}$
#0	1.0	50	**1.5**	2.3	3.1	3.9	4.7	5.4						
#1	1.8	65	2.0	**3.0**	4.0	5.0	6.1	7.1	6.2					
#2	1.8	85	2.6	4.0	**5.3**	6.6	8.0	9.3	8.1	10.0				
#3	5	120	3.2	5.6	7.5	**9.3**	11.3	13.0	10.6	13.2	16.0			
#4	5	160		7.5	10.0	12.5	**15.0**	17.5	15.0	18.7	22.5	26.3		
#5	10	200			12.5	15.6	18.7	**21.8**	20.0	25.0	30.0	35.0	40.0	
#6	10	250				19	23	27	**25.0**	31.2	37.5	43.7	50.0	
#8	20	385				30	36	42	31	**39**	47	55	62	56.2
#10	36	540					51	59	48	60	**72**	84	96	70
1/4	87	1,000							68	84	101	**118**	135	108
5/16	165	1,500							125	156	187	218	**250**	152
3/8	290	2,000								234	280	327	375	**281**
7/16	430	2,500									375	437	500	421
1/2	620	3,000										545	625	562
9/16	620	3,500											750	702
5/8	1,325	4,000												843
3/4	2,400	5,000												985
7/8	5,200	6,000												
1	7,200	7,000												

Table 6-3 (continued)

Nom. size	Seating torque inch-lbs.	Axial holding power (pounds)	Shaft Diameter (Shaft Hardness Rc 15 to Rc 35)											
			Torsional holding power inch-lbs.											
			$\frac{5}{8}$	$\frac{3}{4}$	$\frac{7}{8}$	1	$1\frac{1}{4}$	$1\frac{1}{2}$	$1\frac{3}{4}$	2	$2\frac{1}{2}$	3	$3\frac{1}{2}$	4
# 0	1.0	50												
# 1	1.8	65												
# 2	1.8	85												
# 3	5	120												
# 4	5	160												
# 5	10	200	62											
# 6	10	250	78	94	109									
# 8	20	385	120	144	168	192								
# 10	36	540	169	202	236	270	338							
1/4	87	1,000	312	375	437	500	625	750						
5/16	165	1,500	**468**	**562**	656	750	937	1125	1310	1500				
3/8	290	2,000	625	750	**875**	**1000**	1250	1500	1750	2000				
7/16	430	2,500	780	937	1095	1250	1560	1875	2210	2500	3125			
1/2	620	3,000	937	1125	1310	1500	**1875**	**2250**	2620	3000	3750	4500		
9/16	620	3,500	1090	1310	1530	1750	2190	2620	3030	3500	4370	5250	6120	
5/8	1,325	4,000	1250	1500	1750	2000	2500	3000	**3500**	**4000**	5000	6000	7000	8000
3/4	2,400	5,000		1875	2190	2500	3125	3750	4375	5000	**6250**	**7500**	8750	10000
7/8	5,200	6,000			2620	3000	3750	4500	5250	6000	7500	9000	10500	12000
1	7,200	7,000				3500	4375	5250	6120	7000	8750	10500	12250	14000

Self-tapping Screws

Self-tapping screws are very popular in mass assembly operations. They are equipped with a special type of hardened thread that extends the entire distance to the head. Ordinary screwdrivers (slotted, socket, or Phillips) can be used with these fasteners. Power tools with proper torque adjustment can also be used. Since the thread is hardened and partially tapered, the self-tapping screw cuts and forms its own thread in the material being used. These screws can be removed without damaging the threaded material and used again. If too much torque is applied and the threaded hole is ruined, a slightly larger self-tapping screw can remedy the situation. Since this type of fastener requires no nuts or lock washers, it can be classified as a blind fastener. One of its outstanding advantages is the fact that expensive tapping operations are eliminated. All that is needed is a properly drilled or punched hole of the correct size. These screws are available in many types, depending on the materials in which they are used. They can be used in sheet metals up to $\frac{1}{4}$ in. thick. Some types are ideal for use in molded, extruded, or laminated plastics. Others can be used in aluminum and zinc castings, insulating compositions, ferrous castings, and nonferrous castings. Since these fasteners are inexpensive and easy to use, they have become exceptionally popular.

Figure 6-10 shows a self-tapping screw. Note that the thread extends to the head and is tapered. This particular self-tapping screw is driven by a blade-type screwdriver.

Figure 6-11 shows a Teks® self-drilling fastener. Note how the metal chip is removed in Fig. 6-11(b). The fastener is in its final position in Fig. 6-11(c); it is now securing the two parts together.

A Taptite® thread rolling screw is shown in Fig. 6-12. This type of fastener has a trilobed thread structure. It is driven in a hole different in size

Figure 6-10
Self-tapping screw.
(Courtesy of Parker-Kalon, a division of the Emhart Corporation.)

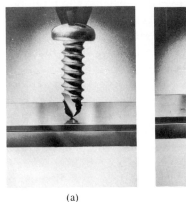

(a)   (b)   (c)

Figure 6-11
Teks® self-drilling fastener (a) in initial position, (b) removing metal chip, (c) in final position.
(Courtesy of Shakeproof, a division of Illinois Tool Works, Inc.)

from the corresponding tapped hole. Since it is a thread rolling screw, no chips are formed—it is rolling a thread and not cutting one.

Nuts and Lock Washers

Nuts and lock washers are necessary components in regular bolted applications. Equipment vibration can loosen fasteners; as a result, they sometimes work completely loose. For this reason, some insurance against loosening must be designed into any machine where vibration presents a problem. A lock washer exerts a force against a nut so that it will not

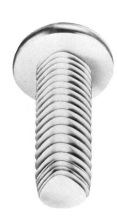

Figure 6-12
Taptite® thread rolling screw.
(Courtesy of Continental Screw Company.)

unthread. If a lock washer can be incorporated into a nut (or bolt), it eliminates one inventory item in a plant and ensures that every assembly *has* a lock washer. The combination of threaded fastener and lock washer is particularly important in mass production, where assembly time is calculated closely.

Some nuts contain a preassembled lock washer that cannot fall loose in handling. Other nuts contain a plastic material (sometimes nylon) that wedges into the threads; this prevents unthreading. Spring-type nuts (discussed in the following section) are also available; these exert a force against the assembled components while permanently engaging the threads of the bolt. Cap or acorn nuts are used where appearance is important. In the acorn type of nut, the top of the nut is closed and rounded. Because of this, the bolt length must be carefully considered. This type is often chrome plated, giving a pleasing appearance to the finished product as well as preventing exposed threads that could snag clothing or cause minor injuries. Many appliances use this type of nut.

Special Fasteners

A designer can choose from hundreds of special fasteners designed for specific purposes. A few of these are presented briefly in this text. Countless new types of fasteners are being developed; thus it is important that a designer keep abreast of current developments.

Figure 6-13 shows a *Flexloc® self-locking nut*, characterized by the six radial slots at the top that do the actual locking. This type of nut has no inserts and is constructed of one piece of metal. It can also be used as a stop nut.

Nylok® cup screws are illustrated in Fig. 6-14. A standard Nylok fastener is shown on the bottom; a Nylok strip-type fastener is shown on the top. The latter is particularly useful with thin materials, or when it is not

Figure 6-13
Flexloc® lock nut.
(Courtesy of SPS Technologies.)

Figure 6-14
Nylok® cap screws.
(Courtesy of Nylok Fastener
Corporation.)

possible to know at what point on the screw a nut will be seated. Nylon inserts are used in many other types of threaded fasteners besides cap screws. Nylok fasteners are useful for sealing and regulating, as well as locking under severe vibration conditions. It should be noted that these fasteners are reusable.

An *ANCO® nut* provides positive locking action by means of a locking pin. Figure 6-15(a) shows the nut; Fig. 6-15(b) shows how the locking pin travels between the bolt threads as the nut is turned. The finished bolt-nut assembly is shown in Fig. 6-15(c); note the position of the pin on the final assembly. The ANCO nut is used with ordinary wrenches and can be reused. The manufacturer lists a breakaway torque of 22 in-lb for the $\frac{1}{2}$-in. size (first removal). Available sizes range from $\frac{1}{4}$ in. to 4 in. They can be used with regular or high-strength bolts in applications where shock or severe vibration can occur.

A few typical *Palnut® fasteners and lock nuts* are shown in Fig. 6-16. Figure 6-16(a) shows a regular lock nut. This can be used alone as a

Figure 6-15
ANCO® nut. (a) The nut. (b) The locking pin. (c) The bolt-nut assembly.
(Courtesy of Bethlehem Steel Corporation.)

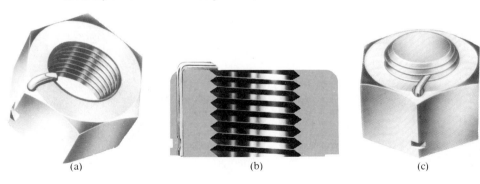

(a)   (b)   (c)

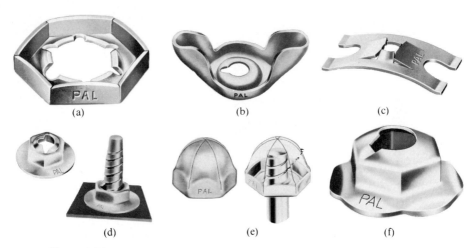

(a) (b) (c)

(d) (e) (f)

Figure 6-16
(a) Palnut® regular-type lock nut (for threaded members). (b) Wing-type Palnut® lock nut. (c) Flat rectangular Pushnut® fastener (for unthreaded studs). (d) Palnut® washer-type self-threading nut, showing threads formed on unthreaded stud. (e) Palnut® acorn-type self-threading nut, showing how threads are formed on unthreaded stud. (f) Washer-type Palnut® self-threading nut.
(Courtesy of the Fasteners Division of TRW, Inc.)

load-carrying nut for light-duty work, or in conjunction with a regular nut to provide locking action. A wing-type nut is shown in Fig. 6-16(b). The Pushnut® fastener shown in Fig. 6-16(c) is used on unthreaded studs and provides fast hand assembly. Self-threading nuts are shown in Figs. 6-16(d), (e), and (f). These nuts are made of heat-treated spring steel and form threads on unthreaded studs.

Figure 6-17 shows a *flange Whiz-lock®* *nut*. The flange on the nut is helpful in gapping oversize holes; the locking teeth prevent unthreading.

Figure 6-17
Flange Whiz-lock® nut.
(Courtesy of MacLean-Fogg Lock Nut Company.)

Figure 6-18
(a) Sems® terminal. (b) Sems® grounding terminal.
(Courtesy of Shakeproof, a division of Illinois Tool Works, Inc.)

Because of the locking teeth, more torque is required to remove this type of fastener than to seat it. Locking teeth are also available on nuts and screws.

Figure 6-18(a) shows a *Sems® terminal* and Fig. 6-18(b) shows a *Sems® grounding terminal*. These units are characterized by a preassembled lock washer on some type of screw. Many head and washer combinations are available. A somewhat similar arrangement is found in the Keps® unit (not shown), where a washer is assembled to a nut. Such units ensure that a washer is used at all times; it is impossible for a careless operator to forget the important lock washer.

Thousands of sizes and types of *Speed Nuts®* and *Speed Clips®* are available for use when rapid assembly is needed. Some of these are shown in Fig. 6-19. The nut-type fastener is self-locking and vibration resistant. The clip-type fastener requires no tools to assemble; it is pushed over a stud. Push-on speed nuts can be removed easily and reused. This type of fastener is popular in the mass assembly operations of the automotive and appliance fields.

Figure 6-20 is a photograph of a standard *Dzus®* quarter-turn self-locking fastener. This fastener, used for easy panel removal, has become popular in the aircraft industry. A stud and grommet are permanently placed in the removable panel. The stud has a spiral cam at the end that engages a permanently mounted (usually riveted) spring in the frame or other part being joined. A quarter-turn of the stud locks the assembly. This fastener is also extensively used, for instance, in the transportation, electronics, and communication industries where frequent access must be made to the "inside" components.

Figure 6-19
Speed Nuts® and Speed Clips®.
(Courtesy of Engineered Fasteners, a division of the Eaton Corporation.)

Figure 6-20
Quarter-turn self-locking fastener.
(Courtesy of Dzus Fastener Company.)

Security Fasteners

A number of special fastener heads have been developed to deter burglaries, unauthorized entry, and vandalism. It is true that nothing can be made tamper-proof, but the need for special tools does have a delaying and discouraging effect.

Figure 6-21 shows a few of the options available. One common means of deterring entry is the one-way screw head, often used on wood screws and metal self-tapping screws. These fasteners are secured by means of an ordinary screwdriver blade. The blade, however, will not "catch" if it is turned counterclockwise. The blade has no bearing surface in that direction; instead, it follows a spiral ramp away from the unit. This type is shown in Fig. 6-21(a). Two of the most common spanner types are shown in Figs. 6-21(b) and 6-21(c); these require special wrenches for fastening or unfastening. The drilled spanner type can be used in small sizes only. The slotted spanner type is adaptable to larger sizes. Figure 6-21(d) shows the pentagon-head type, which requires a socket-head tool designed to accommodate the pentagon shape.

Figure 6-21
Security fastener heads.

(a) One-way type

(b) Drilled spanner type

(c) Slotted spanner type

(d) Pentagon head type

Spacing of Bolts for Wrench Clearance

Care must be exercised in spacing the bolts (or nuts) in any assembly. If this is not done, it may be impossible to tighten the connections without resorting to special wrenches. Sometimes it is desirable to have bolts spaced as close together as possible, particularly when sealing a gasket whose effectiveness depends on the proper pressure being exerted by bolted connections. Spacing can be closer if socket wrenches are used. Table 6-4 shows minimum dimensions for A (the bolt-center distance) and B (the edge distance) if socket-wrench assembly is employed. The values are based on an outside socket diameter of twice the bolt diameter plus $\frac{1}{4}$ in., which approximates commercial socket wrenches. The tabular values also assume American Standard regular hexagon bolt heads.

Table 6-4
**Bolt Spacing for Socket-Wrench Clearance
(American Standard, Regular, Hexagon Head)**

Bolt Diameter	Bolt Centers (A)	Edge Distance (B)
$\frac{1}{4}$*	0.582	0.375
$\frac{5}{16}$	0.714	0.438
$\frac{3}{8}$*	0.810	0.500
$\frac{7}{16}$*	0.907	0.563
$\frac{1}{2}$*	1.039	0.625
$\frac{9}{16}$	1.171	0.688
$\frac{5}{8}$*	1.267	0.750
$\frac{3}{4}$	1.495	0.875
$\frac{7}{8}$	1.7245	1.000
1	1.952	1.125
$1\frac{1}{8}$	2.180	1.250
$1\frac{1}{4}$	2.409	1.375
$1\frac{3}{8}$	2.637	1.500
$1\frac{1}{2}$	2.865	1.625
$1\frac{5}{8}$	3.093	1.750
$1\frac{3}{4}$	3.322	1.875
$1\frac{7}{8}$	3.550	2.000
2	3.778	2.125
$2\frac{1}{4}$	4.235	2.375
$2\frac{1}{2}$	4.692	2.625
$2\frac{3}{4}$	5.148	2.875
3	5.605	3.125

*Nut sizes are larger than bolt heads for these sizes; table is based on bolt heads only.

If it is necessary to use open-end wrenches, both dimensions *A* and *B* must be larger. Formulas for these dimensions can be found in many handbooks. The use of offset box-type wrenches offers a compromise between the requirements of the socket and open-end types.

When hexagonal connections are used with hydraulic or pneumatic lines, open-end wrenches must be used. This is because the connections must be made with the hose or tubing in position, thus blocking out socket or box wrenches.

Spacing must be considered early in the design. If this is not done, at least two parts (the bolted pieces) will have to be redesigned; in some cases, many other parts will be affected as well.

◆ *Example* _____

Sixteen $\frac{3}{4}$-in.-diameter American Standard hexagonal bolts (and nuts) are to be used to connect a plate with a housing. If socket wrenches are to be used for assembly, what is the minimum spacing between bolt centers? How close can the outside bolts be to an adjacent housing?

Solution. From Table 6-4, the minimum center distance between bolts (dimension *A*) is 1.495 in. For simplicity, this would probably be made 1.5 in. Dimension *B* in the table is 0.875 in.

_____

Torque Application for Screw-type Fasteners

It is essential that the proper torque be applied when seating any type of screw fastener. Torque wrenches are wrenches with extra long handles and indicators—often dials—that show the amount of torque being applied in such units as in-lbs or N · mm.

If too much torque is applied in making bolted connections, the bolt threads can strip (shear). Excessive torque applied to self-tapping screws can ruin the newly formed threads, a condition that can only be corrected by using a larger size screw and cutting new threads. The holding power, either axial or torsional, of a setscrew depends upon the prescribed seating torque. If the setscrew is not properly seated, the holding power cannot be ensured; too much torque, on the other hand, can ruin the fastener.

Pins

Dowel pins are used either to secure two parts together for machining or to help sustain the shear load in rotating parts. These pins are usually installed by pressing or hammering them into place. If dowel pins sustain the shearing loads, then the clamping fasteners (such as bolts) are not unduly loaded. Pins are also used to secure (somewhat permanently) gears and similar components to shafts. In such a use, two shearing areas obviously result, since the pin extends completely through the shaft and hub. The use of such pins usually requires close-toleranced reamed holes. Plain pins can be straight or tapered.

Figure 6-22(a) shows a *Groov-Pin*®. The locking principle of this pin provides the holding power of a press-fit pin without the need for close tolerances, and the ease of installation of a taper pin without the need for reaming. When a Groov-Pin® is driven into a hole, the material displaced by the grooves is forced to flow back, setting up a powerful locking force. This enables the pin to hold permanently under severe shock and vibration. Some common types, with recommended uses, are shown in Fig. 6-22(b).

Nameplate Pins

Nameplates are simple but decorative and informative, components that must be mounted on most equipment. They must be secured in such a manner that they will not get lost. Escutcheon pins of many types are frequently used. Some of these are drive pins forced into an untapped hole; others have a spiral thread that wedges into the side of an untapped hole. Such fasteners are usually applied by pressing or hammering them into place.

The development of effective structural adhesives has had an effect on nameplate mounting. If adhesives are used, the hole-drilling operations can be eliminated along with the stocking of the mounting pins.

One must give careful consideration to the information placed on nameplates. The information must be readable, and the plate must be placed in a desirable position on the equipment.

Identification of Metric Fasteners

Fastener manufacturers frequently identify metric components by a *property class* identification. The property class indicates the minimum yield stress—an important consideration in selecting adequate fasteners. For example, if the figures 5.6 are used on a bolt, the number 5 indicates one-tenth of the minimum tensile strength in kgf/mm^2, or 50 kgf/mm^2. The decimal and second figure (.6) represent the ratio of the yield strength to tensile strength. Thus, for the designation .6, the minimum yield strength would be 0.6 × 50, or 30 kgf/mm^2.

Certain fastener manufacturers and users prefer to express tensile and yield strength values in megapascal units (1 MPa = 1 MN/m^2). Since one

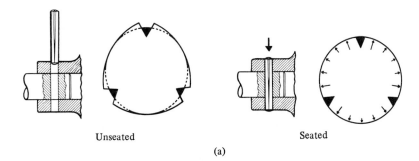

Unseated   Seated

(a)

The basic Groov-Pin, most widely used, has three full-length grooves tapering from maximum diameter at one end to nominal diameter at the other.

Type 1

Grooves cover only half of pin length. Used where holding power is not critical. Speeds assembly because ungrooved portion acts as a pilot.

Type 2

Recommended for severe shock and vibration applications, its full-length parallel grooves give it great holding power. Type 3H is essentially the same, but has pilots on both ends, suiting it for hopper feeds.

Type 3

Similar to Type Two, but taper on grooves is reversed meaning groove end is inserted first. This is particularly good on blind hole applications.

Type 4

Has three oval grooves half the length of the pin, centrally located. Especially suited for quick removal and replacement. Type Five is also well suited for hinge pin applications.

Type 5

(b)

Figure 6-22
(a) Groov-Pin® locking principle. (b) Groov-Pin® types. (Note: Types 24, 3H, and 67 are not shown.)
(Courtesy of Groove-Pin Corporation.)

pascal is equal to 9.81 times the kilogram-force unit (or approximately 10), the value in megapascals is roughly 10 times the kgf/mm^2 value. Essentially, conversions are made by multiplying by 10 since 10^6 appears in both numerator and denominator to change mm^2 to m^2 and to incorporate the prefix mega. Typical class-designation values are 3.6, 4.6, 5.6, 5.8, 6.6, 6.8, 6.9, 8.8, 10.9, 12.9, and 14,9.

Initially, this strength grade was color-coded on the fasteners by the manufacturers. This procedure turned out to be costly and ineffective; the dye sometimes altered performance properties, and color-blind assemblers tended to make mistakes. A number of manufacturers prefer to show the property class by raised numbers on the top of the bolthead and by indentations on the nonbearing face of the nut. It is important that metric fasteners not get confused with the inch type, as the two are definitely *not interchangeable*.

 Retaining Rings

Several types of snap or retaining rings are commercially available. These can be used to secure components on shafts or to secure parts in the bores of housings. Figure 6-23 shows the two basic types. The external type is expanded over a shaft or stud and the internal type is compressed for insertion into a bore or housing. Rings are made of carbon spring steel, stainless steel, or beryllium copper. The use of retaining rings frequently eliminates machined shoulders, collars, rivets, cotter pins, and threaded fasteners. Annular grooves of proper size must be provided to accommodate the rings. The use of special tools can speed assembly and production.

Rings are available in four general types:

1. axial assembly,
2. end-play take-up,

Figure 6-23
Basic external- and internal-type retaining rings. The external type (left) is designed to be expanded for assembly over a shaft, stud, or similar part. The internal ring (right) is compressed for insertion into a bore or housing.
(Courtesy of Truarc® Retaining Rings, a division of Waldes Kohinoor.)

Figure 6-24
A representative collection of Truarc® retaining rings.
(Courtesy of Truarc® Retaining Rings, a division of Waldes Kohinoor.)

3. self-locking,
4. radial assembly.

Figure 6-24 shows a representative collection of Truarc® retaining rings.

 **Metric Fasteners**

Popular sizes of metric fasteners are commercially available. Typical stock items include:

1. cap screws (hex head, socket head, square head, and so on),
2. machine screws (panhead; fillister head, shown on cap screw of Fig. 6-7; round head; flat head; oval head),

3. nuts (hexagon-regular, jam, castellated, nylon stop, weld, wing, dome, ring, spring lock),
4. threaded rods,
5. wing screws,
6. setscrews (flat point, cone point, cup point; square head, slotted head, socket head),
7. studs,
8. washers (flat; lock; spring; Belleville, shown in Chapter 14),
9. retaining rings (external, internal),
10. eyebolts,
11. pins (taper, roll, cotter, threaded),
12. rivets (round head, flat head),
13. shaft collars.

 **Couplings**

Rigid Couplings

Figure 6-25 shows a flanged rigid coupling. Torque is transmitted directly through this type of connector; therefore, the two shafts must have straight, collinear centerlines. This type allows for no linear or angular misalignment. A flange is keyed to each of the connecting shafts; then a series of through bolts is used to connect the flanges. The bolt holes are usually counterbored to prevent the fasteners from projecting beyond the face of the flange; this reduces a potential safety hazard. To calculate the size of through bolt required, the size of the bolt circle must be known, as well as the number of bolts, the allowable stress, and the torque to be transmitted. The following formula can be used if the bolts are arranged in one bolt circle:

$$D = \sqrt{\frac{4T}{\pi S_s r_1 n}} \, , \qquad\qquad (6\text{-}1)$$

where

r_1 = radius of bolt circle (in.),

T = torque (in-lb),

S_s = allowable shearing stress (psi),

n = number of bolts,

D = diameter of bolt (in.).

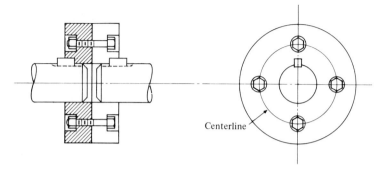

Figure 6-25
Flanged coupling.

◆ *Example* ───────────────────────────

A flanged coupling is designed to use 6 bolts on a 4-in.-diameter bolt circle. The allowable stress in shear is 12,000 psi; the maximum torque is 50 ft-lb. Find the diameter of bolt required.

Solution.

$$D = \sqrt{\frac{4T}{\pi S_s r_1 n}} = \sqrt{\frac{4(50)(12)}{\pi(12,000)(2)(6)}} = 0.073 \text{ in. (root diameter).}$$

Another type of rigid coupling is the sleeve coupling illustrated in Fig. 6-26. This is merely a sleeve that connects the two shafts; the shafts must be correctly aligned, since there is no flexibility to the coupling. The sleeve is then fastened with setscrews to the two shafts. The amount of torque that can safely be handled depends on the holding power of the setscrews. Sometimes a drive pin (pin with grooved slots) is used; in that case, a hole is drilled through the sleeve and shaft. Figure 6-27 shows this application. The

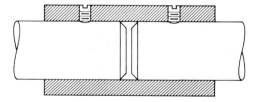

Figure 6-26
Sleeve coupling: setscrew.

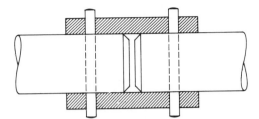

Figure 6-27
Sleeve coupling: drive pin.

size of the pin can be calculated from the following equation:

$$D = \sqrt{\frac{2T}{\pi r S_s}}, \qquad\qquad (6\text{-}2)$$

where

D = pin diameter (in.),

r = shaft radius (in.).

◆ *Example* ――――――――――――――――――――――――
Find the pin diameter for a sleeve-type coupling where the transmitted torque is 400 in-lb and the shaft diameter is 1 in. Assume that the working stress in shear for the pin is 14,000 psi.

Solution.

$$D = \sqrt{\frac{2T}{\pi r S_s}} = \sqrt{\frac{2(400)}{\pi(0.5)(14{,}000)}} = 0.191 \text{ in.}$$

◆ *Example* ――――――――――――――――――――――――
Based on a drive-pin design stress of 80 MN/m², find the power capacity of a 30-mm shaft connected to a sleeve coupling similar to that shown in Fig. 6-27. The shear pin diameter is 4 mm and the shaft speed is 150 rpm.

Solution.

$$T = \frac{\pi r S_s D^2}{2} = \frac{\pi(0.015)(80)(10)^6(0.004)^2}{2} = 30.16 \text{ N} \cdot \text{m};$$

$$kW = \frac{2\pi NT}{1000(60)} = \frac{2\pi(150)(30.16)}{1000(60)} = 0.474 \text{ kW}.$$

Shear Pins

A mechanical type of safety device is the shear pin, similar in function to the familiar electrical fuse. Such a pin is used in a sleeve-type coupling and elsewhere. It is deliberately designed with a smaller factor of safety than the other components of the product. This means that the allowable shear stress for the other components is lower, and the pin will fail before the other parts do. Conveying equipment frequently uses this safety device to prevent part failures that might occur if the system became overloaded. Propellers on outboard motors frequently use a shear pin to prevent driveshaft failure if the propeller strikes an obstacle. To keep part sizes compatible, shear pins are sometimes made of a weaker material than the other components. For example, steel parts are often pinned with a shear pin made of brass, bronze, or some other weaker material. The design of shear pins is covered in Chapter 16 under "Safety Devices."

Flexible Couplings

Because of manufacturing tolerances in the fabrication and mounting of parts, perfect shaft alignment is almost impossible to achieve. When two shafts are connected with rigid couplings, any misalignment (linear or angular) has detrimental effects on several parts of the assembly. Some of these effects are (1) reversed flexure on shafts, (2) unexpected and excessive loads on bearings, and (3) excessive vibration. Flexible couplings can eliminate some of these problems, provided the amount of misalignment is not too great. Commercial couplings are available, most of which consist essentially of three main parts. One of two end parts is keyed or setscrewed to each of the shafts; a central or "floating" part then serves as a connecting link between the end parts. The middle component is made of hardened rubber or some other flexible material. This type of coupling is illustrated in Fig. 6-28; it will correct for both linear and angular misalignment, provided the variations are not too great.

Figure 6-29 shows a type of coupling suitable for linear misalignment, although a certain amount of angular error can also be handled. This device is called an Oldham-type coupling. The end sections are rigidly attached to the shafts. Both end sections contain a rectangular slot that faces the center section; the "floating" central section is provided with a rectangular projection on each side that engages the slots in the end pieces. The two rectangular projections are at right angles to each other. This "floating" section will slide in the grooves of the end pieces as the shaft rotates. The

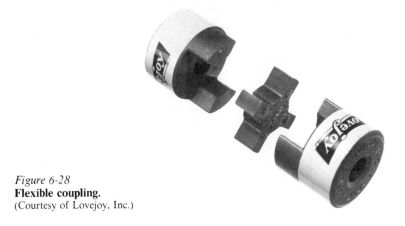

Figure 6-28
Flexible coupling.
(Courtesy of Lovejoy, Inc.)

middle section is usually made of a material unlike that of the end sections; nylon or one of the sintered powdered metals (containing a lubricant) is often used. Use of a suitable material for the "floating" section can do much to reduce the amount of friction caused by the sliding action. Variations of the basic Oldham-type coupling can be used. One type has a projection on one end piece and a rectangular groove on the other end section. The middle section then has a groove on one face and a projection on the other.

Commercial flexible couplings are rated in terms of torque capacity; in addition, the user should carefully study shaft sizes, tolerances, and methods of attachment.

Universal Joints

In certain applications, a definite amount of angular displacement is required because of the nature of the machine. In this case a universal joint

Figure 6-29
Oldham-type coupling.

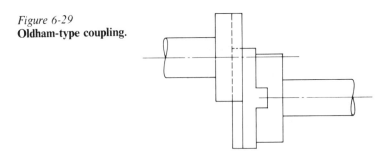

Figure 6-30
Universal couplings.

can be used. Figure 6-30 shows a typical universal. In using the universal joint or any other flexible coupling, it must be remembered that any misalignment whatsoever causes a cycle error. If the driving shaft moves a certain number of degrees, the driven shaft does not have the same angular displacement. The amount of variation depends on the amount of misalignment. If a double universal joint is used, or two universal joints with a "floating" shaft between them, the cycle error can be eliminated provided the yokes of the two universals are properly oriented. In automotive applications, a long drive shaft is used between the two universal joints. Figure 6-31 shows a constant-velocity universal joint. In this device, steel balls operate in races oriented so that the plane of contact always bisects the shaft angle. In using regular universal joints, provision must be made for a small amount of axial displacement of one of the shafts. Excessive angles between shafts should be avoided. Generally, 30 deg is considered a good value for the maximum, although the type of arrangement, the application, and the rotational speed can affect the maximum value selected. Universals might be considered as rigid couplings; however, they are not included in that section because they are generally not used to connect collinear shafts. Instead, they are used where a deliberate angle is desired.

Figure 6-31
Constant-velocity universal joint; bell-type joint on left and disc-type joint on right.
(Courtesy of Con-Vel, a division of the Dana Corporation.)

Figure 6-32
Torque-limiter coupling.
(Courtesy of Morse Chain, a
division of Borg-Warner
Corporation.)

Torque-Limiter Coupling

Figure 6-32 shows a coupling that serves a dual purpose. Besides being a coupling, it is an overload device that will slip when the preset torque is exceeded. The coupling connection is made via a double-strand roller chain that connects two sprockets. For a description of the torque-limiting function of this coupling, refer to the section in Chapter 10 entitled "Torque Limiter." This particular type of coupling allows for a small amount of shaft misalignment. Angular misalignment of $\frac{1}{2}$ deg and parallel misalignment of 0.010 to 0.020 in. can be handled. The double-strand roller chain is easy to connect; one pin makes the connection.

 Keys

A key is a rigid connector between a shaft and the hub of another component such as a pulley, gear, or cam. Its purpose is to prevent relative rotation between the two parts. If a key is to be used, a keyseat must be provided in the shaft and a keyway in the hub of the other part. A keyseat weakens the shaft; thus, the design stress for the shaft has to be reduced as indicated in Chapter 5. Sharp corners on the keyseat also introduce "stress raisers" or stress concentrations that must be minimized as much as possible. Several types of keys are available to a designer; the application often dictates which variety is most suitable. In some instances, a tight fit is needed between the key and both the shaft and hub. In other cases, a tight fit is needed between the key and the shaft, but a loose fit between the key and the hub. Sometimes a feather key is used. This is a key that is bolted into the shaft, but is allowed to slide freely in the other part.

Flat and Square Keys

Table 6-5 lists the standard recommendations for square and flat keys when used with various sizes of shafts. When two parts are keyed together, the key is subjected primarily to shear. The equation for any shear of this type is $S_s = P/A$; the area in shear is the width of the key at the midsection times the length of the key. Therefore, since the width is selected according to the shaft diameter, the real calculation usually turns out to be the length of the key. The widths listed in Table 6-5 are recommendations; if a designer chose a smaller width of key than recommended, the length would be abnormally long if the shaft were designed to carry the desired torque. On the other hand, if a larger key width were chosen, the shaft would be greatly weakened by the loss of cross-sectional area and would have to be redesigned.

Woodruff Keys

Machine tools often use the Woodruff key. Table 6-6 lists standard dimensions for this type. The shear area for the Woodruff key is the area at the top of the shaft. Nominal dimensions (width times length) can sometimes be used to calculate stress without introducing too much error. One outstanding advantage of the Woodruff key is the fact that it cannot

Table 6-5
American Standard Square and Flat Keys

Shaft Diameter	Square Key Width	Flat Key Width × Height
$\frac{1}{2}$ to $\frac{9}{16}$	$\frac{1}{8}$	$\frac{1}{8} \times \frac{3}{32}$
$\frac{5}{8}$ to $\frac{7}{8}$	$\frac{3}{16}$	$\frac{3}{16} \times \frac{1}{8}$
$\frac{15}{16}$ to $1\frac{1}{4}$	$\frac{1}{4}$	$\frac{1}{4} \times \frac{3}{16}$
$1\frac{5}{16}$ to $1\frac{3}{8}$	$\frac{5}{16}$	$\frac{5}{16} \times \frac{1}{4}$
$1\frac{7}{16}$ to $1\frac{3}{4}$	$\frac{3}{8}$	$\frac{3}{8} \times \frac{1}{4}$
$1\frac{13}{16}$ to $2\frac{1}{4}$	$\frac{1}{2}$	$\frac{1}{2} \times \frac{3}{8}$
$2\frac{5}{16}$ to $2\frac{3}{4}$	$\frac{5}{8}$	$\frac{5}{8} \times \frac{7}{16}$
$2\frac{7}{8}$ to $3\frac{1}{4}$	$\frac{3}{4}$	$\frac{3}{4} \times \frac{1}{2}$
$3\frac{3}{8}$ to $3\frac{3}{4}$	$\frac{7}{8}$	$\frac{7}{8} \times \frac{5}{8}$
$3\frac{7}{8}$ to $4\frac{1}{2}$	1	$1 \times \frac{3}{4}$
$4\frac{3}{4}$ to $5\frac{1}{2}$	$1\frac{1}{4}$	$1\frac{1}{4} \times \frac{7}{8}$
$5\frac{3}{4}$ to 6	$1\frac{1}{2}$	$1\frac{1}{2} \times 1$

All dimensions are in inches. ASA B 17.1 (1943).

Table 6-6
American Standard Woodruff Keys

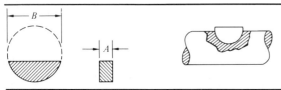

Key Number	Nominal Size, $A \times B$	Depth of Slot in Shaft	Key Number	Nominal Size, $A \times B$	Depth of Slot in Shaft
204	$\frac{1}{16} \times \frac{1}{2}$	0.1718	808	$\frac{1}{4} \times 1$	0.3130
304	$\frac{3}{32} \times \frac{1}{2}$	0.1561	809	$\frac{1}{4} \times 1\frac{1}{8}$	0.3590
305	$\frac{3}{32} \times \frac{5}{8}$	0.2031	810	$\frac{1}{4} \times 1\frac{1}{4}$	0.4220
404	$\frac{1}{8} \times \frac{1}{2}$	0.1405	811	$\frac{1}{4} \times 1\frac{3}{8}$	0.4690
405	$\frac{1}{8} \times \frac{5}{8}$	0.1875	812	$\frac{1}{4} \times 1\frac{1}{2}$	0.5160
406	$\frac{1}{8} \times \frac{3}{4}$	0.2505	1008	$\frac{5}{16} \times 1$	0.2818
505	$\frac{5}{32} \times \frac{5}{8}$	0.1719	1009	$\frac{5}{16} \times 1\frac{1}{8}$	0.3278
506	$\frac{5}{32} \times \frac{3}{4}$	0.2349	1010	$\frac{5}{16} \times 1\frac{1}{4}$	0.3908
507	$\frac{5}{32} \times \frac{7}{8}$	0.2969	1011	$\frac{5}{16} \times 1\frac{3}{8}$	0.4378
606	$\frac{3}{16} \times \frac{3}{4}$	0.2193	1012	$\frac{5}{16} \times 1\frac{1}{2}$	0.4848
607	$\frac{3}{16} \times \frac{7}{8}$	0.2813	1210	$\frac{3}{8} \times 1\frac{1}{4}$	0.3595
608	$\frac{3}{16} \times 1$	0.3443	1211	$\frac{3}{8} \times 1\frac{3}{8}$	0.4065
609	$\frac{3}{16} \times 1\frac{1}{8}$	0.3903	1212	$\frac{3}{8} \times 1\frac{1}{2}$	0.4535
807	$\frac{1}{4} \times \frac{7}{8}$	0.2500			

All dimensions are in inches. The key numbers indicate the nominal key dimensions. The last two digits give the nominal diameter in eighths of an inch; the digits preceding the last two give the nominal width in thirty-seconds of an inch. ASA B 17f (1930; R 1955).

possibly be removed (slip out accidentally) unless the shaft and the keyed-on member are separated. A rule of thumb for selecting a Woodruff key is to find one with a width of approximately one-fourth the shaft diameter and a radius approximately one-half the shaft diameter. The shear stress should then be calculated using the (approximate) shear area.

Pratt and Whitney Keys

The Pratt and Whitney key sizes are listed in Table 6-7. Like the Woodruff key, this key cannot be removed unless the shaft and keyed-on member are separated. The area in shear is approximately the width times the length. The lengths of Pratt and Whitney keys can vary, but they should always be at least twice as long as wide.

Table 6-7
Pratt and Whitney Keys

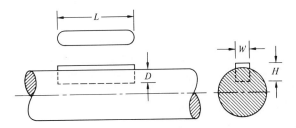

Key No.	L	W	H	D	Key No.	L	W	H	D
1	$\frac{1}{2}$	$\frac{1}{16}$	$\frac{3}{32}$	$\frac{1}{16}$	22	$1\frac{3}{8}$	$\frac{1}{4}$	$\frac{3}{8}$	$\frac{1}{4}$
2	$\frac{1}{2}$	$\frac{3}{32}$	$\frac{9}{64}$	$\frac{3}{32}$	23	$1\frac{3}{8}$	$\frac{5}{16}$	$\frac{15}{32}$	$\frac{5}{16}$
3	$\frac{1}{2}$	$\frac{1}{8}$	$\frac{3}{16}$	$\frac{1}{8}$	F	$1\frac{3}{8}$	$\frac{3}{8}$	$\frac{9}{16}$	$\frac{3}{8}$
4	$\frac{5}{8}$	$\frac{3}{32}$	$\frac{9}{64}$	$\frac{3}{32}$	24	$1\frac{1}{2}$	$\frac{1}{4}$	$\frac{3}{8}$	$\frac{1}{4}$
5	$\frac{5}{8}$	$\frac{1}{8}$	$\frac{3}{16}$	$\frac{1}{8}$	25	$1\frac{1}{2}$	$\frac{5}{16}$	$\frac{15}{32}$	$\frac{5}{16}$
6	$\frac{5}{8}$	$\frac{5}{32}$	$\frac{15}{64}$	$\frac{5}{32}$	G	$1\frac{1}{2}$	$\frac{3}{8}$	$\frac{9}{16}$	$\frac{3}{8}$
7	$\frac{3}{4}$	$\frac{1}{8}$	$\frac{3}{16}$	$\frac{1}{8}$	51	$1\frac{3}{4}$	$\frac{1}{4}$	$\frac{3}{8}$	$\frac{1}{4}$
8	$\frac{3}{4}$	$\frac{5}{32}$	$\frac{15}{64}$	$\frac{5}{32}$	52	$1\frac{3}{4}$	$\frac{5}{16}$	$\frac{15}{32}$	$\frac{5}{16}$
9	$\frac{3}{4}$	$\frac{3}{16}$	$\frac{9}{32}$	$\frac{3}{16}$	53	$1\frac{3}{4}$	$\frac{3}{8}$	$\frac{9}{16}$	$\frac{3}{8}$
10	$\frac{7}{8}$	$\frac{5}{32}$	$\frac{15}{64}$	$\frac{5}{32}$	26	2	$\frac{3}{16}$	$\frac{9}{32}$	$\frac{3}{16}$
11	$\frac{7}{8}$	$\frac{3}{16}$	$\frac{9}{32}$	$\frac{3}{16}$	27	2	$\frac{1}{4}$	$\frac{3}{8}$	$\frac{1}{4}$
12	$\frac{7}{8}$	$\frac{7}{32}$	$\frac{21}{64}$	$\frac{7}{32}$	28	2	$\frac{5}{16}$	$\frac{15}{32}$	$\frac{5}{16}$
A	$\frac{7}{8}$	$\frac{1}{4}$	$\frac{3}{8}$	$\frac{1}{4}$	29	2	$\frac{3}{8}$	$\frac{9}{16}$	$\frac{3}{8}$
13	1	$\frac{3}{16}$	$\frac{9}{32}$	$\frac{3}{16}$	54	$2\frac{1}{4}$	$\frac{1}{4}$	$\frac{3}{8}$	$\frac{1}{4}$
14	1	$\frac{7}{32}$	$\frac{21}{64}$	$\frac{7}{32}$	55	$2\frac{1}{4}$	$\frac{5}{16}$	$\frac{15}{32}$	$\frac{5}{16}$
15	1	$\frac{1}{4}$	$\frac{3}{8}$	$\frac{1}{4}$	56	$2\frac{1}{4}$	$\frac{3}{8}$	$\frac{9}{16}$	$\frac{3}{8}$
B	1	$\frac{5}{16}$	$\frac{15}{32}$	$\frac{5}{16}$	57	$2\frac{1}{4}$	$\frac{7}{16}$	$\frac{21}{32}$	$\frac{7}{16}$
16	$1\frac{1}{8}$	$\frac{3}{16}$	$\frac{9}{32}$	$\frac{3}{16}$	58	$2\frac{1}{2}$	$\frac{5}{16}$	$\frac{15}{32}$	$\frac{5}{16}$
17	$1\frac{1}{8}$	$\frac{7}{32}$	$\frac{21}{64}$	$\frac{7}{32}$	59	$2\frac{1}{2}$	$\frac{3}{8}$	$\frac{9}{16}$	$\frac{3}{8}$
18	$1\frac{1}{8}$	$\frac{1}{4}$	$\frac{3}{8}$	$\frac{1}{4}$	60	$2\frac{1}{2}$	$\frac{7}{16}$	$\frac{21}{32}$	$\frac{7}{16}$
C	$1\frac{1}{8}$	$\frac{5}{16}$	$\frac{15}{32}$	$\frac{5}{16}$	61	$2\frac{1}{2}$	$\frac{1}{2}$	$\frac{3}{4}$	$\frac{1}{2}$
19	$1\frac{1}{4}$	$\frac{3}{16}$	$\frac{9}{32}$	$\frac{3}{16}$	30	3	$\frac{3}{8}$	$\frac{9}{16}$	$\frac{3}{8}$
20	$1\frac{1}{4}$	$\frac{7}{32}$	$\frac{21}{64}$	$\frac{7}{32}$	31	3	$\frac{7}{16}$	$\frac{21}{32}$	$\frac{7}{16}$
21	$1\frac{1}{4}$	$\frac{1}{4}$	$\frac{3}{8}$	$\frac{1}{4}$	32	3	$\frac{1}{2}$	$\frac{3}{4}$	$\frac{1}{2}$
D	$1\frac{1}{4}$	$\frac{5}{16}$	$\frac{15}{32}$	$\frac{5}{16}$	33	3	$\frac{9}{16}$	$\frac{27}{32}$	$\frac{9}{16}$
E	$1\frac{1}{4}$	$\frac{3}{8}$	$\frac{9}{16}$	$\frac{3}{8}$	34	3	$\frac{5}{8}$	$\frac{15}{16}$	$\frac{5}{8}$

Gib Head Keys

This type of key is tapered $\frac{1}{8}$ in. per foot and contains a head that projects beyond the keyed members. The head is included to allow for easy removal. The taper is provided to prevent axial displacement by tightly securing the two parts.

◆ *Example* ────────────────────────────────

A $1\frac{3}{4}$-in.-diameter steel shaft rotates at 150 rpm. It transmits 15 hp and is keyed to an 8-in.-diameter steel pulley. Find the size of square key required if the allowable shearing stress of the key material is 10,000 psi.

Solution.

$$T = \frac{63,000(\text{hp})}{N} = \frac{63,000(15)}{150} = 6300 \text{ in-lb},$$

$$P = \frac{T}{r} = \frac{6300}{\frac{7}{8}} = 7200 \text{ lb};$$

$$A = Lw = \frac{P}{S_s}, \qquad\qquad\qquad (6\text{-}3)$$

$$\tfrac{3}{8}L = \frac{7200}{10,000},$$

$$L = 1.92 \text{ in.}$$

A $\frac{3}{8}$-in. width is selected from Table 6-5.

Note. The key length should be greater than the length of the hub. From the preceding calculation, the minimum hub length is 1.92 in.; therefore, the key length is probably $2\frac{1}{4}$ in. to $2\frac{1}{2}$ in.

──

Metric Keys

Dimensions for metric keys are expressed in millimeters. Popular, commercially available keys include the square, flat, and Woodruff. Square keys are readily obtainable in the following sizes: 4 × 4, 5 × 5, 6 × 6 and in various lengths ranging from 20 mm to 50 mm. Typical flat key sizes are 6 × 4, 8 × 5, 8 × 7, 10 × 6, 10 × 8, 12 × 6, and 12 × 8. Lengths range from 20 mm to 50 mm in the smaller sizes and 35 mm to 100 mm in the larger sizes. The designer should check distributors' catalogs for specific availabilities. In specifying metric keys, it is customary to use the letter M to denote metric, such as: M6 × 20 square key.

Woodruff keys are made in a number of metric sizes. Again, all dimensions are specified in millimeters. They range in size from 1×1.4 mm to 10×16 mm.

Comparison of Keys and Setscrews

The torque-carrying capacity of keys is far greater than that of setscrews. Setscrews are convenient to use because they are easy to assemble and disassemble. When a combination of key and setscrew is used, the key generally carries all the torque; the setscrew merely prevents axial movement. The holding power of setscrews can be determined experimentally and the results used (cautiously) in predicting performance in other designs. In the case of the dog-point setscrew, the shearing area (area at dog point) can be accurately calculated. In the cone-point type, the area in shear depends on the amount of penetration of the point into the shaft—the greater the penetration, the greater area in shear. The shaft is prepared for taking a large penetration by drilling it slightly before applying the setscrew. The area of a cone point in shear is easy to determine if the depth of penetration is known. For most other points, the holding power depends on whether or not a flat spot has been ground into the shaft. Holding power can also depend on the amount of tightening torque applied to the setscrew in assembling. With such variables to be considered, calculations for setscrews must be carefully evaluated before one is assured of suitable holding ability.

◆ *Example* ────────────────────────────────

A $\frac{3}{8}$-in. dog-point setscrew has a dog-point diameter of $\frac{1}{4}$ in. If the allowable shearing stress of this setscrew is 8000 psi and it is used with a $2\frac{1}{2}$-in.-diameter shaft, find its holding capacity. What would be the holding capacity of a key if it were 1 in. long and had the same allowable stress as the setscrew?

Solution.

$$P = S_s A = \frac{S_s \pi D^2}{4} = \frac{8000(\pi)(0.25)^2}{4} = 393 \text{ lb.}$$

This size of shaft would use a $\frac{5}{8}$-in. square key.

$$P = S_s A = S_s w L = 8000(\tfrac{5}{8})(1) = 5000 \text{ lb.}$$

It is obvious that a key can sustain much more torque than a setscrew. Even if three setscrews were used, the total holding capacity would be only $3(393) = 1179$ lb.

Splines

When a shaft is required to carry torque beyond that obtainable with keys (or when the torque is frequently reversed), one solution is to spline the shaft and the hub of the connected member. Splining of the shaft weakens it, since metal is cut away from the usual outside diameter; provision must be made for this when calculating the diameter. Actually, splines are multiple keys that are made an integral part of the shaft and hub, respectively. Two types of splines have been standardized. One is the ASA involute spline, which is available in five tooth forms; these vary in pressure angle. The pressure angles are 30, 45, 20, 25, and $14\frac{1}{2}$ deg.

The other standardized type is the straight spline. Table 6-8 lists the nominal dimensions for the SAE type. The four-spline and six-spline fittings are standardized in maximum diameters from $\frac{3}{4}$ in. to $1\frac{1}{2}$ in. by $\frac{1}{8}$-in. increments, and then by $\frac{1}{4}$-in. increments up to $2\frac{1}{2}$ in. A 3-in. size is also used. The diameters of the ten-spline fittings range from $\frac{3}{4}$ in. to 6 in., with the same increments up to 3 in., and then by $\frac{1}{2}$-in. increments. For the sixteen-spline fitting, the increments are $\frac{1}{2}$ in. and the shaft diameters run from 2 in. to 6 in. The following equation should be used to obtain the torque capacity of a splined fitting in inch-pounds:

$$T = 1000hLnr_m, \tag{6-4}$$

Table 6-8
SAE Straight Splines, Nominal Dimensions

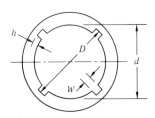

Number of Splines	Permanent Fit		To Slide When not Under Load		To Slide When Under Load		All Fits
	h	d	h	d	h	d	W
4	0.075D	0.850D	0.125D	0.750D	—	—	0.241D
6	0.050D	0.900D	0.075D	0.850D	0.100D	0.800D	0.250D
10	0.045D	0.910D	0.070D	0.860D	0.095D	0.810D	0.156D
16	0.045D	0.910D	0.070D	0.860D	0.095D	0.810D	0.098D

where

L = active length (in.),

n = number of splines,

r_m = mean radius (in.),

h = depth (in.) (see Table 6-8).

◆ *Example* _____

Find the torque-carrying capacity of a splined shaft that is 5 in. in diameter and has 10 splines. The length of contact between the two connected components is 1.5 in.

Solution. From Table 6-8,

$h = 0.045D = 0.045(5) = 0.225$ in.

Then

$$r_m = \frac{D - h}{2} = \frac{5 - 0.225}{2} = 2.388 \text{ in.},$$

$$T = 1000(0.225)(1.5)(10)(2.388) = 8060 \text{ in-lb.}$$

The SAE straight spline is often used in automotive applications. In some of these designs, splines are used for permanent, nonmovable uses. In other designs, they are used where the parts are to slide with no load applied. In still others they are used where the components are to slide under load. It must be remembered that splining requires extra manufacturing operations and thus is more costly than using keys; however, if torque requirements are high, the extra cost is justified.

 Structural Adhesives

Adhesives are becoming more popular in the general area of fasteners. They have been used for many years in various areas of manufacturing; however, their usefulness has now expanded into structural applications. Originally, glues and cements were used only where little strength was required. Now, many aircraft parts are bonded together with adhesives; adhesives have taken the place of riveting and welding on certain subassemblies. Bonded brake linings are commonplace and have good performance ratings.

There are several items a designer should investigate before specifying adhesives. First, the required strength of the bond must be known. This will sometimes force the designer to change the arrangement of the parts slightly to ensure an adequate bonding area. Even more helpful is a change in arrangement that reduces the applied loads on the bonded area. Manufacturing facilities for properly preparing the surfaces for adhesives must be available. It is important to know whether the bonded parts are to be subjected to outside weather conditions or to any chemicals. Some adhesives can withstand such treatment and some cannot. It is essential to know

1. the temperature range in which the assembly will be operating;
2. the type of loading that will be applied to the assembled components; and
3. the available facilities for applying the adhesive.

Certain adhesives require ovens or lamps for curing. Some adhesives require brushing on two surfaces and form the bond upon contact; others can be brushed on one surface. One epoxy adhesive requires that two parts be mixed in equal proportions. Coloring the two mixes aids in determining the quantity of each required. Application can be made as soon as a mixed color is consistent. Curing at room temperatures is desirable from the production standpoint.

Cyanocrylate adhesives are useful for making fast and strong bonds. They require no mixing, no solvent, and little pressure, thus eliminating costly fixtures. Tensile strengths of 5000 psi and overlap shear strengths of 3000 psi are possible with steel substrates. They should not be used where moisture or high temperatures are factors; also, they are limited to small bond-line thicknesses. These adhesives are now being used in subassemblies found in small household appliances, recorders, projection screens, and similar products.

Careful quality control must be maintained when using adhesives for structural purposes. Ultrasonic testing (a nondestructive type of test) is useful in determining whether any voids exist in an adhesive that is bonding two surfaces. X-ray testing is also useful in many cases. "White-room" conditions must often be maintained in production areas where structural bonding with adhesives is performed.

Some adhesives provide shock absorption between the two bonded surfaces. Noise abatement is a property that is especially desirable in bonding metallic surfaces. A nitrile rubber and epoxy combination is used for helicopters; this assists in the general area of damping and flexibility, besides providing for toughness. The aircraft and aerospace industries use adhesives extensively. It is not uncommon to find bonded shear stresses in the 3500 to 4000-psi range.

Many companies manufacture adhesives; the designer has hundreds of types from which to make a selection. A manufacturer's "know how" can be a valuable resource if it is provided with all available information about environmental conditions and plant capabilities. The adhesives available include phenolic epoxies, nylon epoxies, elastomers, and many others. If adhesives are used in structural applications, they must be designed with the same care as rivets and welds.

Materials that can be bonded include such metals as steel, stainless steel, aluminum, copper, and brass. Nonmetals include Teflon®, plastic foams, laminates, glass, and ceramics.

Outstanding advantages of using adhesives include lower manufacturing, sealing, and damping costs. On the negative side, one must use extra care in designing joints and must carefully control manufacturing methods and testing procedures. Also, certain adhesives present a safety hazard with respect to storage and operator use.

In evaluating strength values, one must note the difference between shear strength and peel strength. Peel strength is important where possible failure might be due to peeling. Proper placement of parts that are bonded with adhesives can minimize the peeling effect.

 ## Welding and Weld Design

Soldering, brazing, and welding are all popular shop processes. *Soldering* cannot be considered a welding process; in general, it is a joining operation suitable for electrical connections or for sealing out fluids under low pressures. However, hard solders used in refrigeration systems can withstand high pressures. *Brazing* is a process somewhat more complicated than soldering; it produces a better seal, but the joint will not withstand much load.

Many classifications of welding processes are possible. A complete classification would include such types as forge, gas, thermit, induction, resistance, and arc welding. Most of these could be further subdivided. In this text, we shall deal mainly with resistance and arc welding, since these are the most commonly used in ordinary production processes. Space does not permit a full discussion of all facets of these types; instead, the more popular facets are explored. Welding symbols are illustrated to provide the student with some knowledge of their use. American Welding Society publications, MIL standards, and industry standards provide complete information on the use of symbols. Symbols provide simplification, eliminate some detailing, and provide a working language common to the engineering, inspection, and production departments.

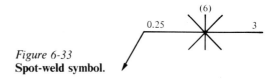

Figure 6-33
Spot-weld symbol.

Resistance Welding

Resistance welding is accomplished by incorporating the work piece into an electric circuit. The resistance of the piece to the flow of electric current produces heat. This heat, with proper pressure, produces the weld. This type of welding has been made popular by the expanding use of sheet-metal processes. There are several types of resistance welds. The three discussed here include spot, seam, and projection welds.

The most common of these is probably the *spot weld*. This is produced by either a bench-type welding fixture or a portable one. The sheets of metal being welded are clamped between the two electrodes and current is passed through the clamped assembly; this forms the weld. Sometimes the locations of the welds are indicated on the drawing and given dimensions in the x and y directions. If the location is not critical, the welding symbol can be used to simplify the subassembly drawing. Figure 6-33 shows a weld symbol for a spot weld. Note that figures can be placed to the left, above, and to the right of the symbol. Each of these is significant. The figure to the left of the symbol can be either the size (diameter) of the weld or the shear strength of the weld. The figure above the symbol shows the number of welds; that to the right indicates the pitch (the distance between weld centers). Weld quality must be checked continually. This is done by subjecting weld samples to testing-machine forces and checking the strength. The size (and strength) of this type of weld is dictated by the size and shape of the electrodes.

Seam welding is a process similar to spot welding. In effect, a seam weld consists of a series of overlapping spot welds. The weld is produced by passing the work (two parts being welded) between two rollers. The electric current is then applied intermittently; this produces the seam weld. Figure 6-34 shows a typical seam-weld symbol. The figure to the left of the symbol

Figure 6-34
Seam-weld symbol.

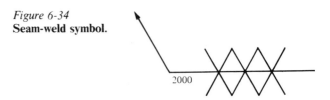

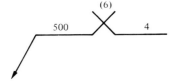

Figure 6-35
Projection-weld symbol.

is either the width of the weld or the shear strength per lineal inch. Certain tanks can be produced by forming sheet metal and then seam-welding. However, a tank produced by seam welding can be used only for low-pressure applications, and requires some additional soldering to make it watertight.

One of the disadvantages of spot welding is the fact that welds are difficult to place where desired. Locating dimensions for the placement of spot welds are often given; however, since the welding tool is awkward to handle, the welder will frequently mislocate some of the welds unless great care is exercised. *Projection welds* are *always* located where desired. This happens because one sheet of the work is prepared in advance by placing dimples at the proper locations. When the two sheets are welded, current passes through the dimples. While this method does ensure proper location, it is more expensive because additional work has to be done. Figure 6-35 shows the welding symbol for projection welds. Figures to the left, above, and to the right of the symbol represent the same items as with the spot weld. Projection welds are desirable when several welds are needed in a limited space. This is particularly true where welded nuts or welded bolts must be used in an assembly.

Arc Welding

Arc welding is a popular type of shop process. Great savings in cost and weight can be made by using common structural shapes that can be cut to size and welded. Machine frames were formerly produced by casting. Such a process involves the use of patterns, molds, and a few finishing processes. Today, most frames are produced by cutting such shapes as channels, angles, rods, and bars to the proper length, then clamping the parts together and welding them. Except for a little grinding in certain places, the welded assembly is easy to prepare for protective and decorative painting. Welded parts must be carefully designed. The designer must analyze forces and stresses carefully before specifying the type and location of welds. Before discussing the various types of arc welds, we will consider the weld symbol and identify each figure or element placed on the basic symbol. Figure 6-36

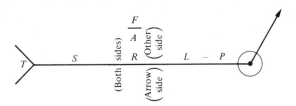

Figure 6-36
Weld symbol and element location.

shows the weld symbol and element location. Figure 6-37 shows the arc- and gas-weld symbols; Fig. 6-38 shows supplementary symbols.

First, one end of the horizontal line contains a pointing arrow, while the other end has a tail. These can be interchanged to facilitate drawing; also, the arrow leader can point up or down as necessary. The letter T represents a specification, process, or other reference. If none is specified, the tail can be omitted. The letter S indicates the size of the weld. Immediately to the right of the size is placed the basic weld symbol. For arc or gas welds, the basic symbol can be placed above or below the reference line, or both. If above, the weld is placed on the side opposite to where the arrow points. If the weld symbol is placed below the horizontal reference line, the arrow points directly to the side where weld metal is to be deposited. If the symbol is placed both above and below the reference line, welding is to be on both sides of the part.

To the right of the weld symbol, L represents the length of the weld and P represents the pitch of the welds. The pitch is the center-to-center distance from weld to weld (this is used where welding is intermittent). In many cases, continuous welding is used. If the part is to be welded on both sides, but staggered, the weld symbols are staggered above and below the horizontal line. Two symbols can be used at the junction of the leader and reference line. A circle around this location indicates that the weld is to extend

Figure 6-37
Arc- and gas-weld symbols.

Type of weld							
Bead	Fillet	Plug or slot	Groove				
			Square	V	Bevel	U	J
⌒	△	⏢	‖	∨	↙	Y	⋃

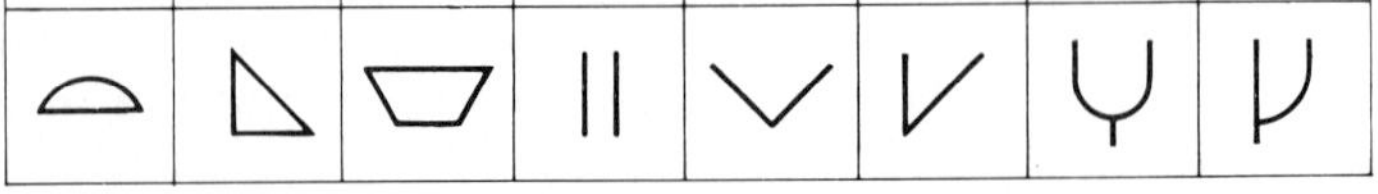

Weld all around	Field weld	Contour	
		Flush	Convex
◯	•	——	◠

Figure 6-38
Supplementary symbols.

completely around the part. A heavy dot implies that the weld is to be made in the field. Sometimes it is difficult to differentiate between shop and field welding. As a general rule, anything done outside one's own plant is considered field welding. Large equipment often has to be fabricated in parts (because of the limited size of railroad freight cars and trailers) and then completed at the site.

In Fig. 6-36, R indicates the amount of root opening, or (for slot and plug welds) the depth of filling. The groove angle is represented by the letter A. The letter F signifies the method of finishing, but not the degree of finishing. The horizontal line between F and A indicates that the contour must be flush with the other metal. The convex contour symbol can be used where flatness is not essential. Many welding standards are available that give complete symbol usage. In this text, only the most common cases are considered.

Probably the most popular type of arc weld is the *fillet weld*. This type is easy to apply and often easy to design. The fillet weld can be used for lap joints and for tee joints. Figure 6-39 shows a welded lap joint using a fillet weld. Parts A and B are subjected to a horizontal force P tending to pull them apart. Fillet welds are applied along edges C and D. An equivalent length of weld along sides E and F could be used; however, welds have greater strength when subjected to transverse shear than when subjected to longitudinal shear. The top view shows an appropriate weld symbol; the weldment is not shown when the symbol is used. This symbol indicates a fillet weld on both sides; the legs of the weld are $\frac{1}{4}$ in. As the fillet symbol is shown on both sides of the horizontal reference line, both sides are to be welded. If both sides have the same size of weld, the size need be shown only once. If sides E and F, as well as C and D, are welded, greater strength will result. This may be important if force P deviates from its horizontal line of action. In any case, it is good practice when welding edges C and D to loop the weld slightly around onto edges E and F. This will provide a slight resistance to turning if force P should deviate in direction. The lapped joint can also be used for plug, slot, grooved, and resistance welds.

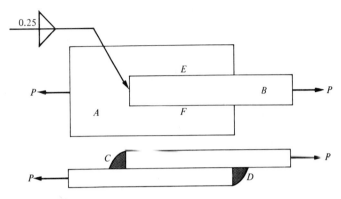

Figure 6-39
Welded lap joint.

Figure 6-40 shows two views of a tee weld. The fillet weld is shown in the front view; it is not necessary to include a front view if the appropriate symbol is given. The symbol shown in the top view indicates that $\frac{1}{4}$-in. fillet welds are used on both sides, that the welds are staggered, and that they are 2 in. long with 2 in. of spacing between welds. In other words, the pitch of the welds is 4 in.

In making weld calculations, the decision between intermittent and continuous welds should be studied with respect to the amount of weld metal deposited. In Fig. 6-40, continuous welding with $\frac{1}{8}$-in. welds would use one-half the weld metal required for intermittent welds made with $\frac{1}{4}$-in. legs. A study of the cross-sectional areas will reveal this difference. The area of a triangle with $\frac{1}{4}$-in. legs is $\frac{1}{32}$ sq in.; the area of a triangle with $\frac{1}{8}$-in. legs is $\frac{1}{128}$ sq in. In addition, it is easier to run a continuous bead.

Butt joints are appropriate for square, V-, bevel, U-, and J-welds. This type of joint is the best for transferring stress from one member to another. When mild steels are being welded, the allowable stress of the weld is the same as that of the two plates that are being welded. All groove welds except the square groove require the preparation of at least one of the two plates being welded. Such extra work adds to the cost of the welding process. It is necessary to specify the root opening (or separation) between the two plates. Figure 6-41 shows a double-V weld with a $\frac{1}{8}$-in. root opening. The symbol for such a weld is shown in the simplified drawing to the right.

Fillet weld sizes are specified by the length of the leg of the fillet. In most cases, the two legs are made equal in length. In making calculations, the throat of the weld is the important consideration. The throat of a weld is

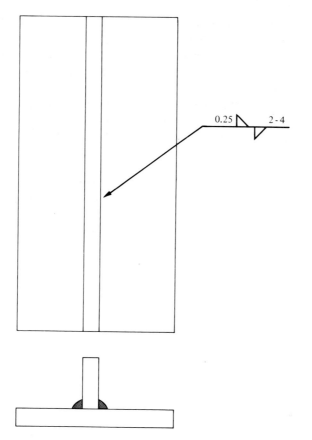

Figure 6-40
Welded tee joint.

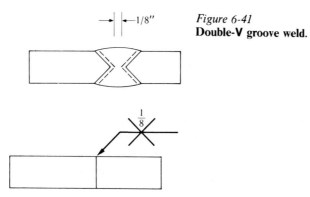

Figure 6-41
Double-V groove weld.

equal to 0.707 times the length of a leg, assuming the weld is made with equal legs (the sine of 45 degrees is 0.707). It is more accurate to use 0.707 times the leg length for longitudinal shear and 0.765 times the leg length for transverse shear. The reasoning behind this is that the throat for transverse shear acts at 67.5 degrees, which is 22.5 degrees away from the 45-degree throat in longitudinal shear. The constant 0.765 is found by dividing 0.707 by the cosine of 22.5 degrees. The following example shows that welds in transverse shear have greater strength than those in longitudinal shear.

◆ *Example* ————————————————————————————

In Fig. 6-39, assume that the allowable shearing stress for the welds is 13,600 psi. Find the load P that can be sustained if welds totaling 4 in. in length are placed at C and D. How does this value compare with the value of P that would be obtained if welds totaling 4 in. were placed at E and F? What is the percentage increase in load-carrying capacity that can be obtained by using welds in transverse shear as compared to longitudinal shear? (In both cases, $\frac{1}{4}$-in. fillet welds are employed.)

Solution. For transverse shear,

$$P = AS_s = 4(\tfrac{1}{4})(0.765)(13,600) = 10,400 \text{ lb.}$$

For longitudinal shear,

$$P = AS_s = 4(\tfrac{1}{4})(0.707)(13,600) = 9620 \text{ lb.}$$

Thus, the percentage increase in load-carrying capacity is

$$\frac{(10,400 - 9620)(100)}{9620} = 8.11\%.$$

Note. To simplify calculations and increase the factor of safety, many designers use the 0.707 constant in all cases.

Allowable Stresses; Load Per Inch

In making calculations, allowable stresses can be used for a particular welding rod; such design stresses are expressed in pounds per square inch. These values are available in manufacturers' catalogs and in welding handbooks. (See Appendix C for reference sources.) For simple loading on welds, it is often simpler to use data in which the allowable loads are expressed in pounds per lineal inch. Table 6-9 lists the allowable loads for fillet welds in shear; these values are given for buildings and bridges. Many mechanical-structural applications are in the category of buildings. Allow-

Table 6-9
Safe Allowable Loads for Fillet Welds (in Shear)

Size of Fillet Weld, in.	Buildings lb / lineal in.	Bridges lb / lineal in.
$\frac{1}{8}$	1200	1100
$\frac{3}{16}$	1800	1650
$\frac{1}{4}$	2400	2200
$\frac{5}{16}$	3000	2750
$\frac{3}{8}$	3600	3300
$\frac{1}{2}$	4800	4400
$\frac{5}{8}$	6000	5500
$\frac{3}{4}$	7200	6600
$\frac{7}{8}$	8400	7700
1	9600	8800

For dynamic or vibrational loads, it may be desirable to reduce unit stress, depending on the severity of the load. (Courtesy of The James F. Lincoln Arc Welding Foundation.)

able loads are based on more or less static conditions. It is desirable to reduce unit stress if there is danger of vibration. Judgment must be exercised in applying such information to certain machine elements. Appropriate codes should always be consulted. Also, much information can be obtained from the American Welding Society.

In many mechanical-structural applications, welding may replace riveting. Table 6-10 lists fillet-weld lengths that could replace rivets. This listing is based on the assumption that an allowable shearing stress of 15,000 psi was used to design the rivets (this 15,000-psi value is commonly specified in codes). Note that an additional $\frac{1}{4}$ in. for starting and stopping the arc should be added to the calculated weld length.

A calculation using values from Table 6-9 appears in the following section.

Welding Structural Angles

When applying longitudinal loads to structural angles, it is desirable to apply the load on a line through the centroid of the angle. In riveted design, the gage line for the rivets passes through the centroid. In welding an angle to a plate, welds are usually placed along the edges of the angle. The length of weld is proportioned between the two sides so that the net effect is to

Table 6-10
Length of Fillet Weld Used to Replace Rivets

Rivet Diameter, in.	Rivet Shear Value at 15,000 lb / sq in.	Length of Fillet Welds (to Nearest $\frac{1}{8}$ in.) "Fusion Code" (Structural), Shielded Arc Welding				
		$\frac{1}{4}$-in. fillet	$\frac{5}{16}$-in. fillet	$\frac{3}{8}$-in. fillet	$\frac{1}{2}$-in. fillet	$\frac{5}{8}$-in. fillet
$\frac{1}{2}$	2950	$1\frac{1}{2}$	$1\frac{1}{4}$	$1\frac{1}{8}$	$\frac{7}{8}$	$\frac{3}{4}$
$\frac{5}{8}$	4600	$2\frac{1}{4}$	$1\frac{3}{4}$	$1\frac{1}{2}$	$1\frac{1}{4}$	1
$\frac{3}{4}$	6630	3	$2\frac{1}{2}$	$2\frac{1}{8}$	$1\frac{5}{8}$	$1\frac{3}{8}$
$\frac{7}{8}$	9020	$4\frac{1}{8}$	$3\frac{3}{8}$	$2\frac{7}{8}$	$2\frac{1}{8}$	$1\frac{3}{4}$
1	11780	$5\frac{1}{4}$	$4\frac{1}{4}$	$3\frac{5}{8}$	$2\frac{3}{4}$	$2\frac{1}{4}$

Note: $\frac{1}{4}$ in. is added to the calculated length of bead for starting and stopping the arc.
(Courtesy of The James F. Lincoln Arc Welding Foundation.)

apply the load through the centroid. The following problem shows how this is accomplished.

◆ *Example*
A 4 in. × 4 in. × $\frac{1}{4}$ in. angle is welded to a $\frac{1}{4}$-in. plate by means of $\frac{1}{4}$-in. fillet welds. Load *P* is applied through the centroid of the angle; the value of *P* is 24,000 lb and the allowable shearing strength of the weld is 2400 lb per lineal inch. From a structural table, distances *a* and *b* are found to be 2.91

Figure 6-42
Angle welded to plate.

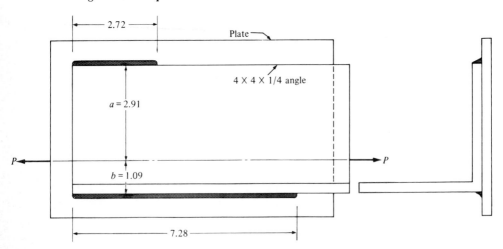

in. and 1.09 in., respectively (refer to Fig. 6-42). Find the length of weld required and how it is to be apportioned between the two sides of the angle.

Solution. The total length of weld required is $24{,}000/2400 = 10$ in. Since this is being distributed along edges 1 and 2, two equations of equilibrium can be written. Thus,

$$L_1 + L_2 = 10 \quad \text{or} \quad L_1 = 10 - L_2$$

and

$$2.91L_1 = 1.09L_2.$$

If we substitute L_1 in terms of L_2 from the first equation, the second equation becomes

$$2.91(10 - L_2) = 1.09L_2;$$
$$29.1 - 2.91L_2 = 1.09L_2;$$
$$4L_2 = 29.1;$$
$$L_2 = 7.28 \text{ in.};$$

and

$$L_1 = 10 - L_2 = 10 - 7.28 = 2.72 \text{ in.}$$

Weld Sizes

Table 6-11 and Figs. 6-43 and 6-44 are taken from *Design of Weldments* (copyright 1963), by permission of the publisher, The James F. Lincoln Arc Welding Foundation. All of these pertain to arc-welding design only. Table 6-11 gives some rule-of-thumb recommendations for selecting the leg size of welds based on various plate thicknesses. Figure 6-43 gives some typical weld situations and corresponding equations.

If the weld is treated as a line and is considered as having no area, the formulas given in Fig. 6-44 are easy to use. It should be noted that the standard formula for vertical shear is

$$S_s = \frac{V}{A}$$

in psi. If the same type of formula is employed, but the weld is treated as a line, the equation becomes

$$F = \frac{V}{L} \tag{6-5}$$

in pounds per inch. A similar situation exists for the standard bending formula:

$$S = \frac{M}{I/c} = \frac{M}{Z}$$

in psi. If this formula is used, but the weld is treated as a line, the equation becomes

$$F = \frac{M}{Z} \tag{6-6}$$

Table 6-11
Rule-of-Thumb Fillet Weld Sizes

Plate Thickness (t)	Strength Design, Full-strength Weld $(w = \frac{3}{4}t)$	Rigidity Design	
		50% of full-strength weld $(w = \frac{3}{8}t)$	33% of full-strength weld $(w = \frac{1}{4}t)$
$\frac{1}{4}$	$\frac{3}{16}$	$\frac{3}{16}$ *	$\frac{3}{16}$ *
$\frac{5}{16}$	$\frac{1}{4}$	$\frac{3}{16}$ *	$\frac{3}{16}$ *
$\frac{3}{8}$	$\frac{5}{16}$	$\frac{3}{16}$ *	$\frac{3}{16}$ *
$\frac{7}{16}$	$\frac{3}{8}$	$\frac{3}{16}$	$\frac{3}{16}$ *
$\frac{1}{2}$	$\frac{3}{8}$	$\frac{3}{16}$	$\frac{3}{16}$ *
$\frac{9}{16}$	$\frac{7}{16}$	$\frac{1}{4}$	$\frac{1}{4}$ *
$\frac{5}{8}$	$\frac{1}{2}$	$\frac{1}{4}$	$\frac{1}{4}$ *
$\frac{3}{4}$	$\frac{9}{16}$	$\frac{5}{16}$	$\frac{1}{4}$ *
$\frac{7}{8}$	$\frac{5}{8}$	$\frac{3}{8}$	$\frac{5}{16}$ *
1	$\frac{3}{4}$	$\frac{3}{8}$	$\frac{5}{16}$ *
$1\frac{1}{8}$	$\frac{7}{8}$	$\frac{7}{16}$	$\frac{5}{16}$
$1\frac{1}{4}$	1	$\frac{1}{2}$	$\frac{5}{16}$
$1\frac{3}{8}$	1	$\frac{1}{2}$	$\frac{3}{8}$
$1\frac{1}{2}$	$1\frac{1}{8}$	$\frac{9}{16}$	$\frac{3}{8}$
$1\frac{5}{8}$	$1\frac{1}{4}$	$\frac{5}{8}$	$\frac{7}{16}$
$1\frac{3}{4}$	$1\frac{3}{8}$	$\frac{3}{4}$	$\frac{7}{16}$
2	$1\frac{1}{2}$	$\frac{3}{4}$	$\frac{1}{2}$
$2\frac{1}{8}$	$1\frac{5}{8}$	$\frac{7}{8}$	$\frac{9}{16}$
$2\frac{1}{4}$	$1\frac{3}{4}$	$\frac{7}{8}$	$\frac{9}{16}$
$2\frac{3}{8}$	$1\frac{3}{4}$	1	$\frac{5}{8}$
$2\frac{1}{2}$	$1\frac{7}{8}$	1	$\frac{5}{8}$
$2\frac{5}{8}$	2	1	$\frac{3}{4}$
$2\frac{3}{4}$	2	1	$\frac{3}{4}$
3	$2\frac{1}{4}$	$1\frac{1}{8}$	$\frac{3}{4}$

*These values have been adjusted to comply with AWS recommended minimums.
(Courtesy of the James F. Lincoln Arc Welding Foundation.)

Type of loading		Standard design formula Stress, lb/in²	Treating the weld as a line Force, lb/in.
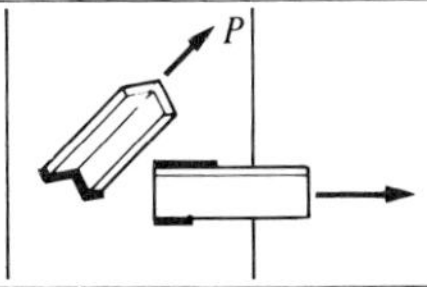	Tension or compression	$S = \dfrac{P}{A}$	$F = \dfrac{P}{L}$
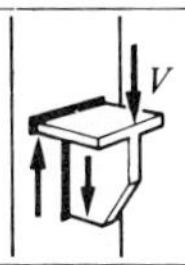	Vertical shear	$S_s = \dfrac{V}{A}$	$F = \dfrac{V}{L}$
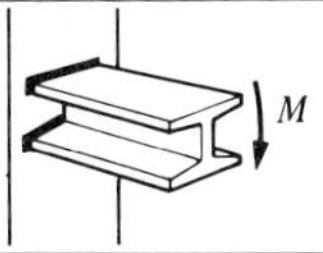	Bending	$S = \dfrac{M}{Z}$	$F = \dfrac{M}{Z}$
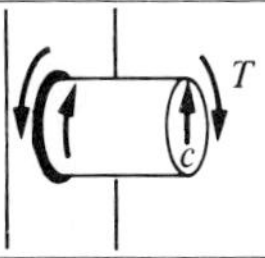	Twisting	$S_s = \dfrac{Tc}{J}$	$F = \dfrac{Tc}{J}$

Figure 6-43
Properties of a weld treated as a line.
(Courtesy of the James F. Lincoln Arc Welding Foundation.)

in pounds per inch. Similar situations exist for tension, compression, and twisting.

Figure 6-45 shows the centroidal distance for some of the weld configurations that are treated as a line. The location of the center of gravity of the weld is a necessary factor in weld calculations.

In treating a weld section as a line, it must be remembered that the width of the line or weld is considered to be zero; thus, "area" is expressed in inches and "moment of inertia" is expressed in in³ instead of in⁴. A dimensional analysis is helpful in making this type of calculation.

In a fillet weld, the allowable load per inch is as follows:

$$F_{\text{allow.}} = (0.707w)(13,600) = 9600w.$$

This is based on an allowable shear stress of 13,600 psi. The above value of F is useful in working any problems where the weld is treated as a line.

The following example shows how a weld can be calculated by treating it as a line.

Figure 6-44

Properties of a weld treated as a line.

(Courtesy of the James F. Lincoln Arc Welding Foundation.)

Outline of welded joint	Bending (about axis $x-x$)	Twisting
$x ---\|--- x \quad d$	$Z = \dfrac{d^2}{6}$	$J = \dfrac{d^3}{12}$
$x -\|---\|- x \quad d$	$Z = \dfrac{d^2}{3}$	$J = \dfrac{d(3b^2 + d^2)}{6}$
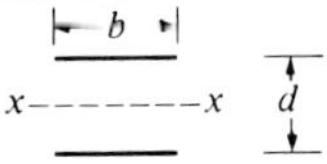 $x ------- x \quad d$	$Z = bd$	$J = \dfrac{b^3 + 3bd^2}{6}$
$x --- x \quad d$	$Z = \dfrac{4bd + d^2}{6} = \dfrac{d^2(4b + d)}{6(2b + d)}$ top · bottom	$J = \dfrac{(b + d)^4 - 6b^2 d^2}{12(b + d)}$
$x ----- x \quad d$	$Z = bd + \dfrac{d^2}{6}$	$J = \dfrac{(2b + d)^3}{12} - \dfrac{b^2(b + d)^2}{2b + d}$
$x -\|-\|- x \quad d$	$Z = \dfrac{2bd + d^2}{3} = \dfrac{d^2(2b + d)}{3(b + d)}$ top · bottom	$J = \dfrac{(b + 2d)^3}{12} - \dfrac{d^2(b + d)^2}{b + 2d}$
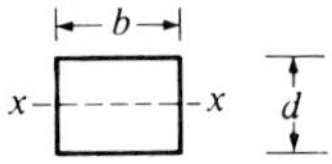 $x ----- x \quad d$	$Z = bd + \dfrac{d^2}{3}$	$J = \dfrac{(b + d)^3}{6}$
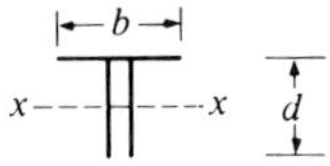 $x --- x \quad d$	$Z = \dfrac{2bd + d^2}{3} = \dfrac{d^2(2b + d)}{3(b + d)}$ top · bottom	$J = \dfrac{(b + 2d)^3}{12} - \dfrac{d^2(b + d)^2}{b + 2d}$
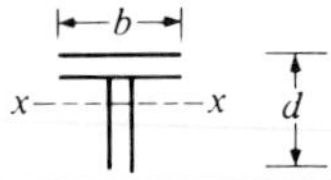 $x --- x \quad d$	$Z = \dfrac{4bd + d^2}{3} = \dfrac{4hd^2 + d^3}{6b + 3d}$ top · bottom	$J = \dfrac{d^3(4b + d)}{6(b + d)} + \dfrac{b^3}{6}$
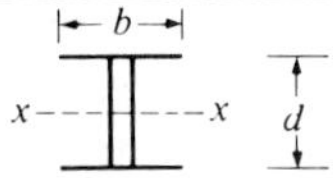 $x --- x \quad d$	$Z = bd + \dfrac{d^2}{3}$	$J = \dfrac{b^3 + 3bd^2 + d^3}{6}$
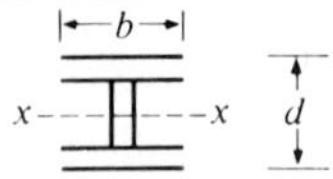 $x --- x \quad d$	$Z = 2bd + \dfrac{d^2}{3}$	$J = \dfrac{2b^3 + 6bd^2 + d^3}{6}$
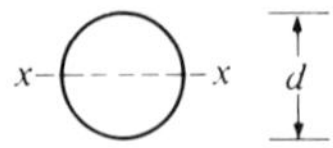 $x ----- x \quad d$	$Z = \dfrac{\pi d^2}{4}$	$J = \dfrac{\pi d^3}{4}$
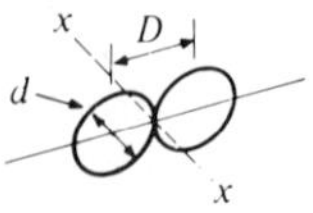 $x \quad D \quad d$	$Z = \dfrac{\pi d^2}{2} + \pi D^2$	

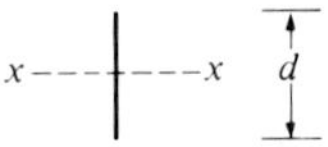

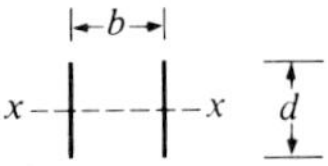

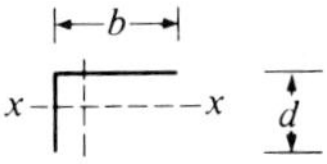

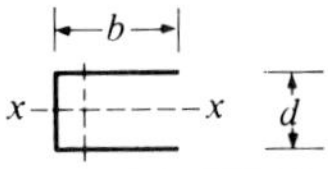

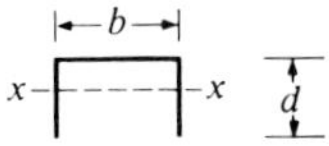

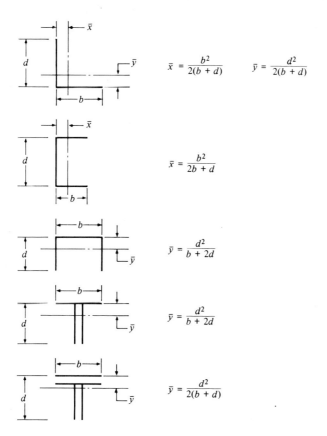

Figure 6-45
Centroids for welds treated as lines.

◆ *Example* ─────────────────────────────

Find the fillet-weld size for the welded assembly shown in Fig. 6-46; consider the weld as a line.

Solution.

$$\bar{x} = \frac{b^2}{2b + d} = \frac{(4)^2}{2(4) + 10} = 0.89 \text{ in.;}$$

$$J = \frac{(2b + d)^3}{12} - \frac{b^2(b + d)^2}{2b + d} = \frac{(8 + 10)^3}{12} - \frac{16(14)^2}{18} = 312 \text{ in}^3.$$

Area of weld treated as line is 18 in. Maximum combined forces are at

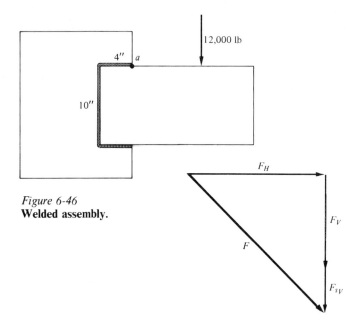

Figure 6-46
Welded assembly.

point *a*. Twisting (horizontal component) is

$$F_H = \frac{Tc}{J} = \frac{(12)(12,000)(5)}{312} = 2310 \text{ lb/in.}$$

Twisting (vertical component) is

$$F_V = \frac{Tc}{J} = \frac{(12)(12,000)(3.11)}{312} = 1440 \text{ lb/in.}$$

Vertical shear is

$$F_{s_V} = \frac{P}{A} = \frac{12,000}{18} = 667 \text{ lb/in.}$$

Resultant force on weld is

$$F = \sqrt{(2310)^2 + (1440 + 667)^2} = 3130 \text{ lb/in.}$$

The required leg size of fillet weld is thus

$$w = \frac{\text{actual force}}{\text{allowable force}} = \frac{3130}{9600} = 0.326 \text{ in.}$$

Therefore, a $\frac{3}{8}$-in. weld is the desired size.

 **Summary**

Each type of fastener serves a specific purpose. Some fasteners, such as rivets, form permanent connections; in such cases, the fastener is destroyed in the process of disassembly. Other fasteners are desirable because they can easily be removed; this is particularly important if maintenance is necessary.

Fastener costs must be considered from three standpoints:

1. initial cost,
2. assembly costs,
3. disassembly costs.

The second item is particularly important if mass production methods are employed. The third item is important from two aspects: ease of disassembly, and reusability of the fastener.

The appearance of fasteners is important from the industrial-design viewpoint. Finishes can add to or detract from the appearance of the end product.

Loads on fasteners must be calculated to ensure proper functioning. Screw fasteners can be subjected to tensile loads, to shear across the shank, or to shear across threaded section. Components should be designed so that undue loads are not placed on the securing parts. Careful consideration of the forces on brackets and similar parts can prevent trouble.

Assembly difficulties should be explored early in the design. Blind fasteners are needed where one does not have access to both sides of an assembly. Spacing of fasteners is also important.

Setscrews are used to prevent axial movement or rotation of one part relative to another. A setscrew has a limited holding capacity; this capacity depends on the type of point selected and the tightening torque. *Safety* setscrews are required in most applications; these have no protruding head. Keys are more effective than setscrews in preventing relative rotation of two parts, because the shearing area is much greater for a key than for a setscrew. Keys are often used in conjunction with setscrews. The key carries the torsional load, and the setscrew prevents axial movement. Other considerations besides shearing strength should go into the selection of a key. Such items as ease of removal and the possibility of slipping loose should be considered. Splines are more effective than keys in transmitting higher torque values.

Pins serve several purposes. A pin placed in a hole completely through a hub and shaft assembly can serve the same purpose as a key. Such connections have two shearing areas; the torsional capacity of the assembly is easy to compute. Dowel pins are often used to hold parts in position for machining operations; in addition, they are often used to carry shear loads in torsional applications, thus preventing threaded fasteners from receiving these loads. Shear pins are mechanical safety devices.

Retaining rings are members of the spring family that are used for fastening purposes. They are easy to assemble and disassemble; they can restrain large axial loads. Collars with setscrews are used for the same type of application; however, axial restraint is limited to the holding power of the setscrews.

Rigid couplings are simple in construction and low in cost; they are used to connect two shafts that are perfectly aligned. A universal joint is used to connect two shafts that are slightly misaligned angularly. Two universal joints used together with a connecting link can correct for parallel misalignment. Oldham-type couplings are also useful for correcting parallel misalignment of shafts.

Flexible connectors are also used to join two aligned shafts. Some types provide for a small amount of:

1. parallel misalignment,
2. angular misalignment,
3. end play.

All flexible connectors dampen vibration and reduce noise level. In most mechanical devices, this feature is of the utmost importance.

Structural adhesives play a prominent role in design. Great structural strength can be developed through their use, and component preparation costs can be substantially reduced.

Most of the components covered in this chapter are stock or "off-the-shelf" items. This simplifies design and tends to reduce manufacturing costs. A designer must rely heavily on manufacturers' literature concerning ratings, empirical formulas, and application factors.

By using as few sizes of fasteners as possible in a complex assembly, fewer stockroom items are necessary; this reduces inventory parts and overall costs.

Welding is a joining process that produces permanent connections. Welded construction results from joining relatively inexpensive structural shapes by the proper welding techniques. Machine frames and other structural systems can be produced at lower overall costs; these assemblies are lighter in weight than those made by casting processes.

Questions for Review

1. What advantages do studs have over cap screws? What disadvantages?
2. How does a machine screw differ from a through bolt?
3. What is the purpose of a washer face on a hexagon nut and on a bolt?

4. How does spot facing differ from counterboring? What is the purpose of each?

5. What is a blind fastener? Give two examples of assemblies where blind fastening is necessary.

6. What is the pitch of a $\frac{1}{4}$-in.–20 UNC-2A screw thread? How does it differ from a $\frac{1}{2}$-in.–20 UNF-2A thread?

7. Why are UNF threads frequently used in automotive applications in preference to UNC threads?

8. Differentiate between the three classes of screw threads.

9. Why does a turnbuckle have left-hand threads on one end and right-hand threads on the other?

10. What is a safety setscrew? What advantages and disadvantages does it have over other types?

11. Why are thicker collars needed when cone-pointed setscrews are used in place of flat-pointed setscrews?

12. In setscrew applications, what is meant by the term "seating torque"?

13. How are retaining rings assembled and removed?

14. What advantages do retaining rings have over collars secured with setscrews?

15. What advantages do structural adhesives have over rivets as means of assembly? What disadvantages?

16. Explain why a regular universal joint does not transmit constant velocity.

17. What undesirable features does an Oldham-type coupling possess?

18. What advantages do flexible couplings have over rigid couplings?

19. What advantages do Woodruff keys have over square or rectangular keys? What disadvantages?

20. What advantages do splines have over keys? What disadvantages?

21. How does a Pratt and Whitney key differ from a regular square or flat rectangular key? What advantages does it have?

22. Explain the differences between the functions of a dowel pin and a shear pin.

23. List three applications of shear pins. Explain why a shear pin is used in each case.

24. Differentiate between the shearing strength and the peeling strength of a structural adhesive. If a structural adhesive were used to secure the plate shown in Fig. 6-47, would the peel strength or the shear strength govern the design?

25. What is a torque wrench? In what applications would its use be essential? Why?

26. What are self-tapping screws? What limitations do they have? List some typical assemblies where self-tapping screws are desirable, and explain why.

27. Does the construction of self-tapping screws differ when intended for use in various materials? Why?

28. What is the primary purpose of a plain washer?

29. What outstanding advantage does projection welding have over spot welding? What is its biggest disadvantage?

30. A $\frac{1}{2}$-in. steel plate is to be placed centrally and welded to a large $\frac{1}{2}$-in. steel plate with $\frac{3}{8}$-in. fillet welds. Welding is to be intermittent; each weld is to be 3 in. in length; 8 in. of space is to be left between the welds. Draw a weld symbol that includes all this information.

31. Draw a weld symbol for 6 spot welds placed on 2 in. centers. The size of each weld is to be $\frac{1}{4}$ in.

32. Are threaded fasteners in the Unified National Coarse series interchangeable with coarse metric-threaded fasteners?

33. A metric thread is designated as M5 × 1.25. What is its nominal size? What is its pitch?

34. What metric thread is close in size to a $\frac{3}{8}''$-UNC-16 thread?

Problems

1. An application similar to that shown in Fig. 6-1 uses $\frac{3}{4}$-in.-diameter rivets. The applied load is 1000 lb and the distance L is 18 in.; $d_1 = d_2 = d_3 = d_4 = 1.41$ in. Find the maximum induced stress. Which rivets are subjected to the greatest stress?

2. In Fig. 6-2, distance a is 3 in. and distance b is 7 in. A force of 600 lb acts at a distance of 18 in. from the wall. Find the direct shear on the bolts and the maximum tensile pull on the upper bolts. Find the minimum nut thickness needed if $\frac{3}{8}$-in. bolts (UNC) are used and the allowable shearing stress is 6000 psi.

3. A 2.5-in.-diameter shaft uses a straight spline with six splines and a contact length of 1.25 in. Assuming it is designed to slide while under load, what is its power-transmission capacity at 100 rpm, in horsepower?

4. A $\frac{1}{4}$-in. cone-pointed setscrew is spotted one-half the length of the cone. The cone angle is 118 degrees, and the allowable shearing stress of the

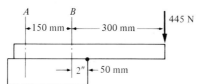

Figure 6-47
Problem 9.

setscrew material is 10,000 psi. Find the holding power of the setscrew when it is securing a pulley to a $\frac{1}{2}$-in. shaft.

5. A rectangular collar is secured to a vertical post by means of two dog-pointed setscrews. If the allowable shearing stress for the setscrew material is 20,000 psi and the collar supports a load of 1250 lb, what size of setscrew should be used?

6. A pulley is secured to a $\frac{1}{2}$-in. steel shaft by means of a cup-pointed setscrew. The shaft transmits $\frac{3}{4}$ hp and rotates at 1200 rpm. What size of setscrew is required for such an application?

7. Two $\frac{1}{4}$-in. cup-pointed setscrews are used to secure a collar to a $\frac{9}{16}$-in.-diameter shaft. What is the total axial holding power? How many horsepower could be safely transmitted if the assembly were to rotate at 1000 rpm? Assume a factor of safety of 4 in both cases.

8. A pulley with a 45-mm.-diameter hub is secured to a 25-mm-diameter shaft by a 6-mm pin placed diametrically through the assembly. If the allowable shearing stress for the pin is 40 MN/m^2 and the assembly rotates at 300 rpm, how much power can the assembly handle?

9. In Fig. 6-47, find the maximum bolt tension caused by eccentric loading. If the 150 mm dimension were changed to 250 mm, what effect would this have on the value? Assume that the plate is rigid.

10. In Fig. 6-48, what is the maximum bolt tension caused by eccentric loading? How can it be reduced? Are bolts A and B subjected to any shearing action? Assume that the bracket is rigid.

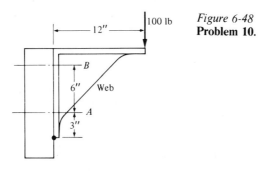

Figure 6-48
Problem 10.

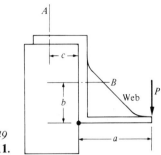

Figure 6-49
Problem 11.

11. Assume that the bracket in Fig. 6-49 is rigid. What advantages would such a bracket have with respect to the bolted fasteners? What disadvantages?

12. A turnbuckle is equipped with $\frac{1}{4}$-in.–20 screw threads. The thread engagement on each end is $\frac{3}{8}$ in. If the allowable stress is 8000 psi in tension and 6000 psi in shear, what is the maximum load that can safely be handled for the arrangement shown in Fig. 6-50?

13. In Fig. 6-51, $\frac{3}{4}$-in.-diameter structural rivets are used to make the connection. Using Table 6-10 and an allowable rivet shear stress of 15,000 psi, find how much load P can be sustained. Also, using Table 6-10, find the length of $\frac{3}{8}$-in. fillet weld that could replace the rivets. How would the weld metal be distributed along the two edges?

14. In Fig. 6-51, what total length of $\frac{1}{4}$-in. weld, properly positioned, could replace five $\frac{5}{8}$-in.-diameter structural rivets having an allowable stress (shear) of 15,000 psi? Use Table 6-10. What load P could be sustained?

15. A $4 \times 4 \times \frac{1}{2}$-in. structural angle is welded to a $\frac{1}{2}$-in. steel plate. The weld lengths are properly distributed along the two edges and the load

Figure 6-50
Problem 12.

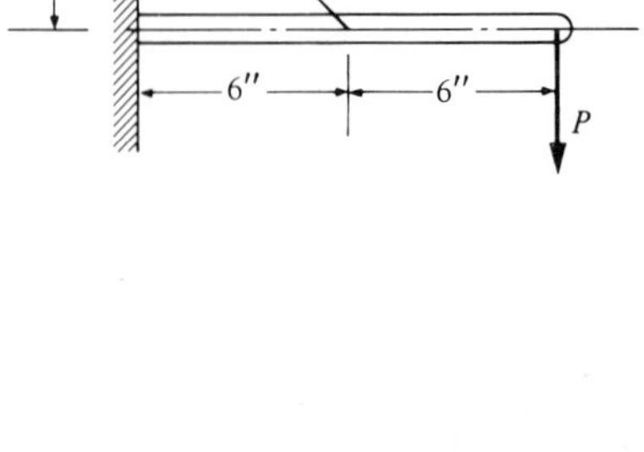

6"

6" 6"

Turnbuckle

P

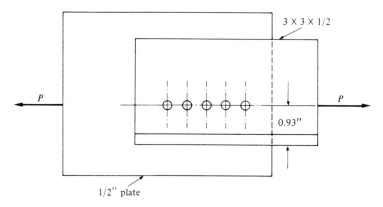

Figure 6-51
Problems 13 and 14.

acts through the centroid. How much load can be handled by this assembly if the total length of weld is 4.5 in., a $\frac{3}{8}$-in. fillet weld is used, and the weld is replacing $\frac{5}{8}$-in.-diameter structural rivets with an allowable shear stress of 15,000 psi? Use Table 6-10.

16. A 3-in. cylinder is fillet-welded to a flat vertical plate. If it is subjected to a load of 2000 lb acting 6 in. from the vertical plate and the allowable load is 9600 lb/in., what size of weld is needed? Treat the weld as a line and use the information from Figs. 6-44 and 6-45.

17. A cylinder similar to that in Problem 16 is subjected to a twisting moment of 40,000 in-lb. What size of weld is needed if the allowable load is 9600 lb/in? Treat the weld as a line.

18. A pulley is keyed to a 24-mm steel shaft with a metric M6 × 20 square key. If the hub length is 20 mm and the allowable shearing stress for the key material is 100 MN/m², how much torque can be transmitted? Express the answer in kN · m units.

19. A metric bolt has a property class identification of 6.8 embossed on the head. Find the minimum tensile and yield stresses in kgf/mm². Also, express the answers in MPa units.

20. Two 32-mm shafts are connected by drive pins in an arrangement similar to that shown in Fig. 6-27. Find the required diameter for these pins if the allowable shearing stress for the pins is 100 MN/m² and the shafts transmit 300 N · m of torque.

21. A heavy-duty shaft coupling similar to that shown in Fig. 6-25 is to be secured with 25-mm bolts at a distance of 150 mm from the shaft center. The shaft transmits 4330 kW of power at a speed of 1200 rpm. If the allowable shearing stress for the bolts is 100 MN/m², how many bolts are required?

22. In an application similar to that shown in Fig. 6-27 a 4-mm pin secures a sleeve to a 30-mm shaft that rotates at 400 rpm and transmits 0.8 kW of power. How much shearing stress is induced in the pin?

23. A metric M4 × 15 square key is used with a 16-mm shaft. If the allowable shearing stress is 50 MN/m², how much torque can the assembly handle? The rotational speed of the shaft is 600 rpm.

24. A gusset plate is riveted to two structural angles. An applied load of 0.44 MN acts along the centroidal axis; the diameter of each rivet is 19 mm. If the allowable shear stress for the rivets is 100 MN/m², how many rivets are required?

25. In Fig. 6-52, find the force acting on rivet #2 of the eccentrically loaded plate.

Figure 6-52
Problem 25.

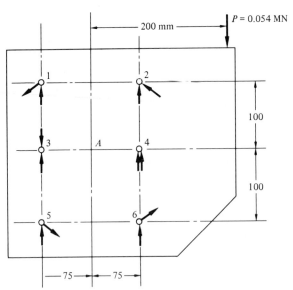

BELTING

7

Introduction

Over the years, a popular method of transmitting power from one shaft to another has been by pulleys and belts. Before the advent of individual electric-motor drives for each machine, central lineshafts with an elaborate array of pulleys for operating each machine were used. In those days, practically all belts were made of leather. Today, belting is still popular; however, various types of materials and pulleys are used. Typical types of drives are open-belt, crossed-belt, serpentine, and quarter-turn. Figure 7-1 illustrates some of these types.

When compared with geared drives and chain drives, belting has certain advantages and disadvantages. A belt drive will tend to slip when an overload is applied to the system. This slipping, while not desirable, is preferable to a complete failure, which could happen if gear teeth were stripped or a roller chain were to break. Unless the overload persisted for an extended period of time, the system would return to normal with only a small amount of wear; failure would not necessarily occur. Belts can absorb quite a large amount of sudden shock. Except for a positive-drive belt, no

Figure 7-1
Belt drives.

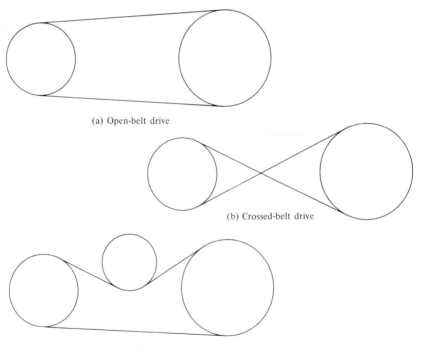

(a) Open-belt drive

(b) Crossed-belt drive

(c) Serpentine drive

belt will deliver the theoretical rotational speed that can be calculated. Theoretically, the belt speed should equal the peripheral speed of the pulleys. The following equation indicates the relationship between rotational speed and pulley diameters:

$$\frac{D_1}{D_2} = \frac{N_2}{N_1},$$
(7-1)

where

N = rotational speed (rpm),

D = pulley diameters (in., ft, or mm).

This equation makes no provision for slip or creep. *Slip* is defined as an actual movement (or sliding) of the belt relative to the pulley. *Creep*, however, is a different phenomenon. When a driving pulley is connected by a belt to a driven pulley, the belt tensions on the two sides of the driver differ. The tension stretches the belt; this stretching is released after leaving the driver pulley. Therefore, the pulley is delivering a different amount of belt than it is receiving. This difference is known as belt creep. A designer is concerned only with the total loss of rotational speed, which is partly creep and partly slip. An assumption in the neighborhood of 4% is often reasonable. Since all belts stretch with use, any belt drive should allow for stretching. This can be done by providing an adjustment for one of the pulleys or a weighted idler pulley resting against the belt.

◆ *Example* ⎯⎯⎯⎯⎯⎯⎯⎯⎯⎯⎯⎯⎯⎯⎯⎯⎯⎯⎯⎯⎯⎯⎯⎯⎯

If a driving pulley is 6 in. in diameter and drives a 12-in.-diameter pulley, find the speed of the driven pulley if the driver rotates at 200 rpm. Calculate this for a condition of no slip, then recalculate assuming a total loss (slip and creep) of 4%.

Solution.

$$N_2 = \frac{D_1 N_1}{D_2} = \frac{(6)(200)}{12} = 100 \text{ rpm};$$

$$100 - (0.04)(100) = 96 \text{ rpm}.$$

Pulley-Belt Contact

In designing belt drives, it is frequently necessary to know the number of degrees through which the belt contacts a given pulley. The following equation gives this contact in degrees for either the small or large pulley (it is usually only necessary to know this figure for the smaller of the two

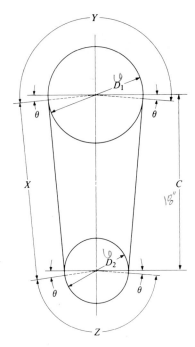

Figure 7-2
Open-belt length.

pulleys):

$$\text{contact (deg)} = 180° \pm 2\,\text{arc sin}\left[\frac{(D_1/2) - (D_2/2)}{C}\right], \qquad (7\text{-}2)$$

where

D_1 = diameter of large pulley (in.),

D_2 = diameter of small pulley (in.),

C = center distance between pulleys (in.).

Length of Belt

If it is necessary to calculate the length of an open belt, the following equations may be used. The first one is an approximation for use when the relative sizes of the two pulleys do not vary too much:

$$L = 2C + 1.57(D_1 + D_2) + \frac{(D_1 - D_2)^2}{4C}. \qquad (7\text{-}3)$$

Referring to Fig. 7-2, we obtain an exact relationship by summing three parts, as indicated in the sketch. Thus

$$L = 2X + Y + Z,$$

where

$$X = \sqrt{C^2 - [(D_1/2) - (D_2/2)]^2},$$

$$Y = \frac{D_1}{2}\left\{\pi + 2\arcsin\left[\frac{(D_1/2) - (D_2/2)}{C}\right]\right\},$$

$$Z = \frac{D_2}{2}\left\{\pi - 2\arcsin\left[\frac{(D_1/2) - (D_2/2)}{C}\right]\right\}.$$

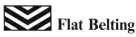 **Flat Belting**

Flat belts are used with pulleys that are either flanged at the edges or crowned in the middle. Flanges ensure that the belt stays on the pulley, but they also damage the edges of the belt. Crowning the pulley is considered better practice. The amount of crowning is greater for leather belting than for duck. In addition, crowning must be greater for low speeds than for high speeds. The amount of crowning varies from $\frac{1}{16}$ in. per foot of pulley face to $\frac{1}{4}$ in. per foot of pulley face.

Friction plays an important part in all belting design. The coefficient of friction between a belt and a pulley varies from approximately 0.3 to 0.5. Surface conditions can alter this value for any given materials (pulley and belt). The ratio of the pull on the tight side of a belt to the pull on the slack side is given by the following equation:

$$\frac{F_1}{F_2} = e^{f\theta}, \tag{7-4}$$

where

F_1 and F_2 are the belt pulls (lb),

f = coefficient of friction,

θ = belt angle of contact (rad),

e = 2.718.

At high speeds (for leather belting, speeds greater than 35 ft/sec), the preceding equation should be modified to include the centrifugal force. The centrifugal force can be found from the following equation:

$$F_c = \frac{12wbtv^2}{g} = \frac{wbtv^2}{2.68}, \tag{7-5}$$

where

w = specific weight of the belt (lb/cu in.)—0.035 for leather belting,

b = belt width (in.),

t = belt thickness (in.),

v = belt speed (ft/sec),

F_c = centrifugal force (lb),

g = acceleration of gravity (ft/sec^2)—approximately 32.2.

For metric units, the equation for the centrifugal force can be written as follows:

$$F_c = \frac{9.81(10)^{-6} wbtv^2}{g},$$

where

w = specific weight of belt material (kg/m^2)—974 for leather,

b = belt width (mm),

t = belt thickness (mm),

v = belt speed (m/s),

F_c = centrifugal force (N),

g = acceleration of gravity—9.8 m/s^2.

◆ *Example* _____

Find F_c for a leather belt with the following characteristics: width = 100 mm, thickness = 6.4 mm, v = 18.3 m/s.

Solution.

$$F_c = \frac{9.81(10)^{-6}(974)(100)(6.4)(18.3)^2}{9.8} = 209 \text{ N}.$$

Note: The specific weight was given in kilogram units; therefore the force must be corrected for newtons by multiplying by 9.81.

_____

The equation for the ratio of the belt pulls then becomes

$$\frac{F_1 - F_c}{F_2 - F_c} = e^{f\theta}.$$

If this equation is inverted and 1 is subtracted from each side, it becomes

$$\frac{F_2 - F_c}{F_1 - F_c} - 1 = \frac{1}{e^{f\theta}} - 1.$$

However, $F_1 = SA = Sbt$, where S = allowable belt strength (psi). Therefore,

$$F_1 - F_2 = (F_1 - F_c)\left(\frac{e^{f\theta} - 1}{e^{f\theta}}\right)$$

$$= \left(Sbt - \frac{wbtv^2}{2.68}\right)\left(\frac{e^{f\theta} - 1}{e^{f\theta}}\right).$$

Thus

$$F_1 - F_2 = bt\left(S - \frac{wv^2}{2.68}\right)\left(\frac{e^{f\theta} - 1}{e^{f\theta}}\right). \tag{7-6}$$

The net belt tension is $F_1 - F_2$. Expressed as an equation, this becomes

$$F_1 - F_2 = \frac{550(\text{hp})}{v}.$$

From this, a usable equation for belt power can be obtained:

$$\frac{550(\text{hp})}{v} = bt\left(S - \frac{wv^2}{2.68}\right)\left(\frac{e^{f\theta} - 1}{e^{f\theta}}\right). \tag{7-7}$$

◆ *Example* _____

A $\frac{1}{4}$-in. belt transmits 30 hp while running on a 2-ft pulley rotating at 600 rpm. The allowable belt stress is 500 psi; the angle of contact is 150 deg. Find the belt width required, assuming the coefficient of friction is 0.4 and the specific weight of the belt material is 0.035 lb/cu in.

Solution.

$$v = \frac{\pi DN}{60}$$

$$= \frac{\pi(2)(600)}{60} = 62.8 \text{ ft/sec};$$

$$e^{f\theta} = 2.718^{(0.4)(150\pi/180)} = 2.85;$$

$$b = \frac{550(\text{hp})}{vt\left(S - \frac{wv^2}{2.68}\right)\left(\frac{e^{f\theta} - 1}{e^{f\theta}}\right)}$$

$$= \frac{550(30)}{62.8(0.25)\left[500 - \frac{(0.035)(62.8)^2}{2.68}\right]\left(\frac{2.85 - 1}{2.85}\right)} = 3.61 \text{ in.}$$

Belt ratings. To simplify matters for the designer, belting manufacturers often rate their product in horsepower per inch of belt width. Two correction factors are applied to these ratings, one for the angle of contact, the other for the type of application. Severe-service applications would reduce the rated horsepower per inch of width, as would an angle of contact of less than 180 deg.

Splicing flat belts. Since flat belting is usually not made endless by the manufacturer, it must be spliced by the user. Several types of connections are available. A tapered belt lap, properly cemented, has approximately 100% the strength of the belt. Machine-laced connections are about 90% efficient; these, however, are sometimes preferred because of the greater work involved in cementing a joint. Other types of connectors can be used, but the strength of the joint is reduced so much that they are seldom used except for emergency repairs.

Pulley sizes. Flat belting is "bent" around a pulley; for this reason, it is good practice to keep the pulley sizes large. However, large pulley sizes can ruin the compactness of a design. If the pulley diameters are too small for the thickness of the belt, cracking will result. Specific recommendations are made in belting manufacturers' catalogs for their particular line of belts. Usually, no pulley should be smaller than 250 times the square of the belt thickness.

Materials for Flat Transmission Belts

Originally, leather was the only belting material for flat belts. Leather is still used, but other materials are common. Such materials include rubber or neoprene (often reinforced with steel wire), duck covered with rubber or neoprene, cotton duck, and tubular fabric. Leather belts are available in single ply (light, medium, heavy), double ply (light, medium, heavy), and triple ply.

Conveyor Belts

Flat belting has an important application in the conveyor field, where the application is to move materials, parts, and products rather than transmit power per se. Typical examples of conveyor-belt uses include handling luggage at airport terminals, moving raw and finished food items in the processing industries, and moving raw materials and parts in the manufacturing industries.

Conveyor belts can be installed horizontally or in the inclined position. In the latter position, the conveyed materials or parts can be ascending or descending. An important consideration in selecting a belt is the condition of the item or items being moved. They could be dry, wet, sticky, abrasive, hot, cold, or saturated with oil or a chemical. Sanitary conditions must be observed in food or drug processing. Belts can be used with or without supporting rollers; this should definitely be taken into account in selecting a belt.

Power requirements can be expressed in horsepower or its metric counterpart, kilowatts. The coefficient of friction is a most important consideration. The coefficient for the pulley side directly affects the power requirement. The pulley side of the belt, therefore, is often given a higher coefficient of friction; the conveying side, in turn, is often textured to provide an adequate gripping surface. As with other flat belts, power capacities are generally specified on a per inch basis. Temperature resistance values are important, depending on whether the equipment is operated continuously or intermittently.

Typical conveyor belt materials are nylon and polyester. Surfaces are coated with urethane, with an elastomer, or with one of many fabrics.

Film Belts

A form of flat belt known as a *film belt* is widely used in tape recorders and in office machines such as photocopiers and duplicators. Most of these belts are made of a thin strip of polyester; some, however, are made of polyimide, a much stronger material. The belt thickness can be as small as 0.5 mil in light duty applications and up to 14.5 mil in heavily used office equipment (a *mil* is equal to 0.001 inches). This type of belt can be used only where power requirements are low but speed is essential. Small pulleys can be used; the thinness of the band prevents heat build-up. Efficiency is high, the belts are durable, and the cost is moderate. Only light loads such as the movement of a sheet of paper, can be handled with a film ribbon. However, it is common to find belt speeds of 100 feet per minute.

 **Round Belting**

Round belting is commercially available in leather. The total length must first be determined, then the ends are fastened with belt hooks. Endless rubber belts, with or without a cord center, are available. Only light loads can be transmitted with this type of belt. It is often used in quarter-turn drives, cross drives, serpentine drives in sewing machines, vacuum cleaners, and light textile machinery. One of its outstanding features is that it can be stretched over the pulleys and snapped into place with very little effort. Round belts usually run from $\frac{1}{8}$ in. to $\frac{3}{4}$ in. in diameter.

Small, endless round belts made of neoprene or rubber are available commercially in metric sizes. Metric belts are often 3 mm in diameter. When not in use, the belt assumes a round shape with an inside diameter of 108–330 mm.

All round belts use pulleys with semicircular grooves, in contrast to the trapezoidal grooves found in V-belt pulleys.

Idler Pulleys

Idler pulleys are sometimes used to regulate belt tension. These are pulleys that are correctly weighted and balanced and then placed in such a position on the back side of the belt that they will impart correct belt tension. It is advisable to use as large a diameter as is compatible with the design. The idler pulley should never be crowned. On short center drives (particularly V-belt drives) a pivoted motor mount can be used to provide correct belt tension.

Angular Drives

Angular drives are possible with most belts. Round belts are particularly adaptable for this type of drive; the quarter-turn drive is quite common. In any angular drive, a guide pulley is placed in the system at an appropriate angle with respect to the other pulleys. It must be so placed that the belt approaching a pulley is directed (or corrected to be) in the proper plane. An unusual angular drive is the mule drive; in this type, the driving and driven pulleys are at right angles to each other, as shown in Fig. 7-3.

Figure 7-3
Mule-drive application.

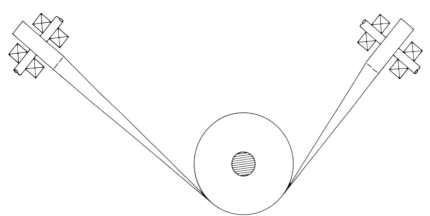

This particular type of drive is used for operating around a corner. It is easy to apply to round and V-belts.

 **V-Belting**

Of the various belt types, the V-belt is the most popular. Some of its advantages are as follows:

1. *Compactness of design.* Short center distances are possible. The center distance between pulleys can be as small as the diameter of the larger sheave.
2. *Smoothness.* This is possible because the V-belt is endless.
3. *Bearing life.* Lower belt tensions, and hence lower bearing loads, are possible. Also, since this type of belt readily absorbs shock, bearing life is lengthened.
4. *Dependability.* When more than one belt is used in a drive, failure of one belt will not cause the entire machine to stop.
5. *Maintenance.* Except for an occasional tightening of the drive (required to correct stretching and seating in the sheave), little maintenance is required.

V-Belt Sizes and Designations

V-belts are essentially trapezoidal in cross section and thus are used with grooved pulleys. Most belts are made of one piece of material in the so-called endless construction. This implies that the machine design must include a means for adjusting belt tension, either through movable pulleys (sheaves) or by moving one of the shafts.

Table 7-1 shows three chief types of V-belts, the designations for the cross sections, and the width of the top of the belt.

Construction of V-Belts

V-belts are very complex in structure. Figure 7-4(a) shows a typical belt. Tensile cords are provided to give great strength in tension, yet allow for flexing. The bottom section of the belt is a rubber or neoprene foundation designed to withstand compression. The outer jacket is a fabric bonded to other parts of the belt. V-belts receive rough treatment; they have to be constructed to withstand speed changes, shock loads, and adverse surrounding conditions.

Table 7-1
V-Belt Sizes

Designation	Top Width (inches)
Conventional or classical	
A	$\frac{1}{2}$
B	$\frac{21}{32}$
C	$\frac{7}{8}$
D	$1\frac{1}{4}$
E	$1\frac{1}{2}$
Narrow wedge	
3 V	$\frac{3}{8}$
5 V	$\frac{5}{8}$
8 V	1
Light duty or fractional horsepower	
2 L	$\frac{1}{4}$
3 L	$\frac{3}{8}$
4 L	$\frac{1}{2}$
5 L	$\frac{21}{32}$

Figure 7-4
V-belt cross section.

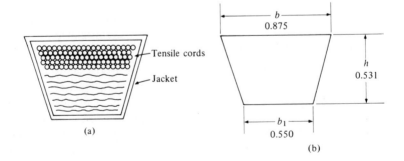

(a)

(b)

Applications and Characteristics

V-belting is found in virtually all industries. The automobile industry uses V-belts to drive fans, water pumps, generators, air-conditioning compressors, power-steering units, and power-brake units. V-belts are used in many applications in the construction field. Industrial uses include lathes, drill presses, compressors, fans, conveyors, among others.

The three most popular types are those listed in Table 7-1. The *conventional* or *classical* type is generally used where more than one belt is required to handle high power requirements. This type is characterized by the single layer of tensile cords; this allows high speeds with short center distances between the sheaves. In multiple-belt drives, one of the belts may fail. However, the equipment will usually continue to operate with one less belt. Instead of replacing only the broken belt, however, it is considered good practice to replace all of them. This is necessary because if the sheaves are adjusted to accommodate the new belt, the others will be too loose, whereas if the drive is adjusted for the old belts, the new one will be overstretched. Belts are sold in matched sets where all the belts are classified as nearly identical.

Narrow wedge V-belts are designed to transmit high power capacities while using less space than the classical type. The three cross sections cover the same capacity range as the five conventional sections. They can be used at up to 6500 fpm without dynamic balancing of the sheaves. The tensile cords are placed near the top of the belt; this frees the sidewall for compression and better gripping. Relative to the conventional types, the narrow-wedge V-belt section has a greater sidewall area.

The *light-duty* or *fractional-horsepower* belts are popular where a single belt is needed in a drive. They are designed to flex around small sheaves without overheating or developing excessive bending stresses. They are used extensively on power lawnmowers and can serve effectively as a clutching device.

Another type of V-belt drive is the *joined V*, which is a banded-together unit designed to replace multiple V-belts. Figure 7-5 shows a Power Band®, which is produced by vulcanizing several V-belts with a reinforced tie band. Such belts are designed so that the high-strength tie band does not touch the sheave. This type is desirable if pulsating or shock loads are present. The danger of whipping belts is eliminated, and the spacing between V-sections conforms exactly to the spacing found in multiple V-belt sheaves. Such a unit also makes matching of belts unnecessary unless a large number of strands requires the use of more than one banded unit. For example, if a 5-strand unit were used in conjunction with a 4-strand unit, the two units should be matched. Banded-together belts are commercially available with 2, 3, 4, or 5 strands per band and in the 3V, 5V and 8V cross sections.

Figure 7-5

Banded-together V-belts.
(Courtesy of the Gates Rubber
Company.)

A *double-*V belt is essentially two classical V-belts molded back to back.
Figure 7-6 shows a Dubl-V® belt cross section. This type is available in four
sizes, designated AA, BB, CC and DD. A Dubl-V can transmit power from
either the top or the bottom of the belt and uses grooved pulleys. It is
particularly useful in serpentine drives, as shown in Fig. 7-7. The pitch
length for this belt is the length at the widest part of the cross section.
Horsepower ratings and arc-of-contact correction factors differ from those
of classical belts. To obtain the length of belt, an accurate mathematical
calculation should be made or an exact layout should be drawn. The driven
load of all sheaves in a drive must not exceed the manufacturer's rating for
the driving pulley.

Link-type V-*belts* are available in the smaller cross sections. One type
uses studs to join the links; another variety couples the links by a tongue-
in-slot method. The link belt can be used where sheave centers are fixed or
where endless belts are impractical; in the latter case, any desirable length
can be constructed by adding the proper number of links. Using this type of

Figure 7-6
Dubl-V® belt.
(Courtesy of Gates Rubber Co.)

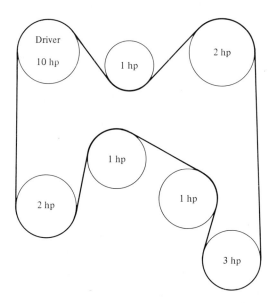

Figure 7-7
Serpentine drive.

belt cuts the cost of stocking various lengths of V-belts and also reduces maintenance costs. Figure 7-8 shows the construction of two types of link belt.

In applications where endless belts are not practical, *open-ended* classical V-belt cross sections are available in roll form; a proper length can be cut and spliced with a single fastener. Because of the fastening method, the belt construction is slightly different. Also, such spliced belts can only be used in lower power applications and at belt speeds under 4000 fpm.

Cog or *notch* V-belts include molded cogs spaced evenly on the bottom side. The cogs are shaped so that the belts flex evenly when used on small

Figure 7-8
(a) Griplink® belt. (b) Griptwist® belt.
(Courtesy of Browning Manufacturing, a division of Emerson Electric Company.)

(a)

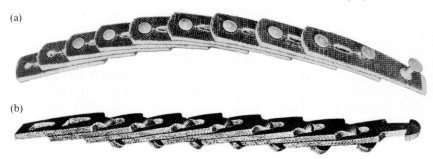

(b)

sheaves; thus bending stresses are reduced. Belts of this type have a longer service life than standard V-belts.

Selection of V-Belt Drives

It is relatively simple to select V-belt drives from manufacturers' catalog information. Properly interpreted, such catalog data provides a mechanical designer with the most economical drive suitable for the existing conditions. As a first step, the problem itself must be defined in terms of the desired ratio between the driving and driven shafts, and any restrictions on center distance must be incorporated. The following discussion outlines the steps in the selection process.

The first step is to find the design horsepower for the application. The design horsepower is the rated horsepower of the driving unit multiplied by a service or application factor based on the driver/driven relationship of the particular application. Table 7-2 shows the range of these factors based on the type of input, the type of driven unit, and the type of operation (normal, intermittent, or continuous). Footnotes on the chart define the types of service in terms of hours of usage.

After the design horsepower has been found, the belt cross section can be found by referring to the selection chart shown in Fig. 7-9. Three cross sections are listed: 3V, 5V, and 8V. A different chart must be used for the classical sections of A, B, C, D, and E or for any other type of V-belt.

Table 7-3 provides the minimum recommended sheave diameters for electric motors, based on the common motor speeds and the motor horsepower. Assume a diameter for the small sheave and calculate its pitch diameter. The pitch diameter can be found by subtracting an appropriate value from the outside diameter in accordance with the following schedule (based on cross sections):

3V	5V	8V
-0.05	-0.10	-0.20

Next, find the diameter of the driven sheave by applying the ratio. Belt speed can be calculated, since the rotational speed for either grooved pulley is known and both diameters are known. The belt length (and center distance, if not already established) must be calculated by Equation 7-3.

The final step is to determine the number of belts required. This process involves correction factors. The arc-of-contact factor A_C can be found in Table 7-4. The difference in the sheave diameters divided by the center distance provides the approximate arc of contact for the smaller sheave, which defines the value for A_C. Two steps are needed to find the belt-length correction factor. First, determine the closest stock belt length and belt

Table 7-2
Service Factors— Ultra-V Drive

| Driven Machine (Note 1) | AC normal torque electric motor (NEMA design A-B) (Note 2) | | AC high torque electric motor (NEMA design C-D) (Note 3) | | |
	Intermittent service (Note 4)	Normal service (Note 5)	Continuous service (Note 6)	Intermittent service (Note 4)	Normal service (Note 5)	Continuous service (Note 6)
Agitators for liquids	1.0	1.1	1.2	1.1	1.2	1.3
Blowers and exhausters						
Centrifugal pumps and compressors						
Conveyors (light duty)						
Fans (up to 10 H.P.)						
Belt conveyors for sand, grain, etc.	1.1	1.2	1.3	1.2	1.3	1.4
Fans (over 10 H.P.)						
Generators						
Laundry machinery						
Line shafts						
Machine tools						
Mixers (dough)						
Positive displacement rotary pumps						
Printing machinery						
Punches-presses-shears (Note 1)						
Revolving and vibrating screens						
Blowers (positive displacement)	1.2	1.3	1.4	1.4	1.5	1.6
Brick machinery						
Compressors (piston) (Note 1)						

Table 7-2 Continued

Driven Machine (Note 1)	AC normal torque electric motor (NEMA design A-B) (Note 2)			AC high torque electric motor (NEMA design C-D) (Note 3)		
	Intermittent service (Note 4)	Normal service (Note 5)	Continuous service (Note 6)	Intermittent service (Note 4)	Normal service (Note 5)	Continuous service (Note 6)
Conveyors (drag-pan-screw)	1.2	1.3	1.4	1.4	1.5	1.6
Elevators (bucket)						
Exciters						
Hammer mills						
Paper mill beaters						
Pulverizers						
Pumps (piston)						
Saw mill and woodworking machinery						
Textile machinery						
Crushers (giratory-jaw-roll) (Note 1)	1.3	1.4	1.5	1.5	1.6	1.8
Mills (ball-rod-tube) (Note 1)						
Hoists (Note 1)						
Rubber calenders-extruders-mills (Note 1)						

Note 1: The driven machines listed above are representative samples only. When one of the sheaves of the drive is used as a flywheel to reduce speed fluctuations and equalize the energy exerted at the shaft or for applications involving impact or jam loads, specially constructed sheaves may be required. Consult the manufacturer.   *Note 2:* Included under this heading are the following electric motors: synchronous and squirrel cage AC normal torque, AC split phase, DC shunt wound and internal combustion engines.   *Note 3:* Included under this heading are the following electric motors: AC high torque, AC hi-slip, AC repulsion, induction, AC single phase series wound, AC slip ring and DC compound wound.   *Note 4:* Intermittent service refers to 3–5 hours of daily or seasonal operation.   *Note 5:* Normal service indicates 8–10 hours of daily operation.   *Note 6:* Continuous service refers to 16–24 hours of daily operation.   *Note 7:* If idlers are used, add the following to the service factor:

Idler on slack side (inside)	None
Idler on slack side (outside)	0.1
Idler on tight side (inside)	0.1
Idler on tight side (outside)	0.2

(Courtesy of T. B. Wood's Sons Company.)

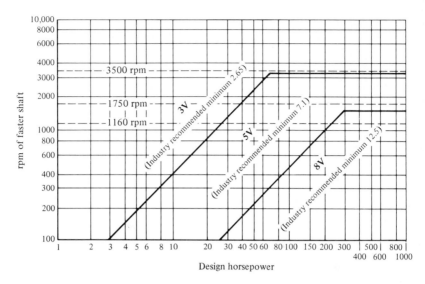

Figure 7-9
Belt selection and outside diameter range of small sheave. This chart is based on rim speeds for cast iron of 6500 f.p.m. Drives may be designed at speeds up to 10,000 f.p.m., but ductile iron sheaves are required.
(Courtesy of T. B. Wood's Sons Company.)

number from Table 7-5, which is a partial listing of Ultra-V® belts. The correction factors can be found in Table 7-6, which lists them by belt number. The next step is to find the basic horsepower rating per belt by referring to Table 7-7, which lists values for 5V belts only. A basic tabular value is first determined by interpolation; to this, an "add on" horsepower is added, based on the ratio. The latter values are found on pp. 293–295.

The corrected horsepower per belt consists of the basic tabular value plus the "add on" amount multiplied by both the arc-of-contact and the belt-length correction factors. To get the required number of belts, the corrected horsepower must be divided into the design horsepower.

It should be noted that this is merely an overview of the procedure. One should carefully read manufacturers' literature and account for all of the details of the selection process in completing the design. For example, many drives will require dynamic balancing, which is not covered in the preceding discussion.

For stock drives, the belt horsepower tables sometimes include a preengineered length-of-arc correction factor, which simplifies the selection process. The following example is based on nonstock selection.

◆ *Example* ─────────────────────────────────────

Find the number of belts required if the input power of 50 horsepower is provided by a squirrel cage normal torque electric motor rotating at 1160 rpm. The driven unit is a compressor that is operated 8 hours per day at 600 rpm. The desired center distance is 75 inches.

Solution.

Service factor = 1.1.   (Table 7-2)

Design horsepower = 50 (1.1) = 55.

Belt cross section: 5V.   (Figure 7-9)

Min. sheave dia. = 8.2 in.   (Table 7-3)

Table 7-3
Minimum Recommended Sheave Diameters in Inches for Electric Motors (NEMA Standards)

Motor Horsepower	Motor RPM			
	870	1160	1750	3500
$\frac{1}{2}$	2.2	—	—	—
$\frac{3}{4}$	2.4	2.2	—	—
1	2.4	2.4	2.2	—
$1\frac{1}{2}$	2.4	2.4	2.4	2.2
2	3.0	2.4	2.4	2.4
3	3.0	3.0	2.4	2.4
5	3.8	3.0	3.0	2.4
$7\frac{1}{2}$	4.4	3.8	3.0	3.0
10	4.4	4.4	3.8	3.0
15	5.2	4.4	4.4	3.8
20	6.0	5.2	4.4	4.4
25	6.8	6.0	4.4	4.4
30	6.8	6.8	5.2	—
40	8.2	6.8	6.0	—
50	8.4	8.2	6.8	—
60	10.0	8.2	7.4	—
75	10.0	10.0	8.6	—
100	12.0	10.0	8.6	—
125	—	12.0	10.5	—
150	—	—	10.5	—
200	—	—	13.2	—
250	—	—	—	—
300	—	—	—	—

(Courtesy of T. B. Wood's Sons Company.)

Table 7-4
Arc-of-contact Correction Factor A_C

$\dfrac{D - d}{C}$	Approximate Arc of Contact on Small Sheave	Correction Factor A_C
.00	180°	1.00
.10	174°	.99
.20	168°	.97
.30	162°	.96
.40	156°	.94
.50	150°	.92
.60	144°	.90
.70	138°	.88
.80	132°	.87
.90	126°	.85
1.00	120°	.83
1.10	114°	.80
1.20	108°	.78
1.30	102°	.75
1.40	96°	.72
1.50	90°	.69

(Courtesy of T. B. Wood's Sons Company.)

$$\text{Ratio} = \frac{1160}{600} = 1.93.$$

Assume 8.25 in. sheave dia. for driver.

Pitch dia. $= 8.25 - 0.10 = 8.15$ in.

Pitch dia. for driven $= 8.15\,(1.93) = 15.7$ in.

Driven sheave dia. $= 15.7 + 0.10 = 15.8$ in.

Belt speed $= \dfrac{\pi}{4}(8.25)(1160) = 2507$ fpm.

$$L = 2C + 1.57\,(D + D) = \frac{(D - d)^2}{4C}. \qquad \text{(Equation 7-3)}$$

$$L = 2(75) + 1.57\,(15.8 + 8.25) + \frac{(15.8 - 8.25)^2}{4(75)}.$$

$L = 187.95$ in.

$A_C = 0.99.$ \qquad (Table 7-4)

Table 7-5
Partial Listing of Ultra-V® Stock Belts (5V and 5VX Belts)

Belt Number	Belt Length	Wt.	Belt Number	Belt Length	Wt.
5VX500	50.0	.6	5VX1600	160.0	1.9
5VX530	53.0	.7	5VX1700	170.0	2.0
5VX560	56.0	.7	5VX1800	180.0	2.1
5VX600	60.0	.7	5VX1900	190.0	2.3
5VX630	63.0	.7	5VX2000	200.0	2.4
5VX670	67.0	.8	5V2120	212.0	2.5
5VX710	71.0	.8	5V2240	224.0	2.7
5VX750	75.0	.8	5V2360	236.0	2.8
5VX800	80.0	.9	5V2500	250.0	3.0
5VX850	85.0	.9	5V2650	265.0	3.2
5VX900	90.0	1.1	5V2800	280.0	3.3
5VX950	95.0	1.1	5V3000	300.0	3.6
5VX1000	100.0	1.2	5V3150	315.0	3.8
5VX1060	106.0	1.2	5V3350	335.0	4.1
5VX1120	112.0	1.3	5V3550	355.0	4.3
5VX1180	118.0	1.4			
5VX1250	125.0	1.5			
5VX1320	132.0	1.6			
5VX1400	140.0	1.7			
5VX1500	150.0	1.8			

(Courtesy of T. B. Wood's Sons Company.)

Nearest stock belt length: #5V1900. (Table 7-5)

$L_C = 1.07.$ (Table 7-6)

Basic horsepower rating per belt = 14.80. (Table 7-7)

Rated hp = 14.80 + 1.26 "add on" = 16.06.

Corrected hp/belt = 16.06 (0.99)(1.074) = 17.01.

No. belts req's $= \dfrac{55}{17.01} = 3.23.$

To be on the safe side, 4 belts should be used.

Sheaves

A mechanical designer has a wide variety of choices in specifying sheaves. Some have single grooves for light drives; others have multigrooves requiring more than one belt. All are designed by the manufacturer to accommodate the standard sizes of V-belts. Cast iron sheaves are available either with finished bores or bores requiring a bushing. Many are made of formed steel or are die cast using various materials that are suitable for die casting.

Multigroove sheaves consume much more axial space and are obviously much heavier than single grooved sheaves. Since a large pulley is a substantial rotating mass of metal, it is important that it be machined accurately and statically balanced.

In ordering a sheave, it is necessary to specify the number of grooves, the V-belt cross section, the pitch diameter, and the bushing bore size (if a bushing is used). Although many pitch sizes are available from which to

Table 7-6
Belt Length Correction Factors L_C for Ultra-V®, 5V Cross Section

Belt No.	Correction Factor L_C	Belt No.	Correction Factor L_C
5V500	.85	5V1400	1.02
5V530	.86	5V1500	1.03
5V560	.87	5V1600	1.04
5V600	.88	5V1700	1.05
5V630	.89	5V1800	1.06
5V670	.90	5V1900	1.07
5V710	.91	5V2000	1.08
5V750	.92		
5V800	.93	5V2120	1.09
5V850	.94	5V2240	1.09
		5V2360	1.10
5V900	.95	5V2500	1.11
5V950	.95	5V2650	1.12
5V1000	.96		
5V1060	.97	5V2800	1.13
5V1120	.98	5V3000	1.14
		5V3150	1.15
5V1180	.99	5V3350	1.16
5V1250	1.00	5V3550	1.17
5V1320	1.01		

(Courtesy of T. B. Wood's Sons Company.)

Table 7-7
Ultra-V ® Horsepower Ratings

| RPM of Faster Shaft | Basic Horsepower Rating per Belt | | | | | | | | | | | | | | | | |
| | Small sheave outside diameter | | | | | | | | | | | | | | | | |
	7.00	7.10	7.50	8.00	8.50	9.00	9.25	9.75	10.00	10.30	10.50	10.90	11.00	11.30	11.80	12.00	12.50
435	4.86	4.99	5.48	6.10	6.72	7.33	7.64	8.25	8.55	8.91	9.15	9.64	9.76	10.11	10.71	10.95	11.54
485	5.34	5.48	6.03	6.71	7.39	8.07	8.41	9.08	9.41	9.81	10.08	10.61	10.74	11.14	11.80	12.06	12.71
575	6.19	6.35	6.99	7.79	8.58	9.37	9.76	10.55	10.93	11.40	11.71	12.33	12.48	12.94	13.70	14.01	14.76
585	6.28	6.44	7.09	7.91	8.71	9.51	9.91	10.71	11.10	11.57	11.89	12.51	12.67	13.14	13.91	14.22	14.99
690	7.23	7.42	8.17	9.12	10.05	10.98	11.44	12.36	12.81	13.36	13.72	14.44	14.62	15.16	16.05	16.41	17.29
725	7.54	7.74	8.53	9.51	10.49	11.45	11.94	12.89	13.37	13.94	14.32	15.07	15.26	15.82	16.75	17.12	18.04
870	8.78	9.02	9.95	11.10	12.24	13.38	13.94	15.06	15.61	16.27	16.71	17.59	17.81	18.46	19.53	19.96	21.02
960	9.53	9.79	10.80	12.05	13.29	14.52	15.14	16.35	16.95	17.66	18.14	19.09	19.32	20.02	21.18	21.64	22.78
1160	11.12	11.41	12.60	14.07	15.52	16.95	17.66	19.06	19.76	20.59	21.13	22.22	22.49	23.29	24.61	25.13	26.43
1460	13.30	13.66	15.09	16.84	18.57	20.26	21.09	22.73	23.54	24.50	25.13	26.38	26.69	27.61	29.10	29.69	31.13
1750	15.19	15.60	17.22	19.21	21.15	23.04	23.96	25.77	26.66	27.70	28.38	29.72	30.05	31.03	32.60	33.21	34.70
2850	20.01	20.53	22.53	24.89	27.06	29.04	29.96	31.65	32.41	33.25	33.77	34.70	34.91	35.48	36.23	36.47	36.87
3450	20.79	21.29	23.19	25.29	27.09	28.57	29.18	30.14	30.48	30.76	30.88	30.91	30.88	—			
200	2.46	2.52	2.76	3.06	3.36	3.66	3.81	4.11	4.26	4.44	4.56	4.79	4.85	5.03	5.32	5.44	5.73
300	3.51	3.60	3.96	4.40	4.83	5.27	5.49	5.92	6.14	6.39	6.57	6.91	7.00	7.25	7.68	7.85	8.27
400	4.52	4.63	5.10	5.67	6.24	6.81	7.09	7.65	7.94	8.27	8.50	8.94	9.05	9.39	9.94	10.16	10.71
500	5.48	5.63	6.19	6.90	7.59	8.29	8.64	9.33	9.67	10.08	10.36	10.90	11.04	11.44	12.12	12.39	13.06
600	6.41	6.58	7.25	8.08	8.91	9.72	10.13	10.94	11.35	11.83	12.15	12.79	12.95	13.43	14.22	14.54	15.32
700	7.32	7.51	8.28	9.23	10.17	11.11	11.58	12.51	12.97	13.52	13.89	14.62	14.80	15.35	16.25	16.61	17.50
800	8.19	8.41	9.27	10.34	11.41	12.46	12.98	14.03	14.54	15.16	15.57	16.39	16.59	17.20	18.21	18.61	19.60
900	9.04	9.28	10.23	11.42	12.60	13.76	14.34	15.49	16.06	16.74	17.19	18.09	18.32	18.98	20.09	20.53	21.62

1000	9.86	10.12	11.17	12.47	13.75	15.02	15.66	16.91	17.53	18.27	18.76	19.73	19.98	20.70	21.89	22.37	23.54
1100	10.65	10.94	12.07	13.48	14.87	16.24	16.92	18.27	18.94	19.73	20.26	21.31	21.57	22.34	23.62	24.12	25.37
1200	11.42	11.73	12.95	14.46	15.95	17.42	18.14	19.58	20.29	21.14	21.70	22.81	23.09	23.91	25.26	25.79	27.11
1300	12.16	12.49	13.79	15.40	16.99	18.55	19.32	20.84	21.59	22.49	23.08	24.24	24.53	25.40	26.81	27.37	28.74
1400	12.88	13.23	14.61	16.31	17.99	19.63	20.44	22.04	22.83	23.77	24.38	25.60	25.91	26.80	28.27	28.85	30.27
1500	13.57	13.94	15.40	17.19	18.94	20.67	21.51	23.18	24.01	24.98	25.62	26.88	27.20	28.12	29.64	30.23	31.69
1600	14.24	14.62	16.15	18.02	19.86	21.65	22.53	24.27	25.12	26.12	26.78	28.08	28.41	29.36	30.90	31.51	32.98
1700	14.88	15.28	16.87	18.82	20.73	22.59	23.50	25.29	26.16	27.19	27.87	29.20	29.53	30.49	32.06	32.67	34.16
1800	15.49	15.91	17.56	19.58	21.55	23.47	24.41	26.24	27.13	28.19	28.88	30.23	30.56	31.53	33.11	33.72	35.20
1900	16.07	16.50	18.22	20.30	22.33	24.30	25.26	27.13	28.04	29.10	29.80	31.16	31.49	32.47	34.05	34.65	36.11
2000	16.62	17.07	18.83	20.98	23.06	25.07	26.05	27.94	28.86	29.94	30.64	32.00	32.33	33.31	34.86	35.46	36.88
2100	17.14	17.60	19.42	21.62	23.74	25.78	26.77	28.69	29.61	30.69	31.39	32.74	33.07	34.03	35.55	36.13	37.51
2200	17.63	18.11	19.96	22.21	24.37	26.43	27.43	29.35	30.27	31.35	32.04	33.37	33.70	34.64	36.11	36.67	37.98
2300	18.09	18.58	20.47	22.76	24.94	27.02	28.02	29.94	30.86	31.91	32.60	33.90	34.22	35.13	36.54	37.07	38.29
2400	18.52	19.01	20.94	23.26	25.46	27.55	28.55	30.45	31.35	32.39	33.05	34.32	34.62	35.94	36.82	37.32	38.43
2500	18.91	19.41	21.37	23.71	25.92	28.00	28.99	30.87	31.75	32.76	33.41	34.62	34.91	35.73	36.96	37.41	38.41
2600	19.27	19.78	21.76	24.11	26.32	28.39	29.37	31.21	32.06	33.04	33.65	34.80	35.07	35.83	36.95	37.35	38.20
2700	19.59	20.11	22.10	24.46	26.66	28.71	29.67	31.45	32.28	33.21	33.79	34.86	35.10	35.80	36.79	37.12	37.81
2800	19.88	20.40	22.40	24.76	26.94	28.95	29.88	31.60	32.39	33.27	33.81	34.79	35.01	35.62	36.46	36.73	37.23
2900	20.13	20.65	22.65	25.00	27.16	29.12	30.02	31.66	32.40	33.21	33.71	34.58	34.78	35.30	35.97	36.16	36.45
3000	20.34	20.86	22.86	25.19	27.30	29.20	30.07	31.62	32.31	33.05	33.49	34.24	34.40	34.83	35.30	35.41	35.41
3100	20.51	21.03	23.02	25.32	27.38	29.21	30.03	31.48	32.10	32.76	33.14	33.76	33.89	34.19	34.46	34.47	35.47

—

Table 7-7
(continued)

Basic Horsepower Rating per Belt

Small sheave outside diameter

RPM of Faster Shaft	7.00	7.10	7.50	8.00	8.50	9.00	9.25	9.75	10.00	10.30	10.50	10.90	11.00	11.30	11.80	12.00	12.50
3200	20.64	21.16	23.13	25.39	27.39	29.14	29.91	31.23	31.78	32.35	32.66	33.14	33.22	33.40	33.43	33.34	—
3300	20.73	21.24	23.19	25.40	27.33	28.98	29.69	30.88	31.35	31.81	32.05	32.36	32.41	32.45	—	—	—
3400	20.78	21.28	23.20	25.34	27.19	28.73	29.37	30.41	30.80	31.15	31.30	31.44	31.43	31.32	—	—	—
3500	20.78	21.28	23.16	25.23	26.98	28.39	28.96	29.83	30.13	30.35	30.42	30.35	30.29	—	—	—	—
3600	20.74	21.23	23.06	25.05	26.69	27.96	28.45	29.14	29.33	29.41	29.38	—	—	—	—	—	—
3700	20.65	21.13	22.90	24.80	26.31	27.43	27.84	28.32	28.40	28.34	—	—	—	—	—	—	—
3800	20.52	20.99	22.69	24.48	25.86	26.81	27.12	27.39	27.34	—	—	—	—	—	—	—	—
3900	20.34	20.79	22.42	24.09	25.32	26.09	26.30	26.32	—	—	—	—	—	—	—	—	—
4000	20.11	20.54	22.09	23.63	24.70	25.27	25.36	—	—	—	—	—	—	—	—	—	—
4200	19.50	19.89	21.25	22.49	23.19	23.31	—	—	—	—	—	—	—	—	—	—	—
4400	18.68	19.02	20.16	21.04	21.31	—	—	—	—	—	—	—	—	—	—	—	—
4600	17.65	17.93	18.80	19.28	—	—	—	—	—	—	—	—	—	—	—	—	—
4800	16.39	16.60	17.16	17.17	—	—	—	—	—	—	—	—	—	—	—	—	—
5000	14.90	15.03	15.24	—	—	—	—	—	—	—	—	—	—	—	—	—	—
5200	13.16	13.20	—	—	—	—	—	—	—	—	—	—	—	—	—	—	—
5400	11.17	11.12	—	—	—	—	—	—	—	—	—	—	—	—	—	—	—

Table 7-7
(continued)

| Basic Horsepower Rating per Belt | | | | | | | | "Add-on" Rating | | | | | | | | | | |
| Small sheave outside diameter | | | | | | | | Speed ratio | | | | | | | | | | RPM of Faster Shaft |
13.00	13.20	13.50	14.00	15.00	16.00	18.70	21.20	1.00–1.01	1.02–1.05	1.06–1.11	1.12–1.18	1.19–1.26	1.27–1.38	1.39–1.57	1.58–1.94	1.95–3.38	3.39 & up	
12.13	12.37	12.72	13.31	14.47	15.62	18.68	21.45	.00	.05	.12	.20	.28	.33	.39	.44	.48	.50	435
13.36	13.62	14.01	14.65	15.93	17.19	20.54	23.56	.00	.05	.13	.23	.31	.37	.43	.49	.53	.56	485
15.52	15.82	16.26	17.01	18.48	19.94	23.78	27.22	.00	.06	.16	.27	.36	.44	.51	.58	.63	.66	575
15.75	16.06	16.51	17.27	18.76	20.24	24.13	27.61	.00	.06	.16	.27	.37	.45	.52	.59	.64	.68	585
18.17	18.51	19.03	19.90	21.61	23.29	27.68	31.57	.00	.07	.19	.32	.43	.53	.61	.69	.75	.80	690
18.95	19.31	19.85	20.75	22.52	24.27	28.82	32.82	.00	.07	.20	.34	.46	.55	.65	.73	.79	.84	725
22.07	22.49	23.11	24.14	26.16	28.14	33.23	37.60	.00	.09	.23	.40	.55	.66	.77	.87	.95	1.00	870
23.91	24.35	25.02	26.12	28.27	30.37	35.71	40.21	.00	.10	.26	.44	.60	.73	.85	.96	1.04	1.11	950
27.70	28.20	28.94	30.17	32.55	34.83	40.46	44.94	.00	.12	.31	.53	.73	.88	1.03	1.16	1.26	1.34	1160
32.54	33.08	33.90	35.22	37.72	40.05	45.33	48.79	.00	.15	.39	.67	.91	1.11	1.29	1.46	1.59	1.68	1460
36.12	36.67	37.47	38.75	41.09	43.13	46.98	48.14	.00	.17	.46	.80	1.09	1.33	1.55	1.74	1.90	2.01	1750
37.01	36.96	—	—	—	—	—	—	.00	.28	.75	1.31	1.78	2.16	2.52	2.84	3.09	3.28	2850
								.00	.34	.91	1.58	2.15	2.61	3.05	3.43	3.74	3.96	3450
6.02	6.14	6.31	6.60	7.18	7.75	9.29	10.69	.00	.02	.06	.10	.13	.16	.18	.20	.22	.23	200
8.70	8.86	9.12	9.54	10.37	11.20	13.41	15.43	.00	.03	.08	.14	.19	.23	.27	.30	.33	.35	300
11.26	11.48	11.80	12.35	13.43	14.50	17.35	19.93	.00	.04	.11	.19	.25	.31	.36	.40	.44	.46	400
13.73	13.99	14.39	15.05	16.36	17.66	21.09	24.19	.00	.05	.14	.23	.32	.38	.45	.50	.55	.58	500
16.10	16.41	16.88	17.65	19.17	20.68	24.65	28.19	.00	.06	.16	.28	.38	.46	.53	.60	.65	.69	600
18.39	18.74	19.27	20.14	21.87	23.57	28.01	31.93	.00	.07	.19	.32	.44	.53	.62	.70	.76	.81	700
20.59	20.98	21.56	22.53	24.44	26.31	31.16	35.37	.00	.08	.21	.37	.50	.61	.71	.80	.87	.92	800

Table 7-7
(continued)

Basic Horsepower Rating per Belt — Small sheave outside diameter | **"Add-on" Rating** — Speed ratio

RPM of Faster Shaft	13.00	13.20	13.50	14.00	15.00	16.00	18.70	21.20	1.00–1.01	1.02–1.05	1.06–1.11	1.12–1.18	1.19–1.26	1.27–1.38	1.39–1.57	1.58–1.94	1.95–3.38	3.39 & up
900	22.69	23.12	23.76	24.81	26.88	28.89	34.08	38.50	.00	.09	.24	.42	.56	.68	.80	.90	.98	1.04
1000	24.70	25.16	25.84	26.97	29.18	31.32	36.75	41.28	.00	.10	.27	.46	.63	.76	.89	1.00	1.09	1.15
1100	26.60	27.09	27.82	29.01	31.33	33.57	39.15	43.68	.00	.11	.29	.51	.69	.84	.98	1.10	1.20	1.27
1200	28.40	28.91	29.67	30.92	33.33	35.63	41.28	45.69	.00	.12	.32	.55	.75	.91	1.06	1.20	1.30	1.38
1300	30.09	30.62	31.40	32.69	35.16	37.50	43.10	47.26	.00	.13	.35	.60	.81	.99	1.15	1.30	1.41	1.50
1400	31.65	32.20	33.00	34.31	36.82	39.16	44.59	48.36	.00	.14	.37	.64	.88	1.06	1.24	1.40	1.52	1.61
1500	33.10	33.65	34.46	35.79	38.29	40.59	45.75	48.98	.00	.15	.40	.69	.94	1.14	1.33	1.50	1.63	1.73
1600	34.41	34.96	35.78	37.10	39.57	41.80	46.53	49.06	.00	.16	.42	.74	1.00	1.21	1.42	1.60	1.74	1.84
1700	35.58	36.14	36.95	38.24	40.64	42.75	46.93	48.59	.00	.17	.45	.78	1.06	1.29	1.51	1.69	1.85	1.96
1800	36.61	37.16	37.95	39.21	41.50	43.45	46.92	47.54	.00	.18	.48	.83	1.12	1.36	1.59	1.79	1.95	2.07
1900	37.49	38.02	38.79	39.99	42.12	43.87	46.49	——	.00	.19	.50	.87	1.19	1.44	1.68	1.89	2.06	2.19
2000	38.21	38.72	39.45	40.58	42.52	44.01	45.60	——	.00	.20	.53	.92	1.25	1.51	1.77	1.99	2.17	2.30
2100	38.77	39.25	39.92	40.96	42.66	43.85	——	——	.00	.21	.56	.96	1.31	1.59	1.86	2.09	2.28	2.41
2200	39.16	39.60	40.21	41.13	42.55	43.37	——	——	.00	.22	.58	1.01	1.37	1.67	1.95	2.19	2.39	2.53
2300	39.37	39.76	40.30	41.08	42.16	42.57	——	——	.00	.23	.61	1.05	1.44	1.74	2.04	2.29	2.50	2.64
2400	39.39	39.73	40.18	40.80	41.49	41.43	——	——	.00	.24	.63	1.10	1.50	1.82	2.12	2.39	2.60	2.76
2500	39.22	39.49	39.85	40.28	40.53	——	——	——	.00	.25	.66	1.15	1.56	1.89	2.21	2.49	2.71	2.87
2600	38.85	39.05	39.29	39.51	——	——	——	——	.00	.26	.69	1.19	1.62	1.97	2.30	2.59	2.82	2.99
2700	38.27	38.39	38.50	38.48	——	——	——	——	.00	.26	.71	1.24	1.68	2.04	2.39	2.69	2.93	3.10
2800	37.48	37.51	37.47	——	——	——	——	——	.00	.27	.74	1.28	1.75	2.12	2.48	2.79	3.04	3.22
2900	36.47	36.40	——	——	——	——	——	——	.00	.28	.77	1.33	1.81	2.19	2.57	2.89	3.15	3.33

RPM																		
3000	3.45	3.25	2.99	2.65	2.27	1.87	1.38	.79	.29	.00	—	—	—	—	—	—	—	—
3100	3.56	3.36	3.09	2.74	2.34	1.93	1.42	.82	.30	.00	—	—	—	—	—	—	—	—
3200	3.68	3.47	3.19	2.83	2.42	2.00	1.47	.84	.31	.00	—	—	—	—	—	—	—	—
3300	3.79	3.58	3.29	2.92	2.50	2.06	1.51	.87	.32	.00	—	—	—	—	—	—	—	—
3400	3.91	3.69	3.38	3.01	2.57	2.12	1.56	.90	.33	.00	—	—	—	—	—	—	—	—
3500	4.02	3.80	3.48	3.10	2.65	2.18	1.60	.92	.34	.00	—	—	—	—	—	—	—	—
3600	4.14	3.90	3.58	3.18	2.72	2.24	1.65	.95	.35	.00	—	—	—	—	—	—	—	—
3700	4.25	4.01	3.68	3.27	2.80	2.31	1.70	.98	.36	.00	—	—	—	—	—	—	—	—
3800	4.37	4.12	3.78	3.36	2.87	2.37	1.74	1.00	.37	.00	—	—	—	—	—	—	—	—
3900	4.48	4.23	3.88	3.45	2.95	2.43	1.79	1.03	.38	.00	—	—	—	—	—	—	—	—
4000	4.60	4.34	3.98	3.54	3.02	2.49	1.83	1.05	.39	.00	—	—	—	—	—	—	—	—
4200	4.82	4.55	4.18	3.71	3.17	2.62	1.92	1.11	.41	.00	—	—	—	—	—	—	—	—
4400	5.05	4.77	4.38	3.89	3.33	2.74	2.01	1.16	.43	.00	—	—	—	—	—	—	—	—
4600	5.28	4.99	4.58	4.07	3.48	2.87	2.11	1.21	.45	.00	—	—	—	—	—	—	—	—
4800	5.51	5.20	4.78	4.24	3.63	2.99	2.20	1.26	.47	.00	—	—	—	—	—	—	—	—
5000	5.74	5.42	4.98	4.42	3.78	3.11	2.29	1.32	.49	.00	—	—	—	—	—	—	—	—
5200	5.97	5.64	5.17	4.60	3.93	3.24	2.38	1.37	.51	.00	—	—	—	—	—	—	—	—
5400	6.20	5.85	5.37	4.77	4.08	3.36	2.47	1.42	.52	.00	—	—	—	—	—	—	—	—

Horsepower ratings in the shaded areas imply the sheave rim speed is 6500 fpm or more. Sheaves of special ductile iron construction are required. (Courtesy of T. B. Wood's Sons Company.)

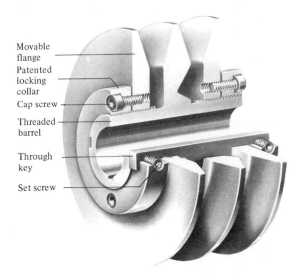

Figure 7-10
Variable-pitch sheave.
(Courtesy of Browning Manufacturing, a division of Emerson Electric Company.)

make a selection, one must put together the right combination of driving and driven sheaves to effect the desired ratio. Figure 7-10 shows the construction features of a variable-pitch sheave. *Exact* pitch diameter can be obtained and held. This particular assembly is designed for the 5- to 10-hp range.

 Synchronous (Timing) Belt Drives

A belt drive using toothed belting in conjunction with grooved pulleys ensures an accurate ratio between the driving and the driven shafts. The ratio is based on the pitch diameters of the grooved pulleys. Timing belts combine the advantages of V-belts with those of chains, which are discussed in Chapter 8.

The tensile members of a timing belt are made of extremely strong, continuous, helically wound fiberglass or steel cords encased in strong, wear-resisting neoprene. The teeth of the belt are usually made of a moderately hard, shear-resistant neoprene compound. The contacting surfaces of the teeth are usually covered with a tough, low-friction nylon duck facing; this protects the teeth in a manner similar to the case hardening of metallic gears.

The standard belt sections are classified according to their pitch, which is the distance measured along the pitch circle from a point on one tooth to the corresponding point on the next tooth. Belt widths vary in accordance with the available pitches. Table 7-8 shows standard belt sections for both single-sided and double-sided timing belts; the latter can be used in serpentine drives.

Figure 7-11 shows typical cross sections for timing belts. Figure 7-12 pictures a complete SURE GRIP® HTD® (high-torque drive) drive. This particular drive is available as a stock item in 8-mm and 14-mm pitches. Stock belt widths in the 14-mm pitches are available up to 170-mm and have ratings approaching 300 hp.

Applications and Advantages

Timing-belt drives are ideal for low speed, high torque applications. As it is a toothed belt, it provides a positive drive, which places it in competition with both chain and gear drives and offers several advantages.

No lubrication is required, and the belts are corrosion resistant. Noise levels are minimal. Toothed belts provide long service and require little attention. Compared with other types of belting, there is virtually no stretching due to wear.

Selection of Timing Belts

Timing belts are selected by a procedure similar to that used for V-belts. The proper ratio must be established; the belt length and center distance must be calculated. The service factor is based on the power supply

Table 7-8
Standard Belt Sections—Timing Belts

Designation	Pitch (inches)
Single-sided belts	
MXL	0.080
XL	0.200
L	0.375
H	0.500
XH	0.875
Double-sided belts	
DXL	0.200
DL	0.375
DH	0.500

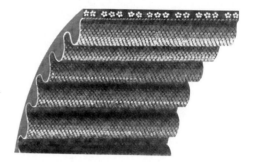

Figure 7-11
Cross sections of timing belts.
(Courtesy of T. B. Wood's Sons
Company.)

Figure 7-12
Sure Grip® HTD® drive.
(Courtesy of T. B. Wood's Sons
Company.)

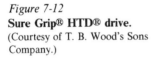

and the type of equipment being driven. The horsepower (or kilowatt) rating is supplied by the manufacturer for the given type of belt (section and width). Two correction factors are used to modify these ratings. One is the belt-width factor; the other is the teeth-in-mesh factor. The latter is the counterpart of the arc-of-contact factor used in selecting V-belt drives.

After the proper belt is selected, the next step is to find an appropriate pulley. This should have the proper pitch so that its grooves will mesh with the teeth on the belt. Both flanged and unflanged pulleys are available. The pulley width must be appropriate for the belt width specified; pulley bores and keyseat sizes must be carefully checked for compatibility with the other components used in the drive.

 Design of a V-Flat Drive

A V-*flat* drive is one where the smaller pulley has a groove and the larger pulley (actually a flywheel or crowned pulley) does not. Existing flywheels or flat pulleys can thus be used at quite a saving in cost. However, the capacity is greatly reduced because the driven pulley contacts only the bottom of the V-*belt* section. This type of drive is most effective when the speed ratio is at least 3:1 and the center distance is relatively small. The driven pulley must be wide enough to handle the required number of belts.

The equation for the ratio of V-belt pulls is as follows:

$$\frac{F_1 - F_c}{F_2 - F_c} = e^{f\theta/\sin(\alpha/2)}, \tag{7-8}$$

where

α = groove angle in degrees.

(For a flat pulley, this would be 180 degrees). In a V-flat drive, it is good practice to see how $e^{f\theta/\sin(\alpha/2)}$ compares for the grooved pulley and the flat pulley. The design should be based on the pulley with the smaller value. If the center distance is no larger than the diameter of the larger pulley, an adequate angle of contact is ensured. This type of drive is usually not suited for large center distances unless an idler pulley is added to increase the angle of contact. The following example illustrates a mathematical approach to V-flat drives.

◆ *Example* —————————————————————————

A V-flat drive is to transmit 20 hp using a 9-in. pitch diameter sheave and a 36-in. flat pulley. The center distance is 36 in. Assume that the coefficient of friction for both belt-pulley conditions is 0.25, and that the allowable tension is 165 lb per belt. The groove angle of the sheave is 36 degrees. The dimensions for the trapezoidal cross section of the belt are shown in Fig. 7-4(b). The specific weight of the belt material is 0.04 lb/cu in. The rotational speed of the 9-in. sheave is 1750 rpm. Find the number of belts required to meet these conditions.

Solution. First, find the centroidal distance for the trapezoidal section, so that the pitch diameter of the belt on the flat pulley can be found:

$$\bar{y} = \frac{h(b_1 + 2b)}{3(b + b_1)} = \frac{0.531[0.550 + 2(0.875)]}{3(0.875 + 0.550)} = 0.286 \text{ in.}$$

Thus the pitch diameter of the large pulley is $36 + 2(0.286) = 36.6$ in.

$$\text{Contact (degrees)} = 180 \pm 2 \arcsin\left[\frac{(D_1/2) - (D_2/2)}{C}\right]$$

$$= 180 \pm 2 \arcsin\frac{18.3 - 4.5}{36}$$

$$= 180 \pm 2(22.5).$$

Thus, the angle of contact for the larger pulley is $180 + 45 = 225$ degrees; for the smaller pulley, it is $180 - 45 = 135$ degrees. For the larger pulley,

$$e^{f\theta/\sin(\alpha/2)} = 2.718^{(0.25)(225\pi/180)/\sin 90} = 2.718^{0.982} = 2.67.$$

For the smaller pulley,

$$e^{f\theta/\sin(\alpha/2)} = 2.718^{(0.25)(135\pi/180)/\sin 18} = 2.718^{1.91} = 6.75.$$

Since 2.67 is less than 6.75, the design should be based on the larger pulley. The area of the belt cross section is

$$h\left(\frac{b + b_1}{2}\right) = 0.531\left(\frac{0.875 + 0.550}{2}\right) = 0.378 \text{ in}^2;$$

$$v = \frac{\pi DN}{60} = \frac{\pi(9/12)(1750)}{60} = 68.7 \text{ ft/sec};$$

$$F_c = \frac{12wAv^2}{g} = \frac{(12)(0.04)(0.378)(68.7)^2}{32.2} = 26.6 \text{ lb};$$

where

$A =$ area in square inches of belt cross section.

$$\frac{F_1 - F_c}{F_2 - F_c} = e^{f\theta/\sin(\alpha/2)} \qquad \text{or} \qquad \frac{165 - 26.6}{F_2 - 26.6} = 2.67.$$

Table 7-9
Coefficient-of-friction Values for Belts Used with Cast Iron or Steel Pulleys

Belt Material	Coefficient of Friction
Oak-tanned leather	0.25–0.30
Mineral-tanned leather	0.40–0.42
Canvas	0.20
Rubber/fabric	0.32–0.35
Rubber	0.30
Woven cotton	0.22–0.25

Therefore $F_2 = 78.4$ lb. Now,

$$F_1 - F_2 = \frac{550(\text{hp})}{v} \; ;$$

$$\text{hp} = \frac{(F_1 - F_2)v}{550} = \frac{(165 - 78.4)(68.7)}{550}$$

$$= \frac{(86.6)(68.7)}{550} = 10.8 \text{ hp/belt.}$$

Therefore, the number of belts required is $20/10.8 = 1.85$, or 2 belts. This design does not include any factor for the type of application (or service) nor any factor for abrupt changes in speed.

Coefficient of Friction Values for Belts

The static coefficient of friction is generally used for all types of belting. However, such friction values vary widely because of such operating conditions as speed changes, total slip, and belt tension. Table 7-9 shows a few static-coefficient-of-friction values.

 **Summary**

Belts provide a useful and inexpensive type of connector in the field of power transmission. Except for toothed belting, exact velocity ratios cannot be maintained. However, a belt drive can handle pulsating or shock loading; this compensates for the disadvantage in the area of velocity ratio. Belts can stretch and they can slip before failure occurs; with other types of power transmission devices, failure occurs suddenly, and this sudden failure means a complete shutdown until costly repairs can be made.

Most belting is available in roll form. In using roll belting, the proper length is obtained and suitable connectors are used to close the loop. However, the designer should consider the type of connector being used; certain of these do not provide full belt strength at the joint.

Standard sizes are available for pulleys and sheaves. Handbooks and manufacturers' catalogs are useful in determining the available sizes. Standard lengths of V-belts are available from stock. To simplify selection, manufacturers compile charts listing the ratio, the diameter of each sheave, and the center distance. Thus, belt-length calculations can be eliminated completely.

Proper tensioning of a belt is essential. To provide for this and to compensate for the stretching of belts, it is advisable to include an adjustment for one of the pulley centers in the design. If this cannot be done easily, an idler pulley is usually placed in the system for tensioning.

Questions for Review

1. Explain how the "crowning" of a pulley keeps a flat belt centered.
2. Explain how a band-saw blade is similar to a belt drive in structure and operation.
3. List several methods used to make belts endless. What disadvantages do connectors have?
4. What advantages do V-belts have over flat belts? What disadvantages?
5. Explain the purpose of the various materials used in constructing a V-belt.
6. How can belt slip be minimized?
7. Explain the purpose of an idler pulley.
8. What is a serpentine drive? Give an application.
9. What difficulties are encountered in designing a quarter-turn belt drive?
10. What difficulties arise in using crossed-belt drives?
11. What devices might use round belting? What advantages does this type have? What disadvantages?
12. What advantages do banded-together V-belts have over multiple V-belt drives?
13. What advantages do V-belt drives have over chain drives?
14. What advantages do V-belt drives have over gear drives? What disadvantages?

15. What are matched sets of V-belts?

16. When one belt of a multiple V-belt drive fails, why should all belts be replaced?

17. What advantages do link-type V-belts have over conventional V-belts?

18. In designing V-belt drives, why should the center of one of the grooved pulleys be made adjustable?

19. What advantages does toothed belting have over regular flat belting? What disadvantages?

20. Explain what is meant by a service factor. In selecting such a factor, is it necessary to know the type of driving unit as well as the driven equipment?

Problems

1. Find the angle of contact on the small pulley for a belt drive with a 72-in. center distance. The pulley diameters are 6 in. and 12 in.

2. Compute the length of belt needed for the drive described in Problem 1, using a formula to approximate the length of open belt needed.

3. An open belt drive has a 6-in. pulley driving a 48-in. pulley. The distance between centers is 96 in. Find the *exact* length of belt needed.

4. In Problem 1, the 6-in. pulley rotates at 600 rpm. Total slip amounts to 4%. Find the rotational speed of the 12-in. pulley.

5. In Problem 1, a $\frac{1}{4}$-in. leather belt with a specific weight of 0.035 lb/cu in. is used. The belt speed is 4000 ft/min. If the maximum allowable belt tension is 300 psi and the belt is 3 in. wide, how much horsepower can it handle? Assume that there is no slip, that the coefficient of friction is 0.4, and that the angle of contact is 180 degrees.

6. For a given belt, a manufacturer gives a horsepower rating of 2.5 hp per inch of width based on a belt speed of 2600 ft/min. The drive is to handle 10 hp; the arc-of-contact correction factor is 0.9. Find the width of belt needed. Assume no application factor. If an application (or service) factor of 1.2 were applied, what would be the belt width? Assume that belt widths are available in 1-in. increments.

7. A $\frac{1}{4}$-in. flat leather belt with a specific weight of 0.035 lb/cu in. is 6 in. wide. The driving pulley is 6 in. in diameter and the driven pulley is 18 in. in diameter. The center distance between pulleys is 56 in. The driving pulley rotates at 1800 rpm. The coefficient of friction is 0.3 for

the driving pulley and 0.4 for the driven pulley. The allowable belt pull is 450 lb. Find the belt capacity in horsepower.

8. What is the belt capacity in Problem 7 if all conditions are the same except that the driving pulley rotates at 500 rpm? (Neglect the centrifugal belt pull.)

9. A $\frac{1}{4}$-in. flat leather belt (specific weight 0.035 lb/cu in.) connects an 8-in. pulley and a 24-in. pulley. The center distance is 48 in. Coefficients of friction are 0.4 for the small pulley and 0.3 for the large pulley. The 8-in. pulley rotates at 1200 rpm. The maximum allowable belt tension is 300 psi. The belt is made endless by a machine-laced connector that makes the joint 90% as strong as the belt material. Find the width of belt needed to handle 25 hp.

10. A V-flat drive is to use belts identical to that shown in Fig. 7-4(b). A 9-in. pitch diameter sheave (36-degree groove angle) is used with a 60-in. flat pulley. The center distance is 60 in. The specific weight of the belt material is 0.04 lb/cu in. The coefficient of friction is 0.4 for the grooved pulley and 0.3 for the flat pulley; 40 hp is to be transmitted by this drive. An application factor of 1.1 is used, the belt speed is 3000 fpm, and the allowable belt tension is 165 lb per belt. Find the number of belts required.

11. A $\frac{1}{4}$-in. round endless belt connects a 1-in.-diameter pulley with a 2-in. pulley. The spacing between pulleys is 4 in. on centers. The allowable stress is 400 psi. If the 1-in. pulley is the driver and rotates at 100 rpm, how much horsepower can be transmitted? Assume that the coefficient of friction is 0.3 for each pulley; neglect centrifugal pull.

12. A $\frac{3}{8}$-in. flat leather belt 12 in. wide is used on a 24-in.-diameter pulley rotating at 600 rpm. The specific weight of the belt is 0.035 lb/cu in. The angle of contact is 150 degrees. If the coefficient of friction is 0.3 and the allowable stress is 300 psi, how much horsepower can it transmit?

13. A $\frac{1}{4}$-in. flat leather belt (specific weight 0.035 lb/cu in.) is used to connect a 10-hp motor to a compressor. The pulley diameter of the driver is 8 in. and the driver rotates at 1200 rpm. The maximum allowable pull for the belt is 80 lb per inch of width. Find the belt width. Assume 180-degree contact and a coefficient of friction of 0.4.

Note. Problems 14 through 16 are to be solved using the manufacturers' tables included in the chapter.

14. A 5V3350 drive belt is connected to a normal torque squirrel-cage

electric motor operating at 1750 rpm. Sheaves for both driver pulley and driven pulley are 12.5 inches in diameter. Find the center distance and the capacity of the belt.

15. A blower is operated 24 hours per day by a 20-hp high-torque motor running at 1750 rpm. The small sheave has a diameter of 7.1 inches; the speed ratio is 2:1. The center distance is 120 inches. Find the belt velocity and the number of belts required.

16. A 30-hp squirrel-cage motor operating at 1160 rpm is to drive a piston pump at 400 rpm. The desired center distance is 80 inches. The pump will operate approximately 4 hours per day. Find the belt speed and the number of belts required.

17. Find the *exact* length for a thin belt that operates on 150-mm and 300-mm pulleys spaced 1.5 m apart on centers. Check your answer with the approximate formula.

18. A 5-mm round belt connects a 20-mm pulley with a 40-mm pulley. The center distance is 150 mm. The 20-mm pulley rotates at 100 rpm and the coefficient of friction of the belt is 0.25. Find the power capacity for this arrangement if the allowable belt stress is 2.6 MN/m^2.

CHAIN DRIVES, HOISTS AND CONVEYORS, ROPES

8

 **Chain Drives**

Chain drives occupy a unique position in the mechanical field. In a sense, they are similar to belt drives; a chain connects sprockets on the driving and driven shafts. The velocity ratio transmitted from one shaft to the other depends on the size of the two sprockets (at the pitch line); unlike that found with belts, the ratio is positive. In belt drives, creep and slip play important roles and must be considered; in a chain drive, which is made up of numerous links, there is a small amount of play in the total length of the chain. This may be desirable in case of small overloads. A chain drive is similar to the open-belt type of drive in that the driving and driven shafts rotate in the same direction.

There are also similarities between chain drives and gear drives. Both types transmit a positive velocity ratio. In a regular spur-gear drive, the driving and driven shafts turn in opposite directions unless an idler is employed. In gear drives employing an annular or ring gear, however, the input and output shafts rotate in the same direction. In a spur-gear drive there is very little contact between meshing gears; therefore, tooth loads are excessive. Since the connection between a chain and its sprocket extends over several teeth, no one tooth is subjected to heavy loads.

Chain drives are often noisy. The silent chain type of drive was developed to counteract this undesirable characteristic. Belt, chain, and gear drives are all dangerous. Belt and chain speeds are high, as is the pitch-line velocity of most gear drives. All should be adequately covered by suitable guards (or completely enclosed) so that no body or clothing parts can possibly be caught in the moving components.

Other comparisons can be made among chain, belt, and gear drives. Gear drives can operate at higher speeds than the others, and are usually more compact. However, chain drives do not require the mounting precision that is so essential for gearing. An advantage of chain drives in comparison to belt drives is that the former do not require tension on the slack side; this leads to better bearing life. In general, chain drives are more compact than the belt type; for a given ratio, the sprockets can be smaller than belt pulleys. For a given load, chain widths are narrower than comparable flat belts. The connecting link on chain drives makes them easy to install—the chain is merely placed over the sprockets and the proper pins are placed in position. In most belt installations, force is needed to place the belt in the proper position; such action is detrimental to mounting bearings. The angle of contact for the driving sprocket can be smaller than the driving pulley for a belt drive. For a chain drive, this angle can be as small as 120 degrees. The center distance can also be small; if a 120-degree angle of contact is provided, the center distance need be no more than just enough to provide clearance for the sprocket teeth. A common center distance is 30 to 50 times

Figure 8-1
Roller-chain assembly.
(Courtesy of Morse Chain, a
division of Borg-Warner
Corporation.)

the pitch. The pitch of a chain is defined as the distance from the center of one roller to the center of the adjacent roller. A film of the proper lubricant in the links of a chain does much to soften any shock loads imposed on it.

Specific applications of chain drives will be discussed later. It might be well to consider one of the classic uses for chain drives—the bicycle. Bicycles invariably utilize chain drives, for several reasons. For one thing, the distance between the pedal sprocket and the driving-wheel sprocket must be small; also, the drive must be positive to ensure proper braking action. The use of a gear train would make the mechanism much more complex and far more expensive.

Roller Chains

Many types of chains are commercially available, each type designed for a specific purpose. Figure 8-1 shows the construction of a roller-chain assembly. A length of roller chain is made up of two main types of links, the roller link and the pin link. The roller link consists of two end plates, two rollers, and two bushings. Alternating with each roller link is a pin link, which consists of two end plates connected with two pins. The spacing between the two end plates is sufficient to accommodate the roller-link assembly. To make a chain endless, a connecting (coupling) link or an offset link is necessary. These links have a removable plate secured with a spring clip or cotter pin. When a chain contains an even number of pitches, the coupler or connection type of link can be used. If, however, a chain contains an odd number of pitches, it is necessary to use an offset link. One end of an offset link contains a pin, while the other contains a roller-bushing combination. A careful look at chain construction will show why such a link is necessary. Silent chains and special types are covered separately in this chapter.

Single-pitch chains are often used to transmit power from one shaft to another. A double-pitch chain is also available at far less cost. The construction of this chain is similar to that of the regular chain, except that the pitch is double. It is popular in conveyor applications, where sprocket center distances are large, operating speeds are low, and horsepower requirements are not great. The outstanding feature of this double-pitch type is its comparatively low cost.

Sprockets

An essential part of any chain drive is the sprocket. A chain drive requires a sprocket on the driving shaft and another on the driven shaft. Chain drives sometimes contain additional sprockets, used for additional power take-offs or for taking up chain slack. When used to take up chain slack, they are known as idlers. Idlers should be mounted on the slack side, not on the tight side. It is desirable to have an adjustment for idler positioning where possible. Figure 8-2 shows a roller-chain sprocket.

Sprocket teeth are standardized to accommodate the roller chain. Various sizes of sprockets may be used, and several types are available. The simplest type is one without a hub. Sprockets are also available with hubs on one side or on both sides. Special arrangements are also possible. Certain of these are used in conjunction with a shear-pin hub; the shear pin is designed to fail in case of an accidental or unexpected overload. Some types can be installed on large lineshafts; these sprockets are made in two parts so that they can be installed without disconnecting the long length of shafting. Smaller sprockets are usually made solid; large ones are sometimes constructed with arms or spokes to reduce weight. In general, the selection of a sprocket depends on the type of application. Figure 8-3 shows the pitch of sprocket teeth.

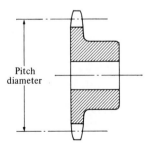

Pitch
diameter

Figure 8-2
Roller-chain sprocket.

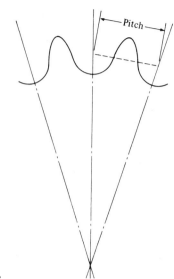

Figure 8-3
Sprocket teeth.

Lubrication

Proper lubrication is necessary to reduce friction and prolong the life of a chain. The American Chain Association defines three basic types of lubrication for chain drives. Type A is the manual or drip type. This type includes oiling by brushing, using a spout can, or directing the lubricant between the link-plate edges with a drip lubricator. Type B includes the oil bath and disc lubrication. With bath lubrication, the lower strand of chain runs through a sump of oil in the drive housing. With disc lubrication, a disc picks up oil from the sump and deposits it on the chain. Type C includes the oil-stream or pressure-spray methods. This is the most thorough type; with these methods, continuous oil is supplied by a circulating pump, through nozzles, to the sprocket side of the chain.

Since chain drives operate in different temperature ranges, it is essential to use an oil of the proper viscosity for operating conditions. It is best to consult chain manufacturers for viscosity recommendations and for the types of oil appropriate for various operating conditions.

Multiple-Strand Roller Chains

For heavy-duty applications, it is often desirable to use multiple-strand roller chains. Double-, triple-, and quadruple-strand chains are commonly used. The use of roller chains in conveyor applications has popularized a variety of special attachment links, usually placed on the pin link. Off-the-shelf attachment links include bent links on one side (or both sides) and straight links on one side (or both sides). Extended pins are also commonly used. Special attachments may have various configurations; each is designed for a specific application. Some are triangular, others resemble channels, and still others provide mounting for slats.

Roller Chain Length

An approximate formula for chain length, which can be used when the center distance is adjustable, is as follows:

$$L = 2C + \frac{N+n}{2} + \frac{(N-n)^2}{4\pi^2 C}.$$ (8-1)

where

C = center distance,

L = chain length,

N = number of teeth (large sprocket),

n = number of teeth (small sprocket).

An approximate value for center distance can be calculated by rearranging the above terms. The chain length has to be converted to a whole number of pitches, preferably an even number. The approximate center distance can be found from Equation 8-2:

$$C = \frac{L - \frac{N+n}{2} + \sqrt{\left(L - \frac{N+n}{2}\right)^2 - 8\frac{(N-n)^2}{4\pi^2}}}{4}.$$ (8-2)

To get a more exact center distance figure, the following equation can be used:

$$C = \frac{L - n\left(\frac{90-\alpha}{180}\right) - N\left(\frac{90+\alpha}{180}\right)}{2\cos\alpha}.$$ (8-3)

where

α = angle of chain with respect to sprocket centerlines.

Horsepower Ratings for Roller Chains*

Table 8-1 is a typical horsepower rating chart for a single-strand standard roller chain. Table 8-1 values are for a $\frac{5}{8}$-in. pitch, number 50 chain. Additional standard sizes of roller chains are available in the following pitches: $\frac{1}{4}$ in., $\frac{3}{8}$ in., $\frac{1}{2}$ in. (2 types), $\frac{3}{4}$ in., 1 in., $1\frac{1}{4}$ in., $1\frac{1}{2}$ in., $1\frac{3}{4}$ in., 2 in., $2\frac{1}{4}$ in., $2\frac{1}{2}$ in., and 3 in. The ASA numbers corresponding to these pitches are: 25, 35, 40, 41, 60, 80, 100, 120, 140, 160, 180, 200, and 240.

A composite horsepower rating chart is an aid in finding the probable chain number for the specific conditions of speed and horsepower for the driving sprocket. If such a chart is not available, one must go directly to the rating tables, such as those in Table 8-1. Tables give specific information based on one strand; also, the table ratings are based on a service factor of one.

Service Factor

The service factor is a judgment factor based on the type of input power and the type of driven load. The listing in Table 8-2 can be used as a guide. To find the design horsepower, the service factor is multiplied by the power to be transmitted. This design horsepower is the figure used in making selections from the horsepower tables.

Multiple-Strand Factors

Since the tables are based on single-strand values, a correction factor must be incorporated if multiple chains are used. The required horsepower-table rating is found by dividing the design horsepower by the multiple-strand factor. If two chains are used, the strand factor is 1.7; if three strands, the factor is 2.5; if four, the factor is 3.3. The following example shows how the horsepower rating tables are used.

The horsepower rating tables are based on approximately 100 pitches and 15,000 hr of service life under full-load conditions. All of the rating information is predicated on the speed of the smaller of the two sprockets, regardless of whether the drive is one of increasing or reducing speed. In addition, the tables are based on horizontally centered shafts. The correct alignment of the sprockets is essential for proper operation. The limiting

---

*Material originally appearing in the *Design Manual—Roller & Silent Chain Drives*, copyright 1955–1968 by American Sprocket Chain Manufacturers, copyright 1974 by American Chain Association, and currently appearing in *Chains for Power Transmission and Material Handling —Design and Applications Handbook*, copyright 1982. Reprinted by permission of the American Chain Association. (This footnote also applies to Tables 8-1 and 8-2.)

Table 8-1

Horsepower Ratings: Standard Single-Strand Roller Chain No. 50 ($\frac{5}{8}$-in. Pitch)

No. of Teeth Small Spkt.	Revolutions per Minute—Small Sprocket																								
	10	25	50	100	200	300	400	500	700	900	1000	1200	1400	1600	1800	2100	2400	2700	3000	3500	4000	4500	5000	5500	6000
9	0.09	0.19	0.36	0.67	1.26	1.81	2.35	2.87	3.89	4.88	5.36	6.32	6.02	4.92	4.13	3.27	2.68	2.25	1.92	1.52	1.25	1.04	0.89	0.77	0.58
10	0.10	0.22	0.41	0.76	1.41	2.03	2.63	3.22	4.36	5.46	6.01	7.08	7.05	5.77	4.83	3.84	3.14	2.63	2.25	1.78	1.46	1.22	1.04	0.90	0.79
11	0.11	0.24	0.45	0.84	1.56	2.25	2.92	3.57	4.83	6.06	6.66	7.85	8.13	6.65	5.58	4.42	3.62	3.04	2.59	2.06	1.68	1.41	1.20	1.04	0.92
12	0.12	0.26	0.49	0.92	1.72	2.47	3.21	3.92	5.31	6.65	7.31	8.62	9.26	7.58	6.35	5.04	4.13	3.46	2.95	2.34	1.92	1.61	1.37	1.19	1.04
13	0.13	0.29	0.54	1.00	1.87	2.70	3.50	4.27	5.78	7.25	7.97	9.40	10.4	8.55	7.16	5.69	4.65	3.90	3.33	2.64	2.16	1.81	1.55	1.34	0
14	0.14	0.31	0.58	1.09	2.03	2.92	3.79	4.63	6.27	7.86	8.64	10.2	11.7	9.55	8.01	6.35	5.20	4.36	3.72	2.95	2.42	2.03	1.73	1.50	
15	0.15	0.34	0.63	1.17	2.19	3.15	4.08	4.99	6.75	8.47	9.31	11.0	12.6	10.6	8.88	7.05	5.77	4.83	4.13	3.27	2.68	2.25	1.92	1.66	
16	0.16	0.36	0.67	1.26	2.34	3.38	4.37	5.35	7.24	9.08	9.98	11.8	13.5	11.7	9.78	7.76	6.35	5.32	4.55	3.61	2.95	2.47	2.11	1.83	
17	0.17	0.39	0.72	1.34	2.50	3.61	4.67	5.71	7.73	9.69	10.7	12.6	14.4	12.8	10.7	8.50	6.96	5.83	4.98	3.95	3.23	2.71	2.31	2.01	
18	0.18	0.41	0.76	1.43	2.66	3.83	4.97	6.07	8.22	10.3	11.3	13.4	15.3	13.9	11.7	9.26	7.58	6.35	5.42	4.30	3.52	2.95	2.52	0	
19	0.19	0.43	0.81	1.51	2.82	4.07	5.27	6.44	8.72	10.9	12.0	14.2	16.3	15.1	12.7	10.0	8.22	6.89	5.88	4.67	3.82	3.20	2.73	0	
20	0.20	0.46	0.86	1.60	2.98	4.30	5.57	6.80	9.21	11.5	12.7	15.0	17.2	16.3	13.7	10.8	8.88	7.44	6.35	5.04	4.13	3.46	2.95	0	
21	0.21	0.48	0.90	1.69	3.14	4.53	5.87	7.17	9.71	12.2	13.4	16.0	18.1	17.6	14.7	11.7	9.55	8.01	6.84	5.42	4.44	3.72	3.18	0	
22	0.22	0.51	0.95	1.77	3.31	4.76	6.17	7.54	10.2	12.8	14.1	16.6	19.1	18.8	15.8	12.5	10.2	8.59	7.33	5.82	4.76	3.99	3.41		
23	0.23	0.53	1.00	1.86	3.47	5.00	6.47	7.91	10.7	13.4	14.8	17.4	20.0	20.1	16.9	13.4	11.0	9.18	7.84	6.22	5.09	4.27	0		
24	0.25	0.56	1.04	1.95	3.63	5.23	6.78	8.29	11.2	14.1	15.5	18.2	20.9	21.4	18.0	14.3	11.7	9.78	8.35	6.63	5.42	4.55	0		
25	0.26	0.58	1.09	2.03	3.80	5.47	7.08	8.66	11.7	14.7	16.2	19.0	21.9	22.8	19.1	15.2	12.4	10.4	8.88	7.05	5.77	4.83	0		
26	0.27	0.61	1.14	2.12	3.96	5.70	7.39	9.03	12.2	15.3	16.9	19.9	22.8	24.2	20.3	16.1	13.2	11.0	9.42	7.47	6.12	5.13			
28	0.29	0.66	1.23	2.30	4.29	6.18	8.01	9.79	13.2	16.6	18.3	21.5	24.7	27.0	22.6	18.0	14.7	12.3	10.5	8.35	6.84	5.73			
30	0.31	0.71	1.33	2.49	4.62	6.66	8.63	10.5	14.3	17.9	19.7	23.2	26.6	30.0	25.1	19.9	16.3	13.7	11.7	9.26	7.58	0			
32	0.33	0.76	1.42	2.66	4.96	7.14	9.25	11.3	15.3	19.2	21.1	24.9	28.6	32.2	27.7	22.0	18.0	15.1	12.9	10.2	8.35	0			
35	0.37	0.84	1.57	2.93	5.46	7.86	10.2	12.5	16.9	21.1	23.2	27.4	31.5	35.5	31.6	25.1	20.6	17.2	14.7	11.7	9.55	0			
40	0.43	0.97	1.81	3.38	6.31	9.08	11.8	14.4	19.5	24.4	26.8	31.6	36.3	41.0	38.7	30.7	25.1	21.0	18.0	14.3	0				
45	0.48	1.10	2.06	3.84	7.16	10.3	13.4	16.3	22.1	27.7	30.5	35.9	41.3	46.5	46.1	36.6	30.0	25.1	21.4	0					

Type A   Type B   Type C

Type A: Manual or drip lubrication; Type B: Bath or disc lubrication; Type C: Oil-stream lubrication.

Table 8-2
Service Factors

	Output conditions		
Input Drive	Smooth	Moderate Shock	Heavy Shock
Internal-combustion engine			
(hydraulic drive)	1.0	1.2	1.4
Electric motor or turbine	1.0	1.3	1.5
Internal-combustion engine			
(mechanical drive)	1.2	1.4	1.7

rpm for each type of lubrication is read from the column to the right of the boundary line. The ratings of multiple-strand chains are greater than those shown in the table.

◆ *Example* ―――――――――――――――――――――――――――――――――

A 10-hp electric motor operates at 1200 rpm and drives equipment with moderate shock at 600 rpm. It is desired to use a double-stranded No. 50 roller chain. Find the number of teeth on each sprocket and the type of lubrication required.

Solution. First, multiply the 10 hp by a service factor of 1.3 obtained from Table 8-2. This gives a design horsepower of 13. Next, find the hp-table rating by dividing the design horsepower by the multiple-strand factor. Thus,

$$\text{hp-table rating} = \frac{\text{design horsepower}}{\text{multiple-strand factor}} = \frac{13}{1.7} = 7.65.$$

Then, scan the 1200 column of Table 8-1; the next reading beyond 7.65 is 7.85. Read horizontally to the left margin from 7.85 and note that a small sprocket of 11 teeth is indicated. Since the driven machine is to operate at half the speed of the motor, the driven sprocket should have twice as many teeth as the smaller sprocket, or 22 teeth. Boundary lines in the table indicate that Type B lubrication—either bath or disc lubrication—is recommended.

Additional consideration should be given to the length of chain and the center distance. For best operation, the length should be approximately 100 pitches. The center distance must be adjusted so that the chain will have an integral number of pitches; an even number is also desirable. The center distance should be such that the angle of chain contact is at least 120 degrees.

Silent Chains

Silent chains are much quieter than roller chains. Chain manufacturers usually produce a somewhat standardized type of silent chain. The joint components differ, but sizes and ratings are standardized to the point of interchangeability. Ratings are given on the basis of horsepower per inch of chain width for the various pitches. Available pitches are $\frac{3}{8}$ in., $\frac{1}{2}$ in., $\frac{5}{8}$ in., $\frac{3}{4}$ in., 1 in., $1\frac{1}{4}$ in., $1\frac{1}{2}$ in., and 2 in. Sprocket teeth for silent chains are similar to those of the roller-chain sprocket, except that the sides are straight. Designations stamped on the link plates indicate the pitch. The complete chain designation is given by the letters SC (indicating silent chain) and one or two digits indicating the pitch in eighths of an inch, followed by two or three digits indicating the chain width in quarters of an inch. For example, the designation SC 814 indicates a silent chain with 8/8 or 1-in. pitch and a chain width of 14/4 or $3\frac{1}{2}$ in.

Sprocket teeth are grooved to allow for the chain link plates. The axial spacing and the widths of the grooves are determined by the particular chain assembly. Figure 8-4 shows a silent-chain drive.

Horsepower rating tables can be found in *Design Manual: Roller and Silent Chain Drives*, published by the American Chain Association, or by contacting any firm that produces silent chains. As with roller chains, a

Figure 8-4
Silent-chain drive.
(Courtesy of Link-Belt Drive
Division, PT Components, Inc.)

service factor must be applied to the specified horsepower to obtain the design horsepower; likewise, the number of links should be an integer. Proper lubrication of this type of drive is another important consideration.

In using silent chains, one must bear in mind that the serpentine type of drive is impossible. Several sprockets can be driven with this type, but not on the reverse side of the chain.

Applications of Chain Drives

Chain drives are used where a positive drive is needed. Chains do not stretch as much as belts and hence require less take-up adjustment. Slipping of belts can generate heat, which can cause explosions; thus, chains are often used under dusty conditions. Chains are not deteriorated by contact with oil or by exposure to the rays of the sun. Chains can be used to convey parts through baking ovens. They are popular components of bicycles, motorcycles, and power lawnmowers. Countless other applications include textile machinery, paper-mill machinery, conveyors, cranes, hoists, dredging equipment, and metal-working and clay-working machinery. In making a selection between gear drives, belt drives, or chain drives, a designer must consider the performance of the equipment and the economics of initial cost, installation, maintenance, and shutdown production losses, as well as the expected life of the driving units.

Light-Duty Chains

Several types of light-duty chains are available to a designer. Bronze or steel sash chain is obtainable in sizes ranging from a metal gage of 0.035 to 0.072 in. Strengths up to 1275 lb are easily obtained. As always, strength values should be divided by a suitable factor of safety before being incorporated in a design. Another light-duty chain is the plumbers' chain, obtainable with brass or steel links. Chains of this type are generally used for supporting small loads or for retaining such parts as electrical-outlet caps.

Figure 8-5 shows a Bead Chain®, which is popularly used in pull-chain lamp sockets and key chains. This is a kinkless chain consisting of hollow balls held together with dumbbell-shaped links. It is commercially available in bead diameters ranging from 0.072 to $\frac{3}{8}$ in. and with approximate tensile strengths of 10 to 185 lb. Certain sizes are made to close tolerances for use in sprocket drives, as shown in Fig. 8-5. Attachments are available in various materials and finishes for use in applications other than drives; this is helpful in designing for appearance.

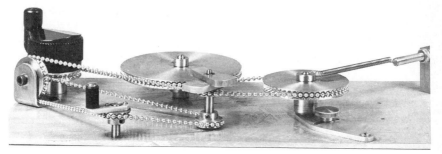

Figure 8-5
Bead Chain® drive.
(Courtesy of the Bead Chain Manufacturing Company.)

Hoisting Chains

Heavy-duty link chains are available in assorted sizes and link shapes. Some of these chains have working load limits of over 50,000 lb. Many are made from open-hearth basic steel wire and then welded. Chain types include straight link, twist link, single jack, and double jack. A designer has quite a large selection of sizes available; specific information should be obtained from manufacturers' catalogs. Sling-chain assemblies complete with ring and grab hook are readily available. When using sling chains, it is desirable to check the angle that the sling makes relative to the direction of the load; such a force analysis can prevent overstressing and possible failure of the chain links. Catalog tables frequently give working loads for sling chains used under various working angles. Figure 8-6 shows a straight-link hoisting chain.

Hoisting Accessories

A chain is useless without certain other parts to form an assembly. One end of a sling chain usually has a circular steel ring or a pear-shaped end ring. A hook is needed at the other end. Several configurations are used, depending on the application of the chain. The parts are usually made of forged steel and are often heat-treated after forging, to provide additional

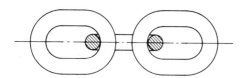

Figure 8-6
Straight-link hoisting chain.

strength and toughness—any failure is serious, and may be physically dangerous. Several different types of hooks are used; these include hoist hooks, grab hooks, slip hooks, and sling hooks. Various sizes are obtainable with safe working loads as high as 25 tons. Sections of chain can be joined in several ways. For light-duty applications, a simple S-hook can be used. Special links that come in sections can be riveted together to join chain sections. A swing link is an assembly that pivots in the middle when opened; chain ends are placed in the link, and the link is then closed to lock together the two chain ends. Working loads for any accessory should be checked carefully before the accessory is used.

 Hoists and Conveyors

As a result of mechanization and automation in modern manufacturing plants, structural systems are often designed around such stock components as hoists, trolleys, and accessories. Ingenuity of design thus reflects one's ability to use stock assemblies with carefully selected structural members so that local codes are satisfied, accepted engineering practices are followed, and the final selection fits into the available space and performs the proper function. Since space is usually at a premium, compactness is essential.

Types of Hoists

The simplest type of hoist is a block-and-tackle arrangement. This usually consists of a fixed multiple-groove pulley and a movable pulley. Either ropes or chains can be used; for this simple type, ropes are convenient. The theoretical mechanical advantage is found by noting the number of ropes supporting the load. Frictional losses are great, and thus efficiency is exceptionally low.

A differential chain hoist of the endless-chain variety is frequently employed where limited use is expected. This is an inexpensive piece of equipment, but frictional losses are high and the mechanical efficiency is only about 35%. After hoisting, the load is held in place by friction in the operating parts. The mechanical advantage is attained by virtue of the fact that the stationary pulley assembly has two different size pulleys. Figure 8-7 shows how the chain is used in conjunction with the pulleys. This type of hoist is available up to capacities of $1\frac{1}{2}$ tons; however, smaller sizes are preferable, since the hand-chain pull at higher loads is excessive. A $\frac{1}{2}$-ton hoist can require a pull of over 100 lb on the hand chain; a $1\frac{1}{2}$-ton unit can require a pull of over 200 lb.

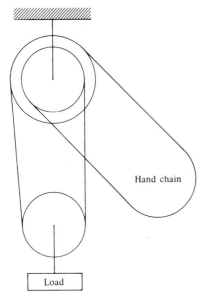

Hand chain

Load

Figure 8-7
Differential chain hoist.

For this endless-chain type of hoist, the ideal mechanical advantage is found by dividing the diametral difference of the pulleys connecting the hand chain into twice the diameter of the larger pulley.

◆ *Example* ───

If the hand chain shown in Fig. 8-7 connects with a 350-mm pulley and a 300-mm pulley, find the theoretical hand-chain pull needed to lift a load of 300 kg. (Ignore all frictional losses.)

Solution. Mechanical advantage (ideal) $= \dfrac{2(350)}{350 - 300} = 14$.

Then,

Pull (theoretical) $= \dfrac{300(9.81)}{14} = 210$ N.

Another type of hand-operated chain hoist is essentially a worm-gear speed reducer. The hand chain operates a worm, and the load-chain sheave is connected to the worm wheel (gear). Efficiency is slightly higher than that of the differential chain hoist. Operation is smooth and the unit is self-locking; thus the load cannot be lowered without moving the hand chain. The loop of chain that operates the worm is made endless for convenience.

Figure 8-8 shows a spur-geared chain hoist. This has an efficiency of approximately 85%; thus, a greater lifting capacity is available for a reasonable pull on the hand chain. A planetary gear train is used to achieve the mechanical advantage. Hoists of this type are available in capacities ranging from $\frac{1}{4}$ to 40 tons. Planetary gear trains are discussed in Chapter 12.

Wire-rope electric hoists are the workhorses in most factory operations. Hand-operated hoists require much physical effort, and usually two operators are needed to move large pieces of material (or equipment) into position. Figures 8-9(a) and 8-10 show two types of electric hoists that are readily available commercially. Such hoists are designed to operate on the flanges of I-beams. Often, beams used in the roof structure of a building are designed to accommodate hoists. Commercial units ranging from $\frac{1}{4}$ ton to 12 tons are manufactured.

When a designer incorporates such equipment in a mechanical-structural project, he or she must check the operating features of the hoist under

Figure 8-8
Spur-geared chain hoist.
(Courtesy of the Hoisting Equipment Division of Eaton Corporation.)

Figure 8-9
**(a) Wire-rope electric hoist
with trolley. (b) Plain trolley.**
(Courtesy of the Hoisting
Equipment Division of Eaton
Corporation.)

consideration. Typical items to be checked include the size of the I-beam, the hoist capacity, provision for overload, lift, minimum radius of curvature for beam, head room, electrical specifications, type and location of operating control, load block and hook size, and trolley type if needed. A permanently located hoist can sometimes be used; this is usually suspended by bolted connections. In checking out beam requirements, the usual flexure and deflection considerations must be used. If the hoist is to move along the beam, calculations should be made under maximum load. This total load must include beam weight, load weight, and the total weight of the hoisting equipment. Provision must also be made for accidental overload.

Figure 8-9(b) shows a plain trolley, designed to operate on the lower flange of an I-beam. Each available size is adjustable to fit several sizes of beams. Various types of hoists can be mounted on trolleys, thus adding flexibility to materials-handling systems. Geared trolleys are also available. These have geared wheels that engage a driving pinion. The driving pinion is driven by means of a chain wheel and hand chain. Thus, the trolley can easily be moved laterally.

Figure 8-10 shows a type of compact electric hoist that does not extend far below the beam (track); this type allows a greater vertical lifting distance.

Conveyors

Materials handling is an important phase of plant operation. It is desirable to move equipment and materials the minimum distance; however, some production processes, particularly assembly lines, require that the parts be moved while the personnel remain in one work station. Various types of conveyors are commercially available; one should consult literature in the materials-handling field for specific information. The chain conveyor is probably the most popular overhead conveyor. Its main function is to move materials or partially assembled parts from one area of an assembly plant to another without the loss of valuable floor space. Special hooks are provided to support the parts being moved. In most cases, the weight of the parts cannot be supported by the chain; a conveyor track does so instead. Chain power should be sufficient to move the conveyor, with loads, and still overcome friction with the track.

Roller conveyors and skate-wheel assemblies are used extensively in the production floor areas. Skate wheels can be used where the parts or materials being moved are flat on the bottom and large enough to contact several of the wheels. The horizontal force needed to move the work depends on rolling resistance and on the friction in the bearings. Most skate

Figure 8-10
Lo-Hed® electric hoist.
(Courtesy of AAI Corporation.)

wheels are mounted with ball bearings to reduce friction. Rollers are much more commonplace. These are usually made of steel tubing with ball-bearing mounts in the ends, and are available in diameters ranging from 1 in. to 5 in. Some are made of materials other than steel, or are plastic-coated to protect the materials or parts being handled. The larger diameters require less horizontal force to move the conveyor. Commercial rollers have ratings up to 10,000 lb per roller.

Rollers are usually mounted in a frame; some of these are expandable to fit various conditions of production. Other rollers are in the form of spring-loaded shafts that can be removed from the frame much as the pin is removed that connects a wristwatch with its band. Also, certain rollers are placed in a spring-loaded frame to somewhat absorb the shock resulting from parts being transferred from one conveyor to another. The frame supporting the rollers is designed in the same manner as a beam with a uniformly distributed load; provision must be made for the holes used to support the roller bearings. In some conveyor systems, live rollers are used. These are rollers that are rotated by belts or chains in pulleys or sprockets. Live rollers are usually used when the product being moved has a flat bottomed surface. However, when a belt is used on the top of the rollers, live rollers must be used; on certain systems, the live roller may take the form of a pulley at the driving end only.

Ball transfer devices are often used to move a product from one conveyor to another. This device is usually made using large, equally spaced balls recessed in a flat plate. Each ball is nested in a hemispherical groove, and small balls of ball-bearing quality are added to separate the large ball from the groove. The small balls are shielded so that they cannot ride out of the groove. Friction losses are low in such a device. They can be calculated in a manner similar to that employed in the study of rolling resistance.

 **Ropes**

Manila Ropes

Ropes made of Manila hemp are popular in marine applications and in certain areas of the construction industry. Block-and-tackle assemblies frequently use this type of rope. Rope sizes range from a nominal diameter of $\frac{3}{16}$ in. to a 4-in. diameter. Manila ropes are commonly available with 3 strands (without center core) and 4 strands (either with or without a core). The strength of a rope varies with age; a factor of safety of 5 is often employed. The tensile strengths range from approximately 420 lb ($\frac{3}{16}$-in.

diameter) to 105,000 lb (4-in. diameter). Ropes should not be used where shock loading is likely to occur. Each strand of rope contains many threads; the number depends on the size of the rope. Before placing any rope in operation, provision should be made to prevent the strands or threads from unwinding.

Certain toys and small electromechanical assemblies use a nylon drive cord. Such a use is only possible for light driving applications. The tensile strength is good, but the sizes used are on the small side; thus, tensile loads on the cord must be kept to a minimum. These drives use nylon cord approximately the size of a fishing line.

Wire-Rope Manufacture

Wire-rope sections are made by twisting wires into a strand and then twisting the strands around a central core. The core can be made of either wire or fiber, although a fiber core is more common. Besides providing support for the strands, a fiber core gives flexibility to the wire rope and aids in internal lubrication. Where high temperatures are encountered, the metal core is necessary; it gives additional strength to the assembly but reduces its flexibility.

There are several rope constructions that should be studied. Figure 8-11(a) shows *regular lay*, the most common construction. In regular lay, the wires are twisted in one direction, the strands in the opposite direction. Thus, the outer wires are approximately parallel to the centerline of the rope. Ropes constructed with the regular lay are less likely to kink and thus are easier to handle. Figure 8-11(b) shows the *Lang lay* construction. In this type, the wires are twisted in one direction and the strands are wound in the same direction. The outer wires run diagonally with respect to the centerline; thus, a greater length of wire is exposed with more area for wearing. Still another possibility is the *reverse lay*, in which the strands are alternately regular and Lang lay. Another classification is based on *right* or *left lay* for the assembled rope, determined by viewing the rope when it is held vertically. If the strands are twisted upward to the right (as when viewing

Figure 8-11
Wire-rope construction. (a) Regular lay. (b) Lang lay.

(a)   (b)

Figure 8-12
**A 6 × 7 regular-lay wire rope
with fiber core.**

external right-hand threads), the rope is considered right lay. If the strands
are twisted upward to the left, the rope is considered left lay. Most wire
ropes are constructed with right lay rather than left lay; very few applications require left-lay rope.

Designation for wire ropes is based on the number of strands and the
number of wires in the strands. Figure 8-12 shows the cross section of a
6 × 7 wire rope (6 strands of 7 wires each). Figure 8-11 shows 6 × 19
construction.

Applications of Wire Ropes

Wire ropes are extensively used in the construction and materials-handling fields. Their uses fall into two general categories, "standing ropes"
and "pulley ropes." Ropes used as standing ropes are not operated over
sheaves. Typical examples include (1) guy strand and rope and (2) highway
guard cable. Ropes used as pulley ropes are rigged over sheaves; hoists and
elevators are typical examples of this use. The following list gives a *few* of
the typical applications, with some common wire-rope designations placed
in parentheses:

Guy (6 × 7)   Hauling (6 × 7)

Hawser (6 × 37)   Hoisting (6 × 19)(6 × 37)(8 × 19)

Elevator (8 × 19)(6 × 19)   Sash cord (6 × 7)

Mooring (6 × 24, 7-fiber cores)   Bridge ropes (6 × 7)(6 × 19)

Highway guard cable (3 × 7)   Drag line rope (6 × 37)

Materials for Wire Ropes

Because of the unusual uses to which they are put, wire-rope materials
require the following properties: high strength in tension, toughness, good
bending-fatigue resistance, and resistance to abrasion. In addition, certain
wire ropes must have good resistance to corrosion. Several rope-wire grades
are used in rope fabrication, including extra-improved plow steel, improved

plow steel, plow steel, and mild plow steel, either with or without a galvanized coating.

Sheaves, Drums, and Fleet Angles

Bending of wire rope can cause serious fatigue problems. For this reason, drum diameters must be large with respect to the size of the outer wires of the rope. Experience and good practice dictate the following relationships between the sheave or drum diameter and the diameter of the outer wires:

800:1   heavily loaded, fast-moving ropes
1000:1   mine hoists
900:1   elevators

These are minimum values; conditions of operation can necessitate either larger or smaller values. It is desirable to use drums with grooves on the winding surface. These grooves (and those on sheaves) should be slightly larger than the rope, to forestall binding. Drum grooves should be placed so as to maintain clearance between ropes, thus reducing abrasion.

The radial pressure on a drum can be computed from the following equation:

$$P = \frac{L}{RD},$$
(8-4)

where

P = pressure (psi),

L = load on rope (lb),

R = tread radius (in.),

D = rope diameter (in.).

It should be noted that the rope diameter is the diameter of the circle that will just enclose all the strands of the rope. Table 8-3 shows maximum radial drum pressures for typical conditions.

Figure 8-13 shows how a fleet angle is defined. The lead sheave that directs wire onto a drum should be placed far enough away from the drum that the fleet angle will not be excessive. Desirable maximum values are $1\frac{1}{2}$ degrees for smooth drums and 2 degrees for grooved drums.

◆ *Example* ───────────────────────────────

A drum has a total traverse travel of 2 ft, or 1 ft on each side of the centerline of the drum. What is the minimum safe distance from the drum

Table 8-3
Maximum Radial Pressure (psi)

Rope construction	Cast Iron	Cast Steel
6 × 19 regular lay	500	900
6 × 19 Lang lay	575	1025
8 × 19 regular lay	600	1075
6 × 37 regular lay	600	1075
6 × 37 Lang lay	700	1225
Flattened strand	800	1450

to the lead sheave, if the drum is grooved?

Solution. For a grooved drum, the maximum fleet angle is 2 degrees. From the right angle formed from the centerline of the drum, the linear distance to the lead sheave is found to be

$$\text{distance} = \frac{1}{\tan 2} = \frac{1}{0.035} = 28.6 \text{ ft.}$$

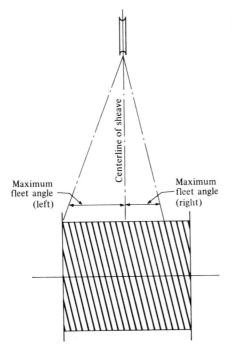

Figure 8-13
Fleet angle.

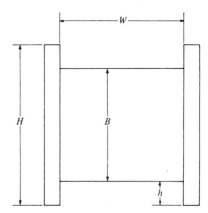

Figure 8-14
Drum dimensions.

When winding a wire rope on a drum, it is important to wind in the proper direction. If this is not done correctly, the coils will tend to spread apart. The next layer will then wedge in the spaces, abrading the rope and shortening its useful life. The correct direction for winding the first layer on a smooth drum depends on the lay of the rope (right or left) and on whether the rope is overwinding or underwinding (that is, whether the rope first contacts the drum surface on the top or bottom of the drum). During overwinding, the top of the drum turns toward the viewer (situated behind the drum) as the rope winds; during underwinding, the top of the drum turns away from the observer.

For right-lay ropes that are overwinding, the rope should travel from left to right. For right-lay ropes that are underwinding, the rope should travel from right to left. These rules are for the first layer only. For left-lay ropes, the directions of travel are reversed. Thus, in overwinding, the path is from right to left; in underwinding, from left to right.

Figure 8-14 defines various parts of a drum for use in the following equation, which gives the length of cable that can be used on a drum.

$$L = \frac{\pi W (H + B)(H - B)}{48 d^2}$$

or

$$L = \frac{0.2618 W h (B + h)}{d^2},$$

(8-5)

where

H = head diameter (in.),

B = barrel diameter (in.),

h = depth of cable (in.),

W = width between flanges (in.),

d = cable diameter (in.),

L = cable length (ft).

Safety Factors

A proper safety factor is essential in using wire ropes. Manufacturers provide the strength of the rope in pounds. The working load is found by dividing this value by a suitable factor of safety. Table 8-4 provides a guide for typical minimum values. Applicable codes must be followed. Judgments for increasing these minimum values should be based on the consequences of failure, the length of rope used, the types of drums and sheaves, the velocity and acceleration of the ropes, the abrasion and corrosion conditions, and the inspection procedures used.

Accessories

Rope thimbles are used to provide protection and support for a looped rope end. These are used in conjunction with wire-rope clips. Figure 8-15 shows a typical wire-rope clip. Basically, this assembly consists of a base, a U-bolt, and nuts. The live or long length of cable should rest against the base; the U-bolt rests against the short end of the rope. The number of clips used depends on the size of the rope. Catalog data from the manufacturer are important in selecting accessories. Rope capacities must be obtained from the same source.

 **Summary**

Chain drives are members of the power-transmission family. Positive velocity ratios are maintained by using appropriate sprockets. Roller chains are more widely used than silent chains. In designing chain drives, one should consider the size needed to handle the required power, the length of chain, the number of pitches, and the type of chain. Correction factors must be applied for service conditions and for multiple-strand operation. Lubrication systems need careful selection to ensure proper chain life.

Table 8-4
Safety Factors

Type of Service	Minimum Safety Factor		
Track cables	3.2		
Guys	3.5		
Mine shafts	8.0 for depths to 500 ft		
	7.0 for depths 500 to 1000 ft		
	6.0 for depths 1000 to 2000 ft		
	5.0 for depths 2000 to 3000 ft		
	4.0 for depths 3000 ft and more		
Miscellaneous hoisting equipment	5.0		
Haulage ropes	6.0		
Overhead and gantry cranes	6.0		
Jib and pillar cranes	6.0		
Derricks	6.0		
Small electric and air hoists	7.0		
Hot ladle cranes	8.0		
Slings	8.0		
Elevators:	**Passenger**	**Freight**	**Dumbwaiter**
Car speed, fpm:			
50	7.50	6.67	5.33
100	7.85	7.00	5.66
150	8.20	7.32	5.98
200	8.54	7.64	6.29
250	8.86	7.92	6.59
300	9.17	8.20	6.88
350	9.47	8.45	7.18
400	9.75	8.70	7.46
450	10.01	8.93	7.74
500	10.25	9.14	8.00
550	10.47	9.32	
600	10.68	9.50	
700	11.00	9.78	
800	11.25	10.02	
900	11.44	10.21	
1000	11.57	10.34	
1100	11.67	10.43	
1200	11.75	10.50	
1300	11.81	10.54	
1400	11.85	10.58	
1500	11.87	10.61	

Figure 8-15
A typical wire-rope clip.

Hoisting ropes and chains are frequently used in maintenance work. Since failure in such applications can be extremely dangerous, one must be careful to select a suitable factor of safety.

Special types of light-duty chains are available. Most of these are used to hoist parts and machinery in light assembly work, although some are useful for light power transmission.

Hoists, conveyors, and transfer devices are the backbone of the materials-handling function. In mass-assembly plants, the handling of bulk materials, small parts, and subassemblies constitutes a major portion of the total operation. The cost of moving materials and parts can amount to 25% of the total manufacturing expense; parts and subassemblies can be moved more than 50 times during the manufacturing function. The design of materials-handling equipment is often outlined by the plant facilities department and further refined by the suppliers of the equipment.

The Engineering Steel Chain Division of the American Chain Association publishes *Applications Handbook*, which outlines the types and components of engineering steel chain. This publication covers specific design information on conveyor, bucket-elevator, drive, and tension-linkage chains, as well as information on sprockets.

Wire ropes are used in the manufacturing, marine, and construction fields. Typical applications include:

1. cranes
2. power shovels
3. derricks
4. skimmers
5. highway guard cables
6. backfillers
7. elevators
8. guy lines
9. dredges

In addition to strength properties, other considerations such as crushing, kinking, and abrasion must be considered in wire-rope selection.

Questions for Review

1. What advantages do chain drives have over belt drives? What disadvantages?

2. How does a roller chain differ from a silent chain?

3. When is an offset link used as a coupler in a roller-chain application? Why?

4. List three types of light-duty chains, with an application for each.

5. How do hoisting chains differ from roller chains?

6. Name three types of hoisting accessories, and explain where each is used.

7. Differentiate between *Lang* lay and *regular* lay. Explain why each of these is used.

8. What is meant by the designation 6 × 37 when classifying wire ropes?

9. What is meant by *independent wire rope core* (IWRC)? How does it differ from other constructions? Why it is used?

10. What advantage does a grooved drum for wire rope have over a smooth drum?

11. What is meant by the term *fleet angle*?

12. What distinctive feature does a ball transfer device have over a skate-wheel conveyor? Where are ball transfer assemblies used?

13. Check the codes dealing with elevator construction and operation in your locality. List their specific requirements.

14. List 10 applications of wire ropes in the construction field. Which applications involve a static condition for the wire rope?

15. Examine the chain drive on a regular bicycle. What ratio exists between the pedal sprocket and the wheel sprocket? How many pitches are in the chain? What provision is made for adjustment?

16. Carefully study the roller-chain horsepower rating tables. How does the horsepower capacity vary with respect to (a) sprocket size, (b) rotational speed, and (c) pitch? What conclusions can be drawn—with respect to sprocket size and speed—about the types of lubrication systems employed?

17. Why is it desirable to have an *even* number of links in a roller-chain application?

Problems

1. A roller-chain drive is operated horizontally by a 15-hp electric motor. The chain, in turn, drives a rotary screen (an application with moderate

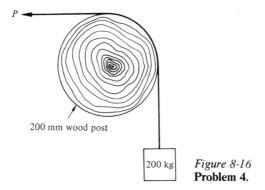

Figure 8-16
Problem 4.

200 mm wood post

200 kg

shock). The speed of the motor is 1200 rpm and the rotating screen operates at 625 rpm. It is desirable to use single-stranded, No. 50 chain. Find the number of teeth on the small and large sprockets and the type of lubrication needed.

2. For conditions similar to those in Problem 1, what size of small sprocket is needed if two strands are used and the motor operates at 900 rpm? What type of lubrication system should be used?

3. A roller-chain application uses three strands of No. 50 chain. The driving sprockets have 25 teeth and rotate at 3000 rpm. What horsepower can be safely handled if the service factor is 1.4?

4. A rope supports a weight of 200 kg and is wrapped $1\frac{1}{4}$ turns around a horizontal, stationary wood post. If the coefficient of friction is 0.3, find the force P needed to prevent slipping. The post is 200 mm in diameter. Express P in newtons. Refer to Fig. 8-16.

5. A 300-lb load is attached to a rope that passes over a bumper and loops one full turn around a vertical post. Referring to Fig. 8-17 for

Figure 8-17
Problem 5.

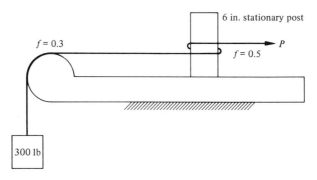

6 in. stationary post

$f = 0.3$

P

$f = 0.5$

300 lb

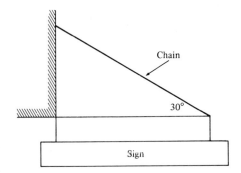

Figure 8-18
Problem 7.

dimensions and for coefficients of friction, find the force *P* needed to prevent downward motion of the load.

6. A block and tackle has an ideal mechanical advantage of 4. If this device uses Manila rope with a breaking strength of 2650 lb, what safe load can be handled if a factor of safety of 5 is to be used?

7. For the sign in Fig. 8-18, the allowable load limit for the link chain is 3300 lb. Neglecting the weight of the other parts, find the limiting weight for the sign. If the chain made an angle of 45 degrees with the horizontal member, what would be the limiting weight of the sign?

8. A guy wire is made of 6 × 7 galvanized iron wire rope with a strength of 7160 lb. Assume a factor of safety of 4. What load *P* could safely be applied to the vertical pole shown in Fig. 8-19, provided the pole remains rigid?

Figure 8-19
Problem 8.

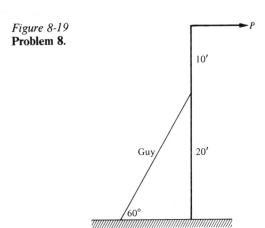

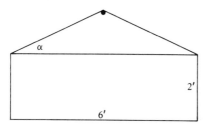

α

2'

6'

Figure 8-20
Problem 9.

9. A 2 ft × 6 ft bulletin board weighs 50 lb and is to be supported with a sash chain having a tensile strength of 600 lb (see Fig. 8-20). What is the minimum angle that the chain can make with the top of the bulletin board if the chain is not to be overstressed? Assume a factor of safety of 5.

10. A hinged boom is supported by two chains as shown in Fig. 8-21. The allowable load on each chain is 3300 lb. Find the total vertical load that can be supported at A. Note that chain loads are determined by the angles made, not by the linear distances; thus, no dimensions are given. (*Hint:* Find the "load value" for one imaginary chain, AE, that could replace AC and AD. Then, three forces are in equilibrium about point A.)

11. Three guy wires are used in conjunction with an 8-m high antenna. They are spaced 120 degrees apart and make an angle of 30 degrees with the horizontal. The maximum horizontal load on the mast is 1300 N and acts at various angles as the wind changes direction. Assume that the 1300-N wind load is concentrated at the midpoint of the antenna and that the guy wires are affixed at the top. Find the maximum guy-wire pull.

12. If the guy wires in Problem 11 make an angle of 60 degrees with the horizontal and the other conditions remained the same, what would be the maximum guy-wire pull?

13. In Fig. 8-22, the allowable load limit for the link chain used in member AB is 14 680 N. How much load can be applied by P?

14. The allowable load for the chain in Fig. 8-23 is 3300 lb. For the arrangement shown, find the permissible sign weight, neglecting the weight of the horizontal member. What problems would be encountered in designing the horizontal member?

15. A hoist uses $\frac{1}{2}$-in. wire rope of 6 × 19 IWRC (independent wire rope core) construction. It weighs 0.46 lb/ft and has a breaking strength of

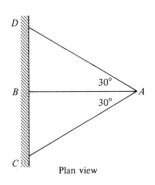

Plan view

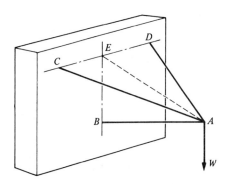

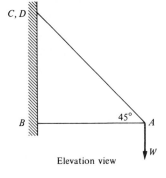

Elevation view

Figure 8-21
Hinged boom: Problem 10.

Figure 8-22
Problem 13.

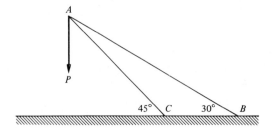

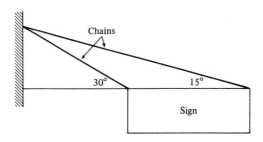

Figure 8-23
Problem 14.

11.5 tons. The single hoisting cable moves a load of one ton at a speed of 480 fpm for a total lift of 800 ft. A factor of safety of 10 is desired. Does this wire rope meet the safety requirement if the load is accelerated for the first 40 ft of the lifting operation?

16. A smooth drum for a wire rope has a diameter of 18 in. and an available width for winding of 18 in. What is the minimum distance from the drum to the lead sheave if a proper fleet angle is to be maintained?

17. A grooved wire-rope drum has a total traverse travel of 3 ft and the lead sheave is 40 ft away. Does this distance provide a proper fleet angle?

18. A winch is equipped with a $\frac{3}{8}$-in.-diameter 6 × 19 wire rope. Assume a factor of safety of 6 and a breaking strength of 6.56 tons. What steady

Figure 8-24
Problem 20.

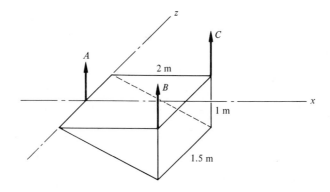

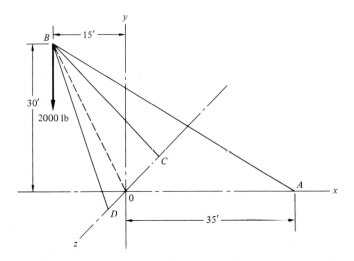

Figure 8-25
Problem 21.

load can be pulled at an angle of 30 degrees with respect to the horizontal?

19. A wire-rope reel has a head diameter of 36 in. and a barrel diameter of 24 in. How much width between flanges is needed to coil 3000 ft of $\frac{1}{2}$-in.-diameter 6 × 19 wire rope?

20. In Fig. 8-24, the triangular prism weighs 600 kg/m³ and is supported from the ceiling by three cables 6.5 m long. Cable A is at the middle of one edge. Find the pull on each cable in newtons.

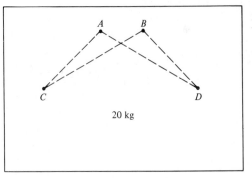

Figure 8-26
Problem 22.

21. Find the "stresses" in the cable and legs of the derrick shown in Fig. 8-25. Distance DO = 5 ft.; distance CO = 10 ft.

22. In Fig. 8-26, wall fasteners at A and B support the 20-kg load on the frame. Cords CAD and CBD each make angles of 45° and 30° with respect to the horizontal. If each cord has a tensile strength of 500 newtons, what is the factor of safety under these conditions?

BRAKES

9

 Materials

Braking action demands much of materials. Stopping a moving member abrades both members, since two materials are sliding relative to each other, and produces a great deal of heat. Thus, brake materials should display good wear characteristics as well as an ability to withstand high temperatures. A good coefficient of friction is necessary to keep operating forces within a reasonable range. Asbestos composition materials facing cast-iron surfaces are frequently used.

    Brakes and clutches are somewhat similar. In the case of a friction clutch, the driving member (with the help of friction) transmits torque to the driven member. On the other hand, brakes, with the help of friction, will stop a moving member or hold a member stationary. In the case of rotating shafts, the torque-holding capacity of the brake must exceed the torque of the shaft in order to be effective. Several types of brakes can be used. Proper selection depends on the type of application, the load to be contained, and the time within which one must brake effectively.

 Analysis of Simple Block Brake

In Fig. 9-1, the following analysis can be made if it is assumed that the pressure between the braking surfaces is uniformly distributed and that angle θ is reasonably small (60 degrees or less). It is unlikely that the distribution of pressure is entirely uniform; thus the following analysis is

Figure 9-1
Block brake.

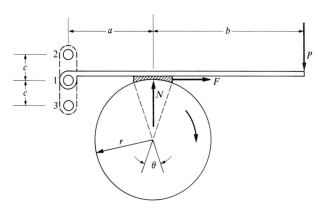

not always 100% correct:

$$F = \frac{T}{r} \quad \text{and} \quad N = \frac{F}{f}. \tag{9-1}$$

For clockwise rotation of the drum, moments about point 1 would be

$$(a + b)P + (0)F = aN.$$

Thus

$$P = \frac{aN}{a + b} = \frac{aF/f}{a + b} = \frac{aT/fr}{a + b} \quad \text{for pivot 1.}$$

Moments about point 2 would be

$$(a + b)P = aN + cF.$$

Thus

$$P = \frac{aN + cF}{a + b} = \frac{(aT/rf) + (cT/r)}{a + b} = \frac{(T/r)[(a/f) + c]}{a + b} \quad \text{for pivot 2.}$$

Moments about point 3 would be

$$(a + b)P = aN - cF.$$

Thus

$$P = \frac{(aT/rf) - (cT/r)}{a + b} = \frac{(T/r)[(a/f) - c]}{a + b} \quad \text{for pivot 3.}$$

If the drum rotates counterclockwise, the direction of F is reversed and the equations become

$$P = \frac{aN}{a + b} = \frac{aF/f}{a + b} = \frac{aT/fr}{a + b} \quad \text{for pivot 1,}$$

$$P = \frac{aN - cF}{a + b} = \frac{(T/r)[(a/f) - c]}{a + b} \quad \text{for pivot 2,}$$

$$P = \frac{aN + cF}{a + b} = \frac{(T/r)[(a/f) + c]}{a + b} \quad \text{for pivot 3.}$$

◆ *Example*

A block brake has a configuration similar to Fig. 9-1, using pivot 2. The drum is 12 in. in diameter, rotates at 100 rpm, and transmits 5 hp. Assume that $a = 6$ in., $b = 12$ in., and $c = 4$ in., and that the coefficient of friction is 0.3. Find the operating force P needed to stop the drum if

1. clockwise rotation is assumed;
2. counterclockwise rotation is assumed.

Solution.

$$T = \frac{63,000(\text{hp})}{N} = \frac{63,000(5)}{100} = 3150 \text{ in-lb};$$

$$P = \frac{(T/r)[(a/f) + c]}{a + b} = \frac{(3150/6)[(6/0.3) + 4]}{6 + 12} = \frac{525(24)}{18} = 700 \text{ lb};$$

$$P = \frac{(T/r)[(a/f) - c]}{a + b} = \frac{(3150/6)[(6/0.3) - 4]}{6 + 12} = \frac{525(16)}{18} = 467 \text{ lb}.$$

The preceding analysis was based on a short shoe (with a small contact angle) and uniform pressure. In the case of *long shoe* brakes, this analysis is not valid because the pressure is not uniform. Frictional wear is proportional to the product of pressure and velocity. Since the velocity is constant for all parts of the shoe, the wear at any given point must be based on the pressure at that point. Therefore, when the contact angle is much larger, such as 90°, corrections must be made in the analysis to deal with the location of the pivot point, the direction of rotation, and the distribution of pressure.

 **Band Brake**

Figure 9-2 is a schematic diagram of a simple band brake. The following equations can be written:

$$T = (F_1 - F_2)r;$$

$$\frac{F_1}{F_2} = e^{f\theta} \qquad \text{or} \qquad F_1 = F_2 e^{f\theta};$$

$$\frac{T}{r} = F_1 - F_2 = F_2 e^{f\theta} - F_2 = F_2(e^{f\theta} - 1)$$

or

$$F_2 = \frac{T}{r(e^{f\theta} - 1)};$$

$$F_1 = F_2 e^{f\theta} = \frac{T e^{f\theta}}{r(e^{f\theta} - 1)}.$$

Taking the arm as a free body, moments can be taken about the pivot point.

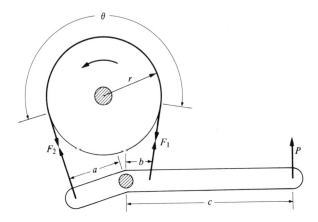

Figure 9-2
Band brake.

Thus, $F_2 a = F_1 b + Pc$. Operating force P would then be

$$P = \frac{F_2 a - F_1 b}{c} = \frac{Ta}{cre^{f\theta}} - \frac{Tbe^{f\theta}}{cr(e^{f\theta} - 1)},$$

$$P = \frac{T}{cr}\left(\frac{a - e^{f\theta}b}{e^{f\theta} - 1}\right). \tag{9-2}$$

The cross-sectional area of the band is found by designing for the tight side (F_1) of the band, which is in tension.

◆ *Example* ──────────────────────────────

Referring to Fig. 9-2, find the operating force required to stop rotation if the drum is 20 in. in diameter, rotates at 100 rpm counterclockwise, and transmits 5 hp. Assume that $a = 6$ in., $b = 1$ in., and $c = 12$ in., and that the coefficient of friction is 0.3. The angle of wrap is 240 degrees.

Solution.

$$T = \frac{63,000(\text{hp})}{N} = \frac{63,000(5)}{100} = 3150 \text{ in-lb};$$

$$e^{f\theta} = 2.718^{(0.3)(240\pi/180)} = 2.718^{1.26} = 3.52;$$

$$P = \frac{T}{cr}\left(\frac{a - e^{f\theta}b}{e^{f\theta} - 1}\right) = \frac{3150}{12(10)}\left(\frac{6 - 3.52}{3.52 - 1}\right)$$

$$= \frac{3150(2.48)}{120(2.52)} = 25.8 \text{ lb}.$$

Note: If $e^{f\theta}b$ is greater than a, the result will be negative and the brake will be self-locking. This could be dangerous in many cases; therefore, care should be taken in selecting dimensions for a and b. The point where self-locking occurs is where the two values are exactly equal.

◆ *Example* ─────────────────────────────

In Fig. 9-3, dimension $a = 6$ in., $b = 20$ in., and $c = 2$ in. The coefficient of friction is 0.25, and the drum diameter is 8 in. Find the operating force P required to balance 400 in-lb of torque.

Solution.

$$\frac{T}{r} = F_1 - F_2;$$

$$\frac{F_1}{F_2} = e^{f\theta} = 2.718^{(0.25)(\pi)}$$

$$= 2.19 \qquad \text{or approximately } 2.2;$$

$$F_1 = e^{f\theta}F_2;$$

$$\frac{T}{r} = e^{f\theta}F_2 - F_2 = F_2(e^{f\theta} - 1).$$

Figure 9-3
Band brake.

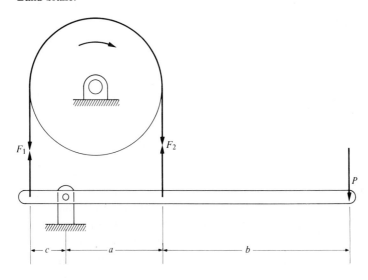

Therefore

$$F_2 = \frac{T}{r(e^{f\theta} - 1)} \quad \text{and} \quad F_1 = \frac{Te^{f\theta}}{r(e^{f\theta} - 1)}.$$

Taking moments about the pivot point, we find

$$F_1c = F_2a - P(a + b).$$

Thus

$$
\begin{aligned}
P &= \frac{F_2a - F_1c}{a + b} = \frac{\dfrac{Tu}{r(e^{f\theta} - 1)} - \dfrac{Tce^{f\theta}}{r(e^{f\theta} - 1)}}{a + b} \\[2ex]
&= \frac{T(a - ce^{f\theta})}{r(e^{f\theta} - 1)(a + b)} \\[2ex]
&= \frac{400[6 - 2(2.2)]}{4(2.2 - 1)(6 + 20)} \\[2ex]
&= \frac{100(1.6)}{1.2(26)} = 5.13 \text{ lb.}
\end{aligned}
$$

Band / Disc Brake

Figure 9-4 illustrates a band/disc brake. This is a heavy-duty brake that combines the desirable features of a band brake with those of a disc brake. This type is extensively used in the agricultural and construction fields, for instance in winches, cranes, and tractors. In addition to the housing, the principal parts are the band, the outside rotor, the inside rotor, and four steel balls. The band and both rotors have brake linings. The four balls are located between the rotors in sloped mating ramps.

The jackshaft is connected to the inside rotor by splines. Both rotors rotate inside the band.

When the brake is operated, the following sequence of events takes place. Pressure is applied to the brake pedal; this tightens the band and slows the outside rotor. The inside rotor continues to rotate at the same speed. The difference in speeds between the two rotors then causes the balls to turn and climb the ramp. As they do, the two rotors change their position and spread apart, bringing the lining surfaces of the rotors in contact with the flat surfaces of the housing to complete the braking operation. When the brake pedal is no longer applied, return springs release the band; when this

Figure 9-4
Band/ disc brake.
(Courtesy of The Bendix
Corporation.)

occurs, the two rotors return to their normal positions. Figure 9-5 is a schematic diagram in which one of the balls is shown.

 Industrial Disc Brakes

This general type of brake is available commercially and has several outstanding features. It can handle high values of kinetic energy and is equally effective for either direction of rotation. Three general types of mounting are used. In one type, the housing is fixed and the disc is placed in a floating position. A second type provides a floating housing and a fixed

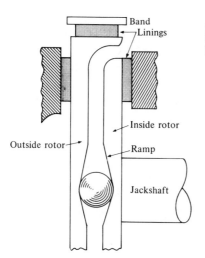

Figure 9-5
Schematic of band/disc brake.

Figure 9-6
Industrial disc brake.
(Courtesy of the Goodyear
Aerospace Corporation.)

disc. A third type has both the disc and housing fixed, but braking is carried out with cylinders on both sides of the disc. The brake can be operated pneumatically, hydraulically, or with a pneumatic-hydraulic booster. (Boosters are described in Chapter 17.) Fail-safe operation is also possible; such a system provides braking by spring action, and release of braking action via hydraulics or pneumatics. Brake adjustment is automatic with hydraulic or pneumatic operation. Brake linings can be easily replaced without disturbing fluid connections.

Figure 9-6 shows one type of available model. This particular brake has a $\frac{3}{8}$-in. disc with an outside diameter of 18-in. Note that the single-stop kinetic-energy capacity is 2,000,000 ft-lb. Return springs provide clearance between the disc and the lining. It is advisable to consult with the manufacturer before incorporating such equipment into a heavy-duty design.

Single-disc or caliper brakes can also be mechanically actuated. This method is not frequently used, however, since the disc-brake principle generally dictates a relatively high axial force, which is difficult to obtain mechanically.

It is important to note that, with many caliper designs, both disc thickness and diameter are "open" within reasonable limitations. This allows further flexibility in both torque and kinetic-energy capability, since the disc is the "torque arm" as well as the heat sink. The model shown in Fig. 9-6 is by no means limited to either this diameter or thickness; this adaptability is a major advantage.

Torque must be absorbed by the friction surfaces, which are located a distance r from the center of the shaft (and disc). The friction force is found by multiplying the coefficient of friction by the normal (operating) force P. However, since most caliper disc brakes have friction surfaces on both sides of the disc, the operating force, exerted parallel to the shaft, will be cut in half. Thus, in equation form, the torque-absorption capacity of the caliper disc brake is:

$$T = 2Prf. \tag{9-3}$$

◆ *Example* ───────────────────────────

A $\frac{1}{4}$-in. caliper disc brake has a disc diameter of 15 in. The friction surfaces are centered 6 in. from the middle. If the coefficient of friction is 0.3 and a force of 50 lb is applied at the friction surfaces, find the horsepower capacity of the brake when operating at 870 rpm.

Solution.

$$T = 2Prf = 2(50)(6)(0.3) = 180 \text{ in-lb}.$$

$$\text{hp} = \frac{NT}{63,000} = \frac{870(180)}{63,000} = 2.49.$$

 Automotive Brakes

One of the largest users of brakes and braking systems is the automotive industry. Passenger vehicles frequently use the expanding-shoe drum brake. The caliper disc brake is also used, in some cases.

Many configurations of the expanding-shoe drum brake are used. Regardless of configuration, the operating principle is the same. Lining material is attached to two stationary brake shoes. When braking is desired, the shoes are forced outward against a rotating brake drum. The actuating force is applied via a wheel cylinder. The wheel cylinder is part of a hydraulic system consisting of a master cylinder and four wheel cylinders (see Fig. 9-7). Sometimes two master cylinders are used; if so, one master cylinder operates two wheel cylinders and the other operates the remaining two. This amounts to two separate hydraulic systems, so that a malfunction in one of the systems still allows braking in two of the four wheels.

In a typical hydraulic system, the master cylinder is operated by depressing the brake pedal, which in turn operates the push rod and piston in the master cylinder and places the hydraulic fluid in the system under high pressure. The high pressure in the wheel cylinders produces a force against the pivoted brake shoes, forcing them against the drum and providing braking action.

Figure 9-8 shows the Duo-Servo® brake. Figure 9-9 shows right-hand self-adjusting brake assemblies. The self-adjustment feature operates when the vehicle brakes while moving in reverse, provided the lining is worn enough to warrant adjustment. The following steps take place in the

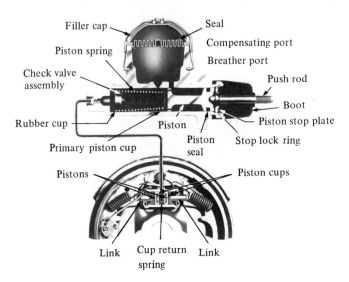

Filler cap

Seal

Compensating port

Piston spring

Breather port

Check valve
assembly

Push rod

Boot

Rubber cup

Piston

Piston stop plate

Primary piston cup

Piston
seal

Stop lock ring

Pistons

Piston cups

Link Cup return Link
spring

Figure 9-7
Brake hydraulic system.
(Courtesy of General Motors Technical Center.)

self-adjustment process:

1. friction forces the front shoe against the anchor pin;
2. the upper end of the rear shoe is forced away from the anchor pin by hydraulic pressure in the wheel cylinder;
3. as this takes place, the upper end of the adjuster lever is prevented from moving by the actuating link attached to the anchor pin;
4. movement forces the lower end of the adjuster lever against the star wheel;

Figure 9-8
Duo-Servo® brake.
(Courtesy of The Bendix
Corporation.)

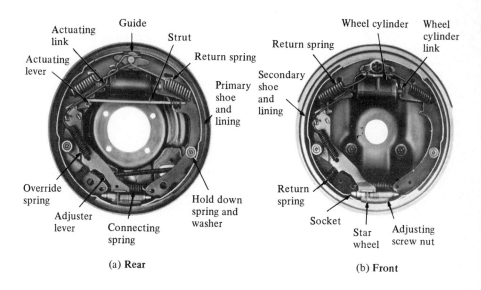

Actuating link
Actuating lever
Guide
Strut
Return spring
Override spring
Adjuster lever
Connecting spring
Hold down spring and washer
Primary shoe and lining
Wheel cylinder
Wheel cylinder link
Return spring
Secondary shoe and lining
Return spring
Socket
Star wheel
Adjusting screw nut

(a) **Rear**

(b) **Front**

Figure 9-9
Right-hand self-adjusting brake assemblies.
(Courtesy of General Motors Technical Center.)

5. if the lining is sufficiently worn, the adjuster lever will cause the screw star wheel to turn one or two teeth, thus making the adjustment.

No adjustment is made when the vehicle is moving forward with brakes applied. This is because the upper end of the rear shoe is forced against the anchor pin by the self-energizing action of the brakes.

To be self-energizing, a brake shoe must be mounted so that there is a tendency for it to move in the same direction as the drum; this produces a wedging action toward the pivot point and increases the braking effect. When drum rotation is reversed, the braking action is reduced. For this reason, automotive brakes are not as effective when the vehicle is moving backwards.

A great deal is expected of braking systems. Two important features are the stopping distance and the sequence of events that occurs during the stopping operation. Wear of brake linings is also important. Since operating conditions vary, it is important to use materials that will exhibit good wear characteristics under many adverse conditions. Still another factor to consider is fade resistance. Brake-lining fade occurs at high temperatures because of a change in the coefficient of friction. Figure 9-10 shows how this coefficient changes with temperature. The brake-pedal force needed for given braking situations is also an important consideration. Since any

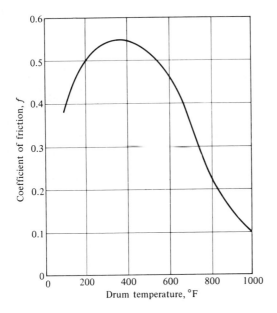

Figure 9-10
Effect of temperature on lining coefficient of friction. In a typical plot, the coefficient gradually rises, peaks out, then falls rapidly as temperature is increased. The rapid fall is referred to as brake-lining fade.
(From *General Motors Engineering Journal*, Vol. 11, No. 4. Copyright 1964 by General Motors Corporation. Reprinted by permission.)

braking application generates heat, a major problem is how to dissipate heat from the brake assemblies after brakes are released.

In automotive applications, it is desirable to have both shoes self-energizing in the front-wheel brakes, since these perform the larger part of the braking operation. In commercial vehicles, the front-wheel and rear-wheel brakes are often of different sizes; often they are also of different types. Both hydraulic and air brakes are extensively used.

The following equations are extracted by permission from *General Motors Engineering Journal*, Vol. 11, No. 4 (copyright 1964 by General Motors Corporation):

1. Heat loss due to convection:

$$q = (h)(t)(\theta)(A_c),$$

where

q = heat loss (Btu),

h = coefficient of heat transfer = $0.21(K/D^{0.4})(v^{0.36}/V_K^{0.6})$,

t = time (sec),

θ = temperature rise at any instant (°F),

A_c = effective convecting area (ft^2),

K = thermal conductivity (Btu/ft^2-sec-°F),

D = brake drum diameter (ft),

v = vehicle speed (mph),

V_K = kinematic viscosity of air (ft^2/sec).

2. Total torque required:

$$B_t = \left[\left(\frac{W_t}{g}\right)(a) - F\right]R_t,$$

where

B_t = total brake torque (ft-lb),

W_t = gross combination weight (lb),

a = deceleration of vehicle (ft/sec^2),

F = vehicle retarding force (lb),

R_t = tire radius (ft).

3. Required line pressure:

$$P = B_t + \frac{(N_1)(S_1)(P_{o1}) + (N_2)(S_2)(P_{o2}) + \cdots}{(N_1)(S_1) + (N_2)(S_2) + \cdots},$$

where

P = line pressure (psi),

N = number of each type of brake,

S = specific torque of brake (ft-lb),

P_o = push-out pressure of brake, or line pressure required to engage lining of drum.

Subscripts 1 and 2 refer to axles one and two.

4. Torque per brake:

$$B = S(P - P_o),$$

where

B = torque per brake (ft-lb).

Electric Brakes

Electric brakes (and clutches) are versatile assemblies that have outstanding advantages over some of their mechanical counterparts. Certain types are obtainable in module form, readily adaptable to standard motor frames. Thus, a designer can use them as building blocks.

In general, electric brakes are divided into two chief types: nonfriction and friction. The *nonfriction* types of brakes are

1. hysteresis,
2. magnetic particle,
3. eddy current.

The hysteresis type is extensively used in electric dynamometers. Magnetic-particle units are often employed in precision servomechanisms; the eddy-current type finds much use in wide-range speed-control servomechanisms.

The *friction* types can be divided into two chief categories: the disc brake and the shoe brake. The disc brake can be of single-disc or multiple-disc construction. A spring-set, solenoid-released shoe brake is often used on elevators, hoists, and cranes to lock the brake in case of power failure. The brake is normally held against its drum by spring action. When the motor is operating, part of the current is shunted to an electromagnetic coil that releases the brake. Thus, when the motor is not operating for any reason, the brake is locked against the drum.

The disc-type electric brake operates by controlling the magnetic force of an electromagnet faced with a suitable friction material. An electric brake can be mounted in several ways. In modular units, the magnet is mounted on the motor frame and the armature is fixed to the rotating shaft. The magnet can be mounted on a foot bracket or a shaft bearing housing.

One unusual feature of electric brakes is that the current can be adjusted using a rheostat to provide instant stopping of the load, gradual deceleration, or a constant slip.

Electric brakes cover a wide range of ratings. The nonfriction types range from $\frac{1}{50}$ to 800 hp. The friction types have ratings as high as 2500 hp.

Handbooks frequently give inertia loads (WR^2) in units of lb-ft^2. In the case of electric brakes, the WR^2-values for the armature must be included. The following formula gives the average brake torque required to stop a given load in a certain time interval:

$$T = \frac{(WR^2)N}{308t}, \tag{9-4}$$

where

$$WR^2 = \text{inertia load (lb-ft}^2),$$

$$T = \text{torque (ft-lb)},$$

$$t = \text{time (sec)},$$

$$N = \text{rotational speed (rpm)},$$

$$W = \text{weight (lb)},$$

$$R = \text{mean radius (ft)}.$$

◆ *Example* _____

Find the average torque required to stop completely an inertia load of 25 lb-ft^2 in 2 sec if the rotational speed is 1750 rpm.

Solution.

$$T = \frac{(WR^2)N}{308t} = \frac{25(1750)}{308(2)} = 71 \text{ ft-lb}.$$

In selecting an electric brake, one can compute the average torque needed. However, after determining this value, one must immediately check a particular brake to be certain that it can properly dissipate the heat that will be generated in stopping the load. Characteristic curves for electric brakes must be studied in much the same manner as one studies motor curves. Manufacturers' catalogs usually provide the necessary curves. Mounting problems must also be taken into consideration.

 Electrically Released Brakes

Figure 9-11 illustrates a variety of permanent-magnet-set, electrically re-leased brakes that automatically engage when electrical power to them is disrupted for any reason. These particular brakes range in static torque-holding capability from 10.5 lb-ft to 400 lb-ft. Brakes of this type are useful in power platforms, paper cutters, manlifts, cranes, power presses, and similar machinery. They are provided to eliminate hazards to operators and equipment.

These brakes are set by the permanent magnetic-flux forces, which attract and hold an iron armature against the magnet pole face. To release the brake, an opposing magnetic field is created by passing current through

Figure 9-11
Electrically released brakes.
(Courtesy of Warner Electric Brake & Clutch Company.)

the electromagnet. This electromagnetic field nullifies the permanent field, releasing the brake armature. Should the current flow be interrupted—either because a safety switch opens or because the power fails—the electromagnetic field collapses and the armature is immediately attracted to the permanent magnet face, setting the brake. Permanent-magnet-set devices are not subject to spring fatigue, which is a common problem of mechanically set brakes; thus, no adjustment is required to maintain uniform torque.

 ## Air Brakes

Air brakes became well known from their use in railway cars, buses, and trailer trucks. Air is used to press the brake shoe against a drum or wheel. Air is supplied by a compressor; however, auxiliary tanks are needed to supply air rapidly when needed. Air brakes operate quickly, but every precaution must be taken to ensure that no leakage occurs in the system.

 ## Summary

Although brakes and clutches serve separate purposes, they are similar in many ways. In most cases, both depend on materials that possess a high coefficient of friction, exhibit good wear characteristics, and are not adversely affected by high temperatures.

A brake converts kinetic energy into heat energy, which is then dissipated to the atmosphere by a part of the brake that serves as a heat exchanger. The ability of a brake to dissipate heat is one of the prime considerations in brake selection. A brake also stops or slows the desired moving part by surface-to-surface friction. It must perform the slowing or stopping action without undue wear on the brake lining. Since the coefficient of friction of lining materials changes as a function of temperature, it is important to consider fade resistance in brakes.

Automotive brakes are frequently self-energizing to assist in the braking operation. Front-wheel brakes are often different in design or in brake-surface area from rear-wheel brakes, because much greater braking action is needed at the front. This ratio can be as large as 3 to 2. Because of the self-energizing action of automotive brakes, braking action while traveling in reverse is not nearly so effective as when moving forward.

Brake assemblies can be actuated by mechanical, electrical, hydraulic, or pneumatic means. The type of actuation selected depends on the type of brake employed and the type of load being handled.

In many applications, one must be concerned with keeping contaminants away from the braking surfaces; mud, grit, or water can ruin a braking system.

In any type of brake design, one should know the torque being handled; the frictional drag can then be designed large enough to contain this torque.

Fittings used in the hydraulic or pneumatic circuits of a braking system must be designed to avoid any possible failure. Any leak makes the brakes inoperative. In electric brakes, the electrical circuits should also be designed to avoid possible failure of the circuit and the brakes.

The operating force needed to apply a brake properly must be considered. If necessary, power assistance can be supplied, either mechanically or hydraulically.

In any hydraulic system, the proper brake fluid must be used. It must have a high viscosity index and should not be adversely affected by elevated temperatures.

Questions for Review

1. What properties should an effective brake material have? List some typical brake materials.

2. Compare the advantages and disadvantages of block brakes and band brakes.

3. In a block brake like that shown in Fig. 9-1, what effect does a decrease in angle θ have on the problem? Explain.

4. Why are single-disc brakes often called *caliper* brakes?

5. How can single-disc brakes be actuated? List three ways.

6. Why do the front-wheel brakes of an automobile handle more than half the load?

7. Explain why automotive brakes are not as effective when applied with the vehicle traveling in reverse.

8. How can automotive brakes be made self-adjusting?

9. What is meant by modular design when referring to electric brakes? What advantages does it have?

10. From the curve of Fig. 9-10, what conclusions can be drawn about maximum brake-drum temperature?

11. Why are brake housings roughly finished and dark in color?

12. What advantages do air brakes have over hydraulic brakes? What disadvantages?

13. What factors are needed to compute the heat loss due to convection in an automotive internal-expanding-shoe brake?

14. What types of brakes are used on elevators and hoists? Why are these types chosen?

15. How does a band/disc brake operate? Give three applications for this type of brake.

16. Explain how the hydraulic system functions on automotive brakes. What advantage does the use of two master cylinders have over the use of one?

Problems

1. For Fig. 9-1, assume that pivot 3 is being used; that $a = 12$ in., $b = 18$ in., and $c = 3$ in.; that the coefficient of friction is 0.25; and that the drum is 20 in. in diameter. What is the operating force required to stop the drum if it rotates at 200 rpm clockwise and transmits 4 hp?

2. Solve Problem 1 for counterclockwise rotation.

3. For Fig. 9-1, assume that pivot 2 is being used; that $a = 0.3$ m, $b = 0.5$ m, and $c = 0.1$ m; that the drum diameter is 0.46 m, and that a force of 550 N is required to contain a clockwise torque of 100 N · m. What is the coefficient of friction between the drum and the brake?

4. For Fig. 9-1, assume that pivot 1 is being used; that the drum rotates counterclockwise; that the drum diameter is 400 mm; that $a = 250$ mm, $b = 500$ mm, and $c = 100$ mm; and that the coefficient of friction is 0.3. What is the torque capacity if an operating force of 225 N stops the drum?

5. In Fig. 9-2, find the value of P for the following conditions: Rotation is counterclockwise; $a = 150$ mm, $b = 50$ mm, and $c = 380$ mm; the coefficient of friction is 0.25; the drum diameter is 450 mm; and the angle of wrap is 230 degrees. The drum rotates at 150 rpm and transmits 140 N · m of torque.

6. In the block brake of Fig. 9-1, assume that $a = 12$ in., $b = 20$ in., and $c = 4$ in.; that $r = 4$ in.; that the coefficient of friction is 0.4; that rotation is clockwise; that pivot 3 is being used; and that a force of 100 lb is applied at P. Under these conditions, how much torque (in-lb) can be contained?

7. In the band brake of Fig. 9-3, $a = 8$ in., $b = 15$ in., and $c = 2$ in. The drum transmits 5 hp and rotates at 200 rpm. The coefficient of friction

between the drum and the band is 0.4. How much force must be applied at *P* to stop the drum from rotating?

8. In Fig. 9-3, the pulls on the tight and slack sides of the band are 800 and 300 lb, respectively. The band is made of steel with an ultimate tensile strength of 80,000 psi. If the band thickness is to be $\frac{1}{16}$ in. and a factor of safety of 8 is required, what band width is needed?

9. In Fig. 9-1, pivot 2 is used; dimensions *a* and *b* are 6 in. and 12 in., respectively; and 800 lb of force is needed to sustain the torque of the drum. If the cross section of the operating lever is 1 in. × 1 in., how much bending stress is induced? How could the arm dimensions be modified to reduce the flexural stress?

10. The band of a band brake has 210 degrees of contact with its drum. By laboratory tests, it is found that the pull on the tight side is 800 lb and the pull on the slack side is 285 lb. What is the coefficient of friction?

11. The inertia load on an electric brake (including the armature) is 32 lb-ft^2 and the drive shaft is rotating at 2000 rpm. What average torque is required to bring this load to a complete stop in 1.2 sec?

12. In Fig. 9-3, the band pulls are 1800 N and 675 N respectively. The allowable tensile stress for the 2-mm-thick steel band is 70 MN/m^2. Find the band width.

13. The band of a band brake has a contact angle of 180 degrees. It is found experimentally that the pulls on the bands are 275 kg and 100 kg respectively under certain operating conditions. Find the coefficient of friction.

14. The drum in Fig. 9-3 is handling 0.5 kW of power and rotates at 100 rpm. Find the force *P* needed to stop rotation if $f = 0.25$, $a = 50$ mm, $b = 150$ mm, and $c = 500$ mm.

15. A brake requires 900 in-lb of torque to stop a shaft operating at 840 rpm in a period of 3.5 seconds. What is the inertia load expressed in lb-ft^2?

16. By experiment, it is found that the pulls on the tight and slack sides respectively of a band-type brake with 180 degrees of drum contact are 1200 N and 400 N. The drum is 400 mm in diameter and rotates at 80 rpm. Find the coefficient of friction and the amount of power that can be absorbed under these conditions.

CLUTCHES

10

A clutch is a mechanical assembly used to connect a driving member with a driven, collinear member. It differs from a coupling in that the driven member in a clutch assembly can be disconnected from the driving member easily and rapidly. Square-jaw couplings and spiral-jaw couplings are exceptions to this general classification. Friction plays an important part in the design of most clutches. Conventional types of clutches include cone, disc, band, spring, centrifugal, electric, and overrunning clutches. From the theoretical standpoint, friction clutches and friction brakes are analyzed in much the same manner.

 **Cone Clutch**

Figure 10-1 shows a diagram of a cone-type clutch. In this figure,

$$\sin \alpha = \frac{P}{N_1}.$$

In Chapter 3, it was determined that $f = F/N$, where F is the frictional resistance. Then

$$N_1 = \frac{P}{\sin \alpha} = \frac{F}{f}.$$

Thus,

$$F = \frac{Pf}{\sin \alpha} \qquad \text{or} \qquad T = \frac{Pfr_m}{\sin \alpha}, \qquad\qquad \textbf{(10-1)}$$

Figure 10-1
Cone clutch.

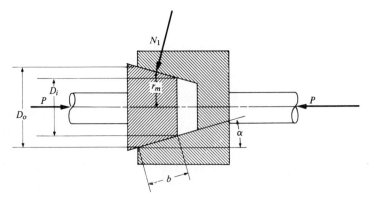

where

$$r_m = \frac{D_m}{2} = \frac{D_o + D_i}{4}.$$

This analysis assumes uniform *wear* and a normal reaction at the midpoint of the contacting surfaces. F is not shown in the sketch, since it acts perpendicularly to the plane of the paper.

If the design of the clutch is based on uniform *pressure*, the equation for the torque becomes

$$T = pb\pi D_m f \left[\frac{2}{3} \left(\frac{r_o^3 - r_i^3}{r_o^2 - r_i^2} \right) \right], \tag{10-2}$$

where

p = pressure (psi).

In either case (uniform wear or uniform pressure) horsepower capacity is found by multiplying the torque by $N/63,000$.

Force P is usually applied to the clutch by a spring. The above analysis is based on the fact that movement (or slipping) is impending; therefore, it is good practice to make the value of P larger than the calculated value. The above equations also assume ideal conditions; when an actual clutch is engaged under load, the faces of the cones tend to slide relative to each other. The total value of P should be at least $N_1(\sin \alpha + f \cos \alpha)$. Values for α should be carefully considered. If this angle is too small, it is difficult to disengage the clutch; if the angle is too large, a greater spring force (P) is required. Cone angles should not be made less than 8 degrees; a good workable figure is 12.5 degrees.

◆ *Example*

A cone clutch has a cone angle of 12.5 degrees and a mean diameter of 20 in. If the clutch is to transmit 15 hp at 900 rpm and the coefficient of friction is 0.3, what axial spring force is required? Calculate on the basis of uniform wear.

Solution.

$$T = \frac{Pfr_m}{\sin \alpha}, \qquad \frac{63,000(\text{hp})}{N} = \frac{Pfr_m}{\sin \alpha},$$

$$P = \frac{63,000(15)(0.216)}{900(0.3)(10)} = 75.6 \text{ lb (minimum required under operation)};$$

or

$$N_1 = \frac{T}{fr_m} = \frac{63{,}000(\text{hp})/N}{fr_m} = \frac{63{,}000(15)/(900)}{0.3(10)} = 350,$$

$$P = N_1(\sin\alpha + f\cos\alpha) = 350[0.216 + 0.3(0.976)]$$
$$= 350(0.509) = 178\ \text{lb (minimum required to engage clutch)}.$$

◆ *Example*

In Fig. 10-1, $D_o = 10$ in., $D_i = 8$ in., the coefficient of friction is 0.3, the capacity is 10 hp at 1000 rpm, and the allowable pressure is 8 psi. Find the face width if the design is based on uniform pressure.

Solution.

$$T = pb\pi D_m f\left[\frac{2}{3}\left(\frac{r_o^3 - r_i^3}{r_o^2 - r_i^2}\right)\right];$$

$$\frac{63{,}000(\text{hp})}{N} = pb\pi D_m f\left[\frac{2}{3}\left(\frac{r_o^3 - r_i^3}{r_o^2 - r_i^2}\right)\right];$$

$$b = \frac{63{,}000(\text{hp})}{Np\pi D_m f\left[\frac{2}{3}\left(\dfrac{r_o^3 - r_i^3}{r_o^2 - r_i^2}\right)\right]}$$

$$= \frac{63{,}000(10)}{1000(8)\pi(9)(0.3)\left[\dfrac{2}{3}\left(\dfrac{125 - 64}{25 - 16}\right)\right]}$$

$$= \frac{63{,}000(10)}{1000(8)\pi(9)(0.3)(4.52)} = 2.05\ \text{in.}$$

Disc Clutch

A very simple type of disc clutch is sketched in Fig. 10-2. The frictional radius is determined as in Chapter 3:

$$N = P, \qquad \frac{F}{f} = P, \qquad F = fP.$$

Therefore,

$$T = Fr_f = fP\left[\frac{2}{3}\left(\frac{r_o^3 - r_i^3}{r_o^2 - r_i^2}\right)\right]. \tag{10-3}$$

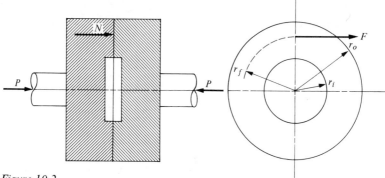

Figure 10-2
Disc clutch.

Horsepower capacity can then be found by multiplying by $N/63,000$. This analysis assumes that uniform *pressure* exists over all the contacting surfaces and that slipping is impending; therefore, it is advisable to make certain that the value of P is larger than the value of N.

If a design is based on *uniform wear*, the torque equation becomes:

$$T = fP\left(\frac{r_o + r_i}{2}\right). \tag{10-4}$$

It is considered good practice to base a design on uniform wear so that the clutch will not slip after the surfaces exhibit some wear. The following example compares results.

◆ *Example* ————————————————————————

A simple disc clutch has elements with outside and inside diameters of 250 mm and 100 mm respectively. The axial force is 1200 newtons. If the coefficient of friction is 0.3, find the torque capacity based on uniform pressure and uniform wear.

Solution. Uniform pressure:

$$T = fP\left[\frac{2}{3}\left(\frac{r_o^3 - r_i^3}{r_o^2 - r_i^2}\right)\right]$$

$$= 0.3(1200)\left[\frac{2}{3}\left(\frac{(0.25)^3 - (0.1)^3}{(0.25)^2 - (0.1)^2}\right)\right] = 66.7 \text{ N} \cdot \text{m}.$$

Uniform wear:

$$T = fP\left(\frac{r_o + r_i}{2}\right) = 0.3(1200)\left(\frac{0.25 + 0.1}{2}\right) = 63 \text{ N} \cdot \text{m}.$$

Multiple-Disc Clutch

A multiple-disc clutch is similar to a single-disc clutch except that it consists of alternate discs (usually of unlike materials) held together by an axial force (often a spring). One set of discs is splined (or keyed) to the driving member, the other to the driven member. If uniform pressure is assumed over all sections of the contacting areas, the equation used for simple disc clutches can be applied, except that the number of *pairs* of contacting surfaces must be considered. If a clutch consists of six steel discs and five bronze discs, the number of pairs of contacting surfaces is one less than the total number of discs. In this case, there would be 10 pairs of contacting surfaces; thus, the torque (or horsepower) capacity of the clutch would be 10 times that of a single-disc clutch with the same size of disc. Figure 10-3 is a diagram of a multiple-disc clutch; other arrangements are possible. It should be noted that both sides of each disc are contacting surfaces.

For a multiple-disc clutch based on uniform pressure, the torque capacity can be found from the following equation:

$$T = fPn\left[\frac{2}{3}\left(\frac{r_o^3 - r_i^3}{r_o^2 - r_i^2}\right)\right]. \tag{10-5}$$

Figure 10-3
Multiple-disc clutch.

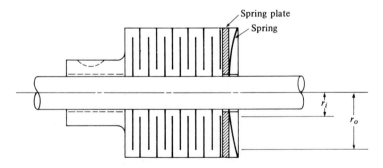

where

n = number of pairs of contacting surfaces.

For uniform wear, the equation is

$$T = fPn\left(\frac{r_o + r_i}{2}\right). \tag{10-6}$$

The following examples illustrate the relative values obtained by assuming uniform pressure and uniform wear.

◆ *Example* ————————————————————

A multiple-disc clutch consists of 10 steel discs and 9 bronze discs. The effective outside and inside diameters are 8 in. and 6 in., respectively. If an axial force of 100 lb is applied and the shaft turns at 1000 rpm, find the horsepower capacity. Assume a coefficient of friction of 0.25 and uniform pressure.

Solution.

$$\text{hp} = \frac{NfPn}{63,000}\left[\frac{2}{3}\left(\frac{r_o^3 - r_i^3}{r_o^2 - r_i^2}\right)\right],$$

where

n = number of pairs of contacting surfaces. Thus,

$$\text{hp} = \frac{(1000)(0.25)(100)(18)}{63,000}\left[\frac{2}{3}\left(\frac{64 - 27}{16 - 9}\right)\right]$$

$$= \frac{(1000)(0.25)(100)(18)}{63,000}\left(\frac{2}{3}\right)\left(\frac{37}{7}\right) = 25.2 \text{ hp.}$$

◆ *Example* ————————————————————

Find the horsepower capacity for the preceding example assuming uniform wear.

Solution.

$$\text{hp} = \frac{NfPn}{63,000}\left(\frac{r_o + r_i}{2}\right)$$

$$= \frac{1000(0.25)(100)(18)}{63,000}\left(\frac{4 + 3}{2}\right) = 25.0.$$

The uniform pressure assumption is valid when clutch lining surfaces are attached to flexible plates.

Band Clutches

Band clutches are similar to band brakes. In the band clutch, a band of steel (sometimes lined with composition material or asbestos) is secured to one of the rotating shafts and tightened by an elaborate mechanism around a drum on the other shaft. Horsepower capacity depends on rotational speed, the size of the drum about which the band is wrapped, the coefficient of friction, the angle of contact, and the pulls developed on the ends of the band. The design of the band is based on the cross-sectional area of the band that is subjected to tension. The larger of the two forces on the ends of the band is used as a basis for design. When this type of clutch is engaged, the band is subjected to shock loading; therefore, a large factor of safety is needed. The following equations are useful in band-clutch design:

$$\frac{F_1}{F_2} = e^{f\theta} \qquad \text{and} \qquad T = (F_1 - F_2)r;$$

thus,

$$\text{hp} = \frac{Nr}{63,000}\left(F_1 - \frac{F_1}{e^{f\theta}}\right), \tag{10-7}$$

where

F_1 = pull on tight side of band (lb),

F_2 = pull on slack side of band (lb),

$e = 2.718$,

θ = angle of contact (rad),

T = torque (in-lb),

hp = horsepower,

N = shaft speed (rpm),

r = drum radius (in.),

f = coefficient of friction.

Spring Clutches

Spring clutches can be divided into two chief types. One is the *expanding* type, in which a spring (helical, with rectangular cross section) is used to connect two round internal surfaces. In this type, the action is such that the spring unwinds against these surfaces, exerting radial force against the drums. (See Fig. 10-4.) This type of clutch is considered an overrunning type

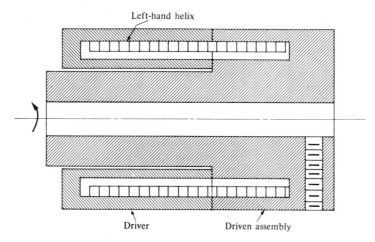

Left-hand helix

Driver   Driven assembly

Figure 10-4
Expanding-type spring clutch.

—that is, the driving member rotates the driven member in one direction only. If the direction of the driver is reversed, the clutch will disengage. Also, if the speed of the driven member exceeds the speed of the driver, the spring will cease to unwind, and the two shafts will disengage. A flat drawing rolled inside a cylindrical mailing tube illustrates the principle involved. If one tried to rotate the drawing inside the tube, it would turn in one direction only; trying to rotate it in the opposite direction would merely lock it tighter in the tube. The expanding type of spring clutch operates in a similar manner.

Figure 10-5 illustrates a *contracting* type of spring clutch. This type carries the spring on the outside of two drums of the same size; one drum is fixed to each of the shafts. Again, one direction of rotation only will connect the two shafts. Rotation in the opposite direction disengages the clutch (and the shafts).

In the expanding type, the spring is subjected primarily to shear; in the contracting type, the spring is subjected primarily to tension.

The horsepower capacity of a spring clutch depends on the amount of force that the spring exerts against the cylindrical drums. This, in turn, depends on (a) the number of coils actively pressing against the drum surfaces and (b) the coefficient of friction between the spring metal and the drum surfaces. The direction of rotation of a spring clutch depends on the direction in which the coils are wound. As the resisting load of the driven member increases, the clutching action is increased, provided the resisting

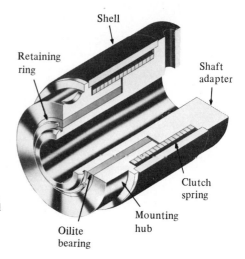

Figure 10-5
Contracting-type spring clutch.
(Courtesy of The Marquette Metal
Products Co., 400 South Main
St., P.O. Box 398, Fountain Inn,
S.C. 29644.)

load does not exceed the capacity of the spring. Appliances such as washing
machines make good use of this type of clutch.

 Centrifugal Clutch (Expanding-shoe Type)

There are several types of rim clutches with internal expanding shoes. The
simplest type consists of a rim or drum connected to the driven shaft, shoes
connected to the driving member, and an intricate system of links to force

Figure 10-6
(a) Straight centrifugal clutch. (b) Packaged clutch with sheave.
(Courtesy of Colt Industries.)

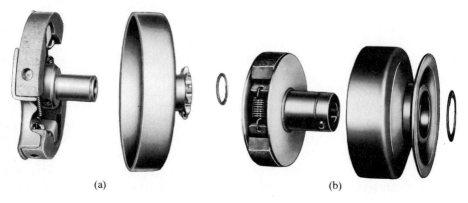

(a)   (b)

the shoes against the inside of the drum. The internal-expanding centrifugal rim clutch is illustrated in Fig. 10-6. In this type, friction shoes are attached to the driving member, but are kept free of the drum by spring action. As the speed of the driving member increases, the shoes are forced against the drum by centrifugal force. As the speed of the driving member increases, the centrifugal force of the shoes against the rim increases. Therefore, the torque capacity increases as the speed is increased. As in any friction calculation, the limiting torque is the product of the normal force (centrifugal or linkage-applied force), the coefficient of friction between the drum and the shoe, the radius of the drum, and the number of brake shoes.

 **Electric Clutch**

The so-called electric clutch is an electromagnetic device. It is somewhat similar in construction to the disc clutch, in that laminations are splined to the driven member. As in the disc clutch, a deliberate amount of axial movement is permitted. Other laminations are alternated and fit over lugs on the driving member. The driving member contains a coil that energizes an electromagnet through the use of a slip ring and brush. The driven shaft contains an armature. When the clutch is disengaged, the outer laminations (and the electromagnet) rotate freely relative to the armature. When the coil is energized, the armature is held tightly against the magnet, the laminations are held tightly together, and the driving and driven members turn as one unit. The holding power of the clutch depends on the number of laminations, the size of the laminations, the coefficient of friction of the lamination material, and the attracting force of the electromagnet. This type of unit is commercially available with capacities up to 2600 ft-lb of running torque. Figure 10-7 shows two types of commercial electric clutches.

 Overrunning Clutch

An overrunning clutch is one that will disengage when, and if, the driven shaft reaches or surpasses the speed of the driving shaft. Several variations of this type are commercially available. Figure 10-8 shows the construction of one type. In this version, the rotor is shaped in such a way that spring-loaded balls or rollers are wedged against the outer rim when the

Armature friction disc

Coil

Magnet

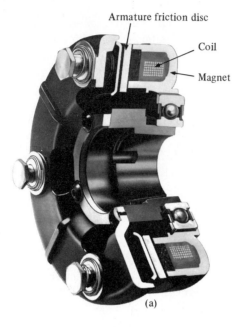

(a)

Figure 10-7
(a) Electric clutch.
(b) Electroclutch.
(Courtesy of Warner Electric
Brake & Clutch Company.)

(b)

Figure 10-8
Overrunning clutch.
(Courtesy of The Hilliard
Corporation.)

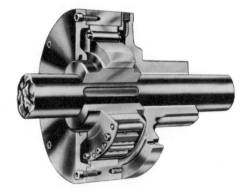

Figure 10-9
Sprag-type overrunning clutch.
(Courtesy of Formsprag, a division
of the Dana Corporation.)

rotor travels in a given direction. If the driver rotates in the opposite direction or if the driven member rotates faster than the driver, the two shafts become disengaged. This device is sometimes used as a silent ratchet; when used as such, continuous indexing is possible.

Another type of overrunning clutch is the sprag type illustrated in Fig. 10-9. The cutaway photograph shows the shape of the sprags and how they are retained. Specially heat-treated, high-strength steel races contribute to high torque-carrying capacities. Torque ratings extend from 4.5 to 150,000 ft-lb; outside diameters are available from 1.250 to 28.750 in. Bore sizes of 0.250 to 12.500 in. can be obtained. Overrunning speeds range from 100 to 3450 rpm. Other varieties of the sprag-type clutch are available in addition to the one illustrated.

 Square-jaw and Spiral-jaw Clutches

Figure 10-10 shows one section of a two-jaw, square-jaw clutch. This section is mounted on one of the shafts, and a similar unit is mounted on the shaft being engaged. Rotation in either direction is possible for this type. It is, however, only possible to engage when the shafts are standing still or are running at the same speed. In the spiral-jaw clutch, one side of the jaw is square and the other is rounded. Thus, it will engage in one direction only; if the direction of rotation is reversed, the shafts will disengage immediately. Starters on light-duty gasoline engines use this type of clutch. Bearing stress

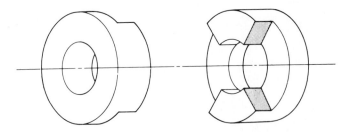

Figure 10-10
Square-jaw clutch.

on the flat surfaces and shearing stress at the roots of the jaws sometimes need to be checked.

Neither of these types depends on friction; thus, there are no heat-generation problems or serious wear problems. These are positive clutches; slipping presents no problem. High-speed engagement is not possible; however, the shape of the jaws can sometimes be modified to permit engagement at certain low speeds and when standing still.

 Ball-spring Overload-release Clutch

Figure 10-11(a) shows a type of clutch that will disengage when the torque reaches a predetermined value. The driving member is engaged with the driven member by two spring-loaded balls, which contact holes in the driving member. When the torque becomes excessive, the balls are forced out of the holes against the spring and the two rotational parts become disengaged. This type is frequently used in power screwdrivers and wrenches. When the driven member (which is connected to the screwdriver) reaches the stalling torque, it ceases to turn. The speedup of the external member and the sound of the balls clicking inform the operator that the screw or nut has been tightened to the desired torque. Various limiting torques can be obtained by replacing the ball-retaining spring with one with a different rating.

◆ *Example* ─────────────────────────
In Fig. 10-11(a), the shaft diameter is $1\frac{3}{4}$ in. The diameter of the balls is $\frac{1}{2}$ in. and the diameter of the sleeve holes is $\frac{3}{8}$ in. Find the maximum torque that can be transmitted before slipping occurs.

Solution. Refer to Fig. 10-11(b):

$$\cos\theta = \frac{\frac{3}{16}}{\frac{1}{4}} = 0.750,$$

$$\theta = 41.4 \text{ degrees};$$

$$R = \frac{P}{\cos\theta} = \frac{10}{\sin\theta}.$$

Therefore,

$$\frac{P}{0.75} = \frac{10}{0.661} \quad \text{and} \quad P = 11.3 \text{ lb.}$$

Figure 10-11
Ball-spring overload-release clutch.

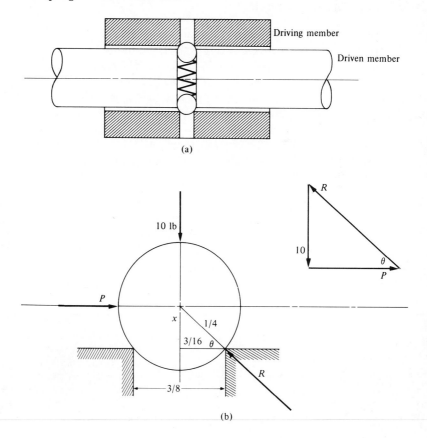

(a)

(b)

The shaft radius is $\frac{7}{8}$ in. To find the torque (maximum) on one ball, multiply P by $(\frac{7}{8} - x)$. The value of x can be found from the following equation:

$$x = 0.25 \sin \theta = 0.25(0.661) = 0.165 \text{ in.}$$

Thus,

$$T = 11.3(0.875 - 0.165) = 8.02 \text{ in-lb.}$$

Since there are 2 balls, the maximum torque would be 2×8.02, or 16.04 in-lb. Any torque above this value would cause the balls to slip over the holes.

 Air-operated Clutches

Figures 10-12 and 10-13 illustrate commercially available air-operated clutches of the rim type or drum type. These industrial clutches (and brakes) are available for a wide range of capacities. The torque ratings for the "CB" Airflex® units range from 1,000 to 260,000 in-lbs of torque; ratings for the "VC" units extend from 27,000 to 1,215,000 in-lbs of torque. In Fig. 10-12, air pressure is applied at the top, causing the friction surfaces to engage the drum for a full 360 degrees. This type is suitable for high-speed cyclic operations as well as for general power transmission service. The "VC" type shown in Fig. 10-13 is built for heavy-duty service and high starting loads.

Figure 10-12
"CB" Airflex® clutch.
(Courtesy of Airflex, a division of the Eaton Corporation.)

Figure 10-13
"VC" Airflex® clutch.
(Courtesy of Airflex, a division
of the Eaton Corporation.)

This type is fully ventilated. Maintenance costs are low, as tube and friction lining are easily replaced in the field.

The Airflex® principle has certain outstanding features. As the tube is constructed of resilient neoprene and cord, a certain amount of parallel and angular misalignment can be tolerated. In addition, because of the resilience, shock loads can be absorbed. The complete and rapid disengagement that occurs when air is released from the flexible actuating tube prevents wear on the friction surface.

 **Torque Limiter**

Figure 10-14 shows a commercial overload limit device somewhat similar to the clutch shown in Fig. 10-3. Several models of this torque limiter are obtainable in capacities that span a large range. The smallest has minimum and maximum torque limits of 5 and 20 lb-ft; the largest has minimum and maximum torque limits of 3150 and 6300 lb-ft. The overload values can be set between these limits for each model.

The limiter shown in Fig. 10-14 has a sintered graphite bronze bushing on which the center member rides. On the left side of the assembly is an integral hub and pressure plate. The darkened sections are friction facings. On the right side of the friction faces is a pressure plate with flats to engage flat surfaces on the hub. Axial force is applied via a disc spring. The load placed on the spring is adjusted by means of three bolts on the right side of the assembly. A ground sprocket can be inserted between the friction

Figure 10-14
Torque limiter.
(Courtesy of Morse Chain, a
division of Borg-Warner
Corporation.)

facings. This sprocket can be part of a chain assembly. The overload value
can be established by means of the adjusting bolts.

The torque limiter is extensively used in the conveyor and materials-
handling fields. Slipping occurs when the system is overloaded; reengage-
ment occurs when the torque returns to a value within the limits of the
setting.

 **Summary**

Clutches and couplings are designed to transmit power from one shaft to
another. Couplings are usually permanent or semipermanent connections,
some of which can correct for small amounts of parallel or angular misalignment. Clutches are designed to disengage a driver quickly from a follower. It
is sometimes difficult to differentiate between a clutch and a coupling. A
typical borderline case is the spiral-jaw or square-jaw clutch (or coupling).

Brakes and clutches also have features in common. Clutches must be
capable of transmitting a certain amount of torque, whereas brakes must be
capable of containing or stopping a given amount of torque, or slowing an
object in a certain period of time. Most brakes and clutches depend on
friction surfaces. Typical clutches include the following:

1. cone,
2. disc (also multiple-disc),
3. band,
4. spring (internal- and external-expanding),
5. centrifugal,
6. electric,
7. overrunning, and
8. overload release.

The method of actuating the clutch should be considered in any clutch design.

Questions for Review

1. What is an overrunning clutch? Explain what conditions are needed to make a clutch overrunning.

2. What is an overload-release clutch? What types are commonly available?

3. Explain how the frictional effect is calculated in a multiple-disc clutch.

4. In a cone clutch, what effect does a larger cone-element angle have on performance? What is undesirable about a small angle?

5. How does a square-jaw clutch differ from a spiral-jaw clutch? List two applications for spiral-jaw clutches.

6. Explain the operation of the following types of clutches: (a) electric, (b) centrifugal, (c) air.

7. Distinguish between braking and clutching functions. What are the principal differences?

8. Explain how an internal-expanding spring clutch differs from an external-expanding spring clutch. List an application for each. Why is rectangular spring wire used?

9. What key properties must friction-clutch materials possess? Why is each property necessary?

10. Explain how an expanding-shoe centrifugal clutch operates. Why does the torque-carrying capacity of this type increase with an increase in rotational speed?

11. Describe the basic principles of an electric clutch.

Problems

1. A square-jaw clutch is made of steel and has the dimensions shown in Fig. 10-15. The allowable stress (shearing) is 10,000 psi. If this unit is designed for shear only, how much torque can this clutch transmit?

2. A cone clutch has an angle of 10 degrees and a coefficient of friction of 0.42. Find the axial force required if the capacity of the clutch is 7 kW at 500 rpm. The mean diameter of the active conical sections is 300 mm.

3. How much torque can a cone clutch transmit if the angle of the conical elements is 10 degrees, the mean diameter of the conical clutch sections

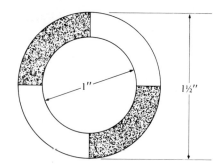

Figure 10-15
Problem 1.

200 mm, and an axial force of 550 N is applied? Assume that the coefficient of friction between the clutch elements is 0.45.

4. In a clutch arrangement similar to Fig. 10-2, the outside diameter is 8 in. and the inside diameter is 4 in. An axial force of 400 lb is used to hold the two parts together. If the coefficient of friction of the mating materials is 0.4, how much torque can the clutch handle?

5. A disc clutch has 6 pairs of contacting friction surfaces. The frictional radius is 2 in. and the coefficient of friction is 0.3. An axial force of 100 lb acts on the clutch. The shaft speed is 400 rpm. How much horsepower can the clutch transmit?

6. A cone clutch has cone elements at an angle of 10 degrees. The clutch transmits 20 hp at a speed of 1000 rpm. The mean diameter of the conical friction sections is 16 in. and the coefficient of friction is 0.3. Find the axial force needed to engage the clutch.

7. In a band clutch, the ratio of the pull on the tight side of the band to that of the slack side is 4/1. The band contacts the drum for 250 degrees. What is the coefficient of friction?

8. A band clutch has an angle of contact of 270 degrees on a 15-in.-diameter drum. The rotational speed of the drum is 250 rpm and the clutch transmits 8 hp. The band is $\frac{1}{16}$ in. thick and has a design stress of 5000 psi. How wide should the band be? Assume a coefficient of friction of 0.4.

9. The angle of contact of a band clutch is 250 degrees. The cross section of the band is $\frac{1}{16}$ in. × 1.5 in. The design stress for the band material is 8000 psi. If the drum is 16 in. in diameter and rotates at 350 rpm, what is the horsepower capacity of the clutch? The coefficient of friction is 0.4.

10. A ball-spring overload-release clutch similar to that shown in Fig. 10-11(a) uses a 1-in.-diameter shaft. The diameter of each ball is 0.4 in.

and the diameter of each sheave hole is 0.3 in. Find the spring force required if slipping is desired at 20 in-lb of torque.

11. A square-jaw clutch similar to that shown in Fig. 10-15 has the following dimensions:

    outside diameter = 40 mm,
    inside diameter = 25 mm.

    If the allowable shearing stress is 100 MN/m², how much torque can the clutch transmit?

12. A simple disc clutch similar to that shown in Fig. 10-2 has an outside diameter of 200 mm and an inside diameter of 100 mm. The coefficient of friction of the clutch materials is 0.4 and force P is equal to 1500 newtons. Find the torque-handling capacity of this arrangement, assuming uniform pressure.

13. Find the power capacity under uniform wear of a cone clutch with a mean diameter of 250 mm if the conical elements are inclined 8 degrees and the axial force is 450 N. The rotational speed of the driver is 200 rpm and the coefficient of friction is 0.2.

14. Determine the power capacity of a cone clutch under uniform pressure and assuming the following conditions: major diameter = 250 mm; minor diameter = 200 mm; length of conical elements in contact = 125 mm; rotational speed = 870 rpm; coefficient of friction = 0.3; and allowable pressure = 70 000 N/m².

15. Find the power capacity under uniform wear of a cone clutch with the following specifications: speed = 870 rpm; length of conical elements in contact = 125 mm; major diameter = 250 mm; minor diameter = 200 mm; coefficient of friction = 0.3; and axial operating force = 500 N.

16. Assuming uniform wear, find the power capacity of a single-disc clutch with an outside and inside diameter of 200 and 100 mm respectively, a rotational speed of 1160 rpm, a coefficient of friction of 0.35, and an axial operating force of 800 newtons.

POWER SCREWS

11

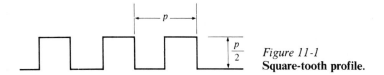

Figure 11-1
Square-tooth profile.

 Types of Threads

Two types of screw threads are popular for translation: the square and Acme threads. Figure 11-1 illustrates the *square* thread. Although it is effective as a translation screw, this type of thread has not been standardized, and it is costly to produce. For design purposes, the depth of thread can conveniently be taken as one-half the pitch. Nominally, the width of the space is equal to the thread thickness. The other popular type of power screw is the 29-degree *Acme* thread; this is a standard thread. Figure 11-2 shows the configuration of this type. The width of space at the pitch line is equal to the thread thickness, or one-half the pitch. (Again, these are nominal proportions.)

Pitch is defined as the distance measured along the pitch line from a point on one thread to the corresponding point on the next thread. For a single thread, the pitch is equal to the lead. It is important to understand the following relationships:

Single thread: lead equals the pitch.
Double thread: lead equals twice the pitch.
Triple thread: lead equals three times the pitch.
Quadruple thread: lead equals four times the pitch.

In the following derivations for translation screws, it is assumed that the load is applied at the midpoint of the thread. It must be remembered that, when a load is raised by means of a thread, the thread is merely a helix wound around a cylinder. When this helix is unwrapped from a cylinder, it forms an inclined plane. The horizontal distance is equal to the circumference (based on mean diameter), and the height of the plane is equal to the lead. A force analysis must be made, and the force at the pitch line of

Figure 11-2
Profile of 29-degree Acme thread.

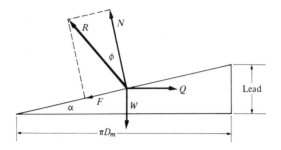

Figure 11-3
Force analysis: translation screw (load being raised).

the thread must be ascertained. In this force analysis, the slope of the incline (determined by the circumference and the lead) and the amount of friction enter into the calculation. It is important to know whether the load is being raised or lowered. The slope of the incline remains constant for both situations, but the effect of friction (frictional resistance) is reversed. When a load is raised, frictional resistance must be overcome by the operating force on the thread; when a load is lowered, friction helps prevent fast movement downward. Figure 11-3 illustrates a section of unwound thread, together with the forces and reactions acting on it.

Equations (Square Threads)

Referring to Fig. 11-3, we can write the following relationships for *raising* a load by means of a square thread

$$\tan \phi = \frac{F}{N} = f,$$

$$\tan \alpha = \frac{\text{lead}}{\pi D_m}.$$

By summing up the horizontal components, we obtain

$$Q = R \sin(\alpha + \phi);$$

by summing up the vertical components, we obtain

$$W = R \cos(\alpha + \phi).$$

Then

$$\frac{Q}{W} = \frac{R \sin(\alpha + \phi)}{R \cos(\alpha + \phi)} \quad \text{or} \quad Q = W \tan(\alpha + \phi),$$

$$\frac{QD_m}{2} = \frac{WD_m}{2} \tan(\alpha + \phi) = T = \text{thread torque}.$$

Substituting the value for the tangent of the sum of two angles,

$$T_f = \frac{WD_m(\tan \alpha + \tan \phi)}{2(1 - \tan \alpha \tan \phi)} = \frac{WD_m(\tan \alpha + f)}{2(1 - f\tan \alpha)},$$

or, by substituting $L/\pi D_m$ for $\tan \alpha$ (where L = lead), the equation can be simplified as follows:

$$T_f = \frac{WD_m}{2}\left(\frac{L + \pi fD_m}{\pi D_m - fL}\right). \tag{11-1}$$

For *lowering* the load by means of a square thread, the following relationships are useful:

$$T_f = \frac{WD_m}{2} \tan(\alpha - \phi) = \frac{WD_m(\tan \alpha - \tan \phi)}{2(1 + \tan \alpha \tan \phi)}$$

or

$$T_f = \frac{WD_m(\tan \alpha - f)}{2(1 + f\tan \alpha)} = \frac{WD_m}{2}\left(\frac{L - \pi fD_m}{\pi D_m + fL}\right), \tag{11-2}$$

where

T_f = frictional torque (in-lb).

Equations (Acme threads)

Similar equations for thread torque can be developed for the 29-degree Acme thread. For raising a load,

$$T_f = \frac{WD_m}{2}\left(\frac{0.968L + \pi fD_m}{0.968\pi D_m - fL}\right); \tag{11-3}$$

for lowering a load,

$$T_f = \frac{WD_m}{2}\left(\frac{0.968L - \pi fD_m}{0.968\pi D_m + fL}\right). \tag{11-4}$$

(0.968 is equal to the cosine of $14\frac{1}{2}$ degrees.)

Equations (Trapezoidal Metric Thread)

A metric counterpart to the Acme power translation screw is the trapezoidal thread as found in DIN 103 standard. Thread dimensions can be found in up-to-date handbooks or by referring to metric thread standards. This is a 30-degree thread; dimensions are given in millimeters. In the following equations, the load W is expressed in newtons (although kilogram-force units can be used), and 10^{-3} is placed in the numerator to produce frictional torque units of newton-meters. Thus, for *raising* a load,

$$T_f = \frac{WD_m}{2}\left(\frac{0.966L + \pi fD_m}{0.966\pi D_m - fL}\right)(10^{-3});$$ **(11-5)**

for *lowering* the load,

$$T_f = \frac{WD_m}{2}\left(\frac{0.966L - \pi fD_m}{0.966\pi D_m + fL}\right)(10^{-3}).$$ **(11-6)**

(0.966 is equal to the cosine of 15 degrees.)

American Standard Threads

The 60-degree thread (V-type) is very seldom used as a translation screw. However, it is sometimes used for small C-clamps. Thread torque is calculated in the same way as for the Acme thread; the following relationships for thread torque are based on the 60-degree angle, which is 30 degrees above the center of the thread and 30 degrees below. For raising a load,

$$T_f = \frac{WD_m}{2}\left(\frac{0.866L + \pi fD_m}{0.866\pi D_m - fL}\right);$$ **(11-7)**

for lowering a load,

$$T_f = \frac{WD_m}{2}\left(\frac{0.866L - \pi fD_m}{0.866\pi D_m + fL}\right).$$ **(11-8)**

 Collar Torque

Collar friction must be overcome as well as thread friction in most power-screw applications. Collar-friction torque is equal to the product of the load, the coefficient of friction, and the frictional radius:

$$T_f = Wr_f f.$$ **(11-9)**

As mentioned in Chapter 3, the frictional radius is

$$r_f = \frac{2}{3}\left(\frac{r_o^3 - r_i^3}{r_o^2 - r_i^2}\right).$$

External Torque

In translation (or power-screw) applications where *both* collar friction and thread friction are involved, the external torque is found by incorporating the collar torque found by the preceding equation into the appropriate equation for thread torque. Thus

external torque = thread torque + collar torque.

The total external torque of a square-threaded power screw (with collar) is

$$Pr = Wr_f f + \frac{WD_m}{2}\left(\frac{L + \pi f D_m}{\pi D_m - fL}\right)$$

if the load is being raised.

◆ *Example* ────────────────────────

A triple-threaded square power screw with a root diameter of 2 in. and 2 threads/in. is used in conjunction with a collar with an outer diameter of 4 in. and an inner diameter of 2.5 in. Find the force necessary to raise a 3000 lb load if the operating force is applied at a radius of 36 in. The coefficient of friction is 0.2 for both threads and collar.

Solution. Pitch = $\frac{1}{2}$ = 0.5 in. Therefore, the lead = 3 × pitch = 3(0.5) = 1.5 in. The outer diameter of the thread = the root diameter + $2(\frac{1}{2} \times \text{pitch}) = 2 + 2(\frac{1}{2})(\frac{1}{2}) = 2.5$ in.;

$$D_m = \frac{2.5 + 2.0}{2} = 2.25 \text{ in.;}$$

$$r_f = \frac{2(r_o^3 - r_i^3)}{3(r_o^2 - r_i^2)} = \frac{2(8 - 1.95)}{3(4 - 1.56)} = 1.65 \text{ in.;}$$

$$Pr = Wr_f f + \frac{WD_m}{2}\left(\frac{L + \pi f D_m}{\pi D_m - fL}\right);$$

$$36P = 3000(1.65)(0.2) + \frac{3000(2.25)}{2}\left(\frac{1.5 + \pi(0.2)(2.25)}{\pi(2.25) - 0.2(1.5)}\right);$$

$$P = 67.8 \text{ lb.}$$

◆ *Example* ────────────────────────

What would the operating force in the previous example be if a single-threaded screw were used?

Solution. Lead = pitch = 0.5 in.

$$Pr = Wr_f f + \frac{WD_m}{2}\left(\frac{L + \pi f D_m}{\pi D_m - fL}\right);$$

$$36P = 990 + 3375\left(\frac{0.5 + 1.41}{7.07 - 0.1}\right);$$

$$P = 53.2 \text{ lb.}$$

◆ *Example*

A single-threaded trapezoidal metric thread has a pitch of 4 mm and a mean diameter of 18 mm. It is used as a translation screw in conjunction with a collar having an outside diameter of 37 mm and an inside diameter of 21 mm. Find the torque (in newton-meters) needed to raise a load of 400 kg if the coefficient of friction is 0.3 for both the threads and the collar.

Solution.

400 kg (force) = 400(9.81) = 3920 newtons;

$$r_f = \frac{2(r_o^3 - r_i^3)}{3(r_o^2 - r_i^2)} = \frac{2(18.5^3 - 10.5^3)}{3(18.5^2 - 10.5^3)} = 14.9 \text{ mm};$$

$$Pr = \frac{Wr_f f}{10^3} + \frac{WD_m}{2}\left(\frac{0.966L + \pi f D_m}{0.966\pi D_m - fL}\right)(10^{-3});$$

$$Pr = \frac{3920(14.9)(0.3)}{10^3} + \frac{3920(18)}{2}\left(\frac{0.966(4) + \pi(0.3)(18)}{0.966\pi(18) - 0.3(4)}\right)(10^{-3});$$

$$Pr = 17.5 + 13.8 = 31.3 \text{ N} \cdot \text{m.}$$

 **Efficiency**

Figure 11-4 is a force-analysis diagram of a translation screw when a load is being lowered. If ϕ is greater than α, a force will be required to lower the load. If α is greater than ϕ, on the other hand, the load will travel down the thread without any force being exerted. If a diagram such as Fig. 11-4 were constructed with accurate angles, the position of R would show the self-locking value. If it is to the right of the line of action of W, force is required to lower the load. If it is to the left of the line of action of W, the load will lower itself without the application of any force.

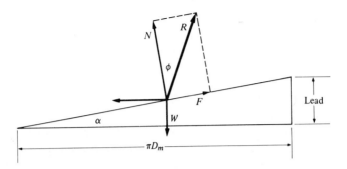

Figure 11-4
Force analysis: translation screw (load being lowered).

Neglecting collar or pivot friction, we can use the following relationships for the efficiency of a translation screw:

$$\text{efficiency} = \frac{\text{torque required to move load without friction}}{\text{torque required to move load with friction}} \times 100$$

$$= \frac{\dfrac{WD_m}{2}\left(\dfrac{L}{\pi D_m}\right)(100)}{\dfrac{WD_m}{2}\left(\dfrac{L + \pi f D_m}{\pi D_m - fL}\right)}$$

or

$$\text{efficiency} = \frac{L(\pi D_m - fL)(100)}{\pi D_m(L + \pi f D_m)} \qquad \text{for raising load,}$$

$$\text{efficiency} = \frac{L(\pi D_m + fL)(100)}{\pi D_m(L - \pi f D_m)} \qquad \text{for lowering load.}$$

These relationships are for square threads only. For angular threads, the thread torque must be modified as described in Eqs. (11-3) through (11-8).

Collar or pivot friction is involved in most translation-screw problems. When this is the case, one must be concerned with the overall mechanical efficiency, not just with the thread friction. To obtain the overall efficiency of an assembly, divide the output work by the input work:

$$e = \frac{WL(100)}{Pr(2\pi)}, \qquad\qquad\qquad \textbf{(11-10)}$$

where

$\quad e$ = efficiency (%),

$\quad L$ = lead (in.),

$\quad P$ = external operating force (lb).

For a given lead, P increases if W is increased. It is obvious that a large lead will increase the efficiency. It is also essential in certain translation-screw applications, such as lifting jacks, C-clamps, and vises, that the efficiency be low enough to make the assembly self-locking. Otherwise, the load will cause the screw assembly to "unwind" and lose its holding power.

◆ *Example* ————————————————————————————

Find the efficiency for the example on page 387.

Solution.

$$e = \frac{WL(100)}{Pr(2\pi)} = \frac{3000(1.5)(100)}{67.8(36)(2)(\pi)} = 29.3\%.$$

Thread Data

Although the square thread has not been standardized, the Sellers type is frequently used. Table 11-1 compares basic thread inforn.ation for the 29-degree standard Acme thread and the Sellers square thread. Not all of the available sizes are listed in this table. Nominal outside diameters and threads per inch are given. The pitch of any thread is found by dividing 1 by the number of threads per inch. For example, if a screw has 16 threads/in., its pitch is $\frac{1}{16}$ in.

Stresses in Translation Screws

Direct *tensile* (or *compressive*) stress can be approximated by considering the root cylinder of the screw. Actually, the thread on this basic cylinder strengthens it; this effect is neglected in the simple equations for tensile or

Table 11-1
Comparison of Acme and Sellers Threads

Outer Diameter	Acme Threads / in.	Sellers Threads / in.	Outer Diameter	Acme Threads / in.	Sellers Threads / in.
$\frac{1}{4}$	16	10	$1\frac{1}{2}$	4	3
$\frac{3}{8}$	12	8	$1\frac{3}{4}$	4	2.5
$\frac{1}{2}$	10	6.5	2	4	2.25
$\frac{5}{8}$	8	5.5	$2\frac{1}{4}$	3	2.25
$\frac{3}{4}$	6	5	$2\frac{1}{2}$	3	2
$\frac{7}{8}$	6	4.5	$2\frac{3}{4}$	3	2
1	5	4	3	2	1.75
$1\frac{1}{4}$	5	3.5	4	2	1.5

compressive stresses that follow:

$$S = \frac{W}{A} = \frac{W}{\pi D_i^2/4} = \frac{4W}{\pi D_i^2},$$

where

D_i = root diameter (in.).

If we neglect the strengthening effect of the thread metal, the direct shearing stress in the root cylinder is

$$S_s = \frac{16T}{\pi D_i^3}.$$

The maximum resultant tensile (or compressive) stress is given by the equation

$$S_{max} = \frac{S}{2} + S_{s_{max}},$$

as stated in Chapter 5. It must be emphasized that in the above equations, only the base cylinder was considered; actual induced stress values would be somewhat smaller.

Shearing stress at the root of the threads of the screw can be found by the equation

$$S_s = \frac{W}{\pi D_i kh},$$

where

h = nut thickness (in.),

k = 0.5 for square thread and 0.63 for Acme thread.

Figure 11-5
Comparison of Acme, square, and 60-degree V-threads at root section.

The equation for shearing stress at the root of the thread on the nut is

$$S_s = \frac{W}{\pi D_o k h},$$

where

D_o = maximum diameter of thread (in.).

Figure 11-5 shows a comparison of the root sections of a 60-degree V-thread with the square and Acme threads.

If the projected area of the threads were used in the analysis, *bearing* stresses could also be considered. The projected area would be $(\pi/4)(D_o^2 - D_i^2)$. In most design situations, it is not necessary to consider bearing stresses.

Buckling stress has to be considered if the translation screw has an excessively long free length. Appropriate column formulas can be used under such circumstances. The slenderness ratio (L/k) determines the necessity for such calculations; L = the effective length of the screw and k = the least radius of gyration. If the L/k ratio is less than 40, the normal compressive stress equation can generally be used, which is $S = P/A$. In determining the k value, one should use the root diameter of the thread, neglecting any strengthening action of the spiraling threads. For the circular cross section, $k = D_i/4$. If L/k is greater than 40, column formulas must be used.

◆ *Example* ————————————————————————

A bench vise has a single square thread with a $\frac{3}{4}$-in. outer diameter. There are 5 threads/in., and the frictional radius of the collar is 1 in. The coefficients of friction are 0.1 and 0.15 for the threads and collar, respectively. If a force of 30 lb is exerted on a 6-in. handle, what force is the jaw of the vise exerting against the work? Find the compressive stress on the base cylinder and the shearing stress at the root of the thread if the jaw of the vise (nut) is 1 in. thick. Neglect the torsional stress in the root cylinder and the strengthening of the cylinder by the screw threads.

Solution. Pitch $= \frac{1}{5} = 0.200$ in. Mean diameter $= 0.650$ in. Root diameter $= 0.750 - 0.200 = 0.550$ in. Thus,

$$\tan \alpha = \frac{lead}{\pi D_m} = \frac{0.200}{(\pi)(0.65)} = 0.098;$$

$$Pr = Wr_f f + \frac{WD_m}{2} \left(\frac{L + \pi f D_m}{\pi D_m - fL} \right).$$

$$(30)(6) = W(1)(0.15) + \frac{0.65W}{2} \left(\frac{0.2 + \pi(0.1)(0.65)}{\pi(0.65) - (0.1)(0.2)} \right).$$

$$180 = 0.15W + 0.065W.$$
$$W = 837.2 \text{ lb.}$$

Hence, the compressive stress is

$$S = \frac{W}{A} = \frac{4W}{\pi D_i^2} = \frac{4(837.2)}{\pi(.550)^2} = 3524 \text{ psi,}$$

and the shearing stress is

$$S_s = \frac{W}{A} = \frac{W}{\pi D_i kh} = \frac{837.2}{\pi(0.550)(0.5)(1)} = 969 \text{ psi.}$$

 Applications of Translation Screws

This screw is used typically in conjunction with a collar for such applications as vises, C-clamps, capstans, vertical screw jacks, scissors jacks, and valve stems (gate valves, water gates, and so on). Collar-friction torque is usually high unless antifriction bearings are used in the collar. In other applications, no collar is used; in these cases, only thread torque is involved. The lead screw on lathes is a typical example of this application. Another example is the drive in some Universal testing machines. In this type of operation, the movable head that subjects a test specimen to tensile or compressive stresses is operated up or down by means of heavy-duty power screws.

Ball-Bearing Screw

Thread friction greatly reduces the overall efficiency in power-screw applications. Figure 11-6 illustrates a ball-bearing screw. In this device

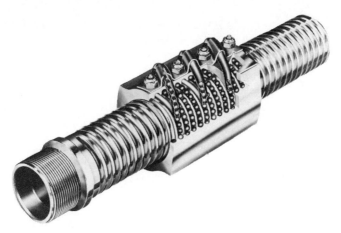

Figure 11-6
Ball-bearing screw.
(Courtesy of Saginaw Steering Gear Division, General Motors Corp.)

(which is really an actuator), a series of steel balls rides between the threads of the screw and the threads of the nut. The threads are shaped to accommodate the balls; in addition, provision must be made to recirculate the balls. Thus rolling friction (a small value) replaces the sliding friction of the usual power screw. With this type of device, efficiencies of over 90% are possible in converting rotation (torque) to translation (thrust). Note that the nut contains a guide for the balls so that they will recirculate externally. The life of this type of screw can be predicted more accurately than can that of the conventional power screw, which undergoes a large and unpredictable amount of wear. Typical applications include automotive steering mechanisms, lens focusing in industrial cameras, and periscopes.

Summary

In general, power (or translation) screws are used to effect a good mechanical advantage. Sometimes they are used in reverse (for convenience) in such applications as self-driving screwdrivers or push drills. When used to gain a mechanical advantage, the lead plays an important role. If the lead is large, linear screw movement per turn of the screw is also large. However, the external operating force must be greater because the large lead produces a greater slope for the load to climb.

To simplify problem solving, the depth of thread for the usual square thread has been taken as one-half the pitch; it is actually $\frac{19}{40}$ of the pitch. For the Acme thread, the depth of thread is actually one-half the pitch plus 0.01

in. Very little error results from simplifying the thread nomenclature in this manner. Anyone wishing extreme accuracy should refer to handbook values when making power-screw calculations.

Efficiency in such applications as hand-powered screw jacks is important. Too high a value prevents the screw from being self-locking; too low a value makes a large operating force necessary.

Questions for Review

1. List the step-by-step procedure for finding the external torque on the handle of a gate valve. What information is needed to solve this type of problem?

2. What advantages does the ball-bearing screw have over the usual translation screws? What disadvantages does it have? List other applications for the recirculating ball type of screw.

3. Discuss the effect of efficiency in power-screw applications. List some examples where high efficiency is desirable; list others where a lower efficiency might be needed.

4. How does a self-driving screwdriver compare to a translation screw?

5. Differentiate between the terms *pitch* and *lead* for any screw thread.

6. How can collar-friction torque be minimized in a capstan application?

Problems

1. The root diameter of a double square thread is 0.55 in. The screw has a pitch of 0.2 in. Find the outside diameter, the number of threads per inch, and the lead.

2. The collar of a capstan has an outside diameter of 8 in. and an inside diameter of 6 in. The total load it supports is 2000 lb. If the coefficient of sliding friction is 0.1, what is the collar-friction torque?

3. A vise is equipped with a 1-in. single square thread, with 4 threads/in. The frictional radius of the collar is $\frac{1}{2}$ in. The coefficient of friction for both the collar and threads is 0.2. How much external torque must be applied to produce a force of 200 lb against the jaws of the vise?

4. A capstan uses a 2-in. single 29-degree Acme thread with 4 threads/in. The frictional radius of the collar is 1 in. and the coefficient of friction is 0.2 for the threads and the collar. How much external force is required at the end of a 40-in. lever to raise a load of 10,000 lb?

5. A capstan is operated with a double square-threaded power screw with a pitch of 1 in. An external force of 50 lb is required at the end of a 50-in. crank to lift a 2-ton load. Find the efficiency.

6. A single-threaded jackscrew has square threads. The screw diameter is 1 in. and there are 4 threads/in. A friction collar has a 2-in. outside diameter and a 1-in. inside diameter. The coefficient of friction is 0.1 for the collar and 0.15 for the threads. How much force must be applied on a 12-in. (radius) jack lever to raise a load of 1000 lb?

7. What is the overall efficiency of the jackscrew of Problem 6?

8. How much force would be required to lower the load of Problem 6, if all other conditions remained the same?

9. How much force would be required to raise the load of Problem 6 if a 29-degree Acme thread were employed and all other conditions of the problem remained the same?

10. A double square-threaded power screw transmits 1800 in-lb of torque and raises a machine head. The head is guided by linear slides with effective antifriction bearings. If the coefficient of friction for the threads is 0.1, find the load that can be raised if the screw has 2 threads/in. and an outside diameter of $2\frac{1}{2}$ in.

11. A 4000-lb load is to be lowered by the double-threaded screw described in Problem 10. What torque must the screw transmit if all other conditions remain the same?

12. If the power screw of Problem 10 rotates at 10 rpm, what is the linear speed of the driven head in feet per minute?

13. A C-clamp has 10 single square threads per inch. The outside diameter is $\frac{1}{4}$ in. and the frictional radius of the collar is 0.3. The coefficient of friction for the threads is 0.2 and the collar-friction coefficient is 0.15. How much force must be exerted on a 2-in. handle (radius) to produce a force of 300 lb between the jaws?

14. A turnbuckle connects two rods having $\frac{3}{8}$-16 NC threads. The thread friction is 0.2. How much external torque is necessary to "stress" the rods to 500 lb if it is assumed that neither rod turns as the turnbuckle is tightened? The mean diameter of the threads is 0.335 in.

15. Find the horsepower required to drive a power screw lifting a load of 4000 lb. A $2\frac{1}{2}$-in. double square thread with 2 threads/in. is to be used. The frictional radius of the collar is 2 in., and the coefficients of friction are 0.1 for the threads and 0.15 for the collar. The velocity of the nut is 10 ft/min.

16. How much torque is required to lower the power screw of Problem 15, if all other conditions remain the same?

17. What is the efficiency of the translation screw (with collar) in Problem 15? Find the compressive stress based on the root cylinder, neglecting the strengthening effect of the threads.

18. Find the shearing stress at the root of the power screw and the shearing stress at the root of the nut thread in Problem 15. Assume that the nut thickness is 1 in.

19. A turnbuckle has M10 × 1.5 threads. The coefficient of friction for the threads is 0.2 and the basic pitch diameter is 9.026 mm. Assuming that neither rod turns as the turnbuckle is tightened, what is the external torque needed to produce a force of 2500 newtons on the rods? (Metric threads are considered V-type.)

20. A gate valve is operated by a handwheel and a trapezoidal metric translation screw. The effective gate area is 0.0183 m² and the water pressure against the gate is 1.67 MPa. The coefficient of friction between the gate and its seat is 0.2. The frictional radius of the collar is 28 mm, and the coefficient of friction is 0.18 for both the collar and the threads. The pitch of the translation screw is 4 mm and the mean diameter is 18 mm. Find the operating torque for the handwheel, expressing your answer in newton-meters.

21. Find the torque required for a C-clamp that must exert a force of 100 lb. Assume the following data: American Standard threads 1/2″-13 UNC; a root diameter of 0.4001 in.; a root area of 0.1257 in.²; a coefficient of thread friction of 0.12; a coefficient of collar friction of 0.25; and a collar frictional radius of 0.25 in.

22. Find the operating torque for a screw-power vertical jack that must lift a load of 450 kg, assuming a trapezoidal metric thread and the

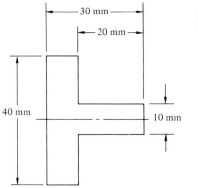

Figure 11-7
Problem 23.

following specifications: OD = 30 mm; pitch = 6 mm; pitch diameter = 27 mm; root diameter = 23.5 mm; coefficient of friction for threads and collar = 0.25; frictional radius of collar = 12 mm; and effective extended length = 230 mm. Also find the compressive, shearing, and maximum stresses for the screw and the slenderness ratio.

23. A C-clamp uses metric trapezoidal threads. Its specifications are as follows: pitch = 5 mm; pitch diameter = 21.5 mm; distance from center of threads to heel of frame = 100 mm; effective frictional radius of the collar = 100 mm; coefficient of friction for threads and collar = 0.15; and applied external torque = 9 N · m. Figure 11-7 shows the dimensions of the T-section for the frame. Under these conditions, find the induced stress (maximum) in the frame.

GEARS

12

Gears provide a positive-ratio drive for transmitting rotary motion from one shaft to another. If the shafts are parallel, any of three types may be used: *spur*, *helical*, or *herringbone* gears. *Spiral* gears are used to connect two shafts that are nonparallel and nonintersecting. *Worm* gears are used where high ratios are desired and where the shafts are nonintersecting and at right angles. *Bevel* gears are often used when two shafts are at right angles to each other and their centerline extensions intersect; however, some bevel gears are at angles other than 90 degrees. *Spiral bevel* gears can be used in the same applications as straight-tooth bevel gears; the spiral bevel gears are capable of higher speeds and quieter operation. *Hypoid* gears are similar to spiral bevel gears, except that the extensions of the centerlines do not intersect. Hypoid gears were originally developed for the automotive rear-axle drive. *Rack-and-pinion* drives are used where the rotary motion of one part must be transformed into translating motion of the other part, or vice versa. The incorporation of some of these types into gear trains constitutes an important phase of design. This chapter deals with the gear tooth design of several types of gears and the design of gear-train mechanisms.

 ## Gear Materials

Various materials are used in manufacturing gears. Usually, the material selected depends on the method used for making the gear and the application to which it will be put. Gears can be cast, cut, or extruded. Typical materials include cast iron, cast steel, plain carbon steel, alloy steel, aluminum, phosphor bronze, laminated phenolics, and nylon.

 ## Spur Gears

Spur gears will be considered first, for several reasons. In the first place, they are the simplest and least expensive type of gear; also, spur-gear definitions are usually applicable to other types. Figure 12-1 shows actual spur-gear tooth sizes, classified by diametral pitch. It is important to understand the following definitions, since they are important factors in the design of any equipment utilizing gears.

Diametral pitch: the number of teeth per inch of pitch-circle diameter. The diametral pitch is usually an integer. A small number for the pitch implies a large tooth size. Meshing spur gears must have the same diametral pitch. The speed ratio is based on the fact that meshing gears may have different-sized pitch circles and hence different numbers of teeth.

Figure 12-1
**Chart of relative sizes of
spur-gear teeth of different
diametral pitch.**
(Courtesy of Barber-Colman Co.)

Circular pitch: the distance from a point on one tooth to the corresponding point on an adjacent tooth, measured along the pitch circle. This is a linear dimension and thus has linear units (usually inches).

Pitch circle: the circle on which the ratio of the gear set is based. When two gears are meshing, the two pitch circles must be exactly tangent if the gears are to function properly. The tangency point is known as the *pitch point.*

Pressure angle: the angle between the line of action (see Fig. 12-2) and a line perpendicular to the centerlines of the two gears in mesh. Pressure angles for spur gears are usually $14\frac{1}{2}$ or 20 degrees, although other values can be used. Meshing gears must have the same pressure angles. In the case of a rack, the teeth have straight sides inclined at an angle corresponding to the pressure angle.

Base circle: a circle tangent to the line of action (or pressure line). The base circle is the imaginary circle about which an involute curve is developed. Most spur gears follow an involute curve from the base circle to the top of the tooth; this curve can be visualized by observing a point on a taut cord as it is unwound from a cylinder. In a gear, the cylinder is the base circle.

Addendum: the radial distance from the pitch circle to the top of the tooth.

Dedendum: the radial distance from the pitch circle to the root of the tooth.

Clearance: the difference between the dedendum and the addendum.

Face width: the width of the tooth measured axially.

Face: the surface between the pitch circle and the top of the tooth.

Flank: the surface between the pitch circle and the bottom of the tooth.

Working depth: the distance that a tooth from a meshing gear extends into the tooth space.

Contact ratio: the average number of teeth in contact.

Formulas

The following basic formulas are useful in simple gear calculations. A careful study of the gear-tooth nomenclature will enable one to develop other equations easily:

$$P = \frac{t}{D}, \tag{12-1}$$

$$P_c = \frac{\pi}{P} = \frac{\pi D}{t}, \tag{12-2}$$

$$a = \frac{1}{P}, \tag{12-3}$$

$$d = \frac{1.157}{P}, \tag{12-4}$$

$$D_B = D \cos \phi, \tag{12-5}$$

where

P = diametral pitch,

t = number of teeth,

D = pitch-circle diameter (in.),

P_c = circular pitch (in.),

ϕ = pressure angle (degrees),

D_B = base-circle diameter (in.),

a = addendum (in.),

d = dedendum (in.).

◆ *Example*

A 40-tooth steel gear has a diametral pitch of 2. The pressure angle is $14\frac{1}{2}$ degrees. Find the circular pitch, addendum, dedendum, clearance, base-circle diameter, and outside diameter of the gear. *Note:* Gear dimensions should usually be figured to 0.001 in. accuracy.

Solution.

$$P_c = \frac{\pi}{P} = \frac{\pi}{2} = 1.571 \text{ in.};$$

$$a = \frac{1}{P} = \frac{1}{2} = 0.500 \text{ in.};$$

$$d = \frac{1.157}{P} = \frac{1.157}{2} = 0.579 \text{ in.};$$

$$\text{clearance} = d - a = 0.579 - 0.500 = 0.079 \text{ in.};$$

$$D = \frac{t}{P} = \frac{40}{2} = 20.000 \text{ in.};$$

$$D_B = D \cos \phi = 20 \cos 14\tfrac{1}{2} = 20(0.968) = 19.363 \text{ in.};$$

$$\text{outside diameter} = D + 2a = 20.000 + 2(0.500) = 21.000 \text{ in.}$$

Additional Gear Formulas

The preceding formulas apply to the American Standard full-depth involute tooth, whether the pressure angle is $14\tfrac{1}{2}$ or 20 degrees. For the American Standard 20-degree stub tooth, the addendum, dedendum, and clearance are as follows:

$$a = \frac{0.800}{P}, \qquad d = \frac{1}{P},$$

$$\text{clearance} = \frac{0.200}{P}.$$

The Fellows stub-tooth system, developed by the Fellows Gear Shaper Company, differs somewhat from the others. In this system, *two* diametral pitches are used, expressed in fractional form: 4/5, 5/7, 6/8, 7/9, 8/10, 9/11, 10/12, and 12/14. The numerator pitch is used for all calculations except those dealing with the height of the tooth. The pitch that appears in the denominator is used in calculating addendum, dedendum, and clearance. The following example shows how these fractionally expressed pitches can be applied.

◆ *Example* ─────────────────────────────

Find the pitch-circle diameter, addendum, dedendum, and clearance for a Fellows stub-tooth gear with a pitch of 4/5. The gear has 20 teeth.

Solution.

$$D = \frac{t}{P} = \frac{20}{4} = 5.000 \text{ in.};$$

$$a = \frac{1}{P} = \frac{1}{5} = 0.200 \text{ in.};$$

$$d = \frac{1.25}{P} = \frac{1.25}{5} = 0.250 \text{ in.};$$

clearance $= d - a = 0.250 - 0.200 = 0.050$ in.

Note that 1.25 is used in the numerator for the dedendum calculation.

 Metric Spur Gears

The metric system for spur-gear calculations involves more than changing inches to millimeters. The basis for calculations in the American inch system is diametral pitch, which is the number of teeth divided by the pitch-circle diameter. In the metric system, the term *module* replaces diametral pitch. The module is found by dividing the pitch-circle diameter by the number of teeth. The module is thus a dimension, expressed in millimeters, equal to the reciprocal of the diametral pitch. The pressure angle is 20 degrees; the tooth form is full depth and thus allows interchangeability.

Formulas

The following formulas are presented to permit basic calculations:

$$m = \frac{D}{t}, \tag{12-6}$$

$$a = m, \tag{12-7}$$

$$b = 1.25\,m, \tag{12-8}$$

$$P_c = m\pi, \tag{12-9}$$

$$C = \frac{m(t_1 + t_2)}{2}, \tag{12-10}$$

$$D_o = D + 2m, \tag{12-11}$$

where

  $m =$ module,
  $a =$ addendum (radial distance from pitch circle to top of tooth),

b = dedendum (radial distance from pitch circle to root of tooth),

t = number of teeth,

D = pitch-circle diameter,

P_c = circular pitch,

D_o = outside diameter of gear,

C = center distance between two meshed gears.

Note that all dimensions are specified in millimeters. The following example shows sample gear calculations in the metric system.

◆ *Example* ────────────────────────────────

A pair of meshing spur gears have a module of 2; the pinion (smaller gear) has 18 teeth and the driven gear has 27 teeth. Find the addendum, dedendum, and pitch-circle diameter for each gear. Also find the circular pitch and the center distance between the axes of the two meshing gears.

Solution.

$a = m = 2$ mm,

$b = 1.25\,m = 1.25(2) = 2.50$ mm,

$D_1 = mt_1 = 2(18) = 36$ mm,

$D_2 = mt_2 = 2(27) = 54$ mm,

$P_c = m\pi = 2(\pi) = 6.2832$ mm,

$$C = \frac{m(t_1 + t_2)}{2} = \frac{2(18 + 27)}{2} = 45 \text{ mm.}$$

───

Note that this discussion has focused on spur gears only. Calculations for internal gears would be similar to the preceding, except that the center distance calculation (Eq. 12-10) would be modified—a minus sign would replace the plus sign.

Gear cutters are commercially available for the following modules: 0.5 to 4.0 by 0.25 increments, 4.0 to 7.0 by 0.5 increments, and 7.0 to 10.0 by 1.0 increments.

Contact Ratio

Continuous action between gear teeth is a necessity and it is desirable to have more than one pair of gear teeth in engagement at a time. The contact ratio can be defined as the average number of teeth in contact when two gears are meshing; it can also be defined as a ratio between the arc of action and the circular pitch. It can be found graphically by determining the angle of action and dividing this value by the pitch angle. The angle of action can be found by noting the place where contact between a pair of teeth first occurs and the place where contact ceases. The pitch angle is the angle projected from the circular-pitch dimension to the center of a gear. The value for the contact ratio can also be determined mathematically.

The action of gear teeth is depicted in Fig. 12-2. The length of the line of action is from point 1 to point 2 in this figure. Point 1 is the point where the addendum circle of the driven gear crosses the pressure line. Point 2 is the point where the addendum circle of the driving gear crosses the pressure line. Both points must lie inside perpendiculars scribed from the pressure line to the gear centers, or the gears will interfere. In this particular example, the pressure angle is 20 degrees. Note that the pressure angle is measured

Figure 12-2
Action of gear teeth.

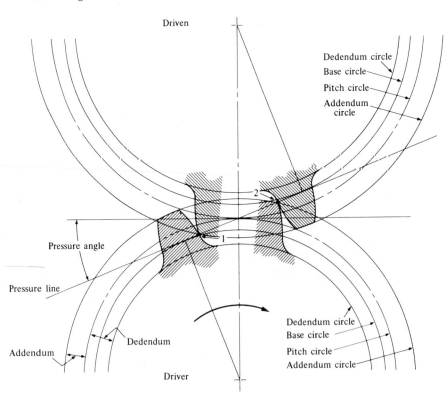

from a perpendicular to the gear centerlines. Forces are transmitted from the driver to the driven along this line. If the driving gear were rotating counterclockwise, the pressure angle would be measured 20 degrees on the opposite side of the above-mentioned perpendicular.

The following equation presents a mathematical approach to determining the contact ratio:

$$\text{CR} = \frac{\sqrt{R_o^2 - R_B^2} + \sqrt{r_o^2 - r_B^2} - C\sin\phi}{P_c\cos\phi}, \tag{12-12}$$

where

R_o and r_o = outside radii for gears (in.),

R_B and r_B = base-circle radii (in.),

C = center distance (in.),

ϕ = pressure angle (deg),

CR = contact ratio.

 **Gear Trains**

A gear train is a system of two or more meshing gears. The simplest type consists of a driver on one shaft engaging a follower on another shaft. If only two gears are used and both are spur gears, the shafts rotate in opposite directions. If one of the gears is an internal gear, the two shafts turn in the same direction. An idler gear is sometimes used; an idler gear does not change the ratio, but merely the direction of rotation. Pitch-circle velocities are the same for all meshing gears in a train.

Ratio

The following general equation applies to simple gear trains, regardless of whether spur gears or internal gears are employed:

$$R = \frac{D_f}{D_d} = \frac{t_f}{t_d} = \frac{N_d}{N_f}, \tag{12-13}$$

where

D = pitch-circle diameter (in.),

t = number of teeth,

N = speed (rpm).

The subscripts f and d represent follower and driver, respectively. Gear trains can have a ratio of unity ($1:1$), less than 1, or greater than 1. The majority of gear trains are reduction units—that is, they have a ratio greater than 1. If all gear centers are fixed, the diameters (or numbers of teeth) can be used in product form in the preceding equation. The following example illustrates this situation.

◆ *Example*
In Fig. 12-3, shaft 1 rotates at 1800 rpm. Find the ratio between shafts 1 and 2 and the ratio between shafts 1 and 3. Then find the ratio between shafts 1 and 4. What is the rotational speed of shaft 4? Gears B and C are locked together and thus rotate at the same speed; this is also true for gears D and E.

Solution.

$$R_{1\text{-}2} = \frac{t_f}{t_d} = \frac{40}{20} = 2,$$

$$R_{1\text{-}3} = \frac{t_{f_1} t_{f_2}}{t_{d_1} t_{d_2}} = \frac{40(36)}{20(18)} = 4,$$

$$R_{1\text{-}4} = \frac{t_{f_1} t_{f_2} t_{f_3}}{t_{d_1} t_{d_2} t_{d_3}} = \frac{40(36)(32)}{20(18)(16)} = 8,$$

$$N_f = \frac{N_d}{R} = \frac{1800}{8} = 225 \text{ rpm}.$$

Figure 12-3
Gear train.

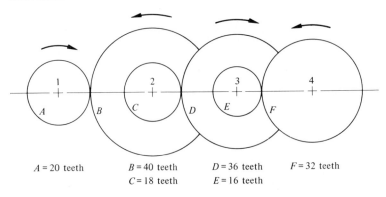

A = 20 teeth   B = 40 teeth   D = 36 teeth   F = 32 teeth
C = 18 teeth   E = 16 teeth

Note: The ratio from shaft 1 to shaft 4 was used in the final step because the output shaft speed was desired. The rotational speed of any of the intermediate shafts can be found by using the ratio to the appropriate shaft. In this problem, even numbers were used for the teeth in each gear to simplify the mathematics. In actual practice, it is desirable to use some odd numbers (even at the expense of not obtaining the exact ratio desired), so that a particular pair of teeth will not always be meshing together. When contact varies from tooth to tooth, the manufacturing discrepancies will disappear and wear will be even on all teeth for both driving and driven gears.

In the preceding example, the rotational direction of each shaft can easily be determined by inspection. If shaft 1 turns clockwise, shaft 2 goes counterclockwise; then shaft 3 would rotate clockwise and shaft 4 would turn counterclockwise. The total ratio in this problem was 8/1. It would therefore be possible to have a train consisting of two gears; if we assumed the driver had 20 teeth, the driven would have to have 8 × 20, or 160, teeth. This one-step ratio has some advantages and some disadvantages. It is obviously simpler than the other train, which requires six gears, four shafts, and eight bearings, as opposed to two gears, two shafts, and four bearings. In addition, provision must be made to connect gear *B* with *C* and *D* with *E*. They could be keyed together (this would require extra components) or they could be made of the same piece of gear material. Either solution would increase the cost of production.

On the other hand, the simpler train would have some very undesirable features. If one gear was exceptionally large, compactness would be difficult to attain. Also, with such a large ratio, action between meshing teeth would present some real problems. The gear teeth would probably interfere; this could be overcome by undercutting the flank of the driver (reducing its strength and load-carrying capacity) or by modifying the tooth profile (making it a nonstandard type). Either solution could greatly increase the cost. In designing gear trains, the designer must consider all possibilities and then make a decision based on overall cost, efficiency, compactness, and general feasibility.

Train Values

For simplicity in evaluating or designing gear trains, it is desirable to use the term *train value*, rather than *ratio*. The train value is merely the reciprocal of the ratio. Thus

$$\text{TV} = \frac{t_d}{t_f} = \frac{D_d}{D_f} = \frac{N_f}{N_d}. \tag{12-14}$$

Then

$$N_f = (\text{TV})N_d.$$

In the previous example, the train value would be

$$\text{TV} = \frac{t_{d_1}t_{d_2}t_{d_3}}{t_{f_1}t_{f_2}t_{f_3}} = \frac{20(18)(16)}{40(36)(32)} = \frac{1}{8}.$$

Then

$$N_f = (\text{TV})N_d = \tfrac{1}{8}(1800) = 225 \text{ rpm}.$$

Reverted Gear Trains

A reverted gear train is one in which the driving shaft is collinear with the output shaft. Figure 12-4 illustrates this type of train. The outstanding

Figure 12-4
Reverted gear train.

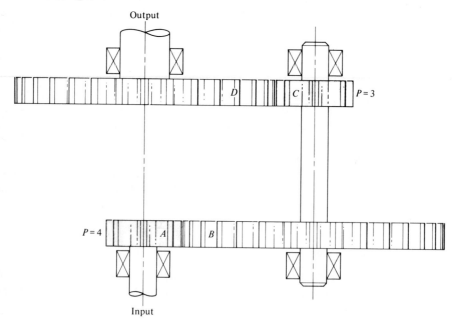

feature of this type of gear train is its compactness. In many cases, the first pair of meshing gears has a different diametral pitch from the other pair (which affects the size). The following example shows how this type of train can be formulated. If all the gears have the same pitch, the problem is much easier to solve.

◆ *Example* ─────────────────────────────────────

The reverted gear train shown in Fig. 12-4 is to have a ratio of 24/1. The minimum number of teeth in any one gear is 16. Gears *A* and *B* have a pitch of 4; gears *C* and *D* have a pitch of 3. Find the number of teeth required in each gear to satisfy these conditions.

Solution. The ratio of 24/1 is accomplished in two stages; therefore, the ideal ratio for each step would be $\sqrt{24}$ /1. However, $\sqrt{24}$ does not give integers, and decimal-fraction ratios tend to complicate the calculation. It is thus best to factor 24 into two factors that are not too far apart numerically. In this case the 24/1 ratio could be factored into a ratio of 6/1 for one set of gears and 4/1 for the other. Since this is a reverted train, the center distances must be equal for each meshing pair. The following equations can be written:

$$TV_{A\text{-}B} = \frac{t_A}{t_B} = \frac{1}{6},$$

$$TV_{C\text{-}D} = \frac{t_C}{t_D} = \frac{1}{4};$$

$$2C = \frac{t_A + t_B}{P_{A\text{-}B}} = \frac{t_C + t_D}{P_{C\text{-}D}}$$

or

$$P_{A\text{-}B}(t_C + t_D) = P_{C\text{-}D}(t_A + t_B).$$

Both diametral pitches are known, but the number of teeth is not known for *any* of the gears in the train. Each side of the following equation can be equated to a least common multiple to solve for tooth values:

$$P_{A\text{-}B}(t_C + t_D) = P_{C\text{-}D}(t_A + t_B).$$

The least common multiple can be found from the product of the following:

1. pitch of gears *A* and *B*,
2. pitch of gears *C* and *D*,
3. the sum of the numerator and denominator of the train value from *A* to *B*,

4. the sum of the numerator and denominator of the train value from *C* to *D*.

The product would be as follows:

$$(4)(3)(1 + 6)(1 + 4) = 420.$$

Then

$$\frac{t_C}{t_D} = \frac{1}{4} \quad \text{or} \quad t_D = 4t_C$$

and

$$P_{A\text{-}B}(t_C + t_D) = 420,$$
$$4(t_C + 4t_C) = 420,$$
$$5t_C = 105.$$

Thus, $t_C = 21$ teeth and $t_D = 4t_C = 4(21) = 84$ teeth. Also,

$$\frac{t_A}{t_B} = \frac{1}{6} \quad \text{or} \quad t_B = 6t_A$$

and

$$P_{C\text{-}D}(t_A + t_B) = 420,$$
$$3(t_A + 6t_B) = 420,$$
$$7t_A = 140.$$

Thus, $t_A = 20$ teeth and $t_B = 6t_A = 6(20) = 120$ teeth. Gears with these tooth numbers would satisfy the 24/1 ratio, provide collinear center-lines, and have at least 16 teeth. If any gear in the preceding calculation had had fewer than 16 teeth, it would have been necessary to multiply all gears by some number that would provide at least 16 teeth for the smallest gear. Similar manipulations can be made by division if the smallest gear has an unusually large number of teeth.

If a train of this type is designed with the same diametral pitch, the calculation is handled in the same manner. However, the least common multiple is much smaller; thus the calculation is somewhat simpler.

In general, reverted trains are used (a) to reduce the size of the speed-reduction unit and (b) to simplify the mounting of other components. Since speed reduction increases the torque, the output shaft will be larger in diameter than the input shaft.

Planetary or Epicyclic Gear Trains

Figure 12-5 shows one type of planetary gear train. Gear 1 is a sun gear, gears 2 and 2' are planet gears, and gear 3 is a fixed-ring gear. For this particular situation, 1 could be considered the driver and the planet carrier 4 could be the driven or follower gear. This would provide slower speed and greater torque-carrying capacity for the driven shaft. If the planet carrier 4 were driving and the sun gear 1 were driven, this would become an overdrive device. Output speed would be increased, but the torque-carrying capacity would be decreased.

From Fig. 12-5, it is obvious that planets 2 and 2' rotate about their axes and revolve about the sun gear—a motion similar to the rotation of the earth on its axis once each day and its revolution about the sun once a year. Planet gears 2 and 2' must be the same size. This type of train can be designed with one planet, with two planets spaced at 180 degrees, or with three planets spaced at 120 degrees. The number of planets has no bearing on the output speed, but the load-carrying capacity increases as the number of planets increases.

Other driving-driven relationships are possible within this same general configuration. If the planet carrier were fixed, the sun gear could rotate freely and the ring gear could be attached to a rotating shaft. This configuration produces no planetary or epicyclic effect; thus, the mechanism becomes a simple gear train.

Another possibility is to have the sun gear fixed, and connect rotating shafts to the planet carrier and the ring gear.

Still another arrangement is possible: the sun gear, ring gear, and planet carrier can all rotate. The rotation of the ring gear is often facilitated by cutting external teeth as well as internal teeth on this gear. The external

Figure 12-5
Planetary gear train.

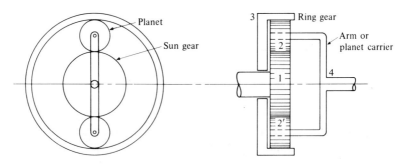

teeth are engaged with a pinion that drives the ring gear. In calculations, however, the meshing of a pinion with the external teeth of the ring gear must be treated as a simple gear train.

Planetary Gear Speeds

A general mathematical relation for calculating rotational speeds in a planetary gear train is as follows:

$$N_s = (TV)N_r + N_c[1 - (TV)], \qquad\qquad (12\text{-}15)$$

where the subscripts s, r, and c represent sun gear, ring gear, and planet carrier, respectively.

The train value is written from the ring (or annular) gear to the sun gear, but the mechanism is treated as a simple gear train, with the arm considered locked. If the sun and ring gears would turn in the same relative direction if the planet carrier was locked, a positive value is assigned to this train value. If the sun gear and ring gear would turn in opposite directions with the planet carrier locked, a negative train value is assigned. In evaluating the preceding equation, care must be exercised in observing algebraic rules for signed numbers. A mistake in sign could lead to a huge mathematical error. The final sign preceding the numerical answer will indicate the direction of rotation of that particular gear. The following example indicates the method used in solving such problems.

◆ *Example* _____

In the planetary gear train of Fig. 12-5, assume that the ring gear is fixed and that the sun gear has 32 teeth, the planet gears have 16 teeth, and the ring or internal gear has 64 teeth. The sun gear rotates at 100 rpm clockwise. Find the rotational speed and direction of the planet carrier.

Solution.

$$TV = \tfrac{64}{16} \times \tfrac{16}{32} = -2,$$
$$N_s = (TV)N_r + N_c[1 - (TV)],$$
$$100 = (-2)(0) + N_c[1 - (-2)],$$
$$3N_c = 100,$$
$$N_c = 33.3 \text{ rpm.}$$

The minus sign signifies that the sun and ring gears turn in opposite directions if the planet carrier is considered fixed (this is an imaginary step).

From the above result, the planet carrier rotates in the same direction as the sun gear; if a minus sign had preceded the 33.3 numerical value, this would have indicated rotation in the opposite direction. Note that this particular train is not too desirable, because the 16-tooth gear engaging a 32-tooth gear will force a given tooth on the 16-tooth gear to contact the same two teeth on the 32-tooth gear during every rotation. This can be avoided by changing one or the other to an odd number; the ratio will be changed slightly, but tooth action will be improved. To design a planetary train, the various terms in the preceding equation can be studied and suitable tooth numbers selected to give a train value that will yield the desired output speed.

◆ *Example* ──────────────────────────────────────

Neglecting gear-tooth strength, design a planetary gear train similar to the one shown in Fig. 12-5. Assume that the annular gear is fixed and has 150 teeth. The rotational speeds of the sun gear and planet carrier are 100 and 40 rpm, respectively.

Solution.

$$N_s = (TV)N_r + N_c[1 - (TV)],$$

$$100 = (TV)0 + 40[1 - (TV)],$$

$$TV = -1.5.$$

The minus sign is significant only to the relative rotation in the planetary train; it can be disregarded in gear-tooth calculations. Since

$$TV = \frac{t_d}{t_f} = \frac{t_3 t_2}{t_2 t_1} = 1.5$$

or

$$t_3 = t_1(TV),$$

then

$$t_1 = \frac{t_3}{TV} = \frac{150}{1.5} = 100 \text{ teeth.}$$

Since the sum of all gear diameters (sun and two planets) must equal the pitch diameter of the ring gear, and since all meshing gears have the same diametral pitch, tooth numbers can also be added diametrically. Both planets must be the same size. Thus,

$$t_3 = t_1 + t_2 + t_{2'} \quad \text{or} \quad t_3 = t_1 + 2t_2;$$

therefore,

$$150 = 100 + 2t_2,$$

$$t_2 = 25 \text{ teeth.}$$

Another configuration of a planetary gear train is shown in Fig. 12-6(a). Two different-sized planets are locked to a shaft that is *revolved* by an arm around the sun gear. The planets are often cut in one piece from the same metal stock, as shown in Fig. 12-6(b). A hole through the center of the gear cluster serves as the bearing for the off-center part of the arm or planet carrier. This type of train can be designed in a manner similar to the previous example using either planet-gear arrangement. The following problem shows how to approach this type.

Figure 12-6
Planetary gear train.

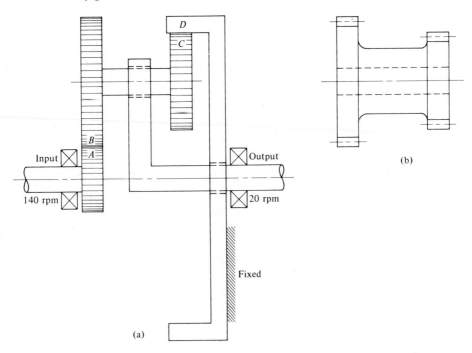

(a)

(b)

Input

140 rpm

Output

20 rpm

Fixed

◆ *Example*

Neglecting the strength of gear teeth, design a planetary train similar to that shown in Fig. 12-6(a) such that the input shaft speed will be 140 rpm and the output rotational speed will be 20 rpm. No gear in the train is to have fewer than 16 teeth.

Solution. Using the general planetary formula, we find

$$N_s = (\text{TV})(N_r) + N_c[1 - (\text{TV})],$$

$$140 = (\text{TV})(0) + 20[1 - (\text{TV})]$$

or

$$\text{TV} = -6.$$

Since the train value is effected in two steps, the ideal arrangement would be to factor 6 into nearly equal factors. Suitable factors are $2/1$ and $3/1$. Using a method similar to that for the reverted simple gear train, we find a suitable least common multiple. Its value is $(3 + 1)(2 + 1)$, if we assume that all gears have the same diametral pitch. Then

$$\frac{t_A}{2} + \frac{t_B}{2} = \frac{t_D}{2} - \frac{t_C}{2} = 12.$$

This equation can also be written without the numerical 2's in the denominator. Then,

$$t_D - t_C = 12.$$

Also,

$$\frac{t_D}{t_C} = \frac{3}{1} \qquad \text{or} \qquad t_D = 3t_C$$

and

$$3t_C - t_C = 12,$$

from which t_C is found to be 6 and t_D is 18. Then,

$$t_A + t_B = 12.$$

Also,

$$\frac{t_B}{t_A} = \frac{2}{1} \qquad \text{or} \qquad t_B = 2t_A$$

and

$$t_A + 2t_A = 12.$$

Therefore, t_A is 4 and t_B is 8.

The smallest gear (A) does not have the required minimum of 16 teeth. If all calculated values are multiplied by 4, the minimum requirements are met. We then have the following values for each of the gears:

t_A = 16 teeth,

t_B = 32 teeth,

t_C = 24 teeth,

t_D = 72 teeth.

There are other combinations that also meet the requirements of the planetary train in terms of speeds and tooth numbers.

Tabulation Method (Planetary Gear Trains)

It is sometimes easier to study the action of a planetary train by a tabulation that involves two imaginary steps. To do this, a table is set up with columns for each gear, identified by name, number, or letter. Then rows are provided for the two imaginary steps and a total. A general format for such a table follows:

	Planet Carrier	Ring Gear	Planet Gear	Sun Gear
1. gears locked	-1	-1	-1	-1
2. carrier locked	0	$+1$	$+4$	-2
3. total	-1	0		-3

The first step is to rotate the entire assembly one turn counterclockwise; this is an imaginary step and assumes that all the gears are locked together as though welded. The turns are then recorded for each component involved. The second step is also an imaginary one. Since the ring gear is fixed, it must have a total value of zero in the tabulation. To obtain a zero value, imagine that it is turned once in a clockwise direction so that the algebraic sum is zero. In this step, the planet carrier is not moving. Thus, a positive value of 1 is entered for the ring gear and a zero value for the planet carrier in step 2. The effects on each of the others must now be determined, both in direction and number of turns. Since the ring gear (with 64 teeth) meshes with the planet gear (with 16 teeth), the planet will turn $+4$. The 16-tooth planet engages the 32-tooth sun gear; thus the relative turns for the sun gear would be $4(\frac{16}{32})$ or 2 turns, but in the opposite direction. (This implies a minus sign.) Next, totals are determined and actual ratios established. The sun gear has a rotational speed of 100 rpm. To find the speed of the planet carrier, we can use the relationship

$$\frac{-1}{N_c} = \frac{-3}{100}.$$

Therefore,

$$N_c = \frac{-100}{-3} = +33.3 \text{ rpm}.$$

The total number of turns for the planet gear was not added, since it had no significance in the problem. In designing this type of train, a careful study of the table columns can be helpful in determining the tooth numbers needed to yield the desirable number of turns for the various gears. It must be remembered, however, that in this type of train, teeth must diametrically total in the same way that diameters would, since all gears that mesh must have the same diametral pitch. For this particular train, the diametral total would be 32 + 16 + 16 = 64.

Not all planetary gear trains require a ring gear. Sometimes a planet travels around a sun gear and operates its arm or planet carrier as a transmitting shaft. Sometimes bevel gears are used in epicyclic gear trains. A notable example of this type is the automotive rear axle. A simple gear train transmits a ratio from drive shaft to rear wheels. However, a bevel-gear differential (a planetary train with a train value of -1) is also incorporated to permit the two rear wheels to operate at different speeds when the vehicle is turning sharply. This fact is obvious when the rear of an automobile is lifted; one turn of a rear wheel will produce one opposite turn on the other wheel.

The following problem shows how the two methods previously described can be used to handle a planetary-gear problem in which the sun gear, ring gear, and planet carrier all rotate.

◆ *Example*

In Fig. 12-7, gear A rotates counterclockwise at 100 rpm. The planet carrier rotates clockwise at 60 rpm. Find the rotational speed of the sun gear.

Solution.

$$N_{100} = (TV) N_{30} = \frac{30}{100}(100) = 30 \text{ rpm clockwise}.$$

Note: This is a simple gear train. For the planetary train,

$$TV = \frac{80}{20}\frac{20}{40} = -2;$$

$$N_s = (TV)N_r + N_c[1 - (TV)]$$
$$= (-2)(+30) + 60[1 - (-2)]$$
$$= -60 + 180 = +120 \text{ rpm}.$$

This value can be checked by the tabulation method as follows:

	Planet Carrier	Sun Gear	Planet Gear	Ring Gear
1. gears locked	−1	−1	−1	−1
2. carrier locked	0	−1	+2	+0.5
3. total	−1	−2		−0.5h3

For the first step, all components are considered locked and are given one turn counterclockwise; thus −1 is recorded for each component. For the second step, the carrier is considered locked and zero is recorded for it. The column total for the planet carrier then shows −1, which represents its speed of +60 rpm. The speed of the ring gear is known to be +30 rpm. Thus, the column total of the ring gear must be equal to one-half that of the planet carrier. To give a total of −0.5, it is necessary to add +0.5 in step 2 to the ring-gear column. Now, if the ring gear is imagined to be turning +0.5 turns, the planet carrier will turn $\frac{80}{20}(0.5)$, which is +2; +2 turns of the planet gear will then turn the sun gear −1 turn. The sun-gear column can now be added up and the result compared to the column total for the planet carrier (or the ring gear, if desired):

$$\frac{-1}{+60} = \frac{-2}{N_s}.$$

Therefore, $N_s = 60(2) = 120$ rpm clockwise.

 **Force Analysis**

A first step in most design projects is a force analysis, quite often in conjunction with a kinematic analysis. If straight-tooth spur gears are used, the total force transmitted from the tooth of the driver gear to the tooth of the driven gear consists of two components. One component is the *tangen-*

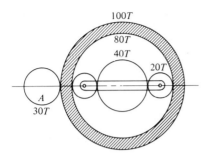

Figure 12-7
Combination of simple and planetary train.

tial force, which is easily determined by the following equation (refer to Fig. 12-8):

$$F = \frac{63,000(\text{hp})}{Nr},$$ (12-16)

where

r = pitch radius of gear (in.),

F = tangential force of gear (lb),

N = rotational speed (rpm).

The *separating force*, which is the other component, is then given by the following equation:

$$Q = F \tan \phi,$$ (12-17)

Figure 12-8
Force analysis: spur and internal gears.

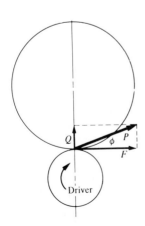

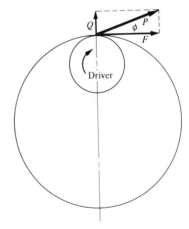

where

ϕ = pressure angle (degrees),

Q = separating force (lb).

Total load, P, is then given by the equation

$$P = \sqrt{F^2 + Q^2} \quad \text{or} \quad P = \frac{F}{\cos \phi}. \tag{12-18}$$

If the gears are of the helical-tooth form, the total load has three components: the tangential force, the separating force, and the *end thrust*. These components are mutually perpendicular to each other. The end thrust is given by the following equation:

$$Q_e = F\frac{\pi D}{L}, \tag{12-19}$$

where

Q_e = end thrust (lb),

L = lead (in.),

D = pitch-circle diameter (in.).

The direction of the end thrust is dictated by the hand of the helix and the direction of rotation of the driving gear. In a meshing pair of helical gears, one gear must have a right-handed helix and the other must have a left-handed helix. The helix angle must be the same for both the meshing gears. Examples of this are shown in the section on pages 441–443.

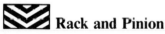 **Rack and Pinion**

A rack can be defined as a gear with an infinitely large pitch-circle diameter; thus, the pitch "circle" becomes a straight line. The teeth of a rack are inclined at an angle equal to the pressure angle ($14\frac{1}{2}$ or 20 degrees). Also, the tooth surfaces are straight, but will mesh properly with the involute surface of a gear. The action between a pinion and a rack is one of transforming rotation into translation. Applications include drill presses, arbor presses, and measuring equipment where the turn of a knob provides a linear adjustment. The line of action between a pinion and a rack starts where the addendum line of the rack crosses the pressure line. Action ceases further down the line of action where the addendum circle of the pinion intersects the pressure line.

Proper operation of a rack and pinion is contingent upon proper placement of the initial line of contact, which must be inside a perpendicu-

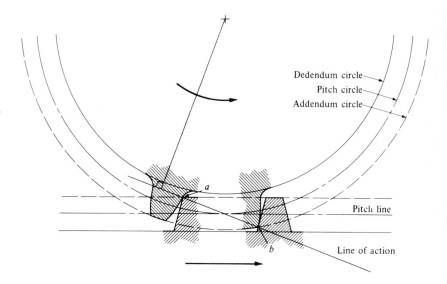

Figure 12-9
Rack-and-pinion tooth action.

lar line from the gear center to the pressure line. Figure 12-9 illustrates this concept. Point *a* is the initial point of contact and *b* is the final point of contact for a given tooth; both points lie along the pressure line and must be on the pitch-point side of the perpendicular from the gear center to the pressure line. The linear velocity of the rack is equal to the pitch-circle velocity of the pinion.

Racks are usually mounted on some type of slide. Friction may present a serious problem unless the slide is adequately lubricated. Linear ball bearings can substantially lower frictional losses.

 **Bevel-Gear Ratios**

In dealing with bevel gears, a ratio can be established by comparing the number of teeth on the follower with the number of teeth on the driver. In comparing pitch angles, however, one must remember that bevel gears are produced by cutting teeth on conical sections. By definition, the pitch angle is the angle formed between the centerline of the gear and the pitch line of a tooth. In Fig. 12-10, the ratio from gear 1 to 2 can be derived in terms of the pitch angles by simple trigonometry and the following simultaneous

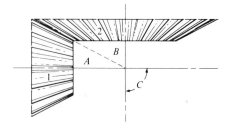

Figure 12-10
Bevel-gear ratios in terms of pitch angles.

equations:

$$R = \frac{\sin B}{\sin A}$$

and

$$A + B = C.$$

The first equation is obvious if one notes that the sides opposite to angles B and A in the right triangles are pitch-circle radii. The second equation can be rearranged to read

$$B = C - A.$$

Thus,

$$R = \frac{\sin (C - A)}{\sin A}.$$

then, clearing the fraction, we obtain

$$R \sin A = \sin (C - A).$$

Then, using the trigonometric equation for the sine of the difference of two angles, we get

$$R \sin A = \sin C \cos A - \cos C \sin A.$$

Transposing, we find

$$R \sin A + \cos C \sin A = \sin C \cos A$$

and

$$\sin A (R + \cos C) = \sin C \cos A.$$

Thus,

$$\frac{\sin A}{\cos A} = \frac{\sin C}{R + \cos C} = \tan A$$

or

$$A = \arctan\left(\frac{\sin C}{R + \cos C}\right). \qquad (12\text{-}20)$$

After angle A is found, B can be found by merely subtracting A from C.

A similar derivation can be made for internal bevel gears. It must be remembered that for internal bevel-gear sets, $C = B - A$. Similar equations can be established. That for angle A is as follows:

$$A = \arctan\left(\frac{\sin C}{R - \cos C}\right). \qquad (12\text{-}21)$$

 Torque Multiplication

It is interesting to evaluate the torque changes that occur when a gear train reduces the speed from an input to an output shaft. For sake of analysis, consider a simple two-gear reduction train under the following assumptions:

1. 100% efficiency in the gear train,
2. pure torsion in the shafts,
3. an input shaft diameter of 1 in.,
4. a reduction ratio of 6/1,
5. an input shaft speed of 1800 rpm,
6. an input horsepower equal to 1.

First, calculate the output shaft size needed to ensure that the induced shearing stress will be equal to that of the input shaft. For the input shaft,

$$T = \frac{63,000(\text{hp})}{N} = \frac{63,000(1)}{1800} = 350 \text{ in-lb.}$$

Then,

$$S_s = \frac{16T}{\pi D^3} = \frac{16(350)}{\pi (1)^3} = 1780 \text{ psi.}$$

To yield the ratio 6/1, the diameter of the output shaft gear must be 6 times that of the input shaft gear. The tangential force at the pitch line is equal for both gears. (This can be considered an action producing a reaction as the driver transmits force to the driven.) If r is the radius of the smaller gear of the train, the pitch-line force is

$$F = \frac{T}{r} = \frac{350}{r} \text{ lb.}$$

For the larger gear in the train, $T = 6r(F) = 6r(350/r) = 2100$ in-lb.

Substituting into the shaft torsion equation, we obtain the minimum size for the output shaft:

$$D = \sqrt[3]{\frac{16T}{\pi S_s}} = \sqrt[3]{\frac{16(2100)}{\pi(1780)}} = 1.82 \text{ in.}$$

Note that the output shaft is much larger than the input shaft. The design is often based on the output shaft; the input shaft is then oversized (and overdesigned) so that there is not a huge difference between the two shaft sizes. The induced stress in the input shaft is greatly reduced by overdesigning.

 ## Design of Spur-gear Teeth

While it is important to design gear trains for .proper kinematic performance, it is equally necessary to make sure that the teeth are strong enough to withstand the applied loads. Gear teeth are similar to cantilever beams. As torque is transmitted from one shaft to another via gears, force is applied along the pressure line. Gear teeth must be strong enough to sustain the applied force, whether a static or dynamic load is applied to the tooth. Gears are stripped when the load exceeds the capacity of the gear. Furthermore, gear installations are expensive; therefore, the rate of wear of the tooth surfaces must be satisfactory. The following equations are presented to serve as a guide in designing proper gear sets. When it is found that preselected preferences do not meet the formula requirements, a designer has several alternatives. If a gear tooth is not strong enough (statically or dynamically), the design can be altered as follows:

1. the diametral pitch can be decreased (this increases the tooth size);
2. the face width can be increased (this increases load-carrying capacity); or
3. the gear material (or hardness) can be changed.

By the same token, if the gear tooth is overdesigned, the designer can change any or all of the previously mentioned factors.

 ## Face Width

Generally speaking, the face width should be kept somewhere between 2.5 times the circular pitch and 4 times the circular pitch. Sometimes the ratio is slightly less than the 2.5 value; if so, exceptionally careful mounting is essential. Unforeseen factors influence gear-set performance; some of these are fluctuating speeds, improper lubrication, poor alignment, high operating temperatures, and severe vibrations. If the face width is excessive for a given size of tooth, any error in alignment may concentrate the load near one end of the tooth, encouraging failure. Sometimes gear teeth are crowned along the width of the gear tooth. The amount of crowning, however, is exceedingly small.

 Beam Strength of Spur Gears

In 1897, Wilfred Lewis developed an equation for the static strength of a spur-gear tooth. The original Lewis equation is as follows:

$$W_b = SP_c by, \qquad\qquad (12\text{-}22)$$

where

W_b = static beam strength (lb),

S = allowable stress (psi),

b = face width (in.),

y = form factor,

P_c = circular pitch (in.).

This equation does not take into account the fact that small errors in tooth proportions can cause higher stresses at high speeds because of fast tooth engagement with resultant shock. Carl Barth developed a speed factor for application to the Lewis equation. The Barth factor is

$$k = \frac{600}{600 + V} \qquad\qquad \text{for metallic gears,}$$

$$k = \frac{150}{200 + V} + 0.25 \qquad \text{for nonmetallic gears,}$$

where

V = pitch-circle velocity (fpm).

Diametral pitch is more commonly used than circular pitch when gears are designed. For this reason, the equation is usually written in terms of diametral pitch and the form factors are modified for the values of diametral pitch. Table 12-1 shows form-factor values (Y) based on diametral pitch (that is, y-values multiplied by π produce the Y-values given in the table). Since horsepower ratings are usually desired, the preceding information can be incorporated into one equation that can be solved for the power values:

$$\text{hp} = \frac{SbYkV}{33,000P},$$ (12-23)

where

S = allowable stress (psi),

Y = form factor based on diametral pitch,

k = speed factor,

b = face width (in.),

V = pitch-line velocity (fpm),

P = diametral pitch.

In using this equation, one should use the allowable flexural endurance stress for the appropriate gear materials. The factor of safety employed becomes a matter of judgment.

Table 12-1
Form Factor (Y) for Use in Lewis Equation

Number of Teeth	$14\frac{1}{2}$-deg Full Depth	20-deg Full Depth	20-deg Stub	Number of Teeth	$14\frac{1}{2}$-deg Full Depth	20-deg Full Depth	20-deg Stub
12	0.210	0.245	0.311	28	0.314	0.352	0.430
13	0.223	0.261	0.324	30	0.317	0.359	0.437
14	0.236	0.277	0.339	34	0.327	0.371	0.447
15	0.245	0.290	0.346	38	0.333	0.384	0.455
16	0.254	0.296	0.361	43	0.339	0.397	0.462
17	0.264	0.302	0.368	50	0.346	0.410	0.474
18	0.270	0.308	0.377	60	0.355	0.421	0.484
19	0.276	0.314	0.386	75	0.361	0.434	0.496
20	0.283	0.321	0.393	100	0.368	0.447	0.505
21	0.289	0.327	0.399	150	0.374	0.460	0.518
22	0.292	0.330	0.405	300	0.383	0.472	0.534
24	0.298	0.337	0.415	Rack	0.390	0.484	0.550
26	0.308	0.346	0.424				

Table 12-2
Allowable Stress (psi)

Material	Hardness (Brinell)	Allowable Stress		
		Spur	Helical	Bevel
Steel	140 min	20,000 to 22,000	20,000 to 22,000	11,000
	180 min	25,000 to 28,000	25,000 to 28,000	14,000
	300 min	35,000 to 40,000	35,000 to 45,000	19,000
	450 min	45,000 to 50,000	45,000 to 60,000	25,000
Cast Iron				
AGMA Grade 20	——	5,000	5,000	2,700
AGMA Grade 30	175 min	8,500	8,500	4,600
AGMA Grade 40	200 min	13,000	13,000	7,000

 AGMA Strength-horsepower Ratings*

Ratings for gears can be classified in very general terms according to strength and durability. Strength is an important consideration, as a gear tooth is subjected to repeated flexure and hence could fail because of fatigue. This consideration is different from the surface failure caused by many types of wear. The allowable horsepower must be calculated for both the pinion and the gear. The smaller of the two values is then used to rate the gear set.

The AGMA formula for power based on strength is as follows (certain symbols have been changed to conform with those used in this text):

$$\text{hp} = \frac{N_p D_p S_a b J K_v K_L}{126{,}000 P K_m K_s K_o K_R K_T}, \tag{12-24}$$

where

N_p = rpm of the pinion,

D_p = pitch diameter of pinion (in.),

S_a = allowable stress for material (psi) (see Table 12-2),

b = face width (in.),

J = geometry factor (see Table 12-3),

---

*Extracted from "AGMA Information Sheet: Strength of Spur, Helical, Herringbone and Bevel Gear Teeth" (AGMA 225.01), with the permission of the publisher, the American Gear Manufacturers Association, 1330 Massachusetts Avenue, N. W., Washington, D.C. 20005.

Table 12-3
**Geometry Factor (*J*), 20-deg Spur, Standard Addendum
(Load Applied at Highest Point of Single-Tooth Contact)**

Number of Teeth for which *J* Is Desired	Teeth in Mating Gear						
	12	17	25	35	50	85	Rack
15	0.25	0.25	0.25	0.25	0.25	0.25	0.25
16	0.25	0.25	0.25	0.25	0.25	0.25	0.25
17	0.29	0.29	0.29	0.29	0.29	0.29	0.29
18	0.30	0.31	0.32	0.32	0.32	0.32	0.32
19	0.31	0.32	0.32	0.33	0.33	0.34	0.36
20	0.31	0.32	0.32	0.34	0.34	0.35	0.37
22	0.32	0.33	0.34	0.35	0.36	0.36	0.38
24	0.33	0.34	0.35	0.36	0.37	0.37	0.39
30	0.35	0.37	0.38	0.39	0.39	0.40	0.42
40	0.38	0.39	0.40	0.41	0.42	0.43	0.45
60	0.40	0.42	0.43	0.44	0.45	0.46	0.48
80	0.42	0.43	0.45	0.46	0.46	0.48	0.50
125	0.43	0.45	0.46	0.47	0.48	0.49	0.52
275	0.44	0.46	0.48	0.49	0.50	0.51	0.54

For other types of gears, refer to charts in AGMA 225.01.

Table 12-4
Life Factor (K_L)

Number of Cycles	Spur and Helical Gears		Bevel gears
	250 to 450 Bhn	Case carburized[†]	Case carburized[†]
1,000	3.0 to 4.0*	2.7	4.6
10,000	2.0 to 2.6*	2.0	3.1
100,000	1.6 to 1.8*	1.5	2.1
1 million	1.1 to 1.4*	1.1	1.4
10 million	1.0	1.0	1.0
100 million	1.0 to 0.9	1.0 to 0.9	1.0

*Use the higher values for higher hardnesses.
[†]55 to 63 R_c

Table 12-5
Load-Distribution Factors (K_m and C_m)

Spur and Helical Gears
(Good Quality Commercial, Accurately Mounted)

Face width, in.	Spur	Helical
0 to 2	1.3	1.2
6	1.4	1.3
9	1.5	1.4
16 and over	1.8	1.7

Bevel Gears
(Good Industrial Quality)

Both members straddle-mounted	One member straddle-mounted	Neither member straddle-mounted
1.00 to 1.10	1.10 to 1.25	1.25 to 1.40

K_v = dynamic factor

$\quad$ = $50/(50 + \sqrt{V})$ for commercial-quality spur gears,

K_L = life factor (see Table 12-4),

$\quad P$ = diametral pitch,

K_m = load-distribution factor (see Table 12-5),

K_s = size factor (usually taken as unity for spur, helical, and herringbone gears),

K_o = overload factor (see Table 12-6),

K_R = factor of safety (see Table 12-7),

K_T = temperature factor (can be taken as unity if gears operate at oil temperatures that do not exceed 160 °F).

Table 12-6
Overload Factors (K_o and C_o)

Power Source	Load on Driven Machine		
	Uniform	Moderate shock	Heavy shock
Uniform	1.00	1.25	1.75
Light Shock	1.25	1.50	2.00
Medium Shock	1.50	1.75	2.25

Table 12-7
Factors of Safety (K_R)

Requirements of Application	K_R
High reliability	1.50 to 3.00
Fewer than 1 failure in 100	1.00 to 1.25
Fewer than 1 failure in 3	0.70 to 0.80

 **AGMA Surface-durability Power Formula***

A direct formula for determining the allowable power considers load, tooth size, and stress distribution. This formula can be stated as follows, with certain symbols changed to conform with those used in this text:

$$\text{hp} = \frac{N_p b C_v I}{126{,}000 C_s C_m C_f C_o} \left(\frac{S_a D_p C_L C_H}{C_p C_T C_R} \right)^2, \tag{12-25}$$

where

D_p = pitch diameter of pinion (in.) (*Note:* for bevel gears, use the value at the large end of the bevel tooth);

N_p = rotational speed of the pinion (rpm);

S_a = allowable contact stress (psi) (see Table 12-8);

b = face width (in.) (use narrowest value of mating gears);

I = geometry factor (see Table 12-9);

C_v = dynamic factor (see Table 12-10);

C_L = life factor (see Table 12-11);

C_H = hardness-ratio factor (see Fig. 12-11);

C_s = size factor (usually taken as unity);

C_m = load-distribution factor (same as K_m) (see Table 12-5);

C_f = surface-condition factor (usually taken as unity if a good surface is developed by processing or run-in);

*Extracted from "AGMA Information Sheet: Surface Durability (Pitting) of Spur, Helical, Herringbone and Bevel Gear Teeth," AGMA 215.01, with the permission of the publisher, the American Gear Manufacturers Association, 1330 Massachusetts Avenue, N.W., Washington, D.C. 20005.

Table 12-8
Allowable Contact Stress (S_a), psi

Material	Minimum Surface Hardness	S_a
Steel (through hardened)	180 Bhn	85,000 to 95,000
	240 Bhn	105,000 to 115,000
	300 Bhn	120,000 to 135,000
	360 Bhn	145,000 to 160,000
	440 Bhn	170,000 to 190,000
Steel (case carburized)	55 R_C	180,000 to 200,000
	60 R_C	200,000 to 225,000
Steel (flame or induction hardened)	50 R_C	170,000 to 190,000
Cast iron		
AGMA Grade 20	——	50,000 to 60,000
AGMA Grade 30	175 Bhn	65,000 to 75,000
AGMA Grade 40	200 Bhn	75,000 to 85,000
Tin bronze, AGMA 2C (40,000 psi minimum tensile strength)		30,000
Aluminum bronze, ASTM B 148-52 (90,000 psi minimum tensile strength)		65,000

Table 12-9
Geometry Factor (I) for Durability Calculations (External Spur Pinion, 20-degree Pressure Angle, Full-Depth Teeth)

	Number of Teeth on Pinion			
Ratio	16	24	30	50 or more
1	0.075	0.077	0.080	0.080
2	0.089	0.099	0.102	0.108
3	0.096	0.108	0.112	0.120
4	0.102	0.114	0.120	0.128
5	0.104	0.118	0.124	0.134
6	0.106	0.121	0.128	0.138
7	0.108	0.124	0.130	0.142
8	0.109	0.125	0.130	0.143
9	0.110	0.125	0.132	0.144
10	0.110	0.126	0.132	0.145

For $14\frac{1}{2}$-degree full-depth teeth, approximate values can be obtained by multiplying the above values by 0.8. For 20-degree stub teeth, approximate values can be obtained by multiplying the above values by 0.95. Geometry factors for bevel, spiral bevel, and helical gearing can be found by referring to the curves in AGMA 215.01.

Table 12-10
Dynamic Factor (C_V and K_V)

Gear Application	Formula
Commercial spur gears	$\dfrac{50}{50 + \sqrt{V}}$
Commercial helical gears	$\dfrac{78}{78 + \sqrt{V}}$
Large-planed spiral bevel gears High-precision helical gears Shaved or ground spur gears	$\sqrt{\dfrac{78}{78 + \sqrt{V}}}$

C_o = overload factor (same value as K_o) (see Table 12-6);

C_p = elastic coefficient (see Table 12-12);

C_T = temperature factor (usually taken as unity when gears operate with oil or the gear blank is less than $250\,°F$);

C_R = factor of safety (see Table 12-13).

In applying the AGMA formulas for strength and durability, it should be remembered that these calculations determine a suitable power rating for the gear set; the service rating is the lowest horsepower obtained. If this rating is not adequate, one can alter one or more of the following:

1. pitch and/or pitch diameter,
2. face width,
3. material and/or hardness.

Past experience with similar gear problems can assist a designer in making good preliminary assumptions for the preceding items to minimize

Table 12-11
Life Factor (C_L)

Cycles	C_L
10^7 and over	1.0
10^6	1.15
10^5	1.30
10^4	1.5

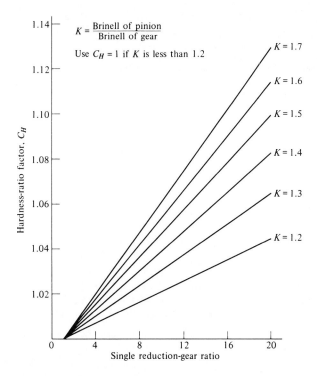

Figure 12-11
Hardness-ratió factor, C_H (for helical gears, based on infinite life).

Table 12-12
Elastic Coefficient (C_p), psi

Pinion Material and Modulus of Elasticity	Gear Material and Modulus of Elasticity			
	Steel 30×10^6	Cast iron, 19×10^6	Aluminum bronze, 17.5×10^6	Tin bronze, 16×10^6
Steel, 30×10^6	2300	2000	1950	1900
Cast iron, 19×10^6	2000	1800	1800	1750
Aluminum bronze, 17.5×10^6	1950	1800	1750	1700
Tin bronze, 16×10^6	1900	1750	1700	1650

Note: These values are for nonlocalized contact. Higher values are needed for bevel gears.

Table 12-13
Factor of Safety (C_R)

Requirements of Application	C_R
High reliability	1.25 or higher
Fewer than 1 failure in 100	1.00

the number of iterations of the procedure. The following example demonstrates the use of these formulas.

◆ *Example* ─────────────────────────────

A spur-gear set has a 3-to-1 ratio. It is to be designed for a life of 10 million cycles. The power source is uniform; there is moderate shock on the driven member. A factor of safety of 2 is to be used. The pressure angle is 20 degrees and the teeth have a standard addendum. The pinion has 21 teeth and rotates at 600 rpm; it is made of steel with a Brinell hardness of 180. The diametral pitch of each gear is 3, and both gears have a face width of 3.5 in. The driven gear is made of AGMA Grade 30 cast iron with a Brinell hardness of 175. Calculate (a) the strength horsepower of the pinion, (b) the strength horsepower of the gear, and (c) the durability horsepower of the gear set. (d) What is the service rating of the gear set?

Solution. (a) First, find the pinion diameter, the pitch-line velocity, and the number of teeth for the gear (needed to get the geometry factor):

$$D = \frac{t}{P} = \frac{21}{3} = 7 \text{ in.};$$

$$V = \frac{\pi DN}{12} = \frac{\pi(7)(600)}{12} = 1100 \text{ fpm.}$$

(*Note:* This value for V is rather high for spur gears; when pitch-line velocity exceeds 1000 fpm, herringbone or helical gears are usually used.) The number of teeth is

$$t_G = 3(21) = 63 \text{ teeth.}$$

From Table 12-3, $J = 0.35$. From Table 12-10,

$$K_v = \frac{50}{50 + \sqrt{V}} = \frac{50}{50 + \sqrt{1100}} = 0.600.$$

We also have the following values:

$S_a = 26{,}500$ psi		from Table 12-2 (average value),
$K_L = 1$		from Table 12-4,
$K_m = 1.4$		from Table 12-5,
$K_s = 1,$		
$K_o = 1.25$		from Table 12-6,
$K_R = 2$		(specified in the problem),
$K_T = 1$		(if we assume that excessive temperatures are not encountered).

Applying the strength-power formula, we obtain

$$\text{hp} = \frac{N_p D_p S_a bJ K_v K_L}{126{,}000 P K_m K_s K_o K_R K_T}$$

$$= \frac{600(7)(26{,}500)(3.5)(0.35)(0.600)(1)}{126{,}000(3)(1.4)(1)(1.25)(2)(1)} = 61.8 \text{ hp}.$$

This is the strength horsepower for the pinion.

(b) A similar procedure is followed for the gear. Thus,

$D = 3(7) = 21$ in.,

$N = 600/3 = 200$ rpm,

$V = 1100$ fpm	(same as pinion),
$J = 0.42$	from Table 12-3,
$S_a = 8500$ psi	from Table 12-2 (average value).

The various K factors are the same as for the pinion. Applying the power formula for the gear, we obtain

$$\text{hp} = \frac{N_G D_G S_a bJ K_v K_L}{126{,}000 P K_m K_s K_o K_R K_T}$$

$$= \frac{200(21)(8500)(3.5)(0.42)(0.600)(1)}{126{,}000(3)(1.4)(1)(1.25)(2)(1)} = 23.8 \text{ hp}.$$

This is the strength horsepower for the gear. It is much lower than that of the pinion because the gear is made of weaker material than the pinion.

(c) Next, the durability horsepower of the gear set is calculated:

$C_v = 0.600$	from Table 12-10,
$S_a = 70,000$ psi	from Table 12-8 (using average value for the gear),
$I = 0.102$	from Table 12-9,
$C_L = 1$	from Table 12-11,
$C_H = 1$	(assumed),
$C_p = 2000$	from Table 12-12,
$C_T = 1$	(if we assume that excessive temperatures are not encountered),
$C_R = 2$	(specified in the problem),
$C_s = 1$,	
$C_m = 1.4$	from Table 12-5,
$C_f = 1$,	
$C_o = 1.25$	from Table 12-6.

Then, applying the durability-power formula for the gear set, we obtain

$$hp = \frac{N_p b C_v I}{126,000 C_s C_m C_f C_o} \left(\frac{S_a D_p C_L C_H}{C_p C_T C_R} \right)^2$$

$$= \frac{600(3.5)(0.600)(0.102)}{126,000(1)(1.4)(1)(1.25)} \left[\frac{70,000(7)(1)(1)}{2000(1)(2)} \right]^2$$

$$= 8.75 \text{ hp}.$$

This is the durability horsepower for the set.

(d) The service-rating horsepower has to be the lowest value obtained in (a), (b), or (c) of the preceding. In this particular problem, 8.75 is the rated horsepower for the set. If such a rating is not satisfactory, changes can be made in tooth size, material, or face width; then horsepower ratings must be recalculated.

In using any type of standard, the designer must use the latest information available. Standards are based on industry-wide experience; improvements are continually being made in manufacturing, design, and application.

 **Gear Failures**

The consequences of gear failure are serious. Failure of one gear in a train will shut down the equipment. In contrast, the failure of one belt in a multiple V-belt drive may not even stop the machine. In a single V-belt drive, it is a simple matter to replace the broken belt with little "down time." Replacing a gear is usually an involved process, even if spare gears are available. Usually, various parts must be removed in order to get at the damaged gear. This process is costly and something to be avoided if at all possible. In spite of careful design, gears will fail. Some of the reasons for failing are as follows:

1. *Overloading:* This is the result of unexpected shock overload. When such a situation occurs, the tooth is "stripped."
2. *Excessive speed:* If speed-control devices are not used, careless operators sometimes operate gear devices beyond their normal speed capacity.
3. *Improper lubrication:* Extreme-pressure lubricants are needed in gear trains, since line contact is made between the two mating gears. The lubrication system should be effective in supplying lubricant to the point where it is needed. An oil must adhere to a gear tooth until it meshes with its mating tooth; if the oil does not adhere, it may never reach the proper location. A filtering device should be provided to keep contaminants away from the components. The cooling effect of the lubricant may be needed to prevent actual welding of the two tooth surfaces. A momentary welding and subsequent pulling apart will damage tooth surfaces, eventually leading to tooth failure.
4. *Outside contaminants:* Any type of grit causes a scoring action on the tooth profiles. Scoring leads eventually to tooth failure.
5. *Improper heat treatment:* Gear teeth are generally surface hardened by a flame hardening process. The hardened surface should be of uniform thickness.
6. *Poor machining:* Tooth profiles must be exact. Proper fillets are needed.
7. *Improper mounting:* Gear centers must be precisely measured. Gears must operate in the proper plane; if they do not, undue forces are applied to the teeth. This situation can lead to early failure.
8. *Poor design:* One should follow the best engineering practice in providing gears with proper diametral pitch, face width, material, and

heat treatment. One should consider carefully the use to which gear sets will be subjected and carry out a careful force analysis before starting the design of the teeth.

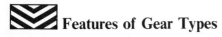

Features of Gear Types

A designer must be concerned not only with the kinematics of gear sets or trains, but also with choosing suitable gear materials for adequate strength and durability. The designer must understand the nomenclature of each type of gear tooth, so that proper center distances can be determined for mounting, as well as the reasons for using the various types available. The purpose of the following sections is to present outstanding features and limitations of different types of gears. Thus, comparisons can be made to aid in choosing the proper gear for a particular application.

External Spur Gears

Figure 12-12 illustrates a simple spur gear. This type is characterized by the straight teeth parallel to the centerline of the bore. Spur gears are used for moderate-speed applications. Mating spur gears produce no undesirable end thrust. Custom-made spur gears can be produced in rather large sizes; outside diameters in excess of 30 ft have been fabricated. Special gears can have face widths of over 5 ft. The larger-sized gears are usually produced in segments. Small sizes of gears can be obtained as "off-the-shelf" parts. In using stock gears, however, one is limited by the available pitches, face widths, tooth numbers, materials, and bores. Thus, if stock gears are used, the design is somewhat restricted. Spur gears can also be produced in miniature sizes.

Figure 12-12
External spur gear.
(Courtesy of Philadelphia Gear
Corp.)

Figure 12-13
Helical gear.
(Courtesy of Philadelphia Gear
Corp.)

Internal Spur Gears

Internal spur gears have several inherent advantages. The tooth space of the internal (or ring) gear resembles the tooth of the pinion in appearance. Ring gears often present mounting problems. However, one of their advantages is the fact that the ring gear forms a kind of safety guard for the set; another advantage is that there is less sliding action and wearing of the teeth, since the pinion and the gear rotate in the same direction. An additional advantage is the shorter center distances possible; this provides compactness. In internal gear sets, the contact ratio (average number of teeth in contact) is larger than in similar external spur-gear sets.

Helical Gears

A helical gear is shown in Fig. 12-13. Helical gears are used to transmit power between parallel shafts. The teeth are cut on a cylinder, but on an angle (the *helix angle*). The helix angle often ranges from 7 to 23 degrees; it can be either left-handed or right-handed. In a gear set, the helix for the pinion and the gear must be of the opposite hand. In addition, the helix angle has to be the same for both mating gears. Helical gearing permits greater load-carrying capacity and higher pitch-line velocities. The face width of a helical gear should be large enough to provide tooth overlap. Helical gears operate more quietly than spur gears. Other than cost of production, the outstanding disadvantage of this type is the great end thrusts they produce. The size of the helix angle determines the amount of axial thrust. Large end thrusts present mounting problems, since they require thrust bearings. The use of double helical gearing eliminates the thrust problem. The double helical gear is characterized by two sets of helical teeth cut in a gear, but of the opposite hand.

In a helical gear, there are two pitches that can be considered. In addition to the regular diametral and circular pitches, one can also consider these pitches in the normal plane. Thus, the following formulas may be developed:

$$\cos \psi = \frac{P_{cn}}{P_c},$$

$$P_{cn} = P_c \cos \psi, \tag{12-26}$$

$$P_n = \frac{P}{\cos \psi}, \tag{12-27}$$

where

P_{cn} = circular pitch in normal plane,

P_n = diametral pitch in normal plane,

ψ = helix angle.

The partial edge view of the helical gear shown in Fig. 12-14(a) defines the circular pitch in the normal plane, as well as the linear circular pitch. The gear shown is a left-handed one; the hand of a helical gear can be found by viewing the gear as if it were a screw thread. The diametral pitch cannot be shown in either plane, since it is merely a ratio and not a dimension.

It should be noted that the circular pitch in the normal plane is the distance from a point on one tooth to a similar point on the adjacent tooth, but measured directly across the tooth. To ensure tooth overlap, point 2 in the figure must be perpendicularly below point 1. The following trigonometric relationship can then be written:

$$\tan \psi = \frac{P_c}{b} \quad \text{or} \quad b = \frac{P_c}{\tan \psi},$$

where

b = face width (in.).

Greater overlap can be obtained by providing enough face width that point 2 will be to the left of a perpendicular from point 1. It is usually desirable to provide more than the minimum value for overlap.

◆ *Example* ———————————————————————————————

A helical gear has a diametral pitch in the normal plane of 3. The helix angle is 23 degrees. Find the diametral pitch, the circular pitch, and the circular pitch in the normal plane. Find the face width needed to place the leading edge of one tooth exactly opposite the trailing edge of its adjacent tooth. If the face width of the gear is 3 in., what is the overlap ratio?

Solution.

$$P = P_n \cos \psi = 3(0.921) = 2.763;$$

$$P_c = \frac{\pi}{P} = \frac{\pi}{2.763} = 1.137 \text{ in.};$$

$$P_{cn} = P_c \cos \psi = 1.137(0.921) = 1.047 \text{ in.};$$

$$b = \frac{P_c}{\tan \psi} = \frac{1.137}{0.425} = 2.675 \text{ in.}$$

Hence, the overlap ratio is $3/2.675 = 1.121$

Axial thrust for helical gears. Figure 12-14(b) illustrates the direction of axial thrust for the possible rotation and helix hand conditions. The thrust is equal to the product of the torque and the tangent of the helix angle divided by the pitch-circle radius:

$$\text{thrust} = \frac{2T \tan \psi}{D}.$$

◆ *Example* ———————————————————————————
A 23-degree helical gear has a pitch-circle diameter of 9.42 in. Find the axial thrust if the gear transmits 1020 in-lb of torque.

Solution. The thrust is

$$\frac{2T \tan \psi}{D} = \frac{2(1020) \tan 23}{9.42} = 91.9 \text{ lb.}$$

Herringbone Gears

Double helical gears eliminate the problem of axial thrust; however, a clearance groove is necessary between the left- and right-handed helical teeth. Continuous-tooth herringbone gearing has no center groove; a special manufacturing process makes this possible. Because of the lack of clearance groove, greater load-carrying capacity is possible. Herringbone gears are often used for heavy loading at moderate speeds. They are useful for continuous service and where a large one-stage ratio is desired. Figure 12-15 shows a continuous herringbone gear.

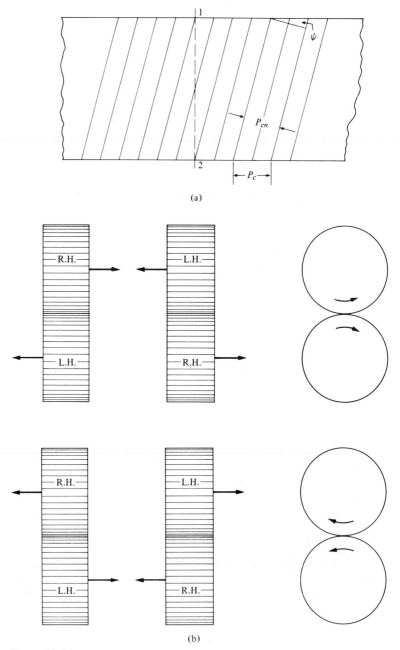

(a)

(b)

Figure 12-14
(a) Pitches for helical gears. (b) Axial thrust for helical gears.

Figure 12-15
Continuous herringbone gear.
(Courtesy of Philadelphia Gear
Corp.)

Bevel Gears

The simplest type of bevel gear is the *straight-tooth bevel gear*. Its teeth are straight and extend toward the axis of the gear. This type of bevel gear is used to connect two shafts that would have intersecting centerlines if extended. If the ratio between the gears is 1/1, the gears are called *miter gears*. Figure 12-16 shows a straight-tooth bevel gear. The velocity ratio in a bevel-gear set is based on the number of teeth in each gear, or the sine of each of the pitch angles.

In bevel gearing, if the pitch angle is less than 90 degrees, the gear is considered an *external* gear; if the pitch angle is greater than 90 degrees, an *internal* bevel gear. If the pitch angle is exactly 90 degrees, the gear is known as a *crown gear* and the teeth lie in a plane instead of in a cone.

A *Zerol gear* is shown in Fig. 12-17. This is a bevel gear with a zero-degree spiral angle; it produces no inward axial thrust. Zerol gears can be used in the same way as straight-tooth bevel gears. Zerol gears operate more quietly than straight-tooth bevel gears and have longer life. There is a small tooth overlap; stress concentrations are eliminated at the tips of the teeth.

Figure 12-18 pictures a *spiral bevel gear*. The advantages of this type over the straight-tooth bevel gear are similar to those of helical gears over spur gears. The teeth are curved and oblique. Spiral angles vary over a wide range (often 20 to 40 degrees); a 35-degree spiral angle is often accepted as a desirable value. Axial-thrust loads are high for spiral bevel gears; this complicates bearing design somewhat. Spiral bevel gears operate quietly and can be used for high-speed operation; their load-carrying capacity is greater than that of straight-tooth bevel gears.

Hypoid gears (Fig. 12-19) were originally developed for automotive use, but they have found other applications. They are similar to spiral-tooth bevel gears, except that the centerlines for the pinion and gear are offset from each other (the pinion centerline can be either above or below that of

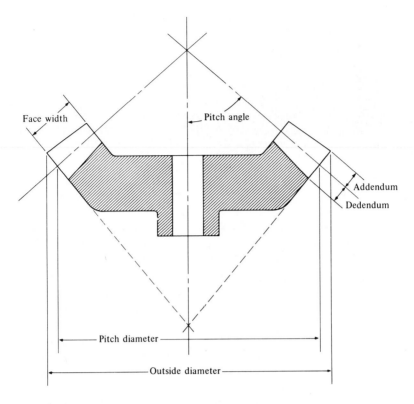

Figure 12-16
Straight-tooth bevel gear.

Figure 12-17
Zerol gear.
(Courtesy of Philadelphia Gear
Corp.)

Figure 12-18
Spiral bevel gear.
(Courtesy of Philadelphia Gear
Corp.)

the gear). Usually, the pinion in a hypoid-gear set is larger than that in the spiral-bevel-gear combination. Thus, greater load-carrying capacity is possible; operation is quiet. Since the centerlines do not intersect, bearings can be mounted on both sides for the pinion and gear; this provides greater rigidity and smoother operation.

Figure 12-20 shows a *worm-gear* set. High ratios can be attained with worm and worm-wheel drives; compactness is easy to obtain with such a combination. The usual practice is to have the worm drive the worm gear. The worm resembles a screw thread in appearance. It can be either left- or right-handed. Also, like a screw, it can be single, double, triple, or quadruple. For example, a double-threaded worm has a lead that is twice the pitch. The greatest ratio is obtained with a single-threaded worm. With small ratios, it is possible to have a gear drive the worm. However, large ratios are usually preferred. With large ratios, it is impossible for the gear to drive the worm. This self-locking feature is desirable in most cases. Worm-gear drives are smooth and quiet. Heat dissipation must often be dealt with by a designer. The efficiency of a worm-gear drive ranges from approximately 20% to 97%. High ratios and small lead angles for the worm produce the lower values of efficiency.

Figure 12-19
Hypoid gear.
(Courtesy of Philadelphia Gear
Corp.)

Figure 12-20
Worm-gear set.
(Courtesy of Philadelphia Gear
Corp.)

To provide maximum contact between the threads of the worm and the teeth of the worm gear, the top of the worm gear tooth is shaped to somewhat envelop the worm; also, the worm itself is shaped to partially envelop the worm gear wheel.

Figure 12-21 shows a Cone-Drive® gear set. Careful examination of this illustration shows how the worm envelops the worm gear wheel and also how the worm gear tooth envelops the worm. The wrap-around or hourglass feature increases load-carrying capacity, since a greater number of teeth are in contact at all times.

Figure 12-22 shows the thrust direction of left- and right-handed worms under indicated directions of rotation.

 **Summary**

Gears are important components in the general area of power transmission. One should carefully evaluate the advantages and disadvantages of gear drives as compared to belt and chain drives before incorporating either into

Figure 12-21
Cone-drive gear set.
(Courtesy of Cone-Drive, a unit
of Ex-Cell-O Corp.)

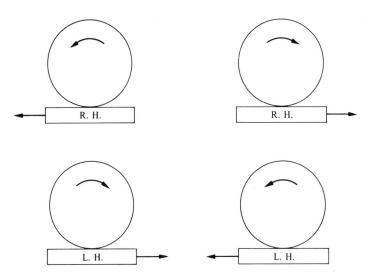

Figure 12-22
Direction of thrust on worms.

a design. Gear drives are expensive, but if properly designed they provide compactness, positive ratios, and long life. See Table 12-14.

In using gear drives, one must determine the type of connection to be made between the driver shaft and the driven shaft. A gear drive can connect shafts whose centerlines are

1. parallel,
2. collinear,
3. perpendicular and intersecting,
4. perpendicular and nonintersecting, or
5. angularly positioned.

Also, rack-and-pinion combinations can be used to convert rotational motion to translational motion or vice versa.

The designer also has a choice of using stock gears or application-engineered gears. Although countless "off-the-shelf" components are available, the designer is then restricted to the available materials, pitches, face widths, tooth numbers, and bores.

Lubrication should be carefully considered in any gear train. Possible causes of tooth failure must be explored; provision should be made to minimize failure wherever possible.

In any simple, compound, or planetary train, the desired motion between input and output has to be studied. Forces are analyzed as in any

Table 12-14
Summary of Gear Types

Type	Applications	Advantages	Disadvantages
External spur	Parallel shafting Moderate speeds	Moderate cost No end thrust	Small contact ratio
Internal spur	Parallel shafting Moderate speeds Same shaft directions	Short centers Large contact ratio Partial safety guard No end thrust	Difficult mounting Expensive
Helical	Parallel shafting High speeds	Quiet operation High load-carrying capacity	End thrust
Herringbone	Parallel shafting Heavy duty	No end thrust Large tooth contact High load-carrying capacity	Expensive
Bevel types			
Straight tooth	Angular drives Moderate speeds	Moderate cost	Difficult mounting
Zerol	Angular drives	Long gear life Smooth and quiet Low stress concentration at tooth tip	Expensive Difficult mounting
Spiral	Right-angle drives High speeds	Good tooth meshing High load-carrying capacity	Expensive Difficult mounting
Hypoid	Right-angle drives Nonintersecting shafts	Mounting rigidity possible High load-carrying capacity	Expensive
Worm gears	Nonintersecting shafts Right-angle drives	High ratios Quiet operation High load-carrying capacity Compact Self-locking possible	Difficult mounting
Rack and pinion	Rotary to linear or linear to rotary	Compact	Difficult mounting Slow speeds Small contact ratio

statics problem. Horsepower requirements are important. Mounting limitations and conditions should also be considered.

In planetary gear trains, shafts can be connected to the sun gear, the ring gear (or the equivalent in bevel gearing), or the planet carrier. All three shafts can rotate; usually, however, one is fixed. The stationary gear has to be either the sun or ring gear; if the planet carrier is fixed, the train becomes a simple gear train.

Gearing is designed on the basis of strength and durability by a process of trial and error based on experience. After preliminary assumptions are made, strength and wear values are calculated. If the assumptions are not suitable, the gear set will be overdesigned or underdesigned. Adjustments are then made by changing any or all of the following:

1. material (or surface hardness);
2. diametral pitch (in effect, this is changing the size of the gear tooth);
3. face width (this somewhat depends on the choice of diametral pitch, since the face width should ideally be somewhere between 2.5 and 4 times the circular pitch and the circular pitch is based on the diametral pitch).

Metric gears are not interchangeable with those made to inch dimensions. Also, metric gears vary slightly among the assorted worldwide standards as used in Germany, Russia, and Japan. The ISO has developed a basic rack and dimensions are normalized for a module of one ($m = 1$). In dealing with metric gear calculations, one must not confuse the module symbol m with the symbol used for the linear measure of distance, m.

Questions for Review

1. Differentiate between the terms *circular pitch* and *diametral pitch*.

2. How does a Fellows stub tooth differ from an American Standard 20-degree stub tooth?

3. What is meant by the term *contact ratio*? Why is it important? Which of the following gear sets would have the larger contact ratio, provided both sets had the same diametral pitch and pitch-circle diameters: (a) a gear set with a pinion driving an external gear or (b) a gear set with a pinion driving an annular gear? Explain why.

4. For a simple gear train with a ratio of approximately 2/1, what advantages would a gear set with 39 and 20 teeth, respectively, have over one with 40 and 20 teeth?

5. Why must spur gears be mounted so that the pitch circles are exactly tangent? What effect would an overlapping of pitch circles have on

gear-set performance? What effect would a gap between the pitch circles of meshing gears have on the action of the gear-train?

6. In a gear-reduction unit, why is the output shaft larger than the input shaft?

7. What is a planetary gear train? How does it differ from a simple or compound train? Why is the term *epicyclic* used to describe this type of train? List some typical applications for planetary gear trains.

8. Explain how the pitch angle of a bevel gear determines whether the gear is an external, an internal, or a miter bevel gear.

9. What advantages do helical gears have over spur gears? What disadvantages?

10. Differentiate between a double helical gear and a herringbone gear.

11. What advantages do gears have over chains in the field of mechanical transmission?

12. In helical gears, how does the circular pitch differ in the normal plane from the linear circular pitch? Which has the greater numerical value?

13. What advantages do annular gear sets have over external spur-gear sets?

14. What steps can be taken to increase the strength- and durability-horse-power ratings of gears?

15. Define the term *module* in metric gears. What relationship does it have to diametral pitch in the American standard system? What units are used to specify module? Diametral pitch?

16. Are American standard spur gears interchangeable with their metric counterparts?

Problems

1. A $14\frac{1}{2}$-degree full-depth involute spur gear has an outside diameter of 8.500 in. and a diametral pitch of 4. Find the circular pitch, the addendum, the dedendum, the pitch-circle diameter, the whole depth of a tooth, the clearance, and the base-circle diameter for this gear.

2. A 20-degree American Standard stub-tooth gear has a total tooth height of 0.9 in. and contains 20 teeth. Find its diametral pitch and the pitch-circle diameter.

3. Two spur gears in mesh have a center distance of 15 in. and a 3/2 ratio. The gears are $14\frac{1}{2}$-degree full-depth involute gears and the diametral

pitch is 4. Find their contact ratio mathematically and then check your result graphically.

4. A simple gear train is to utilize an 80-tooth annular gear. Find the diameter of a suitable pinion and the required center distance. The diametral pitch is 8 and the reduction ratio is 4/1.

5. An 1800-rpm motor is to be connected to a speed reducer that will have an output shaft speed of 90 rpm. The speed reduction is to be accomplished in two stages using gears with at least 20 teeth. Find suitable gear combinations to effect this reduction, neglecting strength and durability.

6. Neglecting strength and durability, find suitable gears for a reverted train similar to that shown in Fig. 12-4. The input shaft rotates at 1800 rpm and the output shaft is to rotate at 90 rpm. The diametral pitch of the gears on the input side is to be 5 and the pitch for the other set is to be 4. All gears must have at least 16 teeth. Why is a reverted gear train desirable for this application?

7. A 26-tooth spur gear has a diametral pitch of 5 and a face width of 2 in. The gears have $14\frac{1}{2}$-degree full-depth involute teeth. Assuming a flexural endurance limit of 60,000 psi and a rotational speed of 600 rpm, find the beam strength by the Lewis equation.

8. A 20-degree spur-gear set has the following specifications:

Pinion	Gear
28 teeth, standard addendum	64 teeth, standard addendum
Steel, Brinell hardness 180	Cast iron, AGMA grade 20
Pitch diameter = 7.000 in.	Pitch diameter = 16.000 in.
Hole diameter = $1\frac{5}{16}$ in.	Hole diameter = $1\frac{9}{16}$ in.
Face width = 3.500 in.	Face width = 3.500 in.
300 rpm	Hub diameter = 5.250 in.

Using the Lewis equation and assuming a flexural endurance limit of 2500 psi, find the horsepower for the gear.

9. Find the durability horsepower for the pinion and gear set of Problem 8. Assume uniform loading on both gears, a factor of safety of 1.5, and a life of 10 million cycles. Use the AGMA procedure for commercial-quality gears.

10. Find the strength horsepower for the pinion and the gear of Problem 8 assuming a life of 10 million cycles, uniform loading on both gears, and a safety factor of 1.5. Follow AGMA practice; assume commercial-quality gears.

11. A 26-tooth helical gear with a helix angle of 15 degrees meshes with a 60-tooth gear of the opposite hand. The smaller gear is made of steel with a Brinell hardness of 265 and it rotates at 100 rpm. The 64-tooth gear is made of steel with a Brinell hardness of 225. The power source is considered in the category of light shock; the 60-tooth gear is connected to equipment that produces moderate shock. The diametral pitch for the gears is 4, and the face width of each gear is 3 times the circular pitch. A factor of safety of 1.6 is to be used. The gears are of commercial quality and are accurately mounted. For a life of 10 million cycles, find the strength horsepower of the 60-tooth gear using AGMA procedures. The geometry factor can be assumed to be 0.60.

12. A straight-tooth bevel pinion with 26 teeth is made of steel with a Brinell hardness of 300. It drives a gear with 50 teeth made of steel with a Brinell hardness of 255. The geometry factor is 0.082 and the loading is uniform. One gear is straddle mounted. Assume the following information:

Pinion	Gear
Factor of safety = 1.4	Face width = 2 in.
Life = 10 million cycles	Dynamic factor = 0.75
Diametral pitch = 4	Elastic coefficient = 2800
Pinion speed = 200 rpm	Hardness ratio = 1.0

Find the durability-horsepower rating for the set.

13. A metric spur gear has an outside diameter of 96 millimeters and a module of 4 mm. Find the number of teeth, the addendum, the dedendum, and the pitch-circle diameter.

14. Two metric spur gears with a ratio of 3/2 and a module of 2.5 mm are engaged at a center distance of 125 mm. Find the number of teeth for each gear.

15. A metric spur-gear set has gears with a module of 0.8 and a 20 degree pressure angle. The driver has 30 teeth and the driven gear has 68 teeth. Find the outside diameter and circular pitch of each gear and the required center distance.

16. A metric spur-gear set has gears with a module of 4 and a 20 degree pressure angle. The driver has 20 teeth and the driven gear has 39 teeth. Find the average number of teeth in contact.

17. In Fig. 12-23, find the rotational speed for the ring gear if the arm turns counterclockwise at 300 rpm and the gear sizes are as follows: $A = 25$

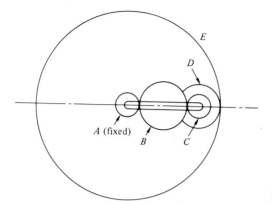

Figure 12-23
Problem 17.

mm, B = 50 mm, C = 22.5 mm, D = 45 mm, and E = 192.5 mm. All gears have the same module.

18. A 20 degree spur-gear pinion with 24 teeth and a diametral pitch of 2 drives a 48-tooth spur gear. The pinion rotates at 1160 rpm and delivers 5 hp. Find the total force transmitted by a tooth on the pinion to a tooth on the gear.

19. In a reverted gear train similar to that shown in Fig. 12-4, gears A and B have a module of 2.5 mm and gears C and D have a module of 2 mm. Find the number of teeth for each gear if no gear is to have less than 15 teeth and the ratio is to be 12 to 1.

CAMS

13

Various motions can be produced by the action of a cam against a follower. Many timing devices are operated by cam action. The purpose of any cam is to produce a displacement of its follower; a secondary follower is often used to produce additional displacement in another location. The most popular type is the plate cam. The cylindrical type is used to transmit linear motion to a follower as the cam rotates. Three-dimensional cams are sometimes used; these provide some unusual follower motions, but also make follower design difficult. The camshaft in the automotive engine illustrates a simple but important application of a plate cam. The cam assemblies in automatic record players illustrate a somewhat more complex application.

Cams are constructed of various materials, including

1. steel,
2. laminated phenolics,
3. nylon, and
4. sintered metals (by the powder metallurgy process).

Cam profiles are accurately constructed by either graphical or mathematical methods. The transition from development drawings to working (shop) drawings can be made in several ways:

1. Make a full-scale template. This is desirable from the manufacturing standpoint, but it will not guarantee accurate cam profiles.
2. Use radial dimensions. This is fairly accurate, but sometimes produces layout problems in the shop.
3. Use coordinate dimensioning. This procedure will ensure accuracy.

In selecting one of these methods, one should consider the function of the cam in terms of desired preciseness.

 Types of Cams

Plate cams are simple to design and easy to fabricate. Figure 13-1 shows a tangential plate cam, which is often used to open and close valves at the correct time. The follower can be moved in various patterns with various rise/fall ratios. Motion should be controlled to avoid abrupt changes in force transmitted from the cam to the follower. One should carefully determine horizontal force components, since these present problems designing the follower assembly guide. Critical reactions occur at points *A* and *B*. These reaction values must be computed. The relative vertical position of point *A* with respect to *B* needs to be raised if the reaction value at *B* is excessive. The position of *B* should be as close to the cam as possible to minimize flexure in the roller-follower support.

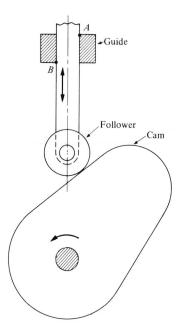

Figure 13-1
Tangential plate cam.

Figure 13-2 shows a *cylindrical cam*. This type produces reciprocating motion in the follower. Again, forces need to be determined and dimensions chosen so as to avoid excessive component sizes. A tapered roller follower is frequently employed; the groove in the periphery of the cam should be shaped to accommodate the follower. This type of cam is expensive to produce. The cylindrical cam has two outstanding features. One is the fact that the cam is positive acting. No outside forces (such as gravity or spring action) are needed to hold the follower against the working surface of the cam. The second feature is the fact that the follower can move through a complete cycle in the course of several revolutions of the cam. For example, it is possible to design the cam so the follower could move from a starting position at the left end to the extreme right position in three revolutions (or more), then the starting position in two revolutions. Other variations are possible.

A *translation cam* is illustrated in Fig. 13-3. In the figure shown, the cam reciprocates horizontally and the follower moves up and down. A pivoted follower can be used with this type. The translation cam can be made positive by providing a guided plate with an inclined slot for the cam; the slot can then engage a pin or roller on a guided vertical reciprocated follower. With the latter type, however, a complete force analysis is a critical phase of the design.

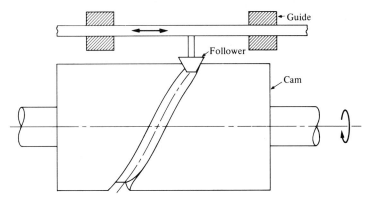

Figure 13-2
Cylindrical cam.

A *face cam* is shown in Fig. 13-4. In this type, the cam rotates and the follower (usually a roller or pin) is guided by a groove cut into the end face of a cylindrical section. Rotation of the cam provides translation of the follower. This type is also positive acting. Production costs for this type of cam are much higher than for a simple plate cam.

A *constant-diameter cam* is illustrated in Fig. 13-5. This is merely a circular plate with the camshaft hole eccentrically located. The amount of eccentricity determines the amount of follower displacement. As the cam rotates, the follower reciprocates. This arrangement is sometimes known as a Scotch yoke mechanism. Follower action is positive; harmonic motion is produced by this type of arrangement.

A simple type of *timing cam* (not shown) can be made by using a thin circular plate notched at appropriate intervals to operate microswitches.

Figure 13-3
Translation cam.

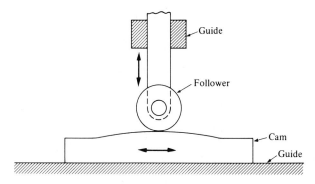

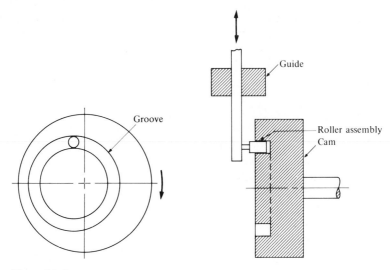

Figure 13-4
Face cam.

Figure 13-5
Positive-action cam (Scotch yoke).

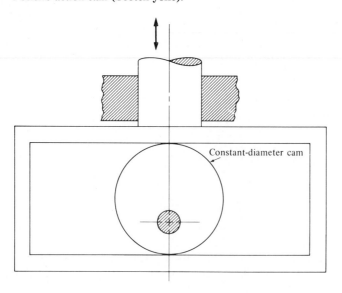

These electrical switches in turn control such components as solenoids and motors. Automatic washing machines and dishwashers usually control the washing, rinsing, and draining cycles by means of such cams. Several cams are mounted side by side, and each cam controls the switches for its particular part of the operation. A timing motor is geared to operate the circular cams so that they make one revolution for the complete process. The 360 degrees of the cam peripheries are then divided so that the intermediate distances between notches provide suitable time intervals to produce the desired results.

 Types of Followers

In general, the follower is considered to be the part that comes in contact with the cam profile. However, when a secondary follower is used, the motion of the secondary follower is dictated by that of the primary follower. For example, a roller follower can be reciprocated by acting against the edge of a pivoted follower.

The simplest type of follower is the reciprocating type that merely moves up and down (or in and out) with the rotation of the cam; the centerline can be either collinear with the cam centerline or offset from it. Contact with the cam can be via a point, a knife edge, a surface, or a roller. Figure 13-6(a) shows a reciprocating knife-edged follower. A flat-faced reciprocating follower is shown in Fig. 13-6(c). If a point or surface is employed for contact, the high normal force can result in abrasion and excessive wear. If the load being transmitted from the cam to the follower is small, the problem is not serious. For example, the operation of a small snap-action switch does not produce cam surface wear. Miniature snap-action electrical switches have actuators with various configurations; some of these are in the form of rounded points or thin metal sections. Miniature three-way valves in air circuits have similar actuators. If cams are used to operate mechanical components directly, a roller is much more effective. Figure 13-6(b) shows a reciprocating roller follower.

Figure 13-7 shows the construction of a cam roller. Cam rollers are commercially available in roller sizes ranging from $\frac{1}{2}$ in. to 6 in. Basic dynamic capacities range from 620 lb to 60,000 lb, based on $33\frac{1}{3}$ rpm and 500 hr of minimum life. Correction factors must be used for any other speed or life values. It should be noted that the cam can be lubricated through an oil hole in the shank or a lubrication fitting in the end of the shank. The cam follower shown in Fig. 13-7 uses small-diameter rollers to reduce friction and increase efficiency. Rolling contact with the cam surface minimizes wear problems. Several mounting arrangements are possible with

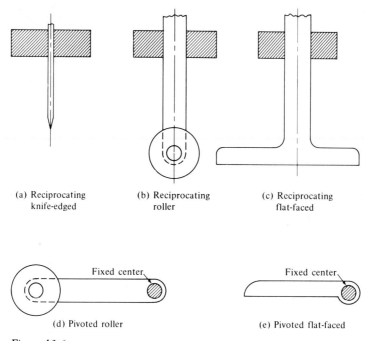

(a) Reciprocating
knife-edged

(b) Reciprocating
roller

(c) Reciprocating
flat-faced

Fixed center

(d) Pivoted roller

Fixed center

(e) Pivoted flat-faced

Figure 13-6
Typical types of cam followers.

Figure 13-7
Camrol® cam follower.
(Courtesy of McGill
Manufacturing Co., Inc.)

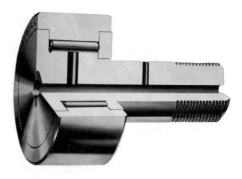

this type of follower. Figure 13-6(d) shows the roller follower mounted on a pivoted arm. A pivoted flat-faced follower is shown in Fig. 13-6(e). As with any flat-faced follower, friction between the follower face and the cam profile must be controlled. Proper lubrication can reduce the effects of friction.

Base and Prime Circles

The *base circle* of a plate cam can be defined as the circle with a radius from the cam center to the profile of the follower at its innermost position. If the cam uses a roller follower, another circle is established from the cam center through the center of the roller; this is often called the *prime circle*. Motion layouts are initiated from the center of the roller on the prime circle. No prime circle is needed for followers that do not use a roller.

 Cam Elements for Plate Cams

As a plate cam rotates, it displaces the follower. To simplify construction, it is best to plot the displacement of the follower first and then transfer each follower position to the appropriate cam element. (This procedure is explained later in the examples showing cam-profile construction.) If a follower reciprocates on a centerline collinear with a cam centerline, the cam elements are radial lines as depicted in Fig. 13-8. The follower is classified as an "in-line" follower whether it is flat-faced, knife-edged, or roller-shaped.

Raising a follower on the surface of a cam is similar to pushing a load up an inclined plane. In the case of the inclined plane, the pushing effort is altered by changing the inclination of the operating force or the slope of the incline. In cam layout work, the term *pressure angle* is an important criterion for design. The pressure angle for a cam assembly with a reciprocating roller follower is the angle formed by a line connecting the cam contact point with the roller center and the cam element (see Fig. 13-9). This angle should be less than 30 degrees, if possible. The horizontal component F_H in Fig. 13-9 causes an undesirable bending moment on the follower stem. The pressure angle θ shown in this figure is obviously too large and produces a large side thrust on the follower guide.

The size of the pressure angle depends on

1. base-circle diameter,
2. size of roller,
3. displacement of follower, and
4. cam motion.

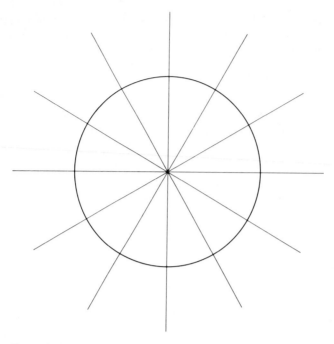

Figure 13-8
Cam elements for in-line follower.

Figure 13-9
Cam pressure angle.

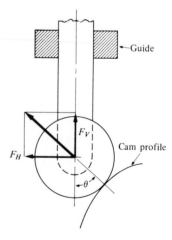

Often, these four items cannot be altered without defeating the cam requirements and space limitations. The value of the horizontal component in Fig. 13-9 can be substantially reduced by offsetting the centerline of the follower. One must be careful to offset the cam elements on the proper side of the cam centerline; this depends on the direction of cam rotation. If the offset is made on the wrong side of the cam centerline, the side thrust becomes greater. Figure 13-10(a) shows the offset cam elements for counter-clockwise cam rotation. The offset is to the right of the cam centerline. An offset circle is constructed with a radius equal to the amount of the offset. Tangential lines are then drawn from the offset circle to form the cam elements as shown. For clockwise cam rotation, a similar procedure is followed—the elements are drawn in the opposite direction, but still tangent

Figure 13-10
Cam elements for reciprocating follower: (a) offset to right, (b) offset to left.

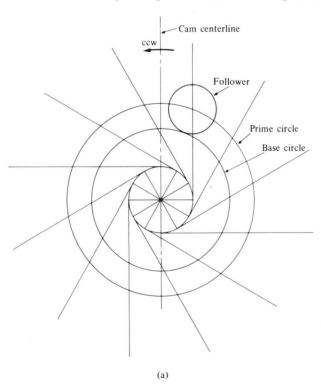

(a)

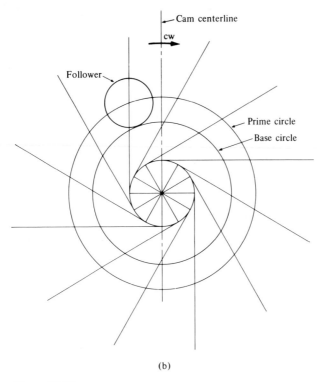

(b)

Figure 13-10
(continued)

to the offset circle (see Fig. 13-10b). In either case, cam elements are merely displaced follower centerlines.

If a pivoted follower is employed, the cam elements become circle arcs with a radius equal to the distance from the pivot point to the center of the roller follower. The procedure for constructing these elements is illustrated in Fig. 13-11. First, a pivot construction circle is made with a radius equal to the distance from the center of the cam to the pivot point. A prime circle is divided into the desired number of increments. In this particular case, 30-degree increments are used. Through each of these positions on the prime circle, arcs equal to the distance from the pivot point to the roller center are scribed to establish center points on the pivot circle. Then, with the same radius, arcs are scribed from each of the displaced pivot points. These circle arcs become the cam elements.

**Types of Motion and Displacement
Diagrams**

Various types of motion can be imparted by a cam to its follower. A displacement diagram can facilitate the study of cam action. The *x*-axis of a displacement diagram represents time or cam increments. These are frequently expressed in degrees of rotation. The *y*-axis usually represents the follower displacement. By making the total length of the *x*-axis equal to the circumference of the prime circle of the cam, one can obtain a fairly good graphical representation of cam action and perhaps foresee abrupt changes in motion and excessive pressure angles. Figure 13-12 shows the graphical construction for *constant velocity*. The *y*-axis is divided into the same number of equal increments as there are divisions in the *x*-axis. Each

Figure 13-11
Cam elements for pivoted follower.

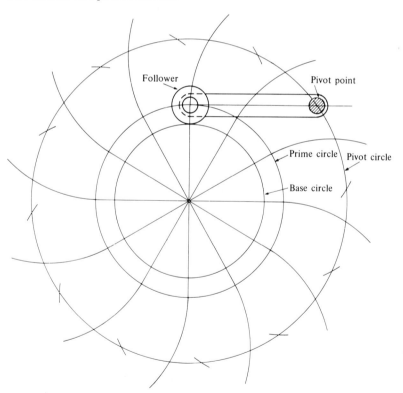

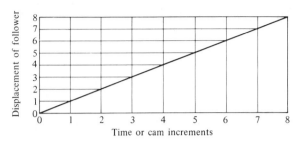

Figure 13-12
Displacement diagram for constant-velocity motion.

vertical division is transferred to the corresponding horizontal cam increment.

Constant velocity (sometimes called *uniform motion*) produces an equal displacement for each time interval throughout the follower travel. When a cam changes the motion from constant velocity to any other type of motion, including dwell, the change is abrupt. Such a change can produce shock loading on the follower.

Harmonic motion avoids sudden changes, since the displacement for each time interval varies in magnitude (see Fig. 13-13). If a point travels in a circular path at a constant speed, the projection of the point on the diameter of the circle moves with harmonic motion. Harmonic motion is also produced when a weight suspended from a helical extension spring is set in motion. In Fig. 13-13, a semicircle is constructed with a diameter equal to the linear displacement. The semicircle is then divided into equal parts, corresponding to the number of time intervals. Each point on the arc of the semicircle is then projected to the diameter. In a cam chart, further projections can be made from each displacement point to the corresponding

Figure 13-13
Displacement diagram for harmonic motion.

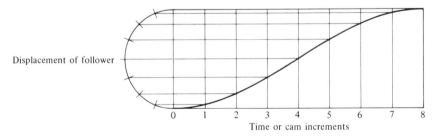

time increment, thus establishing a smooth curve. Note that no abrupt change occurs at the beginning and ending points of the curve. If a cam follower operates on such a curve, severe jolting will not occur.

Still smoother follower action results if *parabolic motion* is employed. In this type of motion, a point is uniformly accelerated for one-half the displacement and then uniformly decelerated for the remainder of the distance. From the motion equation ($s = at^2/2$), if the acceleration is considered constant, time becomes the variable. Substituting 1, 2, 3,... for time units, we obtain distance values as follows:

$$s_0 = \frac{a(0)^2}{2} = 0, \qquad \text{difference} = 1,$$

$$s_1 = \frac{a(1)^2}{2} = 1(a/2), \qquad \text{difference} = 3,$$

$$s_2 = \frac{a(2)^2}{2} = 4(a/2), \qquad \text{difference} = 5,$$

$$s_3 = \frac{a(3)^2}{2} = 9(a/2), \qquad \text{difference} = 7,$$

$$s_4 = \frac{a(4)^2}{2} = 16(a/2), \qquad \text{difference} = 9,$$

$$s_5 = \frac{a(5)^2}{2} = 25(a/2).$$

Note that $a/2$ is a constant; the square of the time intervals becomes 1, 4, 9, 16,.... . If distances are subtracted, the progression becomes 1, 3, 5, 7,.... . This means that the follower will travel three times as far during the second time interval as it did during the first. For the third time increment, the distance traveled is five times as much as during the first time interval.

To construct the cam profile graphically, the displacement is divided into two parts. The first half will have uniformly accelerated motion and the latter uniformly decelerated or retarded motion. If eight time intervals are used, the total displacement is divided in the ratio of 1, 3, 5, 7, 7, 5, 3, and 1. Note this construction in Fig. 13-14. If six time intervals are used, the ratio is 1, 3, 5, 5, 3, and 1. To get true parabolic motion, the number of time intervals must be an *even* number, such as 4, 6, or 8.

Numerous other types of motion can be employed in constructing cam profiles. *Cycloidal motion* is somewhat similar to harmonic motion. The cycloid is produced by a point on a circle rolling along a straight line. In cam applications, it is convenient to make the circumference of the circle equal to the linear displacement of the follower. The generated cycloid is

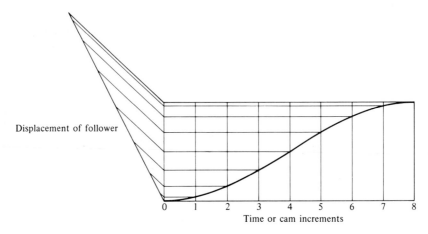

Figure 13-14
Displacement diagram for parabolic motion.

then divided into equal increments in a manner similar to that for harmonic motion, and projections are made to the diameter. Modifications of constant-velocity motion can also be used to good advantage. In general, however, follower action must not be abrupt if the cam operates at high speed. A cam diagram or displacement diagram is helpful to the designer, who can study the transition from one motion to another and make certain that there is not too much abruptness in the cam profile.

Jerk

The preceding curves show displacement for a particular type of motion. Since cams usually employ several different types of motion (including dwell), it is desirable to plot a displacement curve and then plot velocity and acceleration curves directly below the displacement curve in order to get a complete understanding of the cycle.

In high-speed cams with heavy loading, it is important to know the maximum acceleration so as to locate any inertial problems as determined by the equation $F = ma$. Since springs are often used to keep the follower in contact with the cam surface, high acceleration may mean that a spring must exert a high force. The resulting high stress at the point of contact may cause undue wear.

By definition, *jerk* (sometimes called *pulse*) is the time rate of change of acceleration,

$$J = \frac{da}{dt}.$$ (13-1)

Any sudden change in acceleration results in a high value of jerk, which in turn produces a high load on the follower and undesirable shock to the entire assembly. Therefore, if cam speeds and loads are high, a properly designed cam should have a minimum amount of jerk. A large amount of jerk causes vibrations that affect the life of the cam and the follower; also, the noise level becomes high.

The cycloidal motion curve contains no abrupt change in acceleration, which makes it ideally suited for high speed applications.

 **Cam Layout**

Two approaches can be used to construct cam profiles. The displacement diagram with *x*- and *y*-axes can be drawn and the ordinates transferred to a cam layout; or the motion can be constructed directly on the cam layout drawing after the cam elements are established. The following examples show typical cam layouts using some of the previously discussed motions and follower types. To simplify the construction, 30-degree increments of cam rotation are used; in a real-life problem, smaller increments would be used to ensure greater accuracy. If one part of a cam cycle is critical, small increments can be used for that particular part of the layout.

◆ *Example* ──────────────────────────────

Construct the cam profile for a plate cam with an in-line roller follower. The cam is to meet the following specifications (see Fig. 13-15):

1. counterclockwise rotation;
2. diameter of prime circle = 3 in.;
3. roller diameter = $\frac{3}{4}$ in.;
4. displacement = 1 in. in 180 degrees of cam rotation with harmonic motion;
5. dwell for 60 degrees;
6. return in 120 degrees with harmonic motion.

Solution. First, draw the prime circle to size and establish the cam elements radially at 30 degree increments. Consider the prime circle to be the circle that can be drawn through the center of the roller when the cam follower is in its innermost position. Number the cam elements in the opposite direction to that of cam rotation. Note that cam element number 1 will be the first one to contact the roller as the cam starts rotating.

Next, lay out the displacement equal to 1 in. Since the rise is to take six 30-degree increments, a semicircle is constructed with a 1-in. diame-

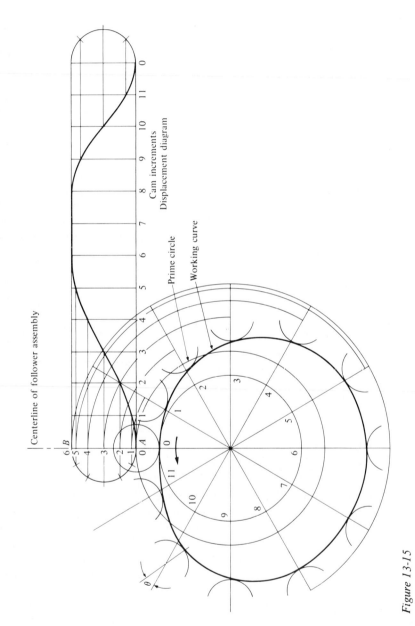

Centerline of follower assembly

Cam increments
Displacement diagram

Prime circle

Working curve

Figure 13-15

Construction of plate cam with in-line roller follower (harmonic-dwell-harmonic).

ter. The arc of this semicircle is divided into six equal parts and projected to the diameter. In this illustration, a cam diagram is constructed; if the *x*-coordinate were made equal to the prime-circle circumference, a true picture of follower climb could be established. The follower positions from *A* to *B* are numbered from 0 to 6 and projected over to the appropriate cam increment. Since the cam dwells for 60 degrees, points 7 and 8 correspond to point 6.

The return motion is also harmonic, but it is effected in 120 degrees of cam rotation. Thus, the displacement distance is used as the diameter of a semicircle, the arc is divided into four equal parts, and projections are made to the diameter—and eventually to the corresponding cam increments on the diagram. For convenience, this construction is given on the right-hand side of the diagram.

Center points for the roller for each cam position can be found by either of two methods: (1) Using dividers, transfer each individual displacement for each cam increment to the radial cam elements (measure from the prime circle). (2) Project all displacement points to the line *AB*; then, using a compass, swing each point on *AB* to the correct cam element. This establishes roller center points based on the center of rotation of the cam.

Through each roller center point, arcs representing the roller must be drawn. A smooth curve tangent to each arc is the working curve of the cam. Note that the tangency point between the cam profile and the roller is not always at the cam element. At position 10, this deviation establishes angle θ, which is the pressure angle. Similar angles can be made at any position.

If one is interested in cam layout only, all construction can be done on line *AB* and the displacement diagram can be eliminated.

◆ *Example* ——————————————————————————

Construct the cam profile for a plate cam that uses a flat-faced follower which is offset $\frac{1}{2}$ in. to the right of the cam centerline, as shown in Fig. 13-16. It is to meet the following specifications:

1. counterclockwise rotation;
2. diameter of prime circle = 3 in.;
3. displacement = 1 in.;
4. follower is to move outward in 120 degrees of cam rotation with parabolic motion; then it is to dwell for 60 degrees;
5. follower is to return in 180 degrees of cam rotation with harmonic motion.

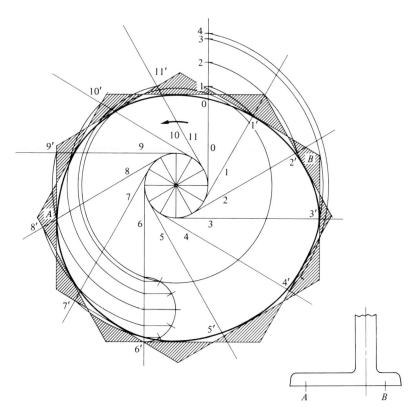

Figure 13-16
Construction of plate cam with offset flat-faced follower.

Solution. Draw a prime circle, then construct an offset circle with a radius of $\frac{1}{2}$ in. Divide the offset circle into 30-degree increments. Draw offset cam elements tangent to the circle (see Fig. 13-16). Since the cam actually rotates counterclockwise, cam elements are numbered clockwise.

The 1-in. follower movement is divided in the ratio of 1, 3, 3, and 1, since there are four 30-degree increments in the 120 degrees of follower rise. Using the center of the cam as the compass point, swing each displacement point to the corresponding cam element; this establishes points 1′, 2′, 3′, and 4′. Since the cam dwells for 60 degrees, points 5′ and 6′ are the same distance from the cam center as 4′. At position 6′, it is convenient to plot the return motion. Since this is harmonic motion for 180 degrees, the 1-in. distance is used for the diameter of a semicircle. This arc is divided into six equal parts, each of which is projected to the diameter and can establish points on the cam elements.

Through each plotted point, draw a line perpendicular to the cam element. This represents the follower centerline; the perpendicular line represents the face of the follower. Each of these lines will become the base of a triangle and the near side of each adjacent triangle. To find all 12 triangles, it is desirable to crosshatch or shade each one as it is formed. A smooth line through the midpoint of each base establishes the working curve for the cam. By measuring the distance from the centerline of the follower to each tangency point, the maximum dimension needed on each side of the centerline can be determined. For added assurance that the cam will never strike any part of the follower except the flat face, it is desirable to add at least $\frac{1}{8}$ to $\frac{1}{4}$ in. to the determined distance. In Fig. 13-16, the maximum distance to the left of the centerline is $9'A$; the required distance to the right is $2'B$.

It is not always necessary for a flat-faced follower to keep its flat face perpendicular to the centerline. It can be at any desired angle; the face is drawn through the plotted points at the desired face angle.

This layout provides the proper motion, but does not consider strength of parts. A force analysis must be done using maximum conditions of loading. Maximum guide reactions occur when the cam strikes the follower at point A. One must also consider bending in the flat-faced follower; this depends on the maximum moment, the working stress of the material selected, and the section modulus for the desired cross section.

◆ *Example* ─────────────────────────────

Construct a plate cam with a pivoted roller follower, as shown in Fig. 13-17. It is to have a 1-in.-diameter roller, a 3-in.-diameter prime circle, and a horizontal pivot point $2\frac{1}{2}$ in. horizontally to the right from point A. The required motion is as follows:

1. rise 2 in. with constant velocity in 210 degrees of counterclockwise cam rotation; then dwell for 30 degrees;
2. return to the starting point in 120 degrees of rotation with constant velocity.

Solution. First, draw the required prime circle and divide it into 30-degree increments. Then establish a pivot circle by scribing a circle concentric with the prime circle, but with a radius equal to the distance from the cam center to the pivot center. Curved cam elements are then drawn with a radius equal to AC; the center of curvature in each case is

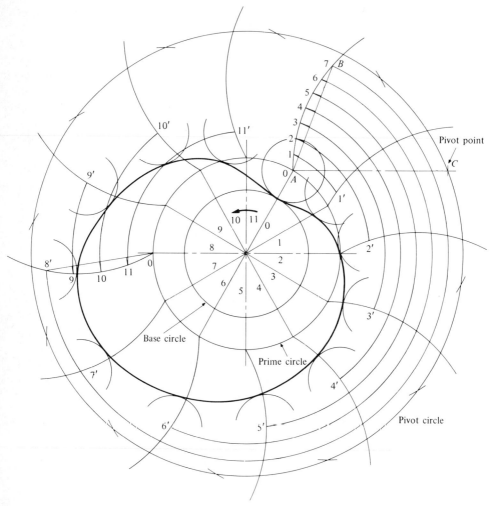

Figure 13-17
Plate cam with pivoted roller follower.

located on the pivot circle and the cam elements extend outward from the prime circle.

The 2-in. displacement is established as the chord of the initial arc and extends from *A* to *B*. Since there are seven 30-degree increments in the 210-degree rise, line *AB* is divided into seven equal parts. Each point is projected to the arc. Next, each numbered point on the arc, such as 1, 2, 3,..., is rotated about the cam center to the appropriate cam element to establish points 1′, 2′, 3′,.... Point 8′ is located the same distance out as point 7', since the cam dwells for 30 degrees. At 8', a similar chord is drawn and divided into four equal parts which are projected from the chord to the arc. Again, using the cam center, each point is rotated to the correct cam element.

Through each plotted point, an arc equal to the roller radius is drawn. A smooth curve tangent to all the roller arcs then provides the working curve for the cam.

Two things can be noted about the cam layout from the working cam profile: (1) it is difficult to establish a smooth curve; (2) the pressure angle is excessive.

To provide a smoother-operating cam, which is essential for high-speed operation, several steps can be taken:

1. Decrease the roller size; this means that a smaller operating load must be used.
2. Increase the base (and prime) circle sizes; this increases the cam size, which could be serious if space is at a premium.
3. Decrease the displacement.
4. Choose a different type of motion for the cam; constant velocity is not an ideal motion.
5. Choose a different location for the pivot point.

In general, point 4 is the item that has the greatest effect on cam performance from the standpoint of smooth operation. If constant velocity is absolutely needed, it should be modified at the beginning and end of a displacement; such motions as harmonic, cycloidal, or parabolic motion can be combined with constant velocity at the beginning and end to smooth the follower operation.

◆ *Example*

Construct the path of the follower for a cylindrical cam with the following specifications:

1. cam diameter = 1.5 in.;
2. direction of rotation as shown in Fig. 13-18(a);
3. horizontal displacement = 2 in.;
4. follower is to move to the left of point *A* in two-thirds of a revolution with harmonic motion, then dwell for one-third of a revolution;
5. follower is to return to point *A* in two-thirds of a revolution with harmonic motion and then dwell for one-third of a revolution.

Solution. Draw orthographic views as shown in Fig. 13-18(a). On the right side view, divide the cam into 30-degree increments. Since rotation is clockwise as viewed from the right side, the cam elements are numbered in a counterclockwise direction. Lay out the total displace-

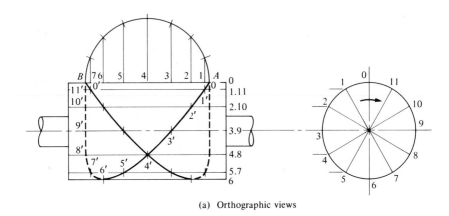

(a) Orthographic views

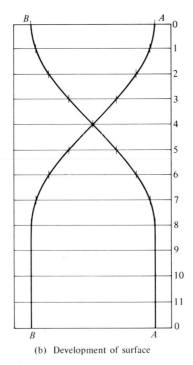

(b) Development of surface

Figure 13-18
Layout of cylindrical cam.

ment from *A* to *B*. Since this distance is covered in two-thirds of a revolution, eight time intervals are needed. Therefore, construct a semicircle with a diameter of 2 in. and divide the arc into eight equal parts. Next, project each of these to the diameter. Cam elements in the front view can be positioned by projecting each one from the right side view. From displacement position 1 on line *AB*, project to cam element 1 on the front view; this locates point 1′. Repeat this procedure for all points from 0 to 8. Since the cam dwells for one-third of a revolution, points 9′, 10′, 11′, and 0′ lie along the same vertical line in the front view.

The return motion is constructed in a similar fashion (to avoid confusion, points are not numbered in the figure). The plotted points represent the centerline of the follower on the periphery of the cylinder. A groove must be cut in the surface of the cylinder to accommodate the follower, not shown here. A smooth curve connecting the plotted points shows the path of the follower. Since the front view is an orthographic representation of the cylinder, points 0′ to 6′ can be seen and hence are shown with a solid line. From 6′ to 0′, the groove is on the back part of the cylinder; these are shown dotted. The return from *B* to *A* is similarly shown.

A development of the surface is helpful and easy to construct. In Fig. 13-18, the circumference of the cam is laid out vertically under the front view for convenience. The circumference is first divided into the proper number of cam increments. Then (starting at *A*) each point in the front view is projected down to the proper element on the development. In one revolution, the follower travels from *A* to *B* in the developed view. For the return motion, the curve is resumed at point *B* on the top of Fig. 13-18(b) and continues until the original starting point *A* is reached. If the developed layout were formed into a cylinder, point *B* at the bottom would correspond to point *B* at the top and point *A* at the bottom to point *A* at the top.

In this cylindrical cam, the entire cycle requires two revolutions of the cam. In some cams of this type, several revolutions are needed for a linear displacement of the follower. If desired, the cam can return the follower in a different number of turns.

Force Analysis

In heavy-duty cam applications for machinery, a thorough force analysis is needed to design the necessary components. Figure 13-19 shows the force is transmitted from cam to follower. In this figure, the centerline of the

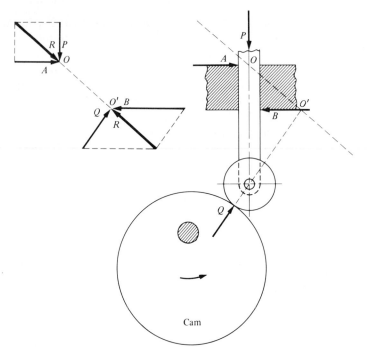

Figure 13-19
Force analysis: cam follower.

follower assembly is offset from the cam centerline. The force P that the cam has to overcome is shown; this is a force or load with known magnitude. The direction of force Q is dictated by line contact between the roller and the cam; its direction is established by drawing a line from the tangency point to the center of the roller. Guide reactions occur at A and B. Treating the follower assembly as a free body, we find that we have a four-force system in equilibrium. The easiest way to solve such a system is to work with two of the forces at a time. The lines of action of P and A intersect at O; then the lines of action of Q and B intersect at O'. Scribe a line from O to O'. Since P has a known value and the directions of A and P are known, a parallelogram can be constructed with the line of action of the resultant along OO'. Find the amount of this resultant. Next, take this value of R and lay it out from O' in the opposite direction. A new parallelogram can be drawn with the known value of R and the direction of B and Q. The value of Q can be scaled; this is the force needed at this particular angle to overcome the load P. The torque needed by the camshaft can be found by multiplying Q by the perpendicular distance from the cam center to the line of action of Q.

In making such an analysis, several other factors should be considered. The distance between A and B has an effect on all values. Also, if sliding friction is to be considered, the reactions at A and B must be sloped downward at the proper friction angle. It should also be noted that this solution is based on this particular position of the cam with respect to the follower. From the design standpoint, the worst condition must be explored. The amount of rolling resistance between the cam and follower is negligible; if the roller contains antifriction bearings, friction at the contact point can be eliminated from the analysis. By making a complete force analysis, the designer is in a favorable position for designing all the components of the cam and follower assembly.

 Cam-follower Relationships

Since a cam imparts a force to its follower, it is essential that the cam profile displace the follower properly, without placing undue loads on it. If a cam is merely operating small electric switches or air pilot valves, a force analysis is not usually needed. In such cases, a careful check on pretravel, operation travel, and overtravel is usually sufficient. Total displacement on such switches, however, must be checked carefully to prevent damage to the unit.

If a cam exerts large forces on its follower, the designer must be careful to avoid impact loading and unnecessary cam surface wear. Figure 13-20 shows some design features that must be avoided. Note the abrupt change in cam profile in Fig. 13-20(a). This gives the follower a jolt that can lead to

Figure 13-20
Poor cam-follower relationships.

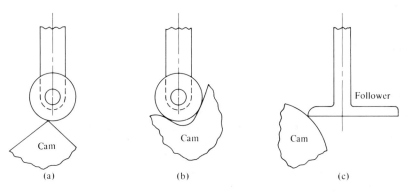

part failure. Figure 13-20(b) shows a situation where the contact between the two components is not continuous. Such a condition usually occurs when the follower is too large with respect to the cam.

One of the difficulties encountered with flat-faced followers is illustrated in Fig. 13-20(c). The cam is engaging the sharp edge instead of the flat face. In laying out a cam with a flat-faced follower, one must determine the amount of face needed and then add $\frac{1}{8}$ to $\frac{1}{4}$ in. for additional safety. A condition such as that shown in Fig. 13-20(c) will produce component failure.

 ## Standard Cams

As with gears, the use of stock or standard cams can save engineering and production time. Standard and semistandard cams are commercially available with preestablished displacements and timing periods. The use of stock cams restricts the design to certain displacements, such as $\frac{3}{4}$-, $1\frac{1}{2}$-, and 3-in. values. Stock cams are designed to operate with needle-bearing roller followers in certain size ranges. A designer who uses such cams has no choice with respect to rise and return motions, although one dwell period or two can be specified. In using these cams, the designer must carefully check the camshaft size and the placement of dowel-pin holes.

Semistandard cams are blanks that can easily be machined into positive-motion face cams. Within reasonable limits, the designer can choose the displacement and minimum follower rise. Blanks are furnished with camshaft and dowel-pin holes; the cam profile can be cut in accordance with the needs of the designer.

Load ratings are given in terms of radial-thrust capacity up to a given cam rotational speed. Standard and semistandard cams are designed for in-line radial follower action, although a limited arc for pivoted followers can be used in certain applications.

 ## Summary

Any cam assembly consists of a cam and a follower. The cam is used to impart motion to the follower or secondary follower. Plate cams can be made of thick metal plate for heavy-loading conditions, or of thin sections of such plastics as laminates and nylon for light-duty applications. Cams can also be produced by the powder metallurgy process. Light-duty applications include timing devices, calculators, sewing machines, record players, and other small appliances. Cylindrical and face cams provide positive follower action, but are expensive to produce.

The forces acting on cams must be analyzed, including the spring force if such a component is used to ensure cam-follower contact. Cam profiles must be studied when large forces are involved to guard against shock loading. Pressure angles should be kept as small as possible to avoid too steep a climb for the follower. If the pressure angle of a tentatively designed cam is greater than 30 degrees, a reduction in value should be considered. This can be done by increasing the cam size, reducing the displacement, or offsetting the follower (for plate cams).

For a complete treatment of cams, consult Harold A. Rothbart, *Cams: Design, Dynamics and Accuracy*, Wiley, 1956.

Questions for Review

1. What disadvantages do flat-faced followers have when compared to roller followers?

2. What steps can be taken to decrease the pressure angle when designing a cam?

3. What steps can be taken to ensure that a cam follower will make continuous contact with a cam profile?

4. What advantage does parabolic motion have over harmonic motion in cam design?

5. List several household appliances that use cams. Also list the type of cam and follower, and the probable motion employed.

6. Where are cams used in machine tools? What purpose do they serve?

7. In engineering practice, the term *jerk* is defined as the time rate of change of acceleration. Explain how excessive jerk can be detrimental to cam actions, and indicate steps that can be taken to minimize jerk.

Problems

1. Graphically lay out a composite displacement diagram as follows:
   a) provide a displacement of 50 mm on the *y*-axis;
   b) provide 8 time intervals on the *x*-axis for the rise and 8 time intervals for the return (use 10 mm spacing for the time intervals);
   c) plot the curve for constant velocity (uniform motion);
   d) superimpose the curve for harmonic motion; and
   e) superimpose the curve for parabolic motion.

2. Draw a displacement diagram with the same dimensions as in Problem 1. Provide the following motion for the rise and return:
   a) rise $\frac{1}{4}$ in. in two time intervals with uniformly accelerated motion;

    b) rise $1\frac{1}{2}$ in. with constant velocity in four time intervals;

    c) rise $\frac{1}{4}$ in. with uniformly decelerated motion in two time intervals; and

    d) return in a similar manner.

3. A plate cam is to have the following specifications:
   a) base-circle diameter of 100 mm;
   b) displacement of 40 mm;
   c) rise full amount in 120 degrees of cam rotation with harmonic motion;
   d) dwell for 120 degrees,
   e) return in 120 degrees with parabolic motion; and
   f) clockwise rotation.
   Plot the working curve for a 25-mm-diameter in-line roller follower.

4. Construct a cam meeting the same specifications as those listed in Problem 3, except that the 1-in.-diameter roller follower is offset $\frac{1}{2}$ in. to the right of the cam centerline.

5. Lay out the cam profile for an in-line flat-faced follower meeting the other specifications of Problem 3.

6. Using the data of Problem 3, construct a cam profile for a knife-edged follower.

7. Plot the path of the follower centerline on the periphery of a cylindrical cam and show the developed surface for a cam with the following characteristics:
   a) diameter of cylinder = 2 in.;
   b) follower is to move 2 in. axially to the right from the starting point with harmonic motion in one revolution;
   c) follower is to dwell for one-quarter of a revolution;
   d) follower is to move 1 in. further to the right with harmonic motion in three-quarters of a revolution;
   e) follower is to return 3 in. to the left in one revolution with parabolic motion.
   Choose the direction of rotation and indicate this on the drawing.

8. Figure 13-21 shows an arrangement for a plate cam with a secondary follower. Pin A is guided along a horizontal centerline by additional links (not shown). The motion of pin A is controlled by the action of the bell crank, which is actuated by a roller follower ($\frac{3}{4}$-in. diameter), which in turn is driven by a plate cam. The plate cam is to rotate counterclockwise. The pin is to move from A to B in 240 degrees of cam rotation with parabolic motion; it is then to return to position A in 120 degrees of rotation with parabolic motion. Displacement is $1\frac{1}{4}$ in.

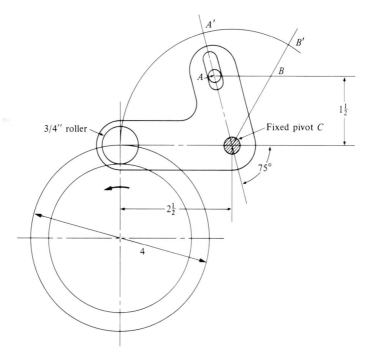

Figure 13-21
Plate cam with secondary follower: Problem 8.

a) Construct the cam profile using 30-degree increments. (*Hint:* After establishing the curved cam elements for a pivoted follower, continue the first one to establish the arc $A'B'$. Lay out the motion of line AB and project each plotted point to the arc $A'B'$. Then, transfer each point to the starting position on the base circle.)

b) Make a force analysis on the bell crank for positions A, B, and the midpoint; assume that the pin must move an 800-lb load. By measurement on the layout, determine the torque required by the cam to move the pin.

c) If a manufacturer's catalog is available, check to see whether a $\frac{3}{4}$-in. needle-bearing roller follower is adequate if the cam rotates at 33 rpm. The follower rotational speed can be found by comparing the circumference of the roller with the periphery of the cam (the latter value can either be estimated or measured with a map measure instrument).

9. Make a timing diagram and cam profile for a timing cam that is nominally 5 in. in diameter. Assume that the cam is driven by a timing motor geared down to one revolution per one-half hour. The cam is to operate a snap-action switch to make and break an electrical circuit (which operates a solenoid) according to the following program:

a) start and circuit closed: 500 seconds,
b) circuit open: 100 seconds;
c) circuit closed: 300 seconds;
d) circuit open: 100 seconds;
e) circuit closed: 300 seconds;
f) circuit open: 100 seconds;
g) circuit closed: 300 seconds; and
h) circuit open: 100 seconds.

   In planning the rise and return distances away from the basic cam, catalog information must be used to obtain such switch specifications as pretravel, overtravel, and the configuration of the switch actuator, which is actually the cam follower.

10. Make a layout for a translation cam that reciprocates horizontally. Divide the 160 mm horizontal travel into 10 mm increments. A 20-mm roller follower is to dwell for the first 40 mm of horizontal travel, rise 50 mm with parabolic motion for 80 mm of travel, and then dwell for the remaining horizontal distance.

SPRING DESIGN

14

 Introduction

Springs are important components for positioning and actuating in the mechanical and electrical fields. Springs can be obtained as off-the-shelf stock parts, or they can be engineered for a specific application. Various types of springs are available; their sizes range from miniature to heavy-duty. In this text, only the conventional types are considered. For unusual applications, a designer should consult a firm specializing in spring manufacture, so that manufacturing capabilities can be determined.

Types of Materials

The majority of springs are made of steel; piano wire is one of the typical types of steel used. Other materials used include phosphor bronze, brass, Monel, silicon bronze, beryllium copper, and others. The properties of these materials must be checked before designing. Coil springs can be made of round or square wire.

Types of Springs

Springs are made in many configurations. The three most popular springs of the coil type are the *extension*, *compression*, and *torsion* types. Figure 14-1 shows an extension spring. In this type, the coils touch each other until a load is applied to the spring. The extension spring is characterized by a loop on each end. Several types of loops can be formed; a few of these are shown in Fig. 14-5.

Figure 14-2 shows a compression spring. When unloaded, the coils are separated; when a load is applied, they move closer together. Four types of

Figure 14-1
Extension spring.

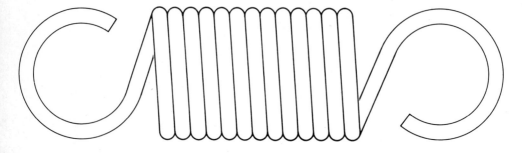

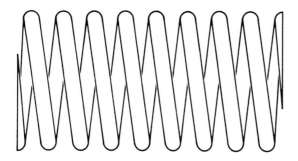

Figure 14-2
Compression spring (ground ends).

ends are used on the compression spring:

1. plain end,
2. ground end,
3. squared end, and
4. squared and ground end.

Figure 14-3 shows a section through a compression spring. Since a compression spring is used with other parts exerting forces on the ends, the spring and the parts should have as much contact as possible; therefore, the spring ends should be flattened. In calculations, one must consider that approximately three-quarters of a turn on each end of a ground compression spring is inactive, since the spring effect is lost from the flattened wire.

Figure 14-3
Section through compression spring.

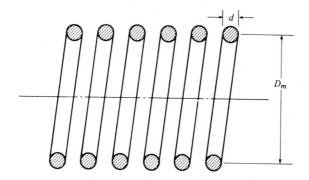

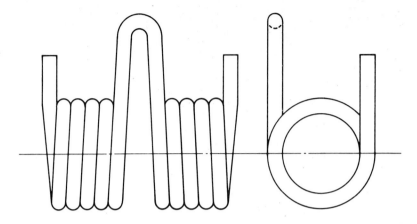

Figure 14-4
Torsion spring (double).

The third classification of coil springs is the torsion type. Figure 14-4 shows a double torsion spring. In this type of spring, the deflection is circular rather than linear. Hinge springs are usually of the torsion type.

Flat springs are available in single-leaf or multiple-leaf form. A single-leaf spring is found in the cursor of most slide rules. Semielliptical types are found in the rear suspension systems of most vehicles. Full-elliptical springs are similar to the semielliptical type, except that the deflection is different and the spring requires more space.

Special springs include the volute spring, Belleville spring, spiral spring, garter spring (see Fig. 16-11), and the Neg'ator® spring.

Before discussing any of these, it is desirable to understand thoroughly the glossary of spring terms that follows.

Glossary of Spring Terms
(for Helical-coil Springs)

Free length: overall length of compression coil spring with no load applied.

Solid length: length of completely compressed compression spring.

Deflection: lengthening (or shortening) of coil spring caused by load.

Spring constant: ratio of load to deflection.

Outside diameter: maximum diameter of coil spring.

Inside diameter: minimum diameter of coil spring (important when the spring must fit over a pin).

Mean diameter: average diameter of coil (important in calculations).

Active number of coils: number of coils used in supporting the load. This does not include the extra three-quarters of a coil at each end of a compression spring that is ground flat.

Winding: the direction of the helix in a coil spring (right hand or left hand).

Spring index: ratio of the mean diameter to the wire diameter in a coil spring.

Stress correction factor: a factor used to correct for high stress caused by curvature on the inside of the coil.

Types of ends: See Fig. 14-5 for common types of loops that can be formed on the ends of an extension spring, such as machine, raised hook, and V-hook.

Figure 14-5
Loops for extension springs. (a) Machine loop. (b) Raised hook. (c) V-hook.

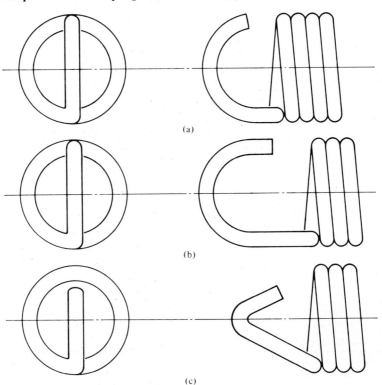

(a)

(b)

(c)

Stress Correction Factor

The stress correction factor can be computed using the following formula developed by Dr. A. M. Wahl:

$$K = \frac{4C - 1}{4C - 4} + \frac{0.615}{C},$$

where

C = spring index = D_m/d,

D_m = mean diameter of coil (in.),

d = wire diameter (in.).

Table 14-1 lists stress correction factors for various spring-index values. The preceding formula can be applied to determine any not shown in the table.

Graphical symbols are frequently used in describing coil springs to avoid the trouble of making an elaborate sketch. Figure 14-6 shows the symbols used to represent an extension spring, a compression spring, and a torsion spring. Spring data are frequently listed adjacent to the symbol in using such sketches.

 Helical-coil Spring Equations
(Round Wire)

The following symbols are used in the equations for helical-coil springs:

P = load (lb),

D_m = mean diameter of coil (in.),

d = wire diameter (in.),

C = spring index = D_m/d,

c = distance from centroid to edge of wire (in.),

J = polar moment of inertia (in^4),

K = stress correction factor,

Table 14-1
Values of Wahl Stress Correction Factors for Round-Wire Extension or Compression Springs

$C = D_m/d$	4	5	6	7	8	9	10	11	12
K (factor)	1.40	1.30	1.25	1.21	1.19	1.16	1.14	1.13	1.12

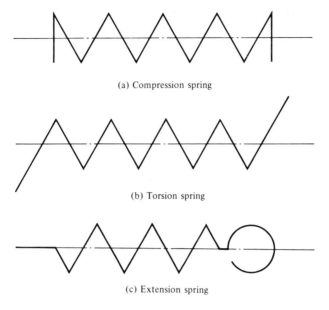

(a) Compression spring

(b) Torsion spring

(c) Extension spring

Figure 14-6
Graphical symbols for springs.

Δ = deflection (in.),

k = spring constant = P/Δ,

G = shear modulus of elasticity (psi),

S_s = shearing stress (psi),

n = number of active coils (coils supporting the load).

Helical coil springs are usually made of round wire and are subjected to a torsional shear stress, a transverse shear stress, and an additional stress caused by the curvature of the wire. The latter two stresses can be corrected by a stress factor applied to the torsional shear stress. If coil springs have static loading only, a factor of safety of 1.5 is generally applied to the torsional yield stress. If the loading is not static, the allowable stress is based on the endurance strength of the material. Several factors must be considered in the case of fatigue loading, such as stress concentration, corrosion fatigue, and surface finish. These factors often affect the life of springs used in such applications as garage doors and appliances such as washing machines and dishwashers.

A load applied to a helical spring must be resisted by the wire. Thus the product of the load and the mean radius of the coil must be equal to the resisting moment of the wire:

$$\frac{PD_m}{2} = \frac{S_s J}{c} = \frac{S_s \pi d^3}{16}.$$

Using the stress correction factor previously discussed, we obtain

$$\frac{PD_m}{2} = \frac{S_s \pi d^3}{16K}.$$

This can be simplified and expressed in terms of P or d as follows:

$$P = \frac{S_s \pi d^3}{8KD_m} \quad \text{or} \quad d = \sqrt[3]{\frac{8KPD_m}{\pi S_s}}. \tag{14-1}$$

An approximate equation for deflection is given as follows:

$$\Delta = \frac{8PD_m{}^3 n}{Gd^4}. \tag{14-2}$$

In the preceding equations, the polar moment of inertia for round wire was used. Similar equations can be developed for square or rectangular wire by using the (J/c)-values for the desired cross section.

The term *spring constant* or *gradient* is useful in comparing springs. It is defined as load divided by deflection:

$$k = \frac{P}{\Delta} = \frac{Gd}{8C^3 n}. \tag{14-3}$$

 ## Series and Parallel Systems

Springs can be used in series or parallel in a manner similar to electrical components. The following equations can be used to compute an "equivalent" spring to take the place of several in series or parallel. These equations can be combined to compute a series-parallel system under certain circumstances. For the series system,

$$\Delta = \Delta_1 + \Delta_2 + \Delta_3$$

and

$$k_e = \frac{1}{(1/k_1) + (1/k_2) + (1/k_3) + \cdots}, \tag{14-4}$$

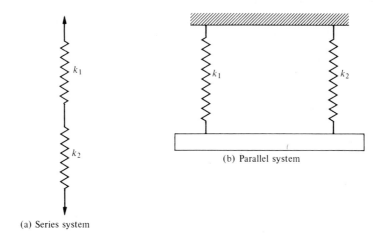

(b) Parallel system

(a) Series system

Figure 14-7
Spring systems.

where

k_e = spring constant (P/Δ) for equivalent spring.

Refer to Fig. 14-7 for series and parallel systems. For the parallel system,

$$\frac{P}{\Delta} = k_e = k_1 + k_2 + k_3 + \cdots \tag{14-5}$$

The following examples should explain how the above equations are used.

◆ *Example* ──────────────────────────

A helical-coil spring has a mean coil diameter of 1 in. and a wire diameter of $\frac{1}{8}$ in. If the shearing stress is 60,000 psi, how much load can it support? How many coils would be needed to restrict the deflection to 1 in. under that load if the shear modulus of elasticity is 12.5×10^6?

Solution.

$$C = \frac{D_m}{d} = \frac{1}{\frac{1}{8}} = 8,$$

$$K = \frac{4C - 1}{4C - 4} + \frac{0.615}{C} = \frac{4(8) - 1}{4(8) - 4} + \frac{0.615}{8} = 1.19.$$

Thus, the spring can support a load

$$P = \frac{S_s \pi d^3}{8KD_m} = \frac{60,000(\pi)(\frac{1}{8})^3}{8(1.19)(1)} = 38.7 \text{ lb.}$$

The number of coils required is obtained as follows:

$$\Delta = \frac{8PD_m^{\,3}n}{Gd^4},$$

$$n = \frac{\Delta Gd^4}{8PD_m^{\,3}} = \frac{1(12,500,000)(\frac{1}{8})^4}{8(38.7)(1)^3}$$

$$= 9.85 \quad \text{or} \quad 10 \text{ coils (active).}$$

Three-quarters of a coil should be added to each end to allow for grinding flat. Thus

$$10 + \tfrac{3}{4} + \tfrac{3}{4} = 11.5 \text{ coils.}$$

◆ *Example* _____

(Refer to Fig. 14-7.) Two extension springs are hooked in series and support a load of 100 lb. One spring has a constant of 50 lb/in. and the other a constant of 100 lb/in. What is the deflection of the load? If these springs were placed in parallel, what would be the deflection?

Solution.

$$k_e = \frac{1}{(1/k_1) + (1/k_2)} = \frac{1}{(1/50) + (1/100)} = 33.3;$$

$$k_e = \frac{P}{\Delta}.$$

Therefore,

$$\Delta = \frac{P}{k_e} = \frac{100}{33.3} = 3 \text{ in.}$$

The deflection for parallel conditions is found as follows:

$$k_e = k_1 + k_2 = 50 + 100 = 150;$$

$$k_e = \frac{P}{\Delta}.$$

Therefore,

$$\Delta = \frac{P}{k_e} = \frac{100}{150} = 0.667 \text{ in.}$$

◆ *Example* _____

A coil spring is to have a spring index of 6 and is to deflect $\frac{1}{2}$ in. under a load of 50 lb. The shear modulus of elasticity is 12,000,000 psi and the spring has 12 active coils. Find the wire diameter and the mean diameter of the coil.

Solution.

$$\Delta = \frac{8PD_m{}^3n}{Gd^4} = \frac{8P(6d)^3n}{Gd^4} = \frac{8P(216)d^3n}{Gd^4} = \frac{1728Pn}{Gd} ;$$

therefore,

$$d = \frac{1728Pn}{G\Delta} = \frac{1728(50)(12)}{\frac{1}{2}(12{,}000{,}000)} = 0.173 \text{ in.}$$

and

$$D_m = 6d = 6(0.173) = 1.038 \text{ in.} \qquad \text{or} \qquad 1\tfrac{1}{16} \text{ in.}$$

◆ *Example* ─────────────────────────────

Find the wire diameter, mean coil diameter, and number of active coils for a compression coil spring that must meet the following conditions:

a) force of 225 N should deflect 125 mm,
b) spring index = 8,
c) shear modulus of elasticity = 78 GN/m²,
d) shearing stress = 400 MN/m².

Solution.

$K = 1.19$ from chart or formula,

$$d = \sqrt[3]{\frac{8KPD_m}{\pi S_s}} = \sqrt[3]{\frac{8(1.19)(225)(8d)}{\pi(400)(10)^6}} ,$$

$$d = \sqrt{\frac{8(1.19)(225)(8)}{\pi(400)(10^6)}} = 0.003\,69 \text{ m or } 3.69 \text{ mm,}$$

$$D_m = 8d = 8(0.003\,69) = 0.0295,$$

$$n = \frac{\Delta Gd^4}{8PD_m{}^3} = \frac{0.125(78)(10^9)(0.003\,69)^4}{8(225)(0.0295)^3} ,$$

$n = 39.1$ active coils; to this, 1.5 coils should be added since this is a compression spring.

───

Spring Constant

The term *spring constant* is commonly used in comparisons between springs. It is defined as the ratio of the load to the deflection. The usual units are pounds for the load and inches for the deflection; thus the spring constant is expressed in pounds per inch. However, any suitable units of force could be used, such as ounces, grams, etc. For the deflection, any desirable units of distance could be used, such as centimeters, millimeters, etc. The units for the spring constant would be altered accordingly.

Synonyms of the term "spring constant" include (1) spring gradient, (2) spring scale, and (3) spring rate. These terms are used interchangeably in the problems at the end of the chapter to familiarize the student with these similar expressions. A careful check of the units should clarify the term.

 Impingement of a Freely Falling Body on Springs

Springs are sometimes used to absorb the momentum of a freely falling body. A typical example of this application is the springs located at the bottom of an elevator to absorb the shock in case of cable failure. If a body falls freely for a distance, then strikes and deflects a spring, it will be resisted by the force of the spring times one-half the spring deflection, as follows:

$$W(d + \Delta) = \frac{P\Delta}{2},$$ (14-6)

where

W = weight (lb),

d = free distance (in.),

Δ = spring deflection (in.),

P = spring force (lb).

◆ *Example* ———————————————————————

A weight of 800 lb falls freely for a distance of 40 in., then strikes and deflects a spring (coil) 10 in. The maximum induced stress is 50,000 psi and the shear modulus of elasticity is assumed to be 12,000,000 psi. Find the wire diameter, the mean coil diameter, and the number of active coils. Assume that the spring index is 7.

Solution.

$$P = \frac{2W(d + \Delta)}{\Delta} = \frac{2(800)(40 + 10)}{10} = 8000 \text{ lb};$$

$$K = \frac{4C - 1}{4C - 4} + \frac{0.615}{C} = \frac{4(7) - 1}{4(7) - 4} + \frac{0.615}{7} = 1.21;$$

$$d = \sqrt[3]{\frac{8KPD_m}{\pi S_s}} = \sqrt[3]{\frac{8(1.21)(8000)(7d)}{\pi(50,000)}}$$

$$= \sqrt[3]{\frac{8(1.21)(8000)(7)}{\pi(50,000)}} = \sqrt{3.46} = 1.86 \text{ in.};$$

$$D_m = 7d = 7(1.86) = 13.02 \text{ in.};$$

$$n = \frac{Gd^4\Delta}{8PD_m^3} = \frac{10(12,000,000)(1.86)^4}{8(8000)(13.02)^3}$$

$$= 10.2 \text{ active coils.}$$

 Leaf Springs

Leaf springs (single leaf) can be used in the cantilever or semielliptical form. Figure 14-8 shows a cantilever leaf spring, which is merely a simple cantilever beam. The stress is thus

$$S = \frac{M}{I/c} = \frac{6M}{bt^2} = \frac{6PL}{bt^2}$$

and the deflection is

$$\Delta = \frac{PL^3}{3EI} = \frac{4PL^3}{bt^3E} = \frac{2SL^2}{3tE},$$

where

b = leaf width (in.),

t = leaf thickness (in.),

S = stress (psi).

Figure 14-8
Single-leaf cantilever spring.

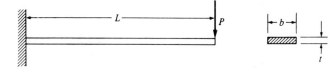

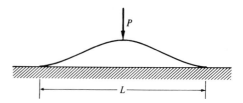

Figure 14-9
Single-leaf semielliptical spring.

The single-leaf semielliptical spring (Fig. 14-9) can be treated as a simple beam loaded at the center. The stress is thus

$$S = \frac{M}{I/c} = \frac{M}{bt^2/6} = \frac{PL/4}{bt^2/6} = \frac{3PL}{2bt^2}$$

and the deflection is

$$\Delta = \frac{PL^3}{48EI} = \frac{PL^3}{4bt^3E} = \frac{SL^2}{6tE}.$$

If a triangular plate is used as shown in Fig. 14-10, uniform stress will prevail throughout the plate. Figure 14-11 shows this same plate, cut and stacked to form a graduated-leaf cantilever spring. Thus

$$S = \frac{6PL}{nbt^2}, \tag{14-7}$$

$$\Delta = \frac{6PL^3}{nbt^3E}, \tag{14-8}$$

Figure 14-10
Triangular plate.

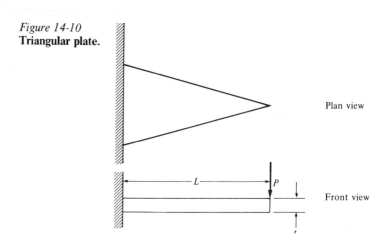

Plan view

Front view

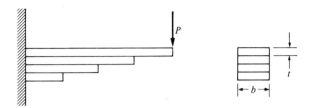

Figure 14-11
Graduated leaves.

where

n = number of leaves.

Semielliptical Leaf Springs

A flat leaf spring can be designed as if it were equivalent to a flat trapezoidal plate, as shown in Fig. 14-12. To save space, however, the plate is sliced into strips, stacked, and banded together in the middle with U-bolts (or some other device). Using deflection formulas for beams with a trapezoidal section, Professor J. B. Peddle derived an equation for the spring load of a semielliptical spring as

$$P = \frac{Snbt^2}{3L}$$

(14-9)

and the deflection as

$$\Delta = \frac{2L^2SK}{tE},$$

(14-10)

Figure 14-12
Flat leaf spring.

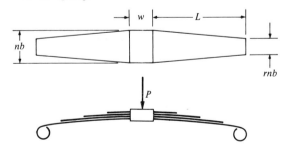

where

  r = number of full-length leaves / total number of leaves,

  w = width of band (in.),

  n = number of leaves.

(For L, see Fig. 14-12.)

In the preceding equation, the value of the constant K can be found from the following equation:

$$K = \frac{1}{(1-r)^3}\left[\frac{1-r^2}{2} - 2r(1-r) + r^2\log_e\frac{1}{r}\right].$$

For a full-elliptical spring, the load equation would be the same, but the deflection value would be doubled. Leaves that are less than full length are usually pointed somewhat and thinned at the ends. The band holding the leaves together deactivates the middle portion of the spring. Sometimes U-bolts are used; they have the same effect as a band. If the leaves are bolted together through the center, the bolt hole weakens the spring; the width of the bolt hole must be considered in making any calculations.

Flexure is the principal stress encountered in this type of spring. However, in automotive applications, swaying can cause additional flexure stresses, and rapid accelerations can cause compression by "driving" through the rear springs.

◆ *Example* ───────────────────────────────

A semielliptical spring is to sustain a load of 800 lb. It consists of one full-length leaf and seven graduated leaves. The leaf width is 2 in. The total length of the spring is 42 in. and U-bolts at the center deactivate 2 in. of length. The allowable stress is 50,000 psi and the modulus of elasticity is 30,000,000 psi. Find the thickness of the leaves and the deflection.

Solution.

$$t = \sqrt{\frac{3PL}{Snb}} = \sqrt{\frac{3(800)(20)}{50,000(8)(2)}} = \sqrt{0.06} = 0.245 \text{ in.};$$

$$K = \frac{1}{(1-r)^3}\left[\frac{1-r^2}{2} - 2r(1-r) + r^2\log_e\frac{1}{r}\right]$$

$$= \frac{1}{\left(1-\frac{1}{8}\right)^3}\left[\frac{1-\left(\frac{1}{8}\right)^2}{2} - 2\left(\frac{1}{8}\right)\left(1-\frac{1}{8}\right) + \left(\frac{1}{8}\right)^2\log_e\frac{1}{\frac{1}{8}}\right]$$

$$= \frac{1}{\left(\frac{7}{8}\right)^3}\left[\frac{\frac{63}{64}}{2} - 2\left(\frac{1}{8}\right)\left(\frac{7}{8}\right) + \frac{1}{64}\log_e 8\right]$$

$$= 1.49[0.492 - 0.219 + 0.0325] = 0.455;$$

$$\Delta = \frac{2L^2SK}{tE} = \frac{2(20)^2(50,000)(0.455)}{(0.245)(30,000,000)} = 2.48 \text{ in.}$$

In all probability, the 0.245-in. thickness would be rounded off to 0.250 in.; this in turn would give a slightly different value for the deflection.

◆ *Example* ────────────────────────────────

A semielliptical leaf spring is used on a 2-wheel trailer. The spring assembly has one full-length leaf and three graduated leaves. The total length between supports is 1.2 m; each leaf has a width of 80 mm and a thickness of 12 mm. U-bolts are used at the center of the assembly, spaced 100 mm apart. If the allowable stress is 330 MN/m² and the modulus of elasticity is 207 GN/m², find the allowable load and the deflection.

Solution.

$$L = 1.2 - 0.1 = 1.1 \text{ m} \qquad \text{and} \qquad r = 1/4;$$

$$P = \frac{Snbt^2}{3L} = \frac{330 \times 10^6(4)(0.080)(0.012)^2}{3(1.1)}$$

$$P = 4608 \text{ N} \qquad \text{or} \qquad \frac{4608}{9.81} = 469.7 \text{ kg};$$

$$K = \frac{1}{(1-r)^3}\left[\frac{1-r^2}{2} - 2r(1-r) + r^2\log_e\frac{1}{r}\right]$$

$$K = \frac{1}{\left(1-\frac{1}{4}\right)^3}\left[\frac{1-\left(\frac{1}{4}\right)^2}{2} - 2\left(\frac{1}{4}\right)\left(1-\frac{1}{4}\right) + \left(\frac{1}{4}\right)^2\log_e\frac{1}{\frac{1}{4}}\right]$$

$$K = 0.428;$$

$$\Delta = \frac{2L^2SK}{tE} = \frac{2(1.1)^2(330 \times 10^6)(0.428)}{0.012(207 \times 10^9)} = 0.138 \text{ m.}$$

────────────────────────────────

When a semielliptical or full-elliptical spring is constructed, some of the shorter leaves are sometimes designed with a different radius of curvature than the main leaf. This causes continuous contact and prevents dirt and grit from entering. Additional clips (rebound clips) are sometimes used in addition to the central band.

Torsion Bars

A simple, yet highly useful, member of the spring family is the torsion bar. Figure 14-13 shows this device in its most elementary form. It is used in a wide variety of applications, from miniaturized components in toys to heavy-duty automotive uses. The spring constant, k, for the torsion bar is the ratio of the torque to the angular deflection. Thus, for a solid round torsion bar,

$$k = \frac{T}{\theta} = \frac{T}{TL/GJ} = \frac{GJ}{L} = \frac{\pi D^4 G}{32L},$$ (14-11)

where

L = length (in.),

G = shear modulus of elasticity (psi),

D = diameter (in.),

J = polar moment of inertia (in^4),

T = torque (in - lb),

θ = angular deflection (rad).

Figure 14-13
Torsion bar.

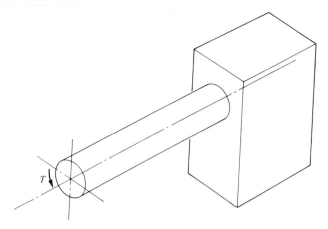

Similar equations can be developed for any cross section by substituting the appropriate relationship for the polar moment of inertia.

◆ *Example* ─────────────────────────────────

A round steel torsion bar is used in a manner similar to that of the bar shown in Fig. 14-13. The bar is $\frac{1}{4}$ in. in diameter and a torque of 150 in-lb is applied 18 in. from the fixed end of the bar. Find the deflection in degrees and the spring constant in in-lb/rad, if the shear modulus of elasticity is 12,000,000 psi.

Solution.

$$\theta = \frac{TL}{GJ} = \frac{32TL}{\pi GD^4} = \frac{32(150)(18)}{\pi(12,000,000)(0.25)^4} = 0.587 \text{ rad};$$

$$\frac{0.587(180)}{\pi} = 33.6 \text{ deg}; \qquad k = \frac{150}{0.589} = 256 \text{ in-lb/rad}.$$

Torsion Springs

Figure 14-14 illustrates a torsion spring. In this type of spring, the force that the spring exerts when deflected is along a circular arc, as is the deflection. The force produces a torque whose value depends on the magnitude of the force and the radial distance from the center of the coil to the point of application. The term *torsion* in a torsion spring refers to the action that the spring produces, not to the stresses within the spring wire. Stresses within the spring wire are flexural, not torsional stresses. (Torsional stresses are present in both extension and compression springs.) Torsional springs are often used in hinges. Of necessity, such springs must fit over a pin or shaft. Sufficient clearance must be allowed both radially and axially. Provision for axial expansion must be made, since more coils appear as the spring is loaded. Radial clearance must be large, since the coil decreases in size when the spring is under load. The load must be applied such that the size of the coil decreases. Though torsion springs can be made with either round or square wire, round wire is preferable. Double torsion springs can be made, but it is more economical to use two single torsion springs of the opposite hand.

Major spring manufacturers produce stock torsion springs in a number of different sizes and with various configurations for the ends where the

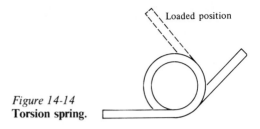
Loaded position

Figure 14-14
Torsion spring.

load is applied. Typical catalog information includes:

wire diameter (in., mm),

outside diameter of coil (in., mm),

position of ends,

deflection (90°, 180°, 270°, 360°),

torque rating (in.-lb, kg-mm or N · m),

radius for applying load (in., mm),

suggested mandrel sizes (in., mm),

protrusion of free ends from centerline (in., mm),

minimum axial space (in., mm).

The rated load can be found by dividing the torque value by the suggested radius where the load should be applied. Torque values are categorized by the amount of the deflection. Intermediate deflections can be prorated. For example, if the torque listing is 9.45 kg-mm for a 90-degree deflection, the value at 45 degrees would be 4.73 kg-mm.

 **Garter Springs**

A special type of extension spring is the garter type, similar to a rubber band holding a set of rolled drawings. One end of a garter spring is formed with smaller coils to screw into the opposite end, forming a garter or circle. This connection is sometimes made by using a short section of a smaller size (diameter and length) as a connecting link. This type of spring is frequently used as the flexible connector in a motion-picture projector. Another application is its use in mechanical oil seals. In this use, the inside diameter of the garter is expanded to fit over a shaft of larger diameter. This stretches the spring, causing a tensile load on the spring and a compressive force against the shaft. This force is transmitted where each coil contacts the shaft. Thus, the force per coil depends on the spring constant (or gradient) and on the number of coils pushing against the shaft. See Fig. 5-13 for a typical application.

 Belleville Springs

Figure 14-15(a) illustrates a special form of spring washer known as the Belleville spring. This type of spring is popular where space is at a premium and the desired deflection is not too great. Belleville springs can be used singly or stacked in three distinct manners. Figure 14-15(b) shows a series stack. Stacking in series increases the deflection in proportion to the number of washers, but the load remains the same as that for a single washer.

Figure 14-15
(a) Belleville spring; (b) series stack; (c) parallel stack; (d) stress constants for Belleville spring washer calculations.
(Courtesy of Spring Manufacturers' Institute, Inc.).

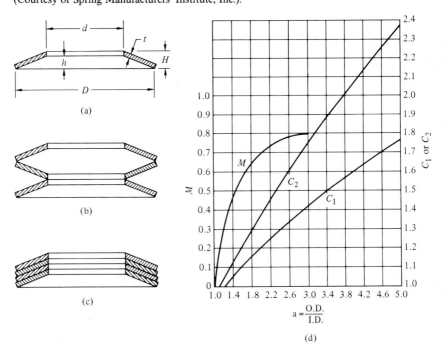

Parallel stacking is shown in Fig. 14-15(c). Parallel stacking increases the load capacity approximately in proportion to the number of washers. However, friction between the washers changes the load-deflection characteristics. The amount of loss is approximately 5% to 10% per washer. Combination series and parallel stacking can also be used.

Belleville washers (springs) must either be installed with a rod through the holes, or the stack must operate within a cylinder. Guides should be hardened, whether internal or external. Plates bearing against the top and bottom of the stack must also be hardened to prevent damage to adjacent parts.

Spring manufacturers provide this type of spring as a stock item. The dimensions shown in Fig. 14-15(a) are expressed in inches or in millimeters. Loads to compress the washer to the flat position ($h = 0$) are given in either pounds or newtons. The following example is given to show the load-deflection characteristics of series and parallel stacks.

◆ *Example* ─────────────────────────────

A Belleville washer has an inside diameter of 3.2 mm and an outside diameter of 8.0 mm. A force of 126 newtons is required to compress the spring 0.55 mm to the flat position. Assuming that six washers are used in series, find the deflection and load capacity for $h = 0$. If the same six washers are stacked in parallel, determine the deflection and the *approximate* load capacity.

Solution.

Series	*Parallel*
Deflection = 6(0.55) = 3.30 mm	Deflection = 0.55 mm
Load = 126 N	Load = (approx.) 6(126)
	= 756 N

Note that the 756 newton answer cannot be correct, since there are frictional losses of 5% to 10% per washer.

───

Often, these types of springs are not compressed to the flat position, but are deflected to 40% of this maximum value. Load values are adjusted accordingly.

The following design formulas, together with the stress constants chart for spring washer calculations, are furnished courtesy of the Spring Manufacturers Institute and are listed as follows:

$$P = \frac{4E\Delta}{M(1 - \mu^2)(D)^2}\left[\left(h - \frac{\Delta}{2}\right)(h - \Delta)t + t^3\right], \tag{14-12}$$

$$S = \frac{4E\Delta}{M(1 - \mu^2)(D)^2}\left[C_1\left(h - \frac{\Delta}{2}\right) + C_2 t\right], \tag{14-13}$$

where

P = load (lb),

S = stress at convex inner edge (psi),

Δ = deflection (in.),

μ = Poisson's ratio (0.3 for steel),

t = thickness (in.),

h = dimension shown in Fig. 14-15 (a),

$a = \dfrac{D}{d}$,

$M = \dfrac{6}{\pi \log_e a}\left(\dfrac{(a - 1)^2}{a^2}\right),$

$C_1 = \dfrac{6}{\pi \log_e a}\left(\dfrac{a - 1}{\log_e a} - 1\right),$

$C_2 = \dfrac{6}{\pi \log_e a}\left(\dfrac{a - 1}{2}\right).$

The following example shows how to determine the load and the induced stress for a given deflection.

◆ *Example* ────────────────────────────────────

A steel Belleville spring washer has the following dimensions: $D = 8$ in., $d = 4$ in., $t = 0.1$ in., $h = \frac{3}{16}$ in., and $H = \frac{1}{4}$ in. For a deflection equal to 40% of the flat position, calculate the load and the stress at the inner edge. Calculate values for the constants, then check your results against the chart in Fig. 14-15 (d).

Solution.

$\Delta = 0.4\,(0.1875) = 0.075$ in.;

$a = \dfrac{D}{d} = \dfrac{8}{4} = 2;$

$$C_1 = \frac{6}{\pi \log_e a}\left(\frac{a - 1}{\log_e a} - 1\right) = \frac{6}{\pi(0.693)}\left(\frac{2 - 1}{0.693} - 1\right) = 1.221;$$

$$C_2 = \frac{6}{\pi \log_e a}\left(\frac{a - 1}{2}\right) = \frac{6}{\pi(0.693)}\left(\frac{2 - 1}{2}\right) = 1.378;$$

$$M = \frac{6}{\pi \log_e a}\left[\frac{(a - 1)^2}{a^2}\right] = \frac{6}{\pi(0.693)}\left[\frac{(2 - 1)^2}{2^2}\right] = 0.689;$$

$$P = \frac{4E\Delta}{M(1 - \mu^2)D^2}\left[\left(h - \frac{\Delta}{2}\right)(h - \Delta)t + t^3\right]$$

$$= \frac{4(30 \times 10^6)(0.075)}{0.689(1 - 0.3^2)(8)^2}$$

$$\times \left[\left(0.1875 - \frac{0.075}{2}\right)(0.1875 - 0.075)(0.1) + 0.1^3\right]$$

$$= 602 \text{ lb};$$

$$S = \frac{4E\Delta}{M(1 - \mu^2)D^2}\left[C_1\left(h - \frac{\Delta}{2}\right) + C_2 t\right]$$

$$= \frac{4(30 \times 10^6)(0.075)}{0.689(1 - 0.3^2)(8)^2}[1.221(0.1875 - 0.0375) + 1.378(0.1)]$$

$$= 72,000 \text{ psi}.$$

In Belleville spring washers, the h/t ratio has an important bearing on the deflection. In the preceding example, the h/t ratio was 2.5. Spring washers with h/t ratios greater than 1.2 should not be used in series. Characteristics of h/t ratios are:

1. constant spring rate: $h/t < 0.4$;
2. nearly constant rate: $h/t < 0.8$;
3. positive decreasing rate: $0.4 < h/t > 1.41$;
4. zero rate over part of deflection: $h/t = 1.5$;
5. positive decreasing rate that becomes zero, then negative and increasing before bottoming out: for $1.4 < h/t > 2.83$.

Other variations of spring washers include instrument types that are light in construction and therefore have small load values. One style is curved in one direction; another is reverse curved to produce a wave appearance.

Constant-force Spring

Figure 14-16 illustrates the four forms available of the Neg'ator® spring; a patented device that exerts a constant force throughout the complete working deflection. Proper prestressing of the strip of spring metal makes this type possible. Some typical applications include (1) maintaining constant pressure for brushes in electric motors, (2) feeding stock in a vending machine, and (3) controlling the force of a cam roller (follower) against the cam. The motor type is used to retract hose reels and steel measuring tapes; multiple motor springs can serve as counterweights in many operations,

Figure 14-16
(a) Neg'ator® extension spring. (b) Neg'ator® motor. (c) Neg'ator® clamp.
(d) Neg'ator® clip.
(Courtesy of AMETEK/Hunter Spring Div.)

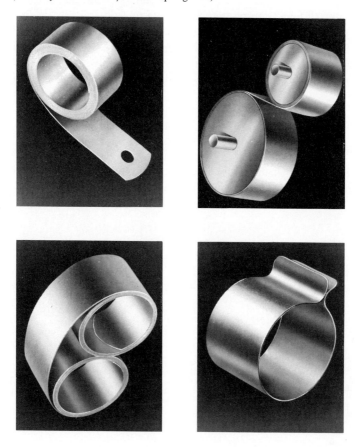

their major advantage being the small amount of space they occupy. Clamps can be used to clamp varying thicknesses of sheets over a wide range. Clips can be used as retaining bands; they are easy to remove. Neg'ator® springs are used in many industrial applications. Design and application data are available from the manufacturer.

 **Volute Springs**

If a decreasing rate of deflection per unit of load is desired, one can use a conical-coil spring made of round wire or a volute spring. The volute spring is made from rectangular wire with the long dimension parallel to the spring axis. It is made in such a manner that each coil fits inside the preceding one. Figure 14-17 shows a volute spring made with rectangular wire. The solid height of the spring is the width of the metal strip from which it is made. This type of spring is usually made so that each coil contacts the preceding one; the sliding friction that results damps out vibrations in the spring. Another outstanding advantage of this type is its compactness.

Stress distributions within the spring are not uniform; this shortens the spring life. In addition, the rubbing action between the coils can cause galling; this produces stress concentration factors that can lead to early failure.

 Retaining Rings

Retaining rings (sometimes called snap rings) are important members of the spring family; they also belong in the fastener category. They are produced with rectangular, square, or circular cross sections and in two general types,

Figure 14-17
Volute spring.

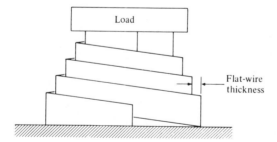

external and internal. The ring is designed to fit (snap) into a circumferential groove. The external type fits into a groove on the outside of a shaft-like member. The internal type fits into a groove cut into the inside of a cylindrical surface.

Retaining rings are inexpensive and can be purchased "off the shelf." Snapping the rings into position is also an inexpensive operation, and their removal is easy. Certain types are merely forced into position with the hands; others require special plier-like tools. However, preparing the groove for the rings does involve additional expense. When the rings are snapped into place, axial movement of the "locked-on" component is prevented. Retaining rings can be used to replace the following fasteners: (1) shoulders (external or internal), (2) collars used in conjunction with setscrews, and (3) cotter pins used with washers.

Calculations for snap rings involve the theory of curved beams, which is beyond the scope of this text. Simplified charts and constants can be found in many spring handbooks. Commercially produced rings are rated in terms of allowable axial thrust; in addition, the manufacturers supply complete recommendations for groove sizes and tolerances. Retaining rings lend themselves well to mass production applications and are therefore extremely popular in these areas. The range of available sizes is large; hence, axial load limitations range from very small to very large values. Figure 6-23 shows two types of retaining rings.

Power Springs

One of the older types of spring is the power spring, frequently called a *clock spring*. It is usually a flat-coil spring. Figure 14-18 shows a typical arrangement. One end of the spring is wound around a shaft; the other end is usually secured to the inside of a retaining drum. In the unwound position, most of the spring rests against the drum. Energy stored after winding can be released as mechanical torque, either through the shaft or through the drum. The drum is often part of a gear. Power springs are inexpensive and convenient to operate.

Ordinary clock springs are power springs. Many toys are powered by this type of spring. Power lawn mowers equipped with recoil (rewinding) starters usually use this type; when one pulls on the starter cord, one is rewinding the spring as well as turning the engine shaft. Other small gasoline engines, such as outboard motors, portable generators, and chain saws, use this principle. Most parking meters use this spring mechanism to assist the timing cycle.

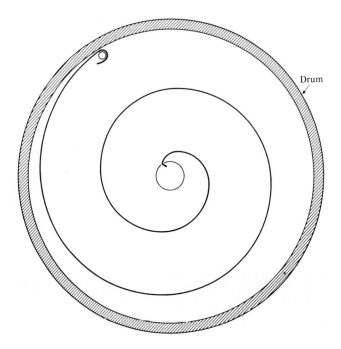

Drum

Figure 14-18
Power spring.

Power springs are also used in such applications as movie cameras and take-up reels such as those found for lubrication hoses in gasoline service stations. The torque varies throughout the entire range; the greatest torque is produced when the spring is fully wound. Sometimes this is an asset, particularly when the initial torque needed to move a resistance is large. The following equation is used in calculations for this type of power spring:

$$M = T = S\frac{I}{c} = S\frac{bt^2}{6},$$

where

    $t =$ thickness (in.),

    $b =$ breadth or width (in.).

This equation is based on a rectangular cross section, the cross section most widely used.

    The shaft on which the spring winds should be carefully designed. If it is too small in diameter, the spring will be subjected to extremely high stress, with failure possibly resulting. If the shaft is too large, the drum will have to be enlarged to provide sufficient space for the spring; otherwise, potential torque will be sacrificed.

 Gradient Analysis

A number of spring-application problems center on determining the gradients or rates required to perform a given function. The basic procedure exploits the equilibrium of forces and reactions using methods similar to those used in finding beam reactions. Either mathematical or graphical methods can be used. The following problems show a few common applications in a step-by-step fashion.

◆ *Example*

A beam is supported by two helical-coil springs as shown in Fig. 14-19. A 200-lb load is applied at P and must deflect the entire beam $\frac{1}{2}$ in. Find the gradient for each of the supporting springs (A and B). Assume that the beam itself is rigid enough that it will have no deflection.

Solution. First, take moments about A:

$$6P = 24R_A \quad \text{or} \quad R_B = \frac{P}{4}.$$

Then, take moments about B:

$$18P = 24R_A \quad \text{or} \quad R_A = \frac{3P}{4}.$$

The gradient at A is then $3P/4$ divided by $\frac{1}{2} = 1.5(200) = 300$ lb/in. and the rate at B is $P/4$ divided by $\frac{1}{2} = \frac{200}{2} = 100$ lb/in. Note that this satisfies the equation for an equivalent spring at P. Thus $k_e = k_1 + k_2$ for springs in parallel; or $200/0.5 = 300 + 100$.

Figure 14-19
Beam with springs in parallel.

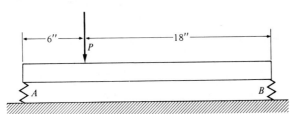

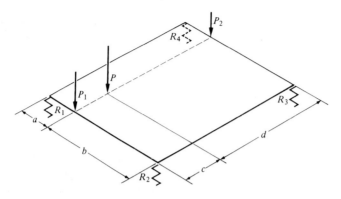

Figure 14-20
Plate with springs in parallel.

The inverse problem, which consists of supporting springs at *A* and *B* with different gradients, is also common. The problem then is to place an operating force *P* (or another spring) so that the beam will remain horizontal with varying values of *P*.

◆ *Example* ————————————————————————————————————

A rigid plate is supported by springs at the four corners. Load *P* is then applied at a position offset from the center of the plate. Find the rate or gradient for each of the compression springs, assuming that the plate must remain horizontal.

Solution. Since the plate must remain horizontal when a load is applied, the deflection for each spring must be equal to the plate deflection. The equation $k_e = k_1 + k_2 + k_3 + k_4$ thus applies; k_e would actually be the rate of an imaginary spring located at *P*. In such a problem, the reactions at the corners constitute the main part of the problem. To provide numerical values, assume the following data: $P = 1000$ lb, $a = 1$ ft, $b = 3$ ft, $c = 1$ ft, and $d = 4$ ft.

As a first step, it is desirable to resolve *P* into two components, P_1 and P_2, at the locations shown. Reactions can then be obtained at each of the four corners. The following equations can be written in terms of letters and numerical values:

$$P_1 = \frac{d}{c + d} P = \frac{4P}{5} = \frac{4(1000)}{5} = 800 \text{ lb},$$

$$P_2 = \frac{c}{c + d} P = \frac{P}{5} = \frac{1000}{5} = 200 \text{ lb},$$

$$R_1 = \frac{b}{a+b}\frac{d}{c+d}P = \frac{3(4)}{4(5)}P = \frac{3P}{5} = \frac{3(1000)}{5} = 600 \text{ lb,}$$

$$R_2 = \frac{a}{a+b}\frac{d}{c+d}P = \frac{1(4)}{4(5)}P = \frac{P}{5} = \frac{1000}{5} = 200 \text{ lb,}$$

$$R_3 = \frac{a}{a+b}\frac{c}{c+d}P = \frac{1(1)}{4(5)}P = \frac{P}{20} = \frac{1000}{20} = 50 \text{ lb,}$$

$$R_4 = \frac{b}{a+b}\frac{c}{c+d}P = \frac{3(1)}{4(5)}P = \frac{3P}{20} = \frac{3(1000)}{20} = 150 \text{ lb.}$$

The rate can then be found by dividing the reaction values by the appropriate deflection.

 Metrication in the Spring Industry

Typical metric information listed in commercial spring catalogs includes:

1. Compression springs
   a) Outside diameter, wire diameter, free length, length under load, and solid height (all designated in millimeters).
   b) Load (expressed in kilograms).
   c) Spring rate (expressed in kilograms per millimeter).

2. Extension springs
   a) Outside diameter, wire diameter, free length, and length under load (specified in millimeters).
   b) Load (denoted in kilograms).
   c) Spring rate (shown in kilograms per millimeter).

3. Torsion springs
   a) Wire diameter, outside diameter, radius at which load is applied, recommended mandrel size, and minimum axial space (shown in millimeters).
   b) Torque (specified in kilogram-millimeters).

Summary

Springs are widely used in the mechanical design field. Applications range from miniaturized assemblies to heavy-duty equipment. Selecting a spring rate or constant is not always easy. Sometimes, holding a load is more important than the deflection; in other cases, the deflection is the critical consideration. It is often prudent to obtain engineering samples of various springs and expose them to actual working conditions before making a final design decision. Selection of materials is also important; corrosion resistance or reactions with certain chemicals or gases could greatly influence the choice.

Springs serve as mechanical elements in two general categories: (1) actuation (or power) and (2) positioning. In the latter application, the spring exerts a force against another part; die springs are examples of this type.

Compression and extension springs are the springs most often used. Deflection is linear in these types. Torsion springs are characterized by angular instead of linear deflection. Flat leaf springs are of the simple beam or cantilever type; their deflection is linear.

The wire used in helical-coil springs can be round, square, or rectangular in cross section. Spring clutches use a rectangular wire to provide adequate radial surface contact. Snap rings also have a rectangular cross section; this provides a large area for axial surface contact with an adjoining part. Belleville washers and snap rings exert their forces through a bending moment.

The garter spring is a special type of extension spring designed to exert radial pressure against a circular part; such springs are found in oil seals. The constant-force spring is used in applications where a predetermined spring force is more important than the deflection; this type has a zero rate. Volute springs require a different force value as the spring reaches various stages of its deflection. Volute springs can have round, square, or rectangular wire cross sections. The hourglass-shaped spring used in cushions and bedsprings is a variation of the volute type.

Questions for Review

1. List three spring applications where the following types are utilized:
   a) compression coil spring (round wire), e) hairspring,
   b) compression coil spring (square wire), f) snap rings,
   c) extension coil spring, g) spring washers,
   d) torsion spring, h) power springs.

2. What advantages does square wire have over round wire in spring design? What disadvantages?

3. What items must be listed in specifying extension springs?

4. What items must be noted in specifying torsion springs?

5. List four applications where nonferrous materials are used in spring design. List also the probable reasons.

Problems

(*Note:* Rates for certain spring problems use S.I. units; others use kg/mm units to conform to spring manufacturer's practice.)

1. An extension coil spring is to elongate 5 in. under a load of 50 lb. The spring index is 8, the modulus of elasticity in shear is 11,500,000 psi, and the maximum induced stress is 90,000 psi. Find the spring gradient, the wire diameter, the mean coil diameter, and the number of active coils required.

2. A load of 50 lb deflects an extension coil spring 8.5 in. What load will deflect it 2.5 in.?

3. Three extension coil springs are hooked in series and support a weight of 150 lb. One spring has a spring rate of 30 lb/in. and the other two have spring rates of 20 lb/in. What will be the total deflection?

4. Four compression coil springs in parallel support a load of 800 lb. Each spring has a gradient of 40 lb/in. How much deflection results?

5. Design a compression coil spring to sustain 200 ft-lb of energy with a deflection of 3 in. Assume that the mean coil diameter is 7 times the wire diameter, the allowable stress is 100,000 psi, and the shear modulus of elasticity is 11,500,000 psi. Find the wire diameter, the mean coil diameter, and the number of active coils.

6. A weight of 100 lb strikes and deflects a compression coil spring. It falls freely for 6 in. and then deflects the spring 6 in. The spring index is 7, the allowable stress is 80,000 psi, and the shear modulus of elasticity is 11,500,000 psi. Find the wire diameter, the mean coil diameter, and the number of active coils.

7. Find the stress on a single-leaf cantilever spring, assuming the following specifications: thickness of leaf $= \frac{1}{8}$ in.; width of leaf $= \frac{1}{4}$ in.; load $= 5$ lb, applied at a distance of 10 in. from a support.

8. What would the deflection in Problem 7 be if the modulus of elasticity were 30,000,000 psi?

9. If the spring in Problem 7 were a graduated-leaf spring (cantilever) with 4 leaves, and all other conditions were the same, what would be the deflection and the induced stress? Assume that the modulus of elasticity is 30,000,000 psi.

10. A compression coil spring has 18 active coils. The wire diameter is 0.063 in. and the mean coil diameter is $\frac{3}{4}$ in. How much load would it take to deflect it $1\frac{1}{2}$ in., if the shear modulus of elasticity were 11,500,000 psi? What weight falling freely for 6 in. would deflect the spring $1\frac{1}{2}$ in.?

11. A semielliptical single-leaf spring similar to that shown in Fig. 14-9 carries a load of 80 lb at its center. Distance $L = 6$ in., the leaf thickness is $\frac{1}{8}$ in., and the width of the leaf is $\frac{1}{2}$ in. The modulus of elasticity is 30,000,000 psi. Find the deflection for this load and the induced stress.

12. A semielliptical leaf spring has one full-length leaf and 3 additional graduated leaves. The total spring length is 54 in.; a 4-in. band is used at the center. The cross section of a leaf is 0.2 in. × 2 in. How much stress is induced in this spring when it is deflected 1 in., if the modulus of elasticity is 30,000,000 psi?

13. Derive an equation for the spring constant of a semielliptical leaf spring with graduated leaves.

14. Derive an equation for the spring constant of a full-elliptical leaf spring with graduated leaves.

15. A semielliptical leaf spring has one full-length leaf and four additional graduated leaves. The leaves are $1\frac{1}{2}$ in. wide. The total spring length is 53 in.; U-bolt connections at the midpoint deactivate 5 in. of length. The spring supports a static load of 500 lb. The working stress is 60,000 psi, and the modulus of elasticity is 30,000,000 psi. Find the leaf thickness and the deflection.

16. A single-leaf cantilever spring has a width of $\frac{1}{2}$ in. and an effective length of 8 in. An 8-oz weight drops freely for 6 in., strikes the spring at the free end, and deflects it 2 in. What thickness of steel spring is necessary if the modulus of elasticity is 30,000,000 psi?

17. A load of 20 kg applied to an extension coil spring provides a deflection of 200 mm. What load will deflect the spring 60 mm?

18. Three extension coil springs are hooked in series and used to support a weight of 70 kg. One spring has a spring rate of 0.408 kg/mm and the other two have spring rates of 0.643 kg/mm. What is the total deflection?

19. Four compression coil springs in parallel support a load of 360 kg. Each spring has a gradient of 0.717 kg/mm. Find the deflection.

20. A semielliptical spring consists of one full-length leaf and four graduated leaves. The total length between supports is 700 mm with a 100-mm band in the center. Each leaf is 7 mm thick and 70 mm wide. A load of 8000 N is applied at the center. Using a stress value of 400 MN/m^2 and a modulus of elasticity of 207 GN/m^2, find the deflection caused by this loading.

21. The outer of two concentric helical compression springs has a rate of 4 N/mm; the inner one has a rate of 3 N/mm. The outer spring is 10

mm longer than the inner one. If the total force on the two springs is 270 N, what force does each spring carry?

22. Three helical-coil compression springs are used in series. A force of 100 newtons deflects the series 50 mm. The rates for two of the springs are 10 N/mm and 5 N/mm. What is the rate for the third spring?

23. An extension coil spring made with round wire is to deflect 25 mm under a load of 20 kg. Find the wire size, mean coil diameter, and active number of coils if the allowable shear stress is 400 MN/m^2 and the shear modulus of elasticity is 78 GN/m^2. Assume a spring index of 7.

24. A 6-mm round bar serving as a torsion bar experiences 12 N · m of torque applied 0.6 m from the fixed end. The transverse modulus of elasticity is 78 GN/m^2. Find the deflection in radians and the spring constant in N · m/rad.

FLYWHEELS

15

A flywheel is a rotating mechanical component designed to store energy and distribute it when needed—in effect, a reservoir of energy analogous to a water storage tower on the roof of a building. Its function is to counteract the fact that most machines do not receive and supply energy in a continuous process. For example, a single-cylinder gasoline engine with a four-stroke cycle supplies energy to the crankshaft only once every two revolutions. Thus the crankshaft gets one good "jolt" of energy every second revolution; the flywheel absorbs excess energy and redistributes it to the crankshaft. Another example of the use of a flywheel is the action of a shearing or punching machine; here, great amounts of energy are needed as the punch (or shear) penetrates the metal. In many pieces of construction equipment, the method of receiving power and using it will determine the need for a flywheel.

 Speed Fluctuations

Certain types of equipment maintain a somewhat stable velocity; others change velocity drastically under load. An ac generator will have a total speed range of plus or minus 0.2%, while a shearing machine will have a speed range of 5 to 15% below normal speed. The coefficient of fluctuation is obtained by writing the speed range in decimal form; thus, the decimal point is moved two places to the left to transform percentage values into fluctuation coefficients. A few typical values for these coefficients are found in Table 15-1. The equations for flywheel design will establish the weight of a flywheel rim depending on the amount of speed regulation desired.

The coefficient of fluctuation can be found from the following equation:

$$C_f = \frac{N_1 - N_2}{N} = \frac{v_1 - v_2}{v}, \tag{15-1}$$

Table 15-1
Typical Values of Fluctuation Coefficients

Type of Equipment	Coefficient
AC generators	0.002 to 0.003
Crushers	0.10 to 0.20
Compressors and pumps	0.03 to 0.05
Speed reducers	0.015 to 0.020

where

C_f = coefficient of fluctuation,

v_1 = maximum rim velocity (fps),

v_2 = minimum rim velocity (fps),

v = mean rim velocity (fps),

N_1 = maximum rotational speed (rpm),

N_2 = minimum rotational speed (rpm),

N = normal or operating rotational speed (rpm).

 Energy of Flywheel

The total kinetic energy of a flywheel is given by the equation

$$E = \frac{Wv^2}{2g},$$

where

W = weight of rim (lb),

g = acceleration of gravity = 32.2 ft/sec^2.

In a flywheel, the change in energy is the important consideration. The following equation is useful in flywheel design:

$$\Delta E - \frac{W}{2g}(v_1{}^2 - v_2{}^2), \tag{15-2}$$

where

ΔE = change in energy (ft-lb).

By factoring, this equation for the change in energy could be written as follows:

$$\Delta E = \frac{W}{2g}(v_1 + v_2)(v_1 - v_2).$$

Since $v = (v_1 + v_2)/2$,

$$v_1 + v_2 = 2v \quad \text{and} \quad C_f = \frac{v_1 - v_2}{v},$$

or

$$v_1 - v_2 = C_f v.$$

Then, by substitution,

$$\Delta E = \frac{W}{2g}(2v)(C_f v).$$

Therefore

$$\Delta E = \frac{W}{g} C_f v^2. \tag{15-3}$$

This equation is based on two assumptions. The first is that speed fluctuations are equally divided above and below the mean velocity. This is generally true except for such equipment as punches and shears, where the operating speed is usually the maximum velocity. In those cases, the preceding shortened equation would not be valid, and one would have to resort to the basic equation, which includes the $(v_1^2 - v_2^2)$ term. The second assumption is that the weight of the rim provides the only flywheel effect. Actually, the weight of the hub and arms contributes from 5 to 10% to this effect, depending on the construction of the flywheel. If we assume 10%, the equations for change in energy would be modified as follows:

$$\Delta E = \frac{1.10W}{2g}(v_1^2 - v_2^2)$$

and

$$\Delta E = \frac{1.10W}{g} C_f v^2.$$

The following examples show some flywheel calculations.

◆ *Example*

A shearing machine requires 1500 ft-lb of energy to shear a specified gauge of sheet steel. The mean diameter of the flywheel is to be 30 in. The normal operating speed of 200 rpm slows to 180 rpm during the shearing process. The rim width is to be 12 in. The weight of cast iron is 0.26 lb/cu in. Find the thickness of the rim, assuming that the hub and arms account for 10% of the rim weight concentrated at the mean diameter.

Solution.

$$v_1 = \frac{\pi DN}{12(60)} = \frac{\pi(30)(200)}{12(60)} = 26.2 \text{ fps},$$

$$v_2 = \frac{\pi DN}{12(60)} = \frac{\pi(30)(180)}{12(60)} = 23.6 \text{ fps},$$

$$\Delta E = \frac{1.10W}{2g}\left(v_1^{\,2} - v_2^{\,2}\right),$$

$$W = \frac{2g(\Delta E)}{1.10\left(v_1^{\,2} - v_2^{\,2}\right)} = \frac{2(32.2)(1500)}{1.10\left[(26.2)^2 - (23.6)^2\right]} = 678 \text{ lb.}$$

The volume of the rim is $678/0.26 = 2608$ cu in. If the cylindrically shaped rim were considered detached and "unwound" into a rectangular solid with a length equal to πD, the thickness of the rim could be found by dividing the volume by the product of this length and the width b of the rim. Thus,

$$t = \frac{\text{volume}}{bL} = \frac{2608}{12(30\pi)} = 2.31 \text{ in.}$$

◆ *Example*

A machine is to be equipped with a cast-iron flywheel 5 ft in diameter and 12 in. wide. The arms and hub of the flywheel are considered equivalent to 5% of the rim weight concentrated at the mean diameter. The change in energy to be handled is 2500 ft-lb and the coefficient of fluctuation is to be 0.03. The running speed of the flywheel is 250 rpm. Find the rim thickness.

Solution.

$$v = \frac{\pi DN}{12(60)} = \frac{\pi(60)(250)}{12(60)} = 65.4 \text{ fps,}$$

$$\Delta E = \frac{1.05W}{g}\left(C_f v^2\right),$$

$$W = \frac{g(\Delta E)}{1.05\left(C_f v^2\right)} = \frac{32.2(2500)}{1.05(0.03)(65.4)^2} = 597 \text{ lb;}$$

thus the rim volume is $597/0.26 = 2296$ cu in. and

$$t = \frac{\text{volume}}{bL} = \frac{2296}{12(60\pi)} = 1.02 \text{ in.}$$

 Flywheel Effect (WR^2)

The equation for the total energy (kinetic) of a flywheel can be written as

$$E = \frac{Wv^2}{2g} = \frac{WR^2\omega^2}{2g} = \frac{WR^2}{2g}\left(\frac{2\pi N}{60}\right)^2 = 0.000171N^2(WR^2), \qquad \text{(15-4)}$$

where

R = radius of gyration (ft),

WR^2 = flywheel effect (lb-ft^2).

This is a quite useful version of the equation, since many handbooks and manufacturers' catalogs tabulate values for WR^2. In using such tables, one should be certain of the units for WR^2. These units are sometimes given as lb-ft^2; in other cases the units are lb-in^2. In the preceding equation, the units are lb-ft^2. The radius of gyration can be found by using the expression $\sqrt{I/A}$; however, a careful dimensional analysis must be made to be certain of the proper units.

The WR^2-value can also be used in a simplified equation to find the torque required to accelerate (or decelerate) a flywheel uniformly from one rotational speed to another. This equation is as follows:

$$T = \frac{0.0391(N_2 - N_1)WR^2}{t}, \tag{15-5}$$

where

T = torque (in-lb),

t = time (sec) required to accelerate (or decelerate)
from N_1 rpm to N_2 rpm.

As in previous equations, the arms, hub, and shaft are often neglected, even though they contribute to the flywheel effect.

In terms of the inside and outside diameters of a flywheel rim (ignoring the effects of the arms and hub), the radius of gyration can be found from the following relationship:

$$R = \sqrt{8.68 \times 10^{-4}\left(D_o^2 + D_i^2\right)},$$

where

R = radius of gyration (ft),

D_o = outside diameter (in.),

D_i = inside diameter (in.).

◆ *Example* _____

A flywheel rim has an inside diameter of 20 in. and an outside diameter of 24 in. The weight of the rim is 215 lb. Neglecting any flywheel effect from the arms and hub, find the torque required to change the rotational speed from 150 rpm to 200 rpm in 10 sec.

Solution.

$$R = \sqrt{0.000868\left[(24)^2 + (20)^2\right]} = 0.920 \text{ ft};$$

$$T = \frac{0.0391(N_2 - N_1)WR^2}{10}$$

$$= \frac{0.0391(200 - 150)(215)(0.920)^2}{10}$$

$$= 35.6 \text{ in-lb.}$$

 ## Flywheel Construction

Flywheels vary in size; large ones can be as large as 20 ft in diameter. For convenience in casting and transporting, the larger sizes are often made in two or more sections. Links are then shrunk into place to ensure high efficiency of the joined sections. When extremely large flywheels are fabricated, multiple sections are used; the number of sections is usually determined by the number of arms used, since each section includes one arm. Small flywheels are usually cast in one piece. The hub diameter is usually 1.8 to 2 times the shaft diameter. Arms are usually elliptical in shape; often the major axis is twice the minor axis. Since the arms serve as cantilever beams, the major axis of the ellipse is placed in the plane of rotation to provide greater flexural strength.

 ## Flywheel Stresses and Hazards

If a flywheel is to be effective, the linear velocity of its rim should be rather high. With high speed—and particularly with large changes in speed—the flywheel is always in danger of bursting or exploding. Extreme care must be exercised to avoid dangerous situations because of the large centrifugal forces present. Speeds and stresses should be carefully checked. Hoop stresses in the rim can be found by considering the rim to be free of the supporting arms. Tensile stresses that tend to pull the two parts of the rim apart can then be calculated. Figure 15-1 shows the action of tensile loads and centrifugal force on one half of a flywheel rim, neglecting the effects of the arms.

Hoop Stress

If the thickness of the rim is relatively small with respect to the mean radius, the centroidal distance $\bar{r}$ can be taken as $2r/\pi$. The centrifugal force tending to separate one half of the rim from the other half is $W\bar{r}\omega^2/g$; this

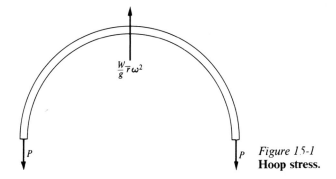

Figure 15-1
Hoop stress.

force is balanced by two P-forces as shown in Fig. 15-1. Thus, the following equations can be written:

$$2P = \frac{W}{g} \bar{r} \omega^2$$

or

$$2SA = \frac{W}{g} \left(\frac{2r}{12\pi} \right) \left(\frac{144v^2}{r^2} \right),$$

$$S = \frac{W}{2Ag} \left(\frac{2r}{12\pi} \right) \left(\frac{144v^2}{r^2} \right) = \frac{w\pi rbt}{2btg} \left(\frac{2r}{12\pi} \right) \left(\frac{144v^2}{r^2} \right),$$

$$S = \frac{12wv^2}{g}, \tag{15-6}$$

where

r = mean radius (in.),

$\bar{r} = 2r/\pi$,

W = total weight (lb) = $w\pi rbt$ for half the rim,

w = lb/cu in.,

v = mean velocity of rim (ft/sec),

ω = angular velocity (rad/sec),

t = rim thickness (in.),

b = rim breadth (in.),

S = stress (psi).

Since cast iron, one of the usual materials for flywheels, weighs 0.26 lb/cu in., this equation can be approximated by $S = v^2/10$.

Bending Stress

In our previous discussion, we assumed that the arms did not restrain the expansion of the rim. If we now assume that the arms do not expand, but allow the rim to extend outward between the arms, a section between the arm supports can be considered a beam fixed at both ends and subjected to a uniformly distributed load equal to the centrifugal force (see Fig. 15-2). The centrifugal force is given by the following equation:

$$P_c = \frac{12wtv^2}{(D/2)g}.$$

The beam span is the length of rim between the arm supports, which is equal to $\pi D/n$; D is the mean diameter of the rim expressed in inches, and n is the number of arms. The maximum moment occurs where the arms join the rim. The section modulus for a rectangular rim would be $bt^2/6$. Incorporating all of this in the flexure formula, we have

$$S = \frac{Mc}{I} = \frac{1}{12}\left[\frac{12wtv^2}{(D/2)g}\right](b)\left(\frac{\pi D}{n}\right)^2\left(\frac{6}{bt^2}\right),$$

$$S = \frac{12wv^2\pi^2D}{gn^2t}. \tag{15-7}$$

Accepted practice is to assume that the arms expand three-fourths of the total radial expansion of the rim. An equation can thus be written for this total stress by combining the hoop stress with the bending stress as follows:

total stress = $\tfrac{3}{4}$(hoop stress) + $\tfrac{1}{4}$(bending stress)

$$= \tfrac{3}{4}\left(12wv^2/g\right) + \tfrac{1}{4}\left(12wv^2\pi^2D/gn^2t\right),$$

$$S = \frac{3wv^2}{g}\left(3 + \frac{\pi^2D}{n^2t}\right). \tag{15-8}$$

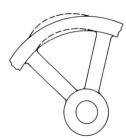

Figure 15-2
Rim action (exaggerated).

The effect of bending can be reduced in one of two ways. If the number of arms is increased, the span is reduced; also, if the rim thickness is increased, the rim performs better as a beam.

 Flywheel Speeds

Rule-of-thumb values for safe flywheel speeds are often taken to be 60 ft/sec for cast-iron wheels and 133 ft/sec for cast-steel wheels. While these values may seem ideal, it must be remembered that thin rims can lead to bending-stress problems. Joint efficiency also enters into the problem. Referring again to the approximate formula for hoop stress, note that the induced stress (hoop stress only) equals the square of the rim velocity in feet per second divided by 10. Thus,

$$v = \sqrt{10S}.$$

If a factor of safety is applied to the ultimate stress and joint efficiency is applied in the case of built-up flywheels, the above formula becomes

$$v = \sqrt{\frac{10S \text{ (efficiency)}}{\text{(factor of safety)}}}.$$

Since the factor of safety is applied under the radical to the stress value, the factor of safety based on velocity is greatly reduced. As a typical example, a factor of safety of 9 applied to the stress becomes a factor of safety of 3 applied to rim velocity. Thus, velocity variations greatly affect stress values in flywheel design.

The following example shows how stress calculations are done for a typical flywheel.

◆ *Example* ——————————————————————

A cast-iron flywheel has a mean diameter of 60 in. The rim is 1 in. thick and 12 in. wide. Cast iron weighs 0.26 lb/cu in. Find the hoop stress, the bending stress, and the total stress, assuming $\frac{3}{4}$ radial expansion of the arms. The rim is supported by six arms. The rim velocity is 65.5 ft/sec at the mean diameter.

Solution. Hoop stress is

$$S = \frac{12wv^2}{g} = \frac{12(0.26)(65.5)^2}{32.2} = 416 \text{ psi.}$$

Bending stress is

$$S = \frac{12wv^2\pi^2 D}{gn^2 t} = \frac{12(0.26)(65.5)^2(\pi)^2(60)}{32.2(6)^2(1)} = 6838 \text{ psi.}$$

Total stress is

$$S = \frac{3wv^2}{g}\left(3 + \frac{\pi^2 D}{n^2 t}\right)$$

$$= \frac{3(0.26)(65.5)^2}{32.2}\left[3 + \frac{(\pi)^2(60)}{(6)^2(1)}\right] = 2020 \text{ psi.}$$

 Rim and Arm Cross Sections

Generally speaking, the rim of a flywheel is rectangular in cross section—sometimes nearly square. Usually, its width is 6 to 15 times its thickness. The cross section through the rim is often built up around the edges and somewhat resembles a channel section. Note that a flywheel sometimes serves a dual purpose in applications where flat belts or V-flat drives are used. As mentioned earlier, an arm is usually elliptical in cross section with the major axis in line with the plane of rotation. Other shapes are possible and are sometimes used. One section resembles an elongated cross; another looks like an I-beam with an unusually large web. Since the arms or spokes serve as cantilever beams, the section modulus becomes an important consideration.

Design of Arms

When designing an elliptical flywheel arm one assumes that each arm carries its share of the load and that bending stresses are induced. Arm stresses are high because of the change in speed of the rim. Arms are often designed for three-fourths of the torsional strength of the shaft. Thus, the moment of an arm can be equated with three-fourths of the twisting moment of the driving shaft. Then,

$$\frac{3}{4}\left(\frac{S_s \pi D^3}{16}\right) = \frac{\pi a^2 b S n}{4},$$

where

a = one-half the major axis,
b = one-half the minor axis,
n = number of arms.

Since the major axis is usually twice the minor axis, this equation becomes

$$\frac{3}{4}\left(\frac{S_s \pi D^3}{16}\right) = \frac{\pi a^2 (a/2) S n}{4}.$$

Solving for a, we obtain

$$a = D\sqrt[3]{\frac{3S_s}{8Sn}} .$$
(15-9)

The major axis is equal to $2a$; the minor axis would be half as long.

Such calculated values are theoretically at the center of the shaft; they are somewhat less at the edge of the hub. The arms usually taper from the shaft center to the rim. Often, the taper is $\frac{1}{4}$ in. per foot.

◆ Example

A 1.75-in. steel shaft carries a rimmed flywheel with six arms. The torsional design stress for the shaft is 6000 psi and the allowable tensile stress for the arms is 5000 psi. Find the major and minor axes for the elliptical arms. The mean diameter of the flywheel is 48 in.; assume no taper for the arms.

Solution.

$$a = D\sqrt[3]{\frac{3S_s}{8Sn}} = 1.75\sqrt[3]{\frac{3(6000)}{8(5000)(6)}} = 0.738 \text{ in.};$$

thus the major axis, $2a$, is 1.476 in. at the center of the shaft;

$b = a/2 = 0.738/2 = 0.369$ in.;

and the minor axis, $2b$, is 0.738 in. at the center of the shaft.

 Solid-disc Flywheels

Not all flywheels consist of hub, arms, and rim. Certain smaller flywheels are made in the form of a solid disc. The weight for such a flywheel is given by the following equation:

$$W = \frac{4g(\Delta E)}{v_1^2 - v_2^2} .$$
(15-10)

A completely solid disc is not practical as a flywheel; in most cases a hole is needed for the shaft. However, the shaft and key add weight in the hole area that approximates the specific weight of the flywheel material.

◆ Example

Find the capacity of a 2-ft-diameter flywheel that weighs 300 lb. The peripheral velocity of the wheel decreases 10% from its rated speed of 60 ft/sec while the machine is performing a shearing operation.

Solution.

$$\Delta E = \frac{W\left(v_1{}^2 - v_2{}^2\right)}{4g}$$

$$= \frac{300\left[(60)^2 - (54)^2\right]}{4(32.2)}$$

$$= 1590 \text{ ft-lb.}$$

◆ *Example*

Find the weight (in kilograms) of a solid-disc flywheel that normally rotates at 500 rpm and that has a diameter of 500 mm. During a forming operation, the speed is reduced 12% and 1400 N · m of energy is expended.

Solution.

$$v_1 = \frac{\pi DN}{60} = \frac{\pi(0.5)(500)}{60} = 13.1 \text{ m/s.}$$

$$v_2 = 0.88(13.1) = 11.5 \text{ m/s;}$$

$$g = 9.81 \text{ m/s}^2;$$

$$W = \frac{4g(\Delta E)}{\left(v_1{}^2 - v_2{}^2\right)(9.81)} = \frac{4(9.81)(1400)}{(13.1^2 - 11.5^2)(9.81)}.$$

Therefore, $W = 142$ kg.

Balancing

Mechanical components such as cams can cause an unbalance that must be corrected. Figure 15-3 shows a cam with more metal on one side of the shaft than on the other. *Static balance* is easy to obtain, however. If the shaft is placed on knife edges, the heavy section will soon rotate to the bottom, as shown in Fig. 15-3. The direction of the unbalance can be found by scribing a vertical line downward from the shaft after positioning the cam-and-shaft assembly on the knife edges. The amount of the unbalance can be determined by measuring cam areas above and below the horizontal centerline through the shaft, assuming that the plate cam has a constant thickness. Irregular areas can be measured with a planimeter. A planimeter will integrate area if a complete periphery of the measured area is traced. After determining the amount of unbalance, weight can be removed from the

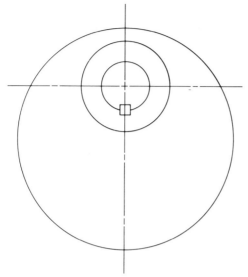

Figure 15-3
Plate cam before static balance.

heavy side by drilling strategically located and sized holes, by adding weights to the light side, or by a combination of both methods.

Figure 15-4 shows a shaft with two identical cams mounted a certain distance apart but 180 degrees out of phase with each other. If this shaft were placed on knife edges, static balance would be indicated. Yet, these weights are not dynamically balanced; shaft rotation would immediately set up dangerous vibrations. Weight adjustments in two locations (or sometimes one) can correct this situation; sometimes weights are added, other times metal is drilled away to reduce weight.

Figure 15-4
Plate cams: statically balanced, dynamically unbalanced.

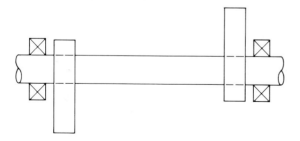

Dynamic Balancing

Several procedures are used to establish dynamic balance. Graphical and mathematical solutions are based on the fact that both force and couple systems must be in equilibrium. For an illustration of such solutions, refer to Fig. 15-5. In this figure, the components are located in planes 1 and 2, and the weights are to be added in the x- and y-planes. The configuration of the mounting assembly most likely dictates the location of the balance weights. Certain distances between the planes have been established and are indicated by linear dimensions. In plane 1, the unbalance of 4 lb is located 3

Figure 15-5
Graphical solution: dynamic balancing.

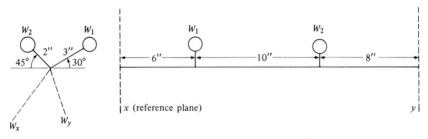

Plane	W, lb	r, in.	a, in.	Wr	Wra
1	4	3	6	12	72
2	5	2	16	10	160
x (reference plane)	W_x	1.5	0	$1.5W_x$	0
y	W_y	1.5	24	$1.5W_y$	$36W_y$

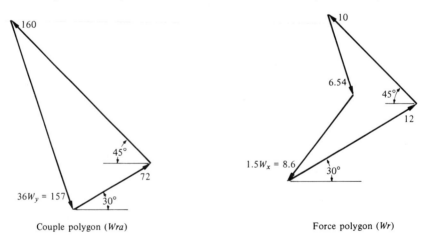

Couple polygon (Wra)

Force polygon (Wr)

in. from the shaft center at a horizontal angle of 30 degrees, as indicated. For the components in plane 2, a weight of 5 lb is located 2 in. from the shaft center and 105 degrees to the left of the other weight, as shown in the left side view. If it is assumed that weights x and y are to be added in planes x and y at radial distances of 1.5 in., the problem is to determine the value of the weights and the relative angles at which they must be placed to attain dynamic balance.

As a first step, it is desirable to tabulate all the known information. The values for Wra and Wr must be represented vectorially. A reference plane must be assumed through the line of action of one of the unknown weights (in this particular case, plane x was chosen). Thus, couple values Wra are based on plane x. The y-plane could have been used if desired; this would have produced different Wra-values. In the x-plane, no Wra-value exists, since dimension a is zero. The next step is to establish a couple polygon using vectors. Three Wra-values comprise this polygon: 72, 160, and the unknown, which is designated $36W_y$. Since these couples must be in equilibrium, the polygon must close. Vectors representing Wra-values of 72 and 160 are constructed at the appropriate angles; the closing line of the polygon represents $36W_y$. This vector scales approximately 157; when this is divided by 36, the resulting weight is approximately 4.36 lb acting in the direction shown by the vector.

The second step is to find the value of weight x. This can be done by constructing a force polygon. Referring again to the table of Fig. 15-5, Wr-values are known for planes 1 and 2; $1.5W_y$ is easily obtained by multiplying the 4.36 lb previously determined by 1.5. Now, the only unknown Wr-value is $1.5W_x$. A force polygon can be constructed by laying out vectors of 12, 10, and 6.54 (the 6.54 value was the result of the preceding multiplication). The vector that closes the polygon will be equal to $1.5W_x$. This scales approximately 8.6; the 8.6 value is divided by 1.5 to yield a W_x value of approximately 5.73 lb. The direction of this vector shows the angle at which the weight must be placed relative to the other weights. The two weights and their angles have now been determined and the assembly should be dynamically balanced. In any graphical solution, the accuracy of the result depends on accurate construction and the use of suitable scales; large-sized polygons contribute to accuracy.

Mathematical Solution for Balancing

Numerical values for the unknown weights can be determined by establishing equations of equilibrium for the Wra- and Wr-values. The desired direction for each weight can then be found by simple trigonometry.

For the couples, let the subscripts H and V stand for "horizontal" and "vertical." Then,

$$36W_{y_H} = 72\cos 30 - 160\cos 45,$$

$$W_{y_H} = \frac{72(0.866) - 160(0.707)}{36} = -1.41 \text{ lb};$$

$$36W_{y_V} = 72\sin 30 + 160\sin 45,$$

$$W_{y_V} = \frac{72(0.500) + 160(0.707)}{36} = 4.14 \text{ lb}.$$

Then

$$W_y = \sqrt{(1.41)^2 + (4.14)^2} = 4.38 \text{ lb},$$

$$\tan\theta = \frac{W_{y_V}}{W_{y_H}} = \frac{4.14}{1.41} = 2.94,$$

$$\theta = 71.2 \text{ degrees}.$$

Similar equations may be written for Wr-values. Thus

$$1.5W_{x_H} = 12\cos 30 - 10\cos 45 + 6.57\cos 71.2.$$

Note that 6.57 was found by multiplying 1.5 by 4.38. Then,

$$W_{x_H} = \frac{12(0.866) - 10(0.707) + 6.57(0.322)}{1.5} = 3.63 \text{ lb}.$$

Also,

$$1.5W_{x_V} = 12\sin 30 + 10\sin 45 - 6.57\sin 71.2$$

$$= \frac{12(0.500) + 10(0.707) - 6.57(0.947)}{1.5} = 4.57 \text{ lb}.$$

Thus

$$W_x = \sqrt{(3.63)^2 + (4.57)^2} = 5.84 \text{ lb},$$

$$\tan\theta = \frac{W_{x_V}}{W_{x_H}} = \frac{4.57}{3.63} = 1.26,$$

$$\theta = 51.6 \text{ degrees}.$$

The mathematical solution and the graphical solution can be used to check each other.

◆ *Example* ―――――――――――――――――――――
Find the value and angular placement of the weights W_x and W_y needed to balance the system shown in Fig. 15-6(a). Assume that W_x and W_y are each centered 80 mm from the centerline of the shaft. Existing positions of W_1 and W_2 are shown in the figure; W_1 = 2 kg and W_2 = 1.5 kg. W_1 is located 75 mm from the centerline and W_2 is 100 mm from the centerline.

Solution. It is usually helpful to tabulate existing data and make rough *Wr* and *Wra* diagrams before starting the mathematical solution. One plane must be established as the reference plane in order to have zero values for one of the *Wr* vectors and one of the *Wra* vectors. In this case, *x* was selected as the reference plane.

Plane	*W*, kg	*r*, mm	*a*, mm	*Wr*	*Wra*
1	2	75	125	150	18 750
2	1.5	100	300	150	45 000
x (ref.)	W_x	80	0	$80W_x$	0
y	W_y	80	375	$80W_y$	$30\,000W_y$

Using *Wra* values, calculate W_y and its angle of action, using horizontal and vertical components.

Figure 15-6
Dynamic balancing example.

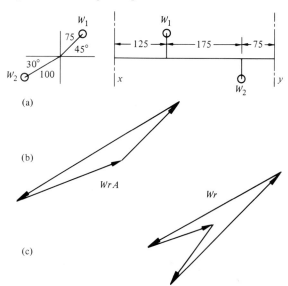

Horizontal: $18\,750 \cos 45 - 45\,000 \cos 30 + 30\,000W_{y_H} = 0$;

$$W_{y_H} = \frac{45\,000 \sin 30 - 18\,750 \sin 45}{30\,000} = 0.857.$$

Vertical: $18\,750 \sin 45 - 45\,000 \sin 30 + 30\,000W_{y_V} = 0$;

$$W_{y_V} = \frac{45\,000 \sin 30 - 18\,750 \sin 45}{30\,000} = 0.308;$$

$$\theta = \arctan \frac{W_{y_V}}{W_{y_H}} = \frac{0.308}{0.857} = 19.77 \text{ degrees};$$

$$W_y = \sqrt{(0.857)^2 + (0.308)^2} = 0.911 \text{ kg.}$$

Then, $80W_y = 72.88$.

Next, using Wr values, calculate W_x and its angle of action.

Horizontal: $150 \cos 45 - 150 \cos 30 + 72.88 \cos 19.77 - 80W_{x_H} = 0$;

$$W_{x_H} = \frac{150 \cos 45 - 150 \cos 30 + 72.88 \cos 19.77}{80} = 0.559.$$

Vertical: $150 \sin 45 - 150 \sin 30 + 72.88 \sin 19.77 - 80W_{x_V} = 0$;

$$W_{x_V} = \frac{150 \sin 45 - 150 \sin 30 + 72.88 \sin 19.77}{80} = 0.696;$$

$$\theta = \arctan \frac{W_{x_V}}{W_{x_H}} = \frac{0.696}{0.559} = 51.23 \text{ degrees};$$

and

$$W_x = \sqrt{(0.559)^2 + (0.696)^2} = 0.892 \text{ kg.}$$

Balancing Machine

Figure 15-7 shows an automatic balancing machine that puts balancing on a production basis. This particular device was designed for balancing automotive crankshafts. With the varying conditions of crankshaft operation, proper balancing is imperative. With such a machine, the operator can easily determine the amount of unbalance and the angular location of the necessary alterations, with no operator judgment required. The balancing machine shown in Fig. 15-7 is used to adjust balance in two planes where weight can be removed (or added). By using proper conversions, the

Figure 15-7
Automatic balancing machine.
(Courtesy of Tinius Olsen Testing Machine Co., Inc.)

diameter and depth of drilled holes for removing the correct amount of metal can be ascertained. With such a system, proper balancing becomes foolproof.

This machine is an electromechanical device in which a velocity pick-up is employed at the planes being used for correcting unbalance. A two-channel amplifier is used; the pick-up signal is rectified and indicated on the *amount meter* as the amount of unbalance. At the same time, the angle channel forms a single pulse that indicates the phase displacement of the pick-up signal with respect to a reference signal generated on the spindle. An *angle meter* thus gives the angular location of the unbalance.

 **Summary**

Flywheels are useful and necessary mechanical components that absorb excess kinetic energy and redistribute this energy to hold speed variations within reasonable values. Because of their shape and size, the best method of fabrication is casting. Shrinkage stresses resulting from the casting process can cause trouble. These stresses are more a production problem than a design problem. Careful inspection of all flywheels is necessary. Better inspection procedures are now possible using effective nondestructive testing methods. X-ray inspection and ultrasonic testing are useful for detecting flaws in the casting process. Centrifugal stresses can be calculated, and stresses due to speed fluctuations can be determined fairly accurately. To be effective, flywheels should operate at fairly high speeds. Flywheel failure, which is often an exploding type of rupture, can thus be exceedingly dangerous. Large factors of safety are generally chosen. The limiting speed (peripheral) is often considered to be 88 fps or 1 mi/min. Higher speeds are sometimes used, but speed variation must be kept to a minimum in such cases.

The balancing of rotating mechanical components is often a complex process. Static balancing is relatively easy; dynamic balancing, on the other hand, can be very difficult. Dynamic balancing is particularly important on high-speed equipment with an excessive speed range. By their very nature, cams are unbalanced components. Thus, provision must be made for adding or removing weight at strategic places in an assembly where cams are used.

Other mechanical parts sometimes contribute a "flywheel effect" to an assembly. The use of heavy pulleys or sheaves can completely eliminate the need for a flywheel—the pulley or sheave is a flywheel in itself. Such effects must be studied carefully by the designer.

Questions for Review

1. What is the purpose of a flywheel?
2. Why are solid-disc flywheels sometimes preferred over those with rim, hub, and arms?
3. Differentiate between *static* and *dynamic* balancing.
4. List all potential stresses induced in flywheels.
5. List five applications where a small coefficient of fluctuation is needed.

Problems

Note: Unless otherwise stated, assume that all flywheels are made of cast iron with a specific weight of 0.26 lb/cu in. or 7600 kg/m³.

1. A flywheel has a mean diameter of 4 ft and a width of 8 in. It is required to handle 2250 ft-lb of kinetic energy at a normal operating

speed of 300 rpm. The coefficient of fluctuation is 0.05. Find the weight and thickness of the rim, assuming that the arms and hub are equivalent to 10% of the total rim weight.

2. A flywheel has a mean diameter of 36 in., a width of 10 in., and a rim thickness of $\frac{3}{4}$ in. Assuming the arms and hub amount to one-eighth of the rim weight and the speed must be kept within 10% of the normal operating speed, find the capacity of the flywheel. The average velocity (at the mean diameter) is 88 fps.

3. A punching machine requires 1800 ft-lb of energy to punch a certain hole. The speed of the machine is reduced from 200 rpm to 180 rpm during the punching operation. The weight of the arms and hub are considered to be 12% of the rim weight, and the mean diameter of the flywheel is 28 in. Find the rim weight required. (*Note:* In such equipment as punches and shears, the rated velocity is the maximum speed.)

4. A shearing machine requires 2000 ft-lb of kinetic energy. The mean diameter of its flywheel is 36 in. The rated speed is 66 fps at the mean diameter. If the coefficient of fluctuation is 0.2, what rim weight should the flywheel have? Neglect the flywheel effect of the arms and hub.

5. If the rim width is 4 times as large as the rim thickness in the flywheel of Problem 4, what will the width and thickness be?

6. A cast-iron flywheel has a mean diameter of 36 in. and a rotational speed (normal) of 400 rpm. Calculate the hoop stress and check with the approximate hoop-stress formula.

7. A 6-ft-diameter flywheel has to absorb 2400 ft-lb of energy and maintain a coefficient of regulation of 0.1. Assume that the arms and hub account for 10% of the rim weight. The mean rim speed (at the mean radius) is 80 fps. Find the cross-sectional area of the rim.

8. A flywheel has a mean diameter of 48 in. and a weight of 800 lb. The speed is to be maintained between 100 and 120 rpm. The arms and hub account for 12% of the rim weight. Find the capacity of this flywheel in ft-lb.

9. A flywheel has a WR^2-value of 595 lb-ft^2. Twelve seconds are required to accelerate the flywheel from 150 to 200 rpm. Find the torque required if the arms and hub are neglected.

10. If the WR^2-value of the arms and hub is neglected, find the possible change in rpm if a torque of 1200 in-lb is applied for 5 sec. The flywheel effect of the rim is 800 lb-ft^2.

11. A flywheel is designed with a maximum hoop stress of 500 psi. Find the approximate peripheral velocity (at the mean radius) of the rim in fps.

12. A cast-steel flywheel has a specific weight of 0.28 lb/in³. The allowable hoop stress is 550 psi. Find the maximum velocity of the rim at the mean radius.

13. What hoop stress is developed in Problem 12 if the rim velocity is 88 fps?

14. A cast-iron flywheel has a mean diameter of 60 in. and a mean rim velocity (at the mean radius) of 88 fps. The rim is 1 in. thick and 6 in. wide. The rim is supported by 6 elliptical arms. Find the total stress, assuming three-quarters radial expansion for the arms.

15. What would be the total stress in Problem 14 if all conditions remained the same except that 4 arms were used instead of 6?

16. A 5-ft-diameter flywheel with 6 arms is driven by a 2-in. shaft designed with a shearing stress of 6000 psi. The allowable tensile stress for the arms is 5500 psi. Find the major and minor axes of the arms, calculated at the shaft center.

17. A 1-ft-diameter flywheel has 4 elliptical arms. A 1-in.-diameter shaft is keyed to this flywheel. The allowable torsional stress for the shaft is 6000 psi; the allowable tensile stress for the arms is 5000 psi. Find the major and minor axes of the arms at the shaft center.

18. A solid-disc flywheel must handle 1000 ft-lb of energy. The diameter of the flywheel is 3 ft; the peripheral velocity varies between 30 and 35 fps. What weight of flywheel is needed?

19. A 2-ft-diameter solid-disc flywheel weighs 200 lb and is designed to handle 800 ft-lb of kinetic-energy change. Its upper speed is 40 fps. Find the coefficient of regulation.

20. List some mechanical components that produce a flywheel effect, either alone or in conjunction with a flywheel.

21. Using vectors, construct Wra and Wr polygons to determine the weights needed in the x- and y-planes of Fig. 15-8. Complete all the blank spaces in the following tabulation.

Plane	W	r	a	θ	Wr	Wra
1	4	3	6	180°	_____	_____
2	5	3	13	30°	_____	_____
3	4	3	18	270°	_____	_____
x (reference)	_____	2	0	_____	_____	_____
y	_____	2	22	_____	_____	_____

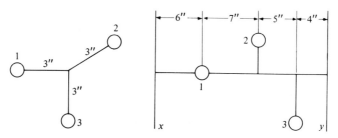

Figure 15-8
Dynamic-balancing problem: Problem 21.

Note that θ_x and θ_y are important values to be determined: weights, radii, and positional location all play an important role in dynamic balancing.

22. Find the rim thickness of a cast-iron flywheel with a width of 200 mm, a mean diameter of 1.2 m, a normal operating speed of 300 rpm, a coefficient of fluctuation of 0.05, and an energy capacity of 3000 N · m. Assume that the hub and arms represent 10% of the rim weight and that the specific weight of cast iron is 7208 kg/m³.

23. A cast-iron flywheel for a blanking press has a mean rim diameter of 1.5 m. Its speed varies from 275 rpm to 250 rpm. The required energy fluctuation is 6500 N · m and the density of the cast iron is 7600 kg/m³. Find the rim dimensions if the arms and hub are neglected and the width is twice the thickness.

24. A solid-disc flywheel for a hole-punching application has a diameter of 0.6 m and must have a capacity of 2150 N · m. Its rated speed is 575 rpm, which decreases 15% during the punching operation. Find the thickness of the flywheel and express your answer in mm.

MISCELLANEOUS MACHINE ELEMENTS AND SPECIAL TOPICS

16

 **Standards**

A standard is the documented result of any particular standardization effort approved by a recognized authority. It provides a common language between the buyer and the seller or the producer and the user. Over 400 organizations write standards in the United States. These include professional societies and trade associations.

The degree of consensus is an important factor in classifying standards. Governmental standards are those developed by the employees of federal, state, or local governing bodies to provide purchasing specifications. Company standards are those developed by groups within a firm to provide purchasing specifications or design procedures. Almost every firm, for example, has a set of drafting-room standards that often interlace with industry-wide practice, yet contain certain stipulations that are appropriate for their mode of operation. Trade associations develop standards relating to certain areas of interest, such as gears, antifriction bearings, roller chains, structural steel, belts, and similar products that are manufactured by several firms specializing in these types of products. Professional societies develop standards that relate to professional practice in such areas as mechanical, electrical, and chemical engineering.

Full consensus standards are those developed by organizations such as the American Society for Testing and Materials (ASTM). ASTM standards are developed by a committee comprised of producers, users, and members representing a general interest. The group is balanced between producers and users to prevent special-interest groups from dominating the development of the finished document. ASTM produces:

1. definition standards,
2. recommended practice standards,
3. method of test standards,
4. classification standards, and
5. specification standards.

The American National Standards Institute (ANSI) is a federation of more than 400 bodies that acts as the national coordinator for standardization in the United States. It is the official U.S. representative to the International Organization for Standardization (ISO).

The National Bureau of Standards does not establish or enforce mandatory standards. Instead it develops measurement methods, instrumentation, and measurement standards for others to use both inside and outside the federal government. NBS cooperates with those producing voluntary consensus standards.

Standards are also classified as *voluntary consensus standards* (basically technological) and *mandatory standards* (basically political). These interlace in many respects. For example, building codes are mandated, but they frequently refer to standards developed by professional societies, ASTM, and others.

In developing a design, one must conform to all mandated codes, follow accepted practice as established by professional societies and trade associations, and conform to the company or agency standards that relate to the project. Therefore, the latest information regarding codes and standards must be readily available.

 # Safety Legislation

Concern for the health and safety of all citizens—at work and away from work—prompted the United States Congress to pass a number of laws related to safety. This legislation provides for both safe operating practices and effective design in order to reduce hazards that lead to deaths and dismemberments.

A brief discussion of the key points in the various acts follows.

Occupational Safety and Health Act

OSHA is a federal act designed to protect workers from dangerous hazards at their employment sites. OSHA enforcement is strict; firms not complying with regulations face heavy fines and a possible closedown of operations. Enforcement officers can check any employer at any time; employee complaints are investigated and can lead to an inspection. The U.S. Department of Labor administers the program. OSHA uses existing federal standards and national consensus standards. The latter are standards that have been adopted and promulgated by nationally recognized organizations such as engineering and professional societies.

The main thrust of this act is to ensure safe and healthful working conditions through effective practices, operations, methods, and similar means. Selected sections of these standards relate to design; most pertain to practices.

The subparts of the Occupational Safety and Health Standards as published in the *Federal Register* are as follows:

  a. General
  b. Adoption and Extension of Established Federal Standards
  c. (Reserved)
  d. Walking-Working Surfaces
  e. Means of Egress
  f. Powered Platforms, Manlifts, and Vehicle-Mounted Work Platforms
  g. Occupational Health and Environmental Control

h. Hazardous Materials
i. Personal Protective Equipment
j. General Environmental Controls
k. Medical and First Aid
l. Fire Protection
m. Compressed Gas and Compressed Air Equipment
n. Materials Handling and Storage
o. Machinery and Machine Guarding
p. Hand and Portable Powered Tools and Other Hand-Held Equipment
q. Welding, Cutting, and Brazing
r. Special Industries
s. Electrical

The following material discusses parts of OSHA's rules and regulations that are closely allied with the design function.

Mechanical power transmission apparatus. OSHA standards concentrate on protecting drives with properly designed guards, maintaining equipment, and establishing orange as a color to reveal dangerous areas. Little is mandated in the way of specific drive-design procedures. Components listed under the general category of power transmission equipment include flywheels, shafting, pulleys, belts, ropes, chains, sprockets, gears, friction drives, belt tighteners, collars, couplings, projecting keys and setscrews, clutches, and other similar parts and assemblies.

Ladders—fixed and portable. Information is provided regarding such design features as materials, rung size, side-rail dimensions, allowable deflections, clearances for fixed ladders, cage sizes for long fixed ladders, and pitches for fixed ladders.

Powered platforms, manlifts, and vehicle-mounted work platforms. Various dimensional stipulations and emergency operation provisions are given. A minimum factor of safety of 10 is required and shall be calculated by the following formula:

$$F = \frac{S \times N}{W},$$

where

S = manufacturer's rated breaking strength of one rope,
N = number of ropes under load,
W = maximum static load on all ropes with the platform and its rated load at any point of its travel.

Machine guards. Details are specified for point-of-operation guarding for machines such as guillotine cutters, shears, alligator shears, power presses, milling machines, power saws, jointers, portable power tools, forming rolls, and calenders. The angle of exposure is specified for the various types of grinding wheels. Also, the size of the mounting flange is given in terms of the grinding-wheel diameter.

Guarding of portable powered tools. Among the features required in portable power tools is the so-called "dead-man control." In this system, the operator must depress the operating trigger when using the tool. If the tool is powered by electricity, the power supply must cut off immediately when the operating switch is released. If the tool is operated by compressed air, release of the operating lever must automatically shut off the air inlet valve.

Electrical machinery. OSHA rules and regulations interlace extensively with various articles and sections of the National Electrical Code with respect to all electrical installations and utilization equipment.

The preceding discussion focuses on very few provisions of the Occupational Safety and Health Act, but is presented so that a student will understand that safety is legislated into design projects. Organizations must have a comprehensive library of standards and codes that apply to any contemplated design project. The *Federal Register* lists requirements for compliance with various standards and codes in addition to supplying pertinent information on OSHA standards.

Consumer Product Safety Act

This legislation was devised to protect the general public from unsafe products. A commission was appointed to study unsafe products, to establish mandatory safety standards, to require warnings on hazardous products, and to ban dangerous products. The commission also can fine or jail offending manufacturers.

Accident rates are high because of faulty appliances or other products used in the home. The following list shows a few of the products that have traditionally been hazardous to consumers and thus require study and corrective action:

bicycles,	stoves,
power lawnmowers,	playground equipment,
power tools,	lighters and fluids,
glass doors,	football helmets,
garage doors,	beds,
soft drink bottles,	shower structures,
space heaters,	cleaning agents.

A number of these products are directly associated with the mechanical design field—particularly bicycles, power lawnmowers, power tools, space heaters, stoves, and playground equipment.

The U.S. Consumer Product Safety Commission issues mandatory safety standards for consumer products that pose unreasonable risks. The Commission seeks voluntary existing standards and welcomes the development of standards from competent sources. In the absence of voluntary input, the Commission develops a suitable standard.

Because it was found that over 1,000,000 children and adults are injured annually in bicycle-related accidents, the Commission considers bicycles the foremost item on their product-hazard index. Studies showed that about 17 percent of the bicycle accidents were directly attributable to mechanical or structural failures. As a result of these studies, the Consumer Product Safety Commission provided a bicycle safety standard.

Let us consider a few of the features included in this initial standard. Note that this standard uses metric units as well as U.S. units for such quantities as distance (millimeters and inches), force (newtons and pounds force), velocity (kilometers per hour and miles per hour), energy (joules and inch-pounds), torque (newton-meters and foot-pounds), and temperature (°Celsius and °Fahrenheit). The complete standard can be found in the *Federal Register*; a few highlights follow:

1. Brakes must be designed to stop within a distance of 4.5 m (15 ft) at a test speed based on the gear ratio.
2. Reflectors must be located on the front of the bicycle, on the pedals, on the back, and on the wheels. For wheel reflectors, manufacturers can choose between reflectorized tires and wide-angle reflectors on the wheel spokes. Reflector color is specified to a degree: front reflectors are essentially colorless, pedal reflectors are essentially colorless or amber, whereas rear reflectors must be red. The optical axis for the front and rear reflectors is specified to be within 5 degrees of the horizontal-vertical alignment of the bicycle when the wheels are tracking in a straight line. To assure that reflectors are not misaligned, a test is provided. The reflectors must withstand a force of 89 N (20 lbf). The location of spoke reflectors is also specified. The color for side reflectors must be colorless or amber for the front wheel and colorless or red for the rear wheel. When retroreflective tire sidewalls are used, the retroreflective material must withstand a test of 50 °C or 120 °F (with tolerances). After 30 minutes at this temperature, it must be determined that the retroreflective material cannot be peeled or scraped away without removal of tire material.
3. The recommended inflation pressure for tires must be indicated on the sidewall. Also, the tire must stay on the rim when inflated to 110

percent of the maximum specified pressure and side-loaded with a test force of 2000 N (450 lbf) under specified test conditions.

4. The handlebars must be designed to allow safe and comfortable control of the bicycle. The ends of the handlebars must have an inside dimension of at least 356 mm (14 in.) and not more than 712 mm (28 in.); also, the ends must be symmetrically located with respect to the longitudinal axis. The ends must be located not more than 406 mm (16 in.) above the seat level at its lowest position when the handlebars are at their highest position. The handlebar stem must be tested for strength and must withstand a force of 2000 N (450 lbf). The ends of the handlebars must be capped or covered.

5. Chain guards must be provided where the chain engages the sprocket for bicycles that are not freewheeling.

6. Handlebar, seat, and stem must be secured with clamps.

7. Locking devices must be provided to secure wheel hubs to the bicycle frame.

8. Fenders must have protected edges. A number of stipulations in this standard relate to mechanical requirements such as sharp edges, integrity (visible fracture), attachment hardware, protrusions, and projections.

9. The fork and frame must meet strength tests as outlined in a formal test procedure that specifies the type of support and the direction of applied forces. For example, the frame is subjected to a load of 890 N (200 lbf) or an energy of at least 39.5 J (350 in-lb); the one that provides the greater load is applied to the fork at the axle attachment point against the direction of the rake in line with the rear wheel axle. After applying the appropriate load, there must be no visible evidence of fracture or deformation of the frame.

In addition to these criteria, there are other requirements that pertain to the seat, to the pedals, to the wheels, and to other parts of the bicycle. So-called sidewalk bicycles must also meet safety requirements; however, these criteria are not quite as rigid as those for bicycles used on roads and highways. Note that standards are constantly reviewed and changed as up-to-date test and performance data become available from recognized manufacturers' organizations and professional societies.

Noise Control Act

Because inadequately controlled noise presents a danger to the health and welfare of the public, the United States Congress enacted the Noise Control Act of 1972. The primary purpose of this act is to control the

Table 16-1
Typical Noise Levels (in decibels)

Noise Source	dB
Riveting operation, steel plates	128
Loud automobile horn	117
Gasoline-powered lawnmower	93
Machine-shop power tools	89
Pneumatic tools	82
Printing press	80
Street traffic	70

emission of noise detrimental to the human environment. Major sources of noise include transportation vehicles and equipment, machinery, appliances, and other products in commerce.

Noise is easy to measure with a sound level meter; the unit is the decibel, abbreviated dB. Sound is usually controlled by isolation. Table 16-1 shows typical decibel values that might be expected from varied sources. Table 16-2 indicates permissible noise exposures as specified in the initial OSHA standard.

Equipment designers can reduce noise levels (and vibration at the same time) by using flexible couplings to connect shafts, by using bonded-rubber mountings to attach noisy machinery to a base, by properly balancing moving parts, and by using other damping materials to prevent sound from being transmitted from one part to another. A muffler will reduce the noise levels of fluid power circuits that use compressed air. The noise produced by

Table 16-2
Permissible Noise Exposures*

Duration per Day, Hours	Sound Level dB (A)[†]
8	90
6	92
4	95
3	97
2	100
$1\frac{1}{2}$	102
1	105
$\frac{1}{2}$	110
$\frac{1}{4}$	115

*OSHA limits as specified in *Federal Register* (10/18/72).
[†]Sound level meters generally have 3 scales; the A scale is one of slow response.

air compressors and automatic metal-forming equipment can be confined in acoustical enclosures. The design of such enclosures involves ingenuity in providing configurations that conform to a machine's profile and that allow the proper placement of sound-absorbing materials.

The noise level of equipment that handles bulk materials can be reduced by bonding a thin layer of sheet lead to the surfaces that come in contact with the bulk materials. Also, a number of plastic materials can be bonded to metal bins and hoppers in order to substantially reduce noise levels.

The Noise Control Act goes beyond factory noise; another concern is general environmental noise. Products in the following categories are greatly affected by the Act: engines, motors, electrical equipment, construction equipment, and transportation equipment.

Uncontrolled noise is a definite health hazard. Values of 180 dB can be lethal, and relatively short exposure to 150 dB can cause deafness. In reviewing decibel intensities, one should note that noise levels progress logarithmically. For example, a reading of 70 dB is 10 times louder than one of 60 dB.

Note that the 90 decibel limit originally established by OSHA was an arbitrary figure, subject to change. European countries consider the limit of 85 dB more realistic. Also, groups other than the federal Occupational Safety and Health Administration also establish noise standards. For instance, the Environmental Protection Agency established an Office of Noise Abatement and Control to safeguard community sound levels. A number of large cities also have noise abatement standards.

Safety Devices

Aside from the factor of safety that should be incorporated in the design of all parts, safety devices are at the disposal of the designer or can easily be fabricated. These safety devices protect the operator and ensure against equipment damage in case of jamming or overloading. Mechanical, electrical, and electromechanical units are used to promote safety in the operation of equipment.

Electrical

Fuses and circuit breakers are useful for breaking an electrical circuit when trouble occurs. To incorporate these devices in the design, a designer must study ratings and mounting conditions. Fuses and circuit breakers should not be confused with interlocks and "panic" buttons. Both of these are installed primarily for the protection of personnel, not equipment. The *interlock* is a switch or plug used to disconnect the source of electricity from the machine when dangerous components are exposed; this type of device often comes into play when someone removes a panel to make adjustments. A qualified repairperson can bypass the interlock or interlocking switch to check circuits; the device is used to protect the unwary operator from dangerous electrical shock. The *"panic" button* is used where numbers of people work around dangerous equipment. These buttons are placed in strategic locations around the equipment where an individual can reach and depress a button without delay to immediately shut off the equipment. Warning lights and buzzers are safety units that are easy to install and easy to understand. Photoelectric assemblies (including light sources) can be used to prevent fingers or hands from being caught in presses, shears, and other metal-working equipment. If the beam of light is broken, movement of the ram is prevented; proper placement of the safety equipment and correct electrical circuitry are the only prerequisites for effective use.

Lockout devices are popular where maintenance crews are needed to service equipment. These units can be commercially purchased or easily designed. In effect, all that is needed is a key for each individual involved in the work. All the keys (or locks, in some cases) are needed to start the equipment operating again. A complex unit secured by several padlocks is placed on the main power switch. Unless all padlocks are removed (indicating that all of the maintenance crew are out of the area), the main power source cannot be connected. Series wiring of operating switches is a time-tested method for ensuring operator safety when using dangerous presses or similar equipment. If two men work together on a heavy press or welding fixture, four series switches are required. To operate the machine, all four switches have to be depressed simultaneously; thus, all four hands are on the switches and not under the ram of the press. This type of system is foolproof, yet simple.

Safety Setscrews

Headless setscrews are mandatory in most industries. Being headless, the setscrews can be placed so that no part protrudes above the surface of the hub, thus reducing the danger of injury from rotating parts. Several types of safety setscrews are available. The headless end can be slotted for use with a screwdriver, or be of the fluted-socket or hexagon-socket type. The latter types require the use of special setscrew wrenches. Such wrenches prevent unequipped (and usually unqualified) personnel from toying with the equipment. They also permit a reasonable tightening torque with good gripping characteristics.

The recessing of fasteners by counterboring or countersinking is a step toward the safer operation of equipment. Whether or not this is done depends on the use of the product; the extra machining operations add greatly to the cost. If counterboring is done, socket-type cap screws rather than hexagon cap screws are often used. This is done to avoid an excessively wide counterbore, which would be necessary to allow tightening with hexagon-socket (external) wrenches. (The internal socket-type wrench used with socket-type cap screws permits the diameter of the counterbore to be just a fraction of an inch larger than the head of the fastener.)

Machine Guards

There are various types of guards that are simple to design and easy to manufacture. The simplest is a railing around the dangerous area. This does not prevent personnel from entering the danger zone, but it does remind them of possible danger. Guards should be placed on individual machines to protect personnel from moving parts. The guard will often be fabricated from expanded metal, wire mesh, or sheet metal. Safety guards are desirable around components such as

1. pulleys and belts,
2. couplings,
3. chains and sprockets,
4. gears,
5. shaft extensions, and
6. cams.

In general, any moving part (rotating or reciprocating) should be guarded if there is any danger of it snagging clothing or parts of the human body. Gears are generally enclosed to allow proper lubrication.*

Point-of-entry guards are used where material is inserted into a machine. The point-of-entry guard is merely a strip of material beyond which the hand, finger, or arm cannot extend. The key point in designing this type of guard is to make certain that human appendages cannot extend into the dangerous area.

Grinding-wheel guards are the most complete type. Since grinding wheels could burst under too great a speed, adequate protection is essential. Grinding wheels are usually enclosed for approximately 270 degrees, and the remaining "open" area is somewhat protected by a shatterproof glass shield between the wheel and the operator. Other precautions are needed in

---

*For additional material in this area, a good source is *The Principles and Techniques of Mechanical Guarding*, Bulletin No. 197, U.S. Department of Labor.

designing grinding wheels; these include the following:

1. Every effort should be made to avoid cracking the wheel. Therefore, the wheel must fit loosely on the spindle.
2. Large flanges must be used to secure the wheel to the spindle. Rubber washers should be used under the flanges to ensure uniform pressure on the grinding wheel.
3. The linear velocity of the edge of the wheel is the most important consideration in grinding-wheel design. In most cases, linear speeds must be kept below 6000 ft/min.
4. In making decisions, the designer should consider the consequences of failure. Adequate exhaust systems should be employed for the general wellbeing of the operator.

Relief Valves

Figure 16-1 shows one type of relief valve. Valves of this type are used to prevent too great a pressure buildup. Most are adjustable over a wide range of "pop-off" pressures. Some are designed for air or other gases; others are for liquids only. In the case of air or steam, the excess fluid is vented into the surrounding atmosphere. In the case of liquids, excess liquid is bypassed back to the reservoir. Check valves are frequently of the spring-loaded ball or swing type. In certain areas of operation, codes specify in detail where and how they must be used.

Shear Pins

The shear pin is the mechanical counterpart of the electrical fuse. Figure 16-2 shows one often-employed type. The purpose of the shear pin is to provide a break between the driving and driven members in case of an

Figure 16-1
Relief valve.
(Courtesy of Bell & Gossett, ITT.)

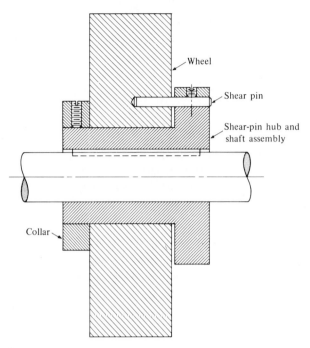

Figure 16-2
Shear-pin application.

overload. Its primary function is to protect valuable equipment, not person-nel directly. The necked-down pin is designed to fail before other compo-nents; thus, it is designed with a slightly lower factor of safety or higher design stress than adjoining components. This type of pin is popular in chain-drive sprockets and conveying equipment, and in similar applications where a jam-up could cause serious trouble. The shear pin is often made of steel when the components are made of cast iron. The pin is inserted into hardened steel bushings and held in place (from lateral movement) by a setscrew. In some applications, brass or bronze are used. The torque-carry-ing capacity of a shear-pin hub can be found from the following equation:

$$T = \frac{S_s \pi D^2 r}{4},$$ $\qquad$ **(16-1)**

where

$\qquad$ r = radius from shaft center to shear-pin center (in.),

$\qquad$ D = necked-down diameter of shear pin (in.),

$\qquad$ S_s = allowable stress for pin material (psi).

Another type of shear pin uses a two-piece key. One part fits into the shaft keyseat, the other into the keyway of the hub. The mating line of the two sections is curved to fit the shape of the shaft. A series of holes is drilled through the two pieces, and an appropriate number of shear pins are inserted to provide shearing and disengagement of the running parts at the desired torque. One advantage of this arrangement is that one two-piece key assembly can be used at various torque levels merely by varying the number of inserted shear pins. The shear pin has a length equal to the height of the two key parts. The strength requirements for keys should be checked before using such an assembly.

Use of Color

Careful specification of colors and color schemes is desirable from the standpoint of operator safety, as is the painting of certain parts and the striping of dangerous areas. Sometimes, embossed or engraved warning signs are also used. Proper placement of warning signs is frequently the responsibility of the designer.

Pastel shades of paint should be used to prevent glare (which could be dangerous). Gray and green are popular colors. Control levers are usually painted with a contrasting color so that they can be easily distinguished.

Remote Control (Electromechanical)

Flexible shafting was mentioned in Chapter 5 as a means of providing remote control. This type shafting, however, has two important limitations. First, the distance between the driving and driven machines is somewhat limited. Second, flexible shafting has an inherent torsional deflection; thus, exact synchronization is impossible. A self-synchronous "master-slave" combination using electrical signals can provide *exact* synchronization. Such an arrangement converts mechanical position into electrical position or electrical position into mechanical position.

Such equipment is useful in working with dangerous materials. Applications include the operation of distant valves and radar operation.

"Dead-Man" Controls

In certain types of equipment, it is desirable and often mandatory to provide control that will automatically shut off if anything should happen to the operator. Such controls are easy to find in either the electrical or mechanical fields. In the electrical field, the usual type is a spring-loaded toggle switch. A force must be applied to the toggle to close the circuit. Thus, when one releases the toggle switch, the circuit opens automatically.

In the mechanical field, the usual type is a valve with an automatic return to the neutral position. Automotive hoists are usually equipped in this manner.

 Indexing Devices

Geneva Mechanism

A traditional device for providing intermittent motion is the Geneva mechanism (see Fig. 16-3). Arm D rotates about its center. As it turns, roller C enters a slot and indexes part B. With four slots, B will rotate one-quarter turn for a full turn of D. After the quarter-turn of motion, roller C will leave the slot and continue around its own center for 270 degrees. As it travels this 270 degrees, part B does not turn. To prevent any motion at all, part A is locked or keyed to crank D and the outside of the Geneva mechanism is shaped to fit exactly the circular arc of A. A small cam follower with needle bearings is often used for the roller. The shape of the locking pad A and the periphery of the Geneva wheel are determined by a kinematic layout. Care must be exercised to ensure that there is no interference of parts.

This type of device provides constant rotation of the driving crank D, but intermittent motion of the driven member B. This mechanism can be provided with any practical number of slots; four, five, or six are common.

Figure 16-3
Geneva mechanism.

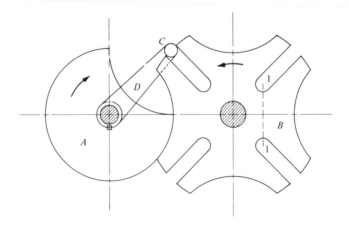

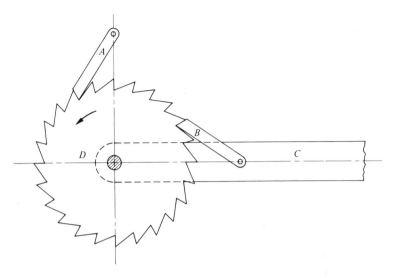

Figure 16-4
Ratchet.

The torque of the driving member must be known. Then, the force at the roller can be found by dividing the torque by the selected radius of the arm *D*. When this force has been determined, a roller of the proper rated capacity can be chosen and the arm *D* checked for strength in bending. Section 1-1 in Fig. 16-3 must be checked when the roller is all the way into the slot. The force exerted by the roller will determine the minimum section required at 1-1. Remember that the force vector at point *C* is always at right angles to the centerline of the arm *D*.

Ratchets

Rachets are useful for indexing. Figure 16-4 shows one type of rachet. Arm *C* has a pawl or detent which engages the teeth of the ratchet wheel *D* when this arm is moved upward. As the arm is lowered, pawl *B* slides past the teeth; when the arm is again raised, detent *B* engages the teeth again. Thus, an up-and-down motion of *C* produces a rotational motion of ratchet wheel *D*. An additional pawl (*A* in Fig. 16-4) is often included to prevent the rachet wheel from reversing direction. The number of indexing positions depends on the number of ratchet teeth. If a greater number of positions is desirable, the number of teeth can be increased. However, increasing the number of teeth means smaller teeth and thus less strength. A greater number of indexing positions can be obtained by using multiple pawls operating from the same pivot as detent *B*, but of different lengths. One additional pawl doubles the number of ratchet positions.

In designing such equipment, care must be taken that the pawl not slip out of the ratchet teeth when the lever is operated. If the mutual perpendicular between the pawl and the ratchet tooth lies between the center of the ratchet wheel and the pivot of the pawl, slipping out cannot occur. The detent should be considered a two-force member and be given sufficient cross-sectional area to withstand the compressive force. Tooth failure usually occurs at the base of the tooth. The area in shear is equal to the circumference of a circle at the base of the teeth divided by the number of teeth and multiplied by the width of the ratchet wheel.

Jack ratchet. Figure 16-5 is a sketch of a simple jack ratchet. Oscillation of the handle engages the upper pawl with the linear teeth; the lower detent prevents the loaded jack from dropping. In this particular type, the detents are designed to withstand the compressive force caused by the weight. Vector analysis is used to determine the amount of this force; then, the arm is considered a free body and the appropriate forces are analyzed. The handle must be designed as a beam. Connecting pins must be designed in double shear. The mechanical advantage of such a jack depends on the distances between the pins and the distance where the force is applied to the handle. Most automotive jacks use a modification of this basic principle. These are designed for both raising and lowering a load. The pawls of the automotive jack are spring-loaded. Also, the vertical, toothed member remains stationary; the frame, detents, and handle all move upward (or downward) as the jack is operated.

Figure 16-5
Simple jack ratchet.

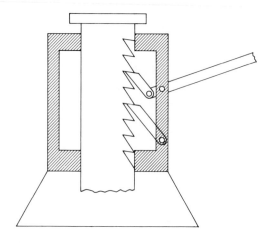

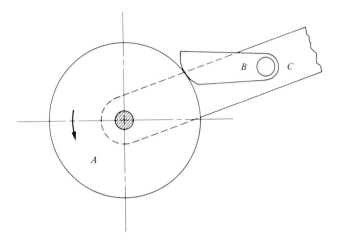

Figure 16-6
Silent ratchet.

Silent ratchet. This type of ratchet can be used for continuous (nondiscrete) indexing. Figure 16-6 shows one arrangement for a silent ratchet. Upward movement of the handle *C* wedges the detent with wheel *A*, producing counterclockwise wheel movement. As the handle is lowered, detent *B* slides past the wheel. Overrunning clutches (discussed in Chapter 10) are actually sophisticated versions of the silent ratchet. High torques cannot be handled by the silent ratchet. Note that friction plays an important role in this type. Careful selection of key dimensions is absolutely necessary for effective operation.

Ratchet applications. In general, ratchets are most useful for indexing light loads. The automotive application (jack ratchets) is an exception to this statement. Ratchets are an important part of many watches and clocks. The ratchet wrench is a convenient and popular device for working in cramped quarters. Ordinary window shades use ratchets. In this unusual application, two detents are placed at 180 degrees to each other and pinned with free fits. The centrifugal force of a fast movement will cause them to spin away from the teeth on the wheel. When the speed is reduced, one of the pawls will drop into its slot by the force of gravity.

The general principles of levers must be considered in determining the operating forces of most ratchets. This elementary concept of mechanics is further complicated in many cases by a force analysis on the operating lever and a study of journal friction (see Chapter 3).

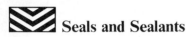

Seals and Sealants

In general, seals are used to keep fluids in a system and contaminants out. The type of sealing depends on what fluids are being used and also whether a static or dynamic seal is desired. The tapering of pipe threads provides an initial sealing step. In any fluid-line connection, it is important to know what fluids are being used and what pressures will be encountered. Commonly used fluids include air, steam, water, oil, and many chemicals. The sealing problem is different for each type of liquid or gas.

Sometimes it is difficult to differentiate between static and dynamic seals. *Static* seals are used for permanent or semipermanent connections. The rubber washer in a garden hose is considered a static seal. The common faucet washer is also considered static, although the seal is broken every time the faucet is opened. The seals used on rotating shafts to retain lubricant and seal out dirt are of the *dynamic* variety. The seals used in hydraulic or pneumatic cylinders around the piston and rod are also of the dynamic type.

Static Seals

Flat gaskets. Simple flat gaskets are often used for seals between two flat surfaces butted together. The surfaces must be flat and smooth. Some typical gasket materials are rubber, compressible cork, asbestos bonded to synthetic rubber, and asbestos cloth spun with brass wire. The materials are available in roll form.

Washers. Washers can be of rubber, synthetic rubber, or fiber composition. The faucet washer is a simple device for sealing; its use, however, is limited to water under low pressure. Rubber washers or "O" rings perform well with water; synthetic rubbers are needed when oil is used as a fluid.

Packings. Packing glands on certain types of pumps and valve stems usually use string packing. This type of packing is available in various materials and shapes. Square flax packing is useful in pumps; round asbestos packing is useful in valve stems. Asbestos wick and candle wick are also used to pack simple valve stems. Hemp and jute are inexpensive packings. Such packings can only be used under low-pressure conditions. In addition, packing glands must be shaped so that tightening will force the packing toward the stem.

Hose clamps. An effective seal can be made between rubber hosing and a tubular section by forcing the hose over a beaded section in the metal part and then clamping the hose in place. Several types of hose clamps are available. Most of them are of the bolted-band type. If leakage occurs, tightening the band will usually stop the leak. A spring clamp is also used. This is less expensive than a bolted-band clamp, but provides no adjustment whatsoever. Automotive radiator hoses and heater hoses are popular applications for clamps.

Metal-to-metal seals. Several types of metal-to-metal seals can be used by a designer. Tapered pipe threads are in this category. Piping compound or "pipe dope" is applied to the male pipe thread. The male and female pipe threads are then tightened, forming the seal. This type of connection is recommended for permanent connections, since frequent coupling and uncoupling will force the threads beyond the effective taper point and thus make sealing impossible. Several types of sealing compounds are available; the choice depends on the fluids used. Typical pipe-thread sizes are listed in Table 17-1.

Another type of all-metal seal is similar in appearance to the garden-hose type of washer; however, the seal is made of a soft metal, such as copper. Annular rings are cut in the faces of the two parts being joined. Force is applied to the sealing ring through the application of torque on a screw thread; this forces the sealing ring snugly into the annular grooves. This type of seal is effective under extreme temperature conditions. The seating (and sealing) of automotive spark plugs is an application of this type.

Flared fittings. Flared fittings can be classified in two different ways. One classification depends on whether the sleeve nut has a male thread or a female thread. The flare of the tubing is the same for either case. The regular flare nut has a female thread; the inverted flare has a male thread on the nut. The inverted type is popular for automotive applications because less space is required. Another way of classifying flared fittings is by the angle of the flare. The SAE type is 45 degrees, whereas the JIC (Joint Industrial Conference) type is 37 degrees. The SAE type is most popular in the automotive field; the JIC type is preferred in industrial applications. Figure 16-7 shows the regular flared fitting; Fig. 16-8 shows the inverted type. The designer must make sure that the parts being connected have the same flare angle.

Flareless fittings. Figure 16-9 shows a common type of compression fitting. This assembly consists of three main parts. A compression sleeve is forced into the recess of an adapter by the screw-thread action of a nut contoured on the inside. This provides a good seal for low to medium pressures. The compression sleeve (ferrule) is made of soft copper when

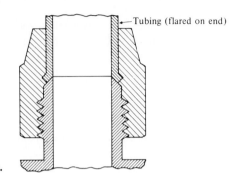

Tubing (flared on end)

Figure 16-7
Regular flared fitting.

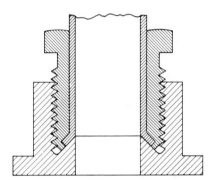

Figure 16-8
Inverted flared fitting.

Figure 16-9
Compression fitting.

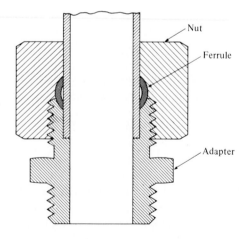

Nut

Ferrule

Adapter

copper tubing is being used. A similar type is popular when plastic tubing (often polyethylene) is used; in the latter case, the sleeve is made of plastic.

Piston Rings

Essentially, piston rings are dynamic seals used in a wide range of operating conditions and for handling many different types of fluids. A piston ring can be made of synthetic rubber in a simple configuration for such an application as a low-pressure air seal under reasonable limits of temperature; the sealing problems for air cylinders are not great. Hydraulic cylinders can have pressures as high as 300 psi; hence, sealing the piston is more complicated. Heavy-duty metallic piston rings are needed in various compressor, pump, and automotive applications. Piston rings in many hydraulic cylinders and pumps are made of Teflon; this material withstands attack from most chemicals, is high in strength, and is usable over a wide temperature range. Internal-combustion engines (gasoline or diesel) demand much of their piston rings because of the high temperatures and varying speeds encountered.

The uppermost compression ring is the one subjected to the highest temperature and pressure during compression and power strokes. Sometimes this ring is chrome-plated on the face. This improves its surface condition and reduces its coefficient of friction. Other engine rings receive somewhat less punishment. Oil-scraper rings are often beveled so that they will pass over the oil during the working stroke; on the return stroke, they collect excess oil. This oil is eventually returned to the crankcase through special holes in the piston.

In designing a piston ring, the exposed annular area is the important consideration. Rings must, of course, be designed against possible failure in shear. The depth and width of the groove are important for proper functioning. The shearing area is equal to the circumference of the piston times the ring thickness; the shearing force is equal to the exposed annular area times the pressure.

The type of application usually dictates the material used in fabricating piston rings. High-grade cast iron with carefully controlled properties of strength and hardness is frequently used. Heat-treating is important in preparing the finished product. If iron is used, it is centrifugally cast; this controls the location of the impurities, which are eventually machined from the ring. Certain pumps use a combination of carbon and bakelite. Bronze is also popular where conditions are not adverse. Piston-ring failure is a serious problem; replacing them is costly. For these reasons, extra care should be used at the start to provide the proper type of ring made of the best material available.

Vibrations and Vibration Control

Vibrations and their control are important factors in most equipment that involves moving parts. The critical speed of shafts was discussed in Chapter 5, and dynamic balancing was covered briefly in Chapter 15. Vibration control is necessary in such equipment as internal-combustion engines, highway vehicles, airborne equipment, marine equipment, delicate instruments, electric motors, compressors, pumps, and transmissions.

The need for shock absorption would be evident if one were to drive an automobile using rigid tires, no springs or shock absorbers, and a seat bolted directly to the frame. Pneumatic tires, rear and front suspension systems, and seat springs are all necessary to provide comfort to the operator or passenger. Vibration control is needed to absorb shock, control noise, and contribute to safety.

Degrees of Freedom

Several vibrational systems can be considered. Figure 16-10 shows respectively, systems with, one, two and three degrees of freedom. The system with one degree of freedom will be discussed briefly; this system is so constrained that it can vibrate in one mode only. The systems with two and three degrees of freedom illustrated in Fig. 16-10 are based on the assumption that only vertical movement is possible. The oscillation of a torsion pendulum about its axis is also considered to be a single degree of freedom.

Figure 16-10
Vibratory systems with one, two, and three degrees of freedom.

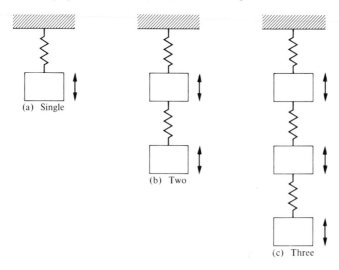

(a) Single

(b) Two

(c) Three

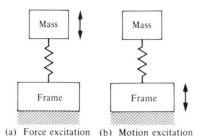

Figure 16-11
Reasons for isolation.

(a) Force excitation (b) Motion excitation

Principal Types of Isolation

Figure 16-11 shows two basic reasons why a system needs to be isolated: force excitation and motion excitation. An example of force excitation is the disturbance created by an internal-combustion engine mounted in a frame for a dynamometer test. The disturbance is in the engine; the absorption system is provided to prevent the frame from vibrating. An example of motion excitation is the disturbance to a bicycle rider created by a rough road. The disturbance is in the frame; spring mounting of the seat isolates the rider from much of the disturbance. Figure 16-12 shows the basic concept of an isolator in a one-degree-of-freedom system. Note that the isolator consists of a spring and some type of damper. The main objective is to convert mechanical energy to heat; this is usually done through friction, although magnetic damping is also used. Friction damping can be obtained with fluid flow (liquid or air) in a cylinder, such as a dashpot.

Mountings to Control Vibration

A flexible coupling (see Fig. 6-28) can be used to isolate one shaft from another. The coupling also performs other functions, such as connecting shafts and allowing for a small amount of angular or linear misalignment between shafts; noise levels are also lower with couplings that have rubber

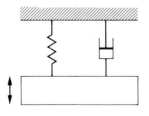

Figure 16-12
Isolation system: spring and dashpot.

Figure 16-13
Plate-form mounting.
(Courtesy of Lord Corp.)

inserts. Figures 16-13 through 16-16 show commercially available absorption units. Many materials are available in sheet or pad form.

Figure 16-13 illustrates a plate-form mounting, a bonded-rubber product that uses the principle of rubber placed in shear. In such a mounting, the plate is attached to a support and the supported component or assembly is attached to the plate by means of a bolt, nut, and lock washer at the center. Two large-diameter washers are placed at the top and bottom. The rubber mounting is provided with snubbing shoulders that cushion any contact between washer and plate that might occur under a shock load. This type of mounting is used for light to medium loads (up to 310 lb). It is often used in appliances and air conditioners.

A tube-form mounting is shown in Fig. 16-14. This is a bonded-rubber shear-type isolator used for steady vibration and intermittent shock under

Figure 16-14
Tube-form mounting.
(Courtesy of Lord Corp.)

Figure 16-15
Center-bonded mount.
(Courtesy of Lord Corp.)

light to heavy loads. The assembly consists of snubbing washers at either end, the supported member on the top, and the usual bolt, nut, and lock washer to hold the assembly together. Load ratings extend to 1500 lb per mounting. The supporting member seats the outside metal cylinder. The metal snubbing washers limit movement under shock or overload.

Figure 16-15 shows a center-bonded mount, often used in the automotive, marine, railroad, and industrial areas. These units are designed to accommodate both U.S. and metric bolts. A cushion of elastomer is bonded to a tubular-steel inner member; the elastomer is designed to extend beyond the inner member, thus accommodating rebound loads and eliminating metal-to-metal bottoming. They provide vibration isolation in all directions as well as good shock protection.

Figure 16-16 illustrates a Flex-Bolt® mounting. This mounting, made with bonded rubber, is available with specially compounded elastomers that perform well under oily conditions. It provides protection from shock and vibration. Simple to mount, it is extensively used in pumps, compressors, engines, and motors.

Damping materials are available in sheet form. In addition, a designer can choose from various available configurations in completing a design.

Figure 16-16
Flex-Bolt® mounting.
(Courtesy of Lord Corp.)

 Designing for Maintenance

Any machine or piece of equipment requires maintenance work from time to time. In some cases, periodic lubrication is needed; in others, parts are replaced at intervals as part of a preventive maintenance program. The designer can reduce maintenance costs by carefully planning in advance. The failure of an inexpensive part can sometimes cause much "down time" and expense because of the labor required to get at the part. Probably the only practical way to prevent such a situation is to design difficult-to-reach parts with a higher factor of safety. The following sections provide a few tips for designers on ways to cut maintenance costs. This is an area where ingenuity and common sense greatly contribute to effective design.

Access Panels and Doors

The internal access needed depends on what the maintenance personnel require. For example, electronic components might be adjusted through a hole only large enough to accommodate an insulated screwdriver. A quick check on screwdriver-blade sizes will reveal the proper opening. However, if the component being adjusted is too far away from the cover, an extra-long screwdriver could be needed. Also, the slots in the parts must be deep enough that the screw-driver blade will not slip out. Another, deceptively obvious, point is that the operator must be able to see what the blade is doing, which implies the presence of a sight hole and adequate lighting for the general area. Figure 16-17 illustrates some of these considerations.

Access panels should be located in a position from which they can be easily removed. A few large fasteners are preferred, whenever possible, to a large number of small ones. The use of thumb screws and special locking fasteners speeds up maintenance work. Access panels should provide ample room for the operator's hand and the necessary tools. Proper wrench clearances must be provided between fasteners and working arcs for wrenches must be provided inside the panel and around the work. Even though ratchet wrenches are helpful in cramped quarters, they are useless unless adequate space is available.

Access doors are used when frequent adjustments or replacements are needed. Hinged doors are often the most desirable. The ideal arrangement is to hinge the door on the bottom or sides. If the door is hinged at the top, anyone working on the equipment has to dedicate a hand, elbow, shoulder, etc., to holding the door open—a procedure not likely to increase productivity. The door must be large enough to permit the introduction and proper use of needed tools. If parts are to be removed through the door, it should be large enough to permit easy removal. If the parts are likely to be heavy, the door opening should be located so that the operator is not subjected to undue physical strain.

Drains and Lubrication Fittings

Drains for oil or water should be located in such a position that wrenching is easily accomplished and the drain flow is in a desirable position. Filling openings for similar fluids should be located so that special equipment is not needed; a filling tube can be helpful if properly located. Lubrication fittings should be located so that oils or greases can be introduced by conventional methods without resorting to special fittings, hoses, or oil-can spouts.

Figure 16-17
Access-hole considerations.

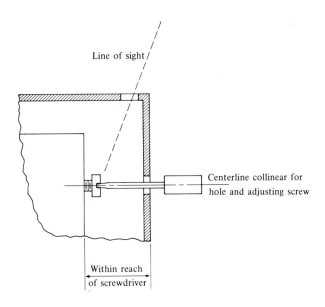

Cable, Wire, and Hose Assemblies

Many electronic circuits use printed circuitry; this often simplifies maintenance, since a complete circuit can be removed and another installed in its place while the defective circuit is taken away for servicing. If wiring is used, proper arrangement of the components can greatly simplify the elaborate wiring harnesses that criss-cross throughout the assembly. Theoretically, only short lengths of wire should be used. This is not always possible in practice, since circuitry is usually complicated and wires cannot always be installed parallel to each other because of unwanted electrical coupling. From the maintenance standpoint, the use of terminal blocks, self-locking connectors, and conventional pin-socket connectors can make defective components relatively easy to disassemble. Terminal blocks require the use of simple hand tools; the other types of connectors can often be unplugged by hand. Color coding can help substantially in saving time and avoiding costly mistakes in reassembly. Most pin-socket assemblies contain guide pins for fast assembly. Figure 16-18 shows a terminal block.

In fluid circuits, hose, tubing, or piping should be designed without unnecessary criss-crossing. A mock-up prior to production can be a great help to the designer. As with electrical circuits, short lengths should be used wherever possible. The use of manifolds (Fig. 16-19) can simplify the design and the assembly operation. Hose, tube, or pipe fittings must be tightened with open-end or open adjustable wrenches; it is impossible to use socket wrenches for tightening or loosening. Therefore, adequate wrench clearance must be provided. Also, the port locations must be arranged so that any hose or tube can be removed without disturbing other connections. Figure 16-20(a) shows a satisfactory arrangement for ports, provided that enough space is allowed between openings to accommodate a wrench. Figure 16-20(b) shows an unsatisfactory arrangement. It would be impossible to disconnect the lines from the interior ports without removing some of the

Figure 16-18
Terminal block.
(Courtesy of Allen-Bradley Co.)

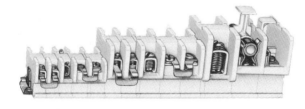

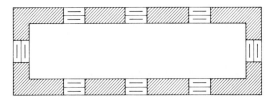

Figure 16-19
Fluid manifold (unneeded ports can be plugged).

outside ones first. The use of quick-disconnect couplings is helpful from the maintenance viewpoint, as well as from the operating point of view. Figure 16-21 shows a unit that seals the fluid in both lines when disconnected.

Check Points

Electrical and mechanical equipment often requires test points. These are trouble-shooting places where test equipment such as meters, gages, etc., can be used to check voltages, pressures, etc., while attempting to pinpoint malfunctions. These check points should be located so that test equipment can be readily positioned and effectively used.

Mechanical Disconnections and Adjustments

If large subassemblies must be removed from equipment, the designer should make the job as easy as possible. Grab bars located in a desirable position with respect to the center of gravity can aid in disassembly; the

Figure 16-20
Port arrangements.

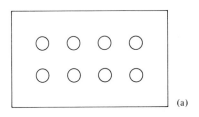

(a)

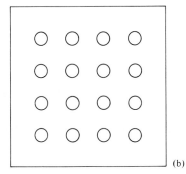

(b)

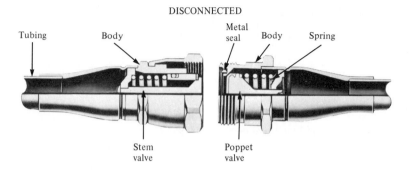

DISCONNECTED

Tubing   Body

Metal
seal   Body   Spring

Stem
valve

Poppet
valve

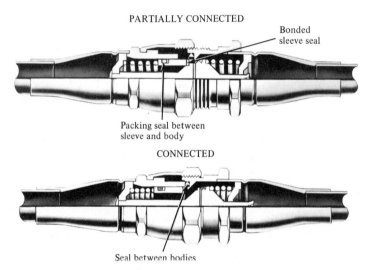

PARTIALLY CONNECTED

Bonded
sleeve seal

Packing seal between
sleeve and body

CONNECTED

Seal between bodies

Figure 16-21
Self-sealing coupling.
(Courtesy of Aeroquip Corp.)

bars should be wide enough to accommodate a hand and deep enough to allow ample hand room.

Mounting bolts should be located in a position from which they can easily be removed. Adequate wrench clearance is necessary. Pulley and sprocket bracket-adjustment bolts for belts and chains should be located in an accessible position for easy adjustment. Bolted shaft couplings are more difficult to remove than couplings secured with splines, keys, and setscrews. Cotter pins are inexpensive and easy-to-remove components; their use can simplify maintenance. For the safety of the mechanic, sharp, protruding edges should be avoided in areas requiring maintenance. It is desirable to have as few bolt and nut sizes as possible so that only a small number of wrenches are needed.

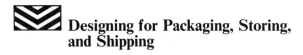 **Designing for Packaging, Storing, and Shipping**

Packing and Packaging

Packing and packaging should be considered during the planning stage. It is difficult to lay out hard-and-fast rules, since different types of equipment require different types of packing. Extra care must be taken in selecting suitable packing materials for delicate instruments. In the case of glass parts, extreme care should be exercised in packing as well as handling. Delicate parts must be isolated with a protective material, such as foam rubber, polystyrene, excelsior (curled shreds of wood), matted paper products, torn strips of paper—or even popcorn. Movable parts must be restrained before shipping. This can be done by properly securing wires, brackets, shipping bolts, or other devices to prevent movement during transport from the manufacturer to the consumer.

Packaging engineers often use a drop test in which the effectiveness of various packing materials is evaluated by dropping the packed product a predetermined distance, then opening the package and examining the contents. Unfortunately, one never knows how much abuse a product will get in the process of storage and transit. The dropping distance is a calculated guess based on observations of handling methods and past experience.

Besides protection against damage by handling, packing often provides protection against corrosion and fumes. Metal parts are often coated with a protective coating of grease or plastic. Some small parts are completely wrapped in waterproof paper or plastic; others are hermetically sealed for long storage under various adverse conditions.

Packaging and Storing

These two terms are interrelated in much the same manner as packing and packaging. In selecting the package size, several factors must be considered. If the product is to be shipped by parcel post, the package cannot exceed the maximum size allowed by postal regulations. The correct carton size can allow the best use of pallets and fork-lift trucks to ease the cost of storage and shipping. The following sizes of pallets are common:

U.S., in.		Metric, mm	
24 × 32	40 × 48	800 × 1 000	1 000 × 1 200
32 × 40	42 × 42	800 × 1 200	1 200 × 1 600
36 × 36	48 × 48	900 × 1 200	1 200 × 1 800
36 × 42	48 × 60	1 000 × 1 000	
32 × 48	48 × 72		

The platform is high enough above the floor to accommodate the fork of a fork-lift truck. In using cardboard cartons, width and length dimensions should be chosen so that alternate tiers can be placed in the opposite direction and exactly fill the available pallet space.

Larger equipment should be bolted directly to a pallet to simplify storage and handling. Thus, overall dimensions should conform to conventional pallet dimensions. Storage space is nonproductive space; thus, as great a quantity as possible should be stored in a given floor space without exceeding allowable floor loads. Proper pallet stacking can cut costs in the storage and handling of finished products.

Shipping

Shipping is the final phase that follows packing and storage. Consideration must be given to the probable shipping methods. One should know the size and weight limitations for shipping by parcel post, air freight, rail, truck, or ship. The methods used to place the product in the hold of a ship, on the railroad boxcar, or in the trailer truck must be considered. For example, if a boxcar is to be loaded by means of an overhead crane or hoist, the clearance between the top of the car opening and the product must be large enough to accommodate the hoisting equipment. If the car is loaded by means of a fork-lift truck, the clearance between the top of the product and the car opening can be smaller.

Excessively large equipment is shipped by dismantling it into readily shipped subassemblies. The finished machine is then assembled at its new site. The subassemblies should be large enough to make maximum use of shipping facilities. If the subassemblies are too small, there will be too many of them; this will complicate on-site assembly. Again, careful attention to these limiting sizes during the design process can lead to more desirable shipping conditions.

Shipping instructions are necessary in the case of fragile equipment, hazardous materials, or unusual tipping dangers. If equipment must be shipped upright, it should be so marked. An unusual center of gravity can sometimes create dangers to personnel and equipment. Provision must be made for such situations. A properly designed and located hoisting eye bolt can help in the handling and placement of certain types of equipment. Hoisting eye bolts are commercially available in shank diameters ranging from $\frac{1}{4}$ in. to $2\frac{1}{2}$ in. Safe working loads are usually specified for each size. Uncrating instructions are also necessary to prevent damage after the equipment arrives at its destination.

 **Control Knobs**

From the industrial-design viewpoint, proper selection of control knobs can enhance the appearance of a product. A designer has many styles and colors from which to choose. Certain knobs are operated by the index finger and thumb. Thus, operating torque is applied to overcome the torque of the rheostat, switch, valve, or whatever component is resisting the movement of the knob. The diameter and finger room needed to supply the external torque must be considered. The knob surface can be plain, knurled, or serrated; in any event, one should be concerned with proper gripping action.

Types

Hundreds of knob configurations are available; most knobs are made of plastic. Figure 16-22 shows an assortment of typical knobs. The skirted types often have markings along the skirt. The use of a skirted knob permits a larger hole opening, which might be necessary for component assembly or adjustment. Note that some bar types are skirted, while others are not. The plain pointed type is usually used when calibrations are placed on the control panel. The bar type and the pointed type permit relatively large amounts of external torque to be applied. Plain knobs can be used with no marking or with a single point marked, often by a dot. Some knobs are equipped with spring clips to hold the unit on its shaft; others are held in position by one or two setscrews. If equipment is subjected to vibrations, a setscrewed knob is usually preferred.

Shaft Types

Figure 16-23 shows cross sections of a few of the various control shafts in common use. Common round shaft sizes are $\frac{1}{8}$, $\frac{3}{16}$, $\frac{1}{4}$, and $\frac{3}{8}$ in. in diameter. Probably $\frac{1}{4}$-in. is the size most often used. The $\frac{1}{4}$-in. size is often ground flat to 0.156, 0.202, or 0.210 in. When referring to Fig. 16-23, one must remember that some shafts are made of steel, while others are made of plastic. Steel shafts are necessary where torque values are relatively high; plastic (low torque) shafts are popular in the electronics field.

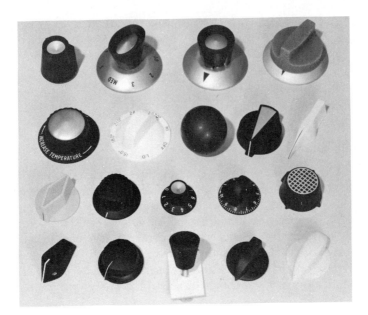

Figure 16-22
Assorted control knobs.
(Courtesy of Rogan Corp.)

Large Knobs

Some equipment uses large-diameter knobs operated by the gripping action of the four fingers and thumb. The larger diameter contributes to the greater operating torque. Again, the diameter must be large enough to fit the hand and the depth great enough to provide adequate finger room.

 Cylinders, Heads, and Plates

Cylinders, heads, and plates constitute a family of mechanical components that have widespread applications. For example, hydraulic cylinders, pneumatic cylinders, and tanks for fluids usually have thin walls; certain gun

Figure 16-23
Shaft extensions for knobs.

barrels have thick walls. Cylinders and tanks are often equipped with flat or dished heads. Utility hole covers and other types of flat plates are subjected to the weight of persons and equipment and must therefore be designed to withstand any forces acting on them. The following sections present key points in the design of these components.

Thin-Walled Cylinders

In the thin-walled cylinder, it is assumed that the stress remains uniform throughout the wall, while it varies in thick-walled cylinders. If the ratio of the inside diameter to the wall thickness is equal to or greater than 15, the cylinder is considered thin-walled.

The circumferential stress of a thin-walled cylinder can be computed by assuming that the two halves of the cylinder are separated from each other by internal pressure (see Fig. 16-24). This is somewhat similar to hoop stress (discussed in Chapter 15). The total force caused by pressure on one half of a cylinder is equal to the resistance offered by the two sides (wall sections) of the cylinder. Thus,

$$pLD = 2tSL \quad \text{or} \quad S = \frac{pD}{2t},$$

and

$$t = \frac{pD}{2S}, \tag{16-2}$$

where

S = tensile stress (psi),

t = wall thickness (in.),

L = length (in.),

p = internal pressure (psi),

D = inside diameter (in.).

Note that the length of the cylinder drops out of the equation and has no effect on the wall thickness.

Longitudinal stress can be found by considering the total pressure. The total force is equal to $\pi D^2 p/4$, applied to a transverse section with an area of πDt. Then,

$$\frac{\pi D^2 p}{4} = \pi DtS \quad \text{or} \quad S = \frac{pD}{4t},$$

and

$$t = \frac{pD}{4S}. \tag{16-3}$$

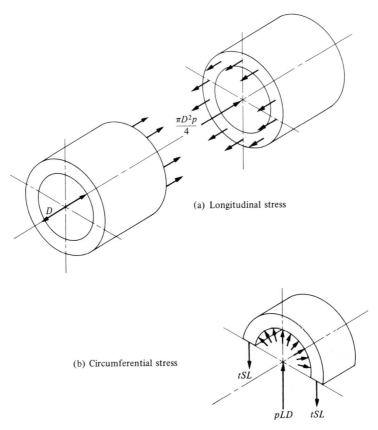

(a) Longitudinal stress

(b) Circumferential stress

Figure 16-24
Stresses in thin-walled cylinders.

Note that the circumferential stress is equal to twice the longitudinal stress; thus, failure would occur by longitudinal rupturing. If riveted joints are used, one must consider the efficiency of the joint.

Note that the inside diameter D was used in these calculations instead of the mean diameter. This does not introduce an appreciable error, since in a thin-walled cylinder the diameter is large relative to the wall thickness.

◆ *Example* ───────────────────────────────

Find the wall thickness of a cylinder with a 12-in. inside diameter; the maximum internal pressure is 200 psi and the allowable tensile stress is 11,000 psi.

Solution.

$$t = \frac{pD}{2S} = \frac{200(12)}{2(11,000)} = 0.109 \text{ in.}$$

It is advisable to check the ratio of the diameter to the computed wall thickness to make certain that this is a thin-walled cylinder. From the above figures, the ratio is obviously greater than 15.

The above discussion refers to internal pressure only. Thin-walled cylinders are sometimes subjected to external pressure. It would appear that the maximum compressive stress would be equal to $pD/2t$; however, this is seldom satisfactory because of nonuniformity of cylinder material and other factors. Extensive experimental work has been done in this area, and empirical formulas have been developed. Such material is beyond the scope of this text.

Thick-Walled Cylinders

In the analysis of thin-walled cylinders, wall stresses are considered uniform throughout because of the large diameter–to–wall thickness ratio. When this ratio is smaller, the circumferential and radial stresses vary throughout the wall section and a different approach is needed. Much research work has been done in this area, and formulas have been developed. Advanced texts and handbooks present these formulas. In using such information, one must note whether the thick-walled cylinder is subjected to internal or external pressure. Gun barrels and heavy-duty cylinders are applications of thick-walled cylinders; high internal pressure dictates heavy wall sections.

Cylindrical Vessels with Uniform Internal Pressure*

The following equations are useful for determining the wall stresses in thick vessels. Figure 16-25 shows the dimensional arrangement and the direction of the stresses. Wall stresses are given in psi units and are

---

*From Roark and Young, *Formulas for Stress and Strain* (5th ed.), Copyright 1975, McGraw-Hill, New York, by permission.

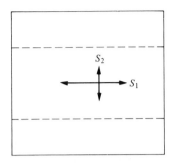 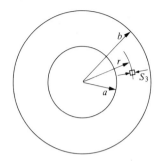

Figure 16-25
Thick-walled cylindrical vessel.

classified as follows:

S_1 = longitudinal,

S_2 = circumferential,

S_3 = radial.

For a thick-walled cylindrical vessel subjected to internal pressure acting in all directions, the stress formulas are as follows:

$$S_1 = p\left(\frac{a^2}{b^2 - a^2}\right), \tag{16-4}$$

$$S_2 = p\left[\frac{a^2(b^2 + r^2)}{r^2(b^2 - a^2)}\right], \tag{16-5}$$

$$S_{2_{max}} = p\left(\frac{b^2 + a^2}{b^2 - a^2}\right) \quad \text{at inner surface,}$$

$$S_3 = p\left[\frac{a^2(b^2 - r^2)}{r^2(b^2 - a^2)}\right], \tag{16-6}$$

$$S_{3_{max}} = p \quad \text{at inner surface,}$$

$$S_{s_{max}} = p\left(\frac{b^2}{b^2 - a^2}\right).$$

◆ *Example* ─────────────────────────────

A steel cylinder with an inside diameter of 2 in. and an outside diameter of 6 in. is subjected to 12,000 psi pressure acting uniformly in all directions.

Find the maximum circumferential stress, the maximum radial stress, and the circumferential stress at the midsection of the wall.

Solution. The maximum circumferential stress is

$$S_{2_{max}} = p\left(\frac{b^2 + a^2}{b^2 - a^2}\right)$$

$$= 12,000\left(\frac{3^2 + 1^2}{3^2 - 1^2}\right)$$

$$= 15,000 \text{ psi.}$$

The maximum radial stress is

$$S_{3_{max}} = p$$

$$= 12,000 \text{ psi.}$$

The circumferential stress at the midsection of the wall is

$$S_2 = p\left[\frac{a^2(b^2 + r^2)}{r^2(b^2 - a^2)}\right]$$

$$= 12,000\left[\frac{1^2(3^2 + 2^2)}{2^2(3^2 - 1^2)}\right]$$

$$= 4880 \text{ psi.}$$

Flat Plates

Flat plates are frequently used in the mechanical field for such things as manhole covers, cylinder heads, pistons, and the sides of rectangular tanks. They can be circular, square, or rectangular in shape.

Several formulas have been developed for the common conditions of loading and supporting. Plate loads can be applied in the following ways, among others:

1. uniformly distributed over the unsupported area,
2. concentrated in the center,
3. concentrated over an *area* in the center.

The supporting function can take several forms, including:

1. freely supported around the periphery,
2. supported in the center,

3. supported around the edge, but fixed (e.g., by bolted or riveted connections),
4. supported along two long edges or two short edges (in the case of rectangular plates).

It is beyond the scope of this text to cover all these situations. The most common application is a uniformly distributed load or pressure over an unsupported area. The plate is often supported around the edge and not rigidly attached.

Plain rectangular flat plates or safety plates with a raised lug pattern are popular for stair treads, platforms, and certain types of manhole covers. It is difficult to determine the stresses in rectangular plates. If it is assumed that the stress is uniform along a diagonal, the following equations can be used to find plate thickness when the plate is subjected to a uniformly distributed load or pressure. If the plate is simply supported along the *four* edges, then

$$t = ab\sqrt{\frac{p}{2S(a^2 + b^2)}} \ . \tag{16-7}$$

If it is fixed at the edges, then

$$t = ab\sqrt{\frac{p}{3S(a^2 + b^2)}} \ . \tag{16-8}$$

In either case,

t = plate thickness (in.),

S = allowable flexural stress (psi),

p = unit load or pressure (psi),

a = width (in.),

b = length (in.).

Simpler versions of these equations can be used for square plates. Floor-plate thicknesses are available in the following sizes: $\frac{1}{8}$, $\frac{3}{16}$, $\frac{1}{4}$, $\frac{5}{16}$, $\frac{3}{8}$, $\frac{7}{16}$, $\frac{1}{2}$, $\frac{5}{8}$, $\frac{3}{4}$, $\frac{7}{8}$, and 1 in. These dimensions do not include the height of the raised pattern.

◆ *Example* ────────────────────────────────

A 24 in. ×36 in. steel plate is made of AISI 1020 steel and is simply supported (by four edges) over an access opening. Find the plate thickness needed if a total load of 1200 lb is to be evenly distributed over the entire plate. Use a factor of safety of 4.

Solution.

$$p = \frac{1200}{(24)(36)} = 1.39 \text{ psi} \quad \text{and} \quad S = \frac{65{,}000}{4} = 16{,}300 \text{ psi};$$

$$t = ab\sqrt{\frac{p}{2S(a^2 + b^2)}}$$

$$= (24)(36)\sqrt{\frac{1.39}{2(16{,}300)\left[(24)^2 + (36)^2\right]}} = 0.130 \text{ in.}$$

Therefore, a $\frac{3}{16}$-in. plate should be used.

◆ *Example* ————————————————————————

Find the induced stress in a 1.25-m-square steel plate fixed at the four edges, assuming that the 500 kg load is evenly distributed and the plate is 5 mm thick.

Solution. Convert 500 kg to newtons.

$$500(9.81) = 4905 \text{ N}.$$

$$p = \frac{4905}{(1.25)^2} = 3139 \text{ N/m}^2 \quad \text{and} \quad a^2 + b^2 = 3.125.$$

$$S = \frac{a^2 b^2 p}{3t^2(a^2 + b^2)} \quad \text{from Eq. (16-8)}$$

$$= \frac{(1.25)^2(1.25)^2(3139)}{3(0.005)^2(3.125)(10)^6} = 32.7 \text{ MN/m}^2.$$

Similar equations can readily be developed for a circular plate with a uniform pressure acting on a surface supported only at the edges. If the plate were to fail, rupture would probably occur across a diameter. If one-half of the plate is considered, the total force due to the pressure is $\pi r^2 p/2$ and can be considered as acting through the centroid of the semicircular half. The section modulus is equal to $Srt^2/3$. The centroidal distance for the total force on one-half of the plate from the center is $4r/3\pi$ and the centroidal distance for the reaction on one-half of the plate is $2r/\pi$. See Fig. 16-26.

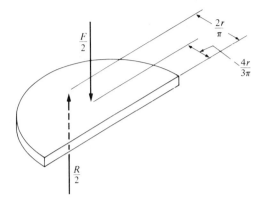

Figure 16-26
Total force and total reaction with centroidal distances for a one-half circular plate under uniform pressure, supported at the edges.

Thus, the distance between the two centroids for the total force and the reaction for one-half of the circular plate is

$$\frac{2r}{\pi} - \frac{4r}{3\pi} = \frac{2r}{3\pi}.$$

Then, considering flexure, we can develop the following equation:

$$M = SZ \quad \text{or} \quad \frac{\pi r^2 p}{2}\left(\frac{2r}{3\pi}\right) = \frac{Srt^2}{3}$$

and

$$r^2 p = St^2 \quad \text{or} \quad t = r\sqrt{\frac{p}{S}}.$$

Since the diameter is usually easier to work with, the plate thickness in terms of diameter becomes

$$t = \frac{d}{2}\sqrt{\frac{p}{S}}. \tag{16-9}$$

◆ *Example* ⎯⎯⎯⎯⎯⎯⎯⎯⎯⎯⎯⎯⎯⎯⎯⎯⎯⎯⎯⎯⎯⎯

A cylinder head is subjected to 200 psi of pressure. The inside diameter of the cylinder is 6 in.; the allowable stress in flexure is 11,000 psi. Assuming uniform support at all parts of the edge, find the plate thickness.

Solution.

$$t = \frac{d}{2}\sqrt{\frac{p}{S}} = \frac{6}{2}\sqrt{\frac{200}{11,000}} = 0.405 \text{ in.}$$

Dished Heads

If cylinder heads are dished, the equation for flat circular plates does not hold. The following equation can be used if the radius to which the head is dished does not exceed the diameter of the head between supporting edges and if the allowable flexural stress includes a factor of safety of at least 5. This equation is also based on the pressure acting on the concave side of the dished head (see Fig. 16-27):

$$t = \frac{5pr}{6S},$$ (16-10)

where

  r = radius to which the head is dished (in.),

  t = head thickness (in.).

If the pressure acts on the convex side of the head, the thickness obtained by the preceding equation must be multipled by $\frac{5}{3}$.

◆ *Example* ─────────────────────────

A cylinder head is dished to a radius of 12 in. and is subjected to pressure on the concave side. The head is $\frac{3}{8}$ in. thick and the allowable stress for the material is 10,000 psi. What is the permissible pressure? For the same thickness, find the permissible pressure if it acts on the convex side of the dished head.

Solution.

$$p = \frac{6St}{5r} = \frac{6(10{,}000)(0.375)}{5(12)} = 375 \text{ psi};$$

$$p_1 = \frac{3p}{5} = 0.60(375) = 225 \text{ psi}.$$

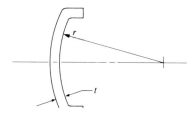

Figure 16-27
Dished head.

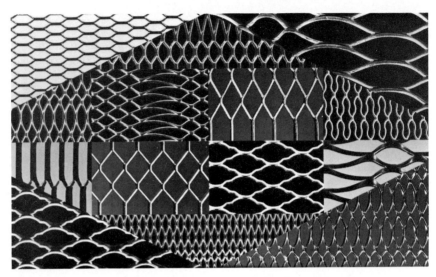

Figure 16-28
Decorative mesh.
(Courtesy of Wheeling Corrugating Co.)

Gratings and Grilles

Machinery in the process industries is large and complex. Trench covers, catwalks, and stair treads are needed in certain areas. If ventilation, heat dissipation, or light is needed, some type of open steel or aluminum grating is often used. If such units are simply supported at the two ends that provide the greatest numerical value for section modulus, the design problem is handled as if the grating were a beam. Assumptions must thus be made for the types and values of loading.

Expanded metal or decorative mesh is often used in grilles, where appearance is important. Some typical configurations are shown in Fig. 16-28.

 Design of Two-Force Members

Many mechanical devices use links or connecting rods that are classified as two-force members. This implies that the loading is axial (either tensile or compressive).

If the loading is tensile, direct stress is applied to the section and the length of the member is not significant. Links or rods loaded in this manner are calculated with direct-stress equations; the allowable tensile stress governs the design. If compressive loads are involved, the members must often be checked as columns.

Columns

Column design is essentially a structural-design problem. However, many two-force links and rods must be checked as columns. Columns can be classified as short, intermediate, or long. If a link or rod is considered a short column, the induced stress is purely compressive and the value for the stress is P/A. In general, if the unsupported length of a column is not greater than 10 times the least transverse dimension, any bending or buckling effect can be neglected. If a column is exceptionally long, failure is usually due to buckling. The capacity of slender columns depends on the stiffness of the material. In the mechanical field, most rods and links fall into a third category between the short and slender classifications, that of intermediate columns.

Slenderness ratio. The slenderness ratio, or L/k ratio, is defined as the ratio of the length to the least radius-of-gyration value. The following gives slenderness ratios that might be expected for different classes of steel columns:

Type of Column	L / k
short	0 to 25
intermediate	25 to 120
slender	over 120

End connections. The method used to restrain the end of a column has a great effect on the allowable or critical load that can be applied axially. Most situations fall into one of the following four cases:

1. both ends hinged;
2. one end hinged, one end fixed;
3. both ends fixed; and
4. one end fixed, one end free.

Case 1 is the most common in the mechanical field.

Euler's formula. The traditional Euler column formula is as follows:

$$\frac{P}{A} = \frac{\pi E}{(L/k)^2}. \qquad\qquad \textbf{(16-11)}$$

The formula can be used in this form if both ends are hinged (Case 1). In Case 2, the effective length is equal to approximately 0.7 times the total length; in Case 3, to 0.5 times the total length; and in Case 4, to twice the total length. Thus, if total-length values are used in Euler's formula, the following modifications must be made:

$$\frac{P}{A} = \frac{2\pi E}{(L/k)^2} \qquad \text{for Case 2,}$$

$$\frac{P}{A} = \frac{4\pi E}{(L/k)^2} \qquad \text{for Case 3,}$$

$$\frac{P}{A} = \frac{0.5\pi E}{(L/k)^2} \qquad \text{for Case 4.}$$

Unfortunately, the area is unknown at the beginning of most design problems; thus the (L/k)-value is also unknown. For this reason, trial-and-error methods are employed. Past design experience plus some calculated guesses based on direct stress and the probable (L/k)-value are helpful in determining suitable final values.

J. B. Johnson formula. This empirical formula, sometimes called the parabolic formula, is useful for finding the critical load for columns. Since it is based on the stress at the yield point and checks out well experimentally, it is often recommended for aeronautical and machine-design work. The critical load, P, is determined from the following relationship:

$$P = AS_y \left[1 - \frac{S_y (L/k)^2}{4c\pi^2 E} \right], \qquad (16\text{-}12)$$

where

S_y = yield stress (psi),
c = constant for end connection
= 0.25 for one free end and one fixed end
= 1 for both ends hinged
= 2 for one fixed end and one free end (but guided)
= 4 for both ends fixed.

All other symbols are the same as for Euler's formula.

◆ *Example* ──────────────────────────────────

Using the J. B. Johnson formula, find the critical load for a 2-in.-diameter steel column with a length of 4.5 ft. This column is a connecting rod with pin connections used in a piece of construction equipment. The ultimate tensile strength is 103,000 psi, and the yield stress is 90,000 psi.

Solution. For a circular (solid) section,

$$k = \sqrt{\frac{I}{A}} = \sqrt{\frac{\pi d^4/64}{\pi d^2/4}} = \frac{d}{4} = \frac{2}{4} = 0.5 \text{ in.}$$

The critical load is then

$$P = AS_y \left[1 - \frac{S_y(L/k)^2}{4c\pi^2 E} \right]$$

$$= \frac{\pi(2)^2}{4}(90,000)\left[1 - \frac{90,000(54/0.5)^2}{4(1)(\pi)^2(30 \times 10^6)} \right]$$

$$= 32,130 \text{ lb.}$$

To be on the safe side, a factor of safety should be applied to the critical load so that the actual applied load will be less than the calculated value of 32,130 lb.

◆ *Example*
A 3 in. × 3 in. × $\frac{1}{2}$ in. steel angle is used as a column in compression. The effective length is 6 ft; both ends are fixed. Using the J. B. Johnson formula, find the critical load for a yield stress of 50,000 psi, assuming a factor of safety of 5. The following information is available for a 3 in. × 3 in. × $\frac{1}{2}$ in. angle:

cross-sectional area = 2.75 in^2,

moment of inertia = 2.20 in^4,

centroidal distance = 0.93 in.

Solution. First, determine the value for k and the L/k ratio:

$$k = \sqrt{\frac{I}{A}} = \sqrt{\frac{2.20}{2.75}} = \sqrt{0.8} = 0.894.$$

Then

$$L/k = 72/0.894 = 80.5;$$

therefore,

$$P = AS_y \left[1 - \frac{S_y(L/k)^2}{4c\pi^2 E} \right]$$

$$= 2.75(50,000) \left[1 - \frac{50,000(80.5)^2}{4(4)(\pi)^2(30 \times 10^6)} \right]$$

$$= 128,000 \text{ lb.}$$

This is the critical load without a factor of safety. Applying the factor of safety of 5, we obtain

$$P = 128,000/5 = 25,600 \text{ lb.}$$

Questions for Review

1. Carefully examine the safety devices of some piece of large equipment. List the mechanical and electrical safety devices and indicate their function.

2. Differentiate between an interlock and a lockout device. Give an example of each.

3. How does a point-of-entry guard differ from a regular machine guard? Give an example of a point-of-entry guard.

4. Name five consumer products that required mechanical design changes in order to conform to the standards established by the Consumer Product Safety Commission.

5. What is meant by a "dead-man" control on electrical or mechanical equipment? Give two examples.

6. What is a relief valve? Name a piece of equipment requiring such a device.

7. How does a Geneva mechanism differ from a ratchet? Give an application of each and explain why each device is used.

8. What advantage does a silent ratchet have over the regular type?

9. Differentiate between a static and a dynamic seal. Give two examples of each.

10. What type of excitation does a tractor engine produce with respect to the engine frame?

11. Examine a modern street bicycle and list all obvious safety features that conform to the U.S. Consumer Product Safety Commission's standard for bicycles. Note any violations of this standard.

12. What considerations should be made in providing access panels for mechanical equipment?

13. Obtain a torque wrench and some sockets for socket wrenches. Reversing the usual procedure, grip the sockets between the thumb and index finger and note the amount of torque required to move each size of socket through a part turn. Tabulate your results, measuring the angle of turn as accurately as possible. What conclusions can be drawn from varying the size of socket? How does this compare with knob-turning torque?

14. Differentiate between a thick-walled cylinder and a thin-walled cylinder. Give an everyday example of each.

15. How does the Consumer Product Safety Act differ from the Occupational Safety and Health Act?

16. If a consumer product produces noise at a level above the Federal standard, what steps can be taken to reduce the noise?

Problems

1. Some equipment is packaged in boxes 18 in. wide, 24 in. long, and 10 in. high. The packages must be kept upright. If the boxes are to be pallet-loaded and stored, what size of pallet would be most desirable? Using a sketch, show how the boxes could be stacked.

2. A thin-walled cylinder is 15 in. in diameter (inside) and contains fluid at a pressure of 150 psi. The wall is 0.125 in. thick. Find the induced stress in the cylinder wall.

3. A cylinder is subjected to an internal pressure of 8000 psi. The inside diameter of the cylinder is 12 in. and the wall is 1.5 in. thick. Determine the maximum stress induced and indicate where this stress is located.

4. A double ratchet is to be used in a mechanism that is to index in 30 positions over a complete revolution. How many ratchet teeth are needed?

5. A single shear pin is used in an application similar to Fig. 16-2. The design shearing stress for the pin material is 18,000 psi. The pin diameter is $\frac{1}{8}$ in. (minimum) and the pin is located 1.50 in. away from

the shaft center. If the operating speed of the shaft is 1200 rpm, how many horsepower are needed to cause pin failure?

6. A flat, circular cylinder head has a diameter of 8 in. and a thickness of $\frac{3}{8}$ in. If the allowable flexural stress is 10,000 psi, how much pressure can it sustain?

7. A cylinder has a dished head with a radius of 10 in. An internal pressure of 200 psi acts on the concave side of the dished head. Find the thickness of the head if the allowable stress is 11,000 psi. If the pressure were applied to the convex side of the head, what would happen to the thickness?

8. A thick-walled cylinder has an inside diameter of 2 in. and an outside diameter of 8 in. The pressure inside the cylinder is 15,000 psi, evenly distributed. Find the circumferential stress at 1-in. increments from the inside of the wall to the outside. What conclusions can be drawn from the results?

9. A steel thick-walled cylinder with an inside diameter of 2 in. is subjected to an internal pressure of 5000 psi. If the maximum circumferential stress is not to exceed 12,000 psi, how thick must the wall be?

10. A steel piston rod in a hydraulic cylinder is 1 in. in diameter and 24 in. in length. Assume that the modulus of elasticity for steel is 30,000,000 psi. Using Euler's formula, find the slenderness ratio and the maximum load that the rod can take in compression. Assume that both ends of the rod are hinged. (*Note:* The radius of gyration of a circular section is equal to $D/4$.)

11. A steel thick-walled cylinder with an inside diameter of 50 mm is subjected to an internal pressure of 400 bars or 40 000 000 Pa. If the maximum circumferential stress is not to exceed 82 MN/m², how thick must the wall be?

12. A thin-walled cylinder has an inside diameter of 280 mm; its maximum internal pressure is 1.6×10^6 Pa. Find the wall thickness if the allowable tensile stress is 80 MN/m².

13. A 30-mm steel rod 1 m long is used as a column with both ends fixed. If the yield stress is 130 MN/m² and $E = 207$ GN/m², find the critical load according to the J. B. Johnson formula. Then, apply a factor of safety of 5.

14. A round vertical steel tank is 6 m in height and has an inside diameter of 3 m. It contains gasoline with a mass density of 725 kg/m³. If the allowable tensile stress is 25 MN/m², how thick must the wall be?

15. Calculate the shear-pin diameter for an arrangement similar to that shown in Fig. 16-2 according to the following conditions: operating torque = 1000 N · m, allowable shearing stress for the pin material = 56 MN/m², and center-to-center distance between pin and shaft centers = 80 mm.

16. A 100-mm-diameter cylinder has a flat head with a thickness of 10 mm. If the allowable stress for the head material is 67 MN/m², how much pressure can be applied?

17. A cylinder with a dished head is subjected to pressure on the convex side of the head. The allowable stress for the head material is 67 MN/m², the radius of the dish is 300 mm, and the head material is 10 mm thick. How much pressure can the head withstand?

POWER UNITS

17

Available Types of Power

In many designs, a final (though preplanned) step consists of selecting a suitable drive for the machine. Several types of power units are available, each of which has both good features and disadvantages. The purpose of this chapter is to survey what is available and suggest how to make an intelligent choice. Each type will be considered and evaluated separately.

Turbines

Turbines are a heavy-duty source of mechanical power. The three main types are (1) hydraulic, (2) steam, and (3) gas. Hydraulic and steam turbines are classified as either impulse or reaction turbines. The reaction type depends on pressure for providing rotary motion to the blades; the impulse type operates the rotor by the direct impingement of the fluid on the blades. The hydraulic type is principally used in the hydroelectric power plant.

Steam and gas turbines are widely used; both are small in comparison to a reciprocating engine of similar capacity. Steam turbines are used to drive alternators, blowers, centrifugal pumps, and ship propellers.

Gas turbines consist of three basic parts: a compressor, a combustor, and a turbine wheel. Gas turbines are used in locomotive, aircraft, and marine applications. Steam power plants frequently use gas turbines as standby equipment. Mobile gas turbines are useful for generating electric power in remote areas. Gas-turbine capacities can range from 25 hp to over 30,000 hp. It is popular in pipeline operations for driving centrifugal pumps, blowers, and related equipment.

Internal-Combustion Engines

Gasoline and diesel engines are classified as internal-combustion engines. In both, combustion takes place within a cylinder and applies force to a piston (power stroke), which in conjunction with a connecting rod and crankshaft changes the linear motion into circular motion (rotation). The entire process is one of transforming chemical energy into mechanical energy. Both engines can use a two-stroke or four-stroke cycle. A stroke is an up or down movement of the piston; thus, two strokes are needed for one revolution of the crankshaft. The sequence of events for the four-stroke engine is as follows:

1. intake (air-fuel mixture in gasoline engines; air in diesel engines),
2. compression,
3. expansion (power),
4. exhaust.

In gasoline engines, the fuel-air mixture is ignited between steps 2 and 3 by means of a carefully timed spark. In diesel engines, fuel is injected between steps 2 and 3. The compression ratio of the diesel is extremely high and the air correspondingly hot; the injected fuel immediately ignites. In the four-stroke cycle, there is one power stroke for every two revolutions of the crankshaft, or one power stroke for every four piston strokes.

The sequence of events in the two-stroke cycle is as follows:

1. intake and compression,
2. expansion (or power) and exhaust.

Therefore, in the two-cycle engine, there are two power strokes for two revolutions of the crankshaft, or one power stroke for every two strokes. Ignition (or fuel injection) occurs after step 1. In this type of cycle, the exhaust and intake steps are combined near the end of the power stroke. In gasoline engines, some of the air-fuel mixture is lost and all the exhaust gases may not be expelled; therefore, the efficiency is not very high. However, this is often offset by the fact that this engine is not nearly as complicated as the four-stroke engine and, therefore, is cheaper to build. The two cycle gasoline engine is often used in small lawn mowers, where initial cost is a more important consideration than operating cost.

Gasoline engines. Gasoline engines are popular, inexpensive, and highly portable power units. They are sometimes used as stationary engines, particularly for generating emergency electric power. Deadly carbon monoxide fumes limit the use of this type of equipment; if used indoors, adequate ventilation must be provided. Gasoline engines are popular for outdoor uses, such as automobiles, lawn mowers, airplanes, pumps, compressors, and some types of farm equipment. In considering this type of engine, it is necessary to evaluate carefully the horsepower requirements, speed requirements, and vibration tolerance of the application. Adequate heat dissipation is important. The number of cylinders in any internal-combustion engine is important; the fewer the cylinders the fewer impulses to the crankshaft.

Diesel engines. Diesel engines are much more efficient than gasoline engines. The diesel engine can have an efficiency of nearly 40%. Diesel engines are frequently used for railroad and marine engines and for driving emergency electric generators. Since the fuel-injection system is highly complex, diesel engines are often more difficult to start than other types. The fuel-injection system replaces the carburetor of the gasoline engine; ignition does not require spark plugs (or distributors). Two-cycle diesel engines are much more popular than two-cycle gasoline engines. The two-cycle diesel is used to power buses, trucks, compressors, and construction equipment. Diesel power is reliable and comparatively low in cost. Like gasoline engines, diesels must be used outdoors or with adequate ventilating systems.

Electric Motors

The electric motor is probably used in a wider variety of applications than any other power unit. Many types are available. Before making a selection, the designer must think carefully about environmental conditions and consider the costs involved in installing the motor, operating it, safeguarding it, and providing for the safety of personnel.

Electric motors are classified in various ways. The designer must decide whether the application requires a general-purpose motor, a special-purpose motor, or a definite-purpose motor. Motors are also classified according to the type of electricity used: A motor can operate on either direct current or alternating current. Alternating current can be single-phase or polyphase. Polyphase alternating current can be two-phase or three-phase. Although one type of current can be converted to another, this process is generally expensive. Therefore, the type of current available should be carefully checked. If polyphase current can be used, the motor will be lighter in weight for a given capacity and will be cheaper to operate. Available voltages and frequencies should also be checked; 60-cycle, 115 and/or 120 volt operation is fairly well standardized in the United States.

The machine on which the motor is being used should be carefully checked, as well as the conditions surrounding the machine. The horsepower requirements and speed must, of course, be noted; if a speed reducer is needed, this will increase the cost. The common speeds available for ac motors are 3600, 1800, 1200, 900, 720, 600, 514, 450, and 400 rpm.

Environmental conditions dictate the type of enclosure needed. Drip-proof or splashproof enclosures are needed to protect the motor against dripping or splashing liquids or other falling objects in the vicinity. Motors must be totally enclosed in dusty areas. Explosion-proof motors are needed in most chemical applications. Ventilation is important in some uses. Ambient temperatures must be considered.

The designer should determine the method for mounting the motor. Mounting lugs (known as foot and flange mountings) are available for either horizontal or vertical mounting. The motor dimensions should be verified and coordinated with the equipment being designed. In some applications, space or allowed weight may be limited. The direction of shaft rotation should be checked.

The type of service that the motor must provide is important. Some are run intermittently; others, continuously. Some driven machines have a constant load; others, a varying load. On some, the speed is constant; on others, it varies.

Performance ratings should be considered. Nameplate (rated) information on a motor should be carefully checked. If the motor will be operated for over an hour at full load, it should be determined whether or not the motor is rated for continuous operation; many units have a rated temperature rise of 40 °C for continuous operation. Service factors can be applied to limit the amount of overload to a certain value; when operated at an overload, however, the temperature rise, efficiency, power factor (in the case of ac), and speed will vary from the rated nameplate values.

Controls are needed in all applications, to protect both operator and equipment. The simplest type is the manual control—basically, an on-off switch. In remote-control operations, magnetic controls use electromagnets to open or close the electrical contacts. It is desirable to have overload and low-voltage protection to break the circuit if the voltage varies more than the usual amount. Short-circuit protection can be supplied by fuses or suitable circuit breakers. Controls are expensive; therefore, their cost must be considered in addition to the cost of the motor.

The following list is a "thought starter" for selecting motors. This list is not complete; instead, it provides comments, characteristics, and applications about some of the commonly used types. The designer should consult motor manufacturers' catalogs before making a selection.

DC Motors

Shunt	constant-speed motor
	rheostats in armature and field circuits are used to regulate speed
	little change in speed between no-load and full-load conditions
	used for duplex and triplex pumps, compressors
Series	does not have a no-load speed
	load must be connected to motor shaft or speed can become dangerous
	high starting torque
	used for hoists, cranes
Compound	has no-load speed
	higher torque than shunt motor
	good for irregular loads such as punches and presses
	used for single-acting reciprocating pumps

AC Motors (polyphase)

Squirrel cage	rugged construction
Induction	constant speed
	no moving electrical contacts
	low starting torque

	used for elevators, reciprocating pumps, compressors, woodworking equipment
Wound rotor (slip ring)	high starting torque
	speed can be controlled
	used for reciprocating pumps, compressors
Synchronous	high efficiency
	not self-starting
	used for constant-speed applications, 20–100 hp and up (compressors and dc generators)

AC Motors (single-phase)

Series	high starting torque
	high speeds with light loads
	universal type (operates on ac or dc)
	used for vacuum cleaners, sewing machines, portable tools
Repulsion	high starting torque
	high speed at light load
	used to start induction motors
Repulsion-induction	high starting torque
	constant running speed
Split-phase	used to start induction motors
	used for refrigerators, air conditioners, freezers, and other compressors with high starting load
Synchronous	constant speed
	used on electric clocks, timers, phonographs

Figure 17-1 shows a cutaway of a dripproof motor. Some of the construction features illustrate and emphasize the way mechanical design enters into electrical-equipment design. Note the sleeve-bearing construction on the left end of the shaft, and the single-shielded ball-bearing construction on the right. Pressed on the shaft ahead of the ball bearing is a bearing washer. The stator is made of heavy steel plate, and the steel foot assembly is welded to the frame ring. Ventilating fans and air guides are used to direct air flow over the core and coils. The air openings are fully dripproof. The terminal box is placed for maximum versatility. The shaft is equipped with a keyseat. Holes are provided in the foot for easy mounting. The foot is finish-milled so that no preparation is needed before the motor is used.

Timing Motors

Many precise timing devices use small synchronous motors, some as slow as $\frac{1}{2}$ rpm. Many are used at a speed of 1 rpm, and others at speeds of 2 to 10 rpm. This type of motor is definitely a subfractional-horsepower type. Ratings are usually given in pound-inches. Usually, these motors have shafts

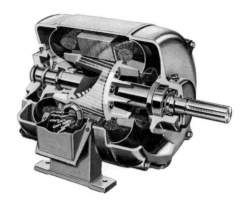

Figure 17-1
Dripproof motor.
(Courtesy of Gould Inc., Electric
Motor Div.)

$\frac{3}{16}$ in. or $\frac{1}{8}$ in. in diameter. The shaft is often $\frac{1}{2}$ in. to $\frac{7}{8}$ in. long. One of the chief uses of such motors is for the automatic sequencing of operations in such appliances as washing machines and dishwashers. The slow speed of 1 rpm is often geared down to even slower operation; in that case, the motor carries a lightweight cam that operates contacts or small snap-action switches. These motors are also used to operate small lightweight displays in showrooms and store windows. These motors are small in size as well as horsepower. Many are only $1\frac{1}{2}$ in. in diameter and 1 in. in length. Manufacturers provide either horizontal or vertical mounting brackets; this expands the possible applications.

Metric Motors

Specifications for metric electric motors use SI units almost exclusively. Frame dimensions and shaft sizes are specified in millimeters; output power capacity is shown in watts or kilowatts; and torque is expressed in newton-meters. However, speeds are generally noted in revolutions per minute rather than in radians per unit of time. The radial and axial load capacities of gear motors are both noted in newtons. The input requirements of metric motors are specified by voltages combined with hertz frequency units.

 Electrical Components

As previously mentioned, electrical controls are used in conjunction with motor operation. A good mechanical designer should be familiar with the applications of electrical components. Much so-called mechanical design is in fact *electromechanical* design. Electrical-equipment handbooks and periodicals should be consulted for specific application data. A general discussion of some commonly used electrical assemblies and parts follows.

Circuit Protection

One of the simplest devices for circuit protection is the fuse. A fuse is designed to melt and break the electrical circuit when a short circuit occurs or when the equipment is overloaded beyond its rated capacity. The two common types are the plug (requiring an appropriate socket) and the cartridge (requiring a holder). Time-lag fuses are used if a temporary overload is planned, such as the starting of a motor-driven compressor. Circuit breakers perform the same function as fuses; they are generally used when circuits are subjected to frequent overloading, and also on large-capacity applications. Circuit breakers can be rapidly reset; thus equipment can be placed in operation promptly after the overload conditions have been corrected. Circuit breakers are more expensive than fuses; however, spares are essential when fusing is used. The cause of the malfunction must be determined and corrected before circuit breakers are reset or fuses are replaced.

Motor Controls

Motor controls are used to provide remote control of a motor and, at the same time, protection for the motor and associated equipment. In addition, controls protect the equipment operator; this is a particularly important requirement because the typical machine operator is not knowledgeable in the field of electricity. A few of the usual controls for single-phase ac operation follow. The simplest type of control is the manual switch; this is usually a snap-type switch with two positions: *on* and *off*. It is usually provided with thermal overload; thus, when an overload is detected, the switch opens. Another familiar type of switch is designated *start* and *stop*; this is pushbutton-operated and contains overload breakers. The breakers are reset manually after an overload. Other types of switches can be solenoid-operated. These are often multiple units designed to operate valves where timing is an important consideration. Solenoid starters are common electrical control assemblies.

Special Switches

Various types of special switches are used in electromechanical products. Several varieties of toggle switches are found. Some are spring-loaded so that the circuit is closed only as long as pressure is applied to the lever.

Figure 17-2
Heavy-duty limit switches.
(Courtesy of Micro Switch, a division of Honeywell, Inc.)

Rotary switches with numerous positions are useful in switching sequences. Float-controlled switches are available; sump pumps use this type. Certain switches are activated by changes in temperature or pressure. Most air compressors are equipped with a switch that shuts off the motor when the tank reaches a preset pressure value. Foot-operated remote-control switches are useful in operating small machine tools. Cam-limit switches are used to carry out operation sequences in such applications as washing machines and dishwashers. When used in conjunction with a timing motor (slow speed, low power), they can operate solenoid valves for controlling water supply and discharge, as well as starting and stopping other components such as agitators. Most of these cam-limit switches can be built into one unit for performing several functions.

Single snap-action switches are available in miniature sizes with various choices of actuators. Figure 17-2 shows one of the types available. This type of switch is often completely sealed against dust and fluids. The designer should carefully study the possible mountings for the switch and design the other components, such as cams, eccentrics, or levers, so that contact will be made when desired. Pretravel and overtravel must be carefully considered with reference to the manufacturer's specifications.

Electrical Accessories

In the electronics field, printed circuitry simplifies construction considerably. Soldering is used for permanent connections; however, it is often desirable to disconnect components from a circuit for testing or replacement. Connector plugs (male and female) are commercially available in various types. Terminal lugs and terminal blocks are often used. Some of these use screwed connections; others use friction and spring-type connection of the male and female parts. While the electrical designer selects most of these parts, the mechanical designer is often responsible for seeing that all the parts fit compactly into the finished assembly; also, the mechanical designer must provide adequate ventilation and heat dissipation. Manufacturers' catalogs, with mounting data, must be carefully studied.

Photoelectric Controls

Photoelectric controls, with appropriate light sources, are used in the production and inspection areas. Originally, this type of control was used primarily for operating automatic doors; when the light beam was interrupted, the door-opening mechanism was started. These same components are now in the following applications:

1. smoke alarm,
2. counting bottles, cartons, or parts,
3. operating water nozzles for automatic car washers,
4. checking part sizes,
5. protecting operator's hands in press operations,
6. sensing marks on cartons for sorting purposes,
7. loop control for proper feed in fabric processes,
8. alarm for break in continuous roll operations,
9. tape coding.

Automation has substantially increased the popularity of photoelectric controls.

Electronic Data Displays

An electro-mechanical designer has modern methods of displaying data effectively and often eliminating bulky meters. This is particularly important if space is limited and if human error in taking readings must be minimized.

Liquid crystal displays, and light-emitting diodes (LEDs) are used extensively in clocks, watches, and hand-held calculators. Crystals of gallium doped with arsenide or phosphide can produce red, green, or amber light. Liquid crystals reflect light and do not produce any. They have the fluidity of liquids and some of the optical properties of crystals. In the digital display, the molecules of liquid crystal align themselves with an applied electric field, thus changing the refractive index of the liquid. A properly positioned strip of Polaroid makes this change visible to the viewer. Some units are now available in the form of a bar-type graph.

Other display units use a dot matrix where a number of dots form the letters or numbers. These can be produced by a number of different techniques. Dot matrix displays are very readable, and hence their application is becoming widespread.

 **Reducers**

Commercial speed-reducer units are available in various ranges of speed. Some are designed for a definite input and output speed; others for a varying speed. Speed-reduction units are available in which the motor is already provided by the manufacturer. This eliminates the problem of providing a shaft coupling, and also makes the assembly more compact. In selecting a reducer or motorized reducer, the designer must check the following:

1. horsepower (or torque) capacities;
2. speed of output shaft;
3. shaft size(s) and provisions for coupling (such as keyseats);
4. location of shaft(s)—vertical and horizontal position; and
5. direction of shaft rotation.

Reducers are designed with spur, helical, herringbone, or worm gears. In some reducers, the output shaft runs through the unit; this provides power take-off on either side. However, the unused shaft might create a safety hazard. Other units provide right-angle drives; in these, the input and output shafts are at right angles to each other. In a speed reducer, torque is multiplied as the speed is reduced. For example, a 1-hp motor turning at 1750 rpm might have 36 in-lb of torque. If this motor is connected to a 10/1 speed reducer, the output shaft of the reducer then rotates at 175 rpm and delivers 360 in-lb of torque. This value is theoretical, however, because friction losses occur in the bearings and the gears.

Figure 17-3 shows a speed reducer designed to give one definite reduction in speed. This type is used if an operating speed other than that produced by the motor or driving device is desired. Since gears are used in this assembly, a positive ratio is ensured.

Stepped, nonpositive ratios may be produced by using cone-type V-pulleys. Stepped pulleys are often used on the driving and driven shaft. The grooved pulleys must be mounted in opposite positions and on parallel shafts. Since these multiple sheaves are connected with a V-belt, the speed of the output shaft can be changed by shifting the belt from one set of grooves

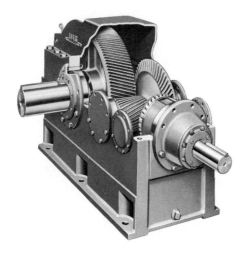

Figure 17-3
Right-angle speed reducer.
(Courtesy of The Falk Corp.)

to another. Drill presses frequently use this simple speed-changing device. For this type of application, the various output speeds are needed because of the varying cutting speeds of different metals and the fact that larger drills need lower rotational speeds to provide the proper linear velocity at the cutting lip.

Variable-speed drives are used when three or four stepped ratios are not adequate. Figure 17-4 shows the construction of the P.I.V.® variable-speed drive; P.I.V. stands for *positive, infinitely variable*. This drive has a unique operating principle. The input and output shafts are connected by a self-pitching chain that connects two pairs of conical wheels. Each wheel has teeth of uniform depth; the tooth width varies from a minimum size toward the center to a maximum size near the outside edge. The links of the chain contain packs of hardened steel slats that slide transversely. A control screw and handwheel unit forces one pair of the conical sections closer together and at the same time separates the other pair. Thus, the self-pitching chain operates on different diameters when the conical wheels are adjusted. By operating the control handwheel, various output speeds can be obtained. Since the device uses a chain, the drive is positive. Other variable-speed units use a belt. This provides an infinite number of speed changes; however, the drive is no longer positive.

Figure 17-5 shows a fractional-horsepower variable-speed device. The name Zero-Max® implies that the speed can be varied in infinitely small decrements from its rated maximum all the way to 0 rpm. Thus, it can also serve as a clutch. Another interesting feature of this unit is that the output shaft always turns in the same direction, regardless of the direction of

Figure 17-4
P.I.V.® mechanism.
(Courtesy of PT Components,
Inc.)

rotation of the input shaft. The operation of this unit is somewhat complex. The output shaft is equipped with four or more one-way (overrunning) clutches. The input shaft contains eccentrics. When the input shaft rotates, the eccentrics operate the connecting rods, which in turn cause the main and control links to move up and down or back and forth (the type of motion depends on the position of the external control lever). Linkage movement is up and down at the zero setting. Either four or eight laminations are used. The eccentrics on the drive shaft are in different positions; thus, the driving strokes on the output shaft overlap, providing smoother operation. When the Zero-Max® unit is operated from an 1800-rpm motor and has four

Figure 17-5
**Fractional-horsepower
variable-speed drive.**
(Courtesy of Zero-Max, a unit
of Barry-Wright.)

laminations, the output shaft receives driving strokes 120 times per second for any speed except zero. Speed regulation and stability are better in the higher part of the range. The speed is continuously variable from zero to one-quarter of the input speed (one-half, on some models).

The manufacturers' ratings and installation and operating procedures should be carefully studied before a particular speed-reduction unit is specified.

Choosing between a stepped speed-reduction unit and a continuously variable unit depends on the type of control desired in a particular project. Proper selection will affect the cost and function of the equipment. Also, one should remember that variable-speed air motors and hydraulic motors are also available for a wide range of speeds. Speed is adjusted by merely adjusting a valve. In general, air motors require a continuous source of air; some of the hydraulic units are self-contained, while others can be remotely operated. Manufacturers' data should be consulted before making any selection.

 **Fluid Power**

Fluid power is becoming popular in manufacturing plants. Automation has fostered greater use of fluids in power applications. Fluid power can be subdivided into two main branches, *pneumatics* (air) and *hydraulics* (oil). Other fluids are sometimes used, but air and oil are the most common. Figure 17-6 shows a simple hydraulic circuit. Oil is pumped from a reservoir through a control valve (protected by a relief valve) to a four-way directional control valve. The four-way valve (which can be hand-operated or solenoid-operated) directs the flow to either end of the cylinder. If oil is directed by this valve into the head end of the cylinder, the oil in the rod end of the cylinder is directed through the four-way valve to the reservoir. When the four-way valve is placed in the opposite position, oil goes to the rod end and is discharged from the head end into the reservoir through the four-way valve. The hydraulic cylinder is the "workhorse" or actuator of the system. It provides linear motion when used as shown; however, if used in conjunction with other components, many types of motion can be specified.

The cylinder shown in Fig. 17-6 is a double-acting cylinder, which requires a port at each end to operate the piston. Some cylinders are single-acting; in these, there is a port in one end only, and the return is carried out by spring pressure. A three-way valve is used instead of a four-way valve. The types of mountings commercially available for cylinders include *clevis*, *foot*, *head*, and *trunnion* mountings.

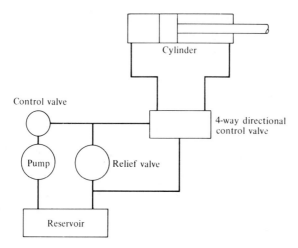

Figure 17-6
Simple hydraulic circuit.

A reservoir is necessary in any hydraulic circuit; it can be separate from the other components or incorporated into the equipment. A reservoir should be equipped with baffle plates for reducing turbulence, a sight glass for checking oil level, and a plug for drainage.

Several types of pumps are available; each has its advantages and limitations. One of the simplest types is the *spur gear* or *herringbone gear pump* for oils under low pressure. The gear pump contains two identical gears; a drive shaft is connected to one of them. Oil enters the inlet port, is carried in the tooth spaces between the gears and housing, and is discharged from a port opposite the intake. In a *centrifugal pump* the inlet port is in the middle of a rotating impeller. The high-speed impeller blades throw the fluid into a volute section that leads to the outlet opening. This type of pump delivers a constant flow, but is not self-priming. It is virtually impossible to overload the centrifugal pump. Several types of *piston pumps* are available. Generally, this type is expensive; however, it is quite efficient and has a wide range of capacity. The basic theory is very simple. As the piston moves in one direction, a ball check valve opens the inlet port. When the piston reverses direction, a ball check valve in the outlet port opens, discharging the liquid. There are several variations of this basic type of pump; some use radially mounted pistons and others axially mounted pistons. Other types of pumps are the lobe type (similar to the gear pump) and the propeller type.

Valves used in fluid-power applications can be classified in several different ways. One common classification is by function. One of the basic types is the *pressure-control valve*. In a sense, this is a safety valve or relief valve. The most common type of pressure-control valve is the spring-loaded ball type. Flow is in one direction only; spring pressure against the ball is then varied to adjust how high the fluid pressure can rise before it is diverted back to the reservoir. Other types of pressure-regulating valves are commercially available. *Flow-control valves* are usually of the needle type. This type is easier to operate than the conventional gate valve, which theoretically could be used for flow control.

Directional control valves are used to direct fluid to the proper lines when needed. The four-way valve shown in Fig. 17-6 directs liquid into the proper end of the hydraulic cylinder. This type often consists of a spool operating in a linear direction to block or open appropriate ports, or of a rotating cylindrical section with appropriate channels directing the flow through the desired ports. Numerous types of directional valves are available. Some are manually operated with two or three positions; others are solenoid-operated. A designer should explore the advantages and limitations of the many components commercially available and select the one best suited to the application. Again, cost must be considered as well as function.

Figure 17-7 shows a pneumatic circuit similar to the hydraulic one. The air supply comes from a compressor. Before the air enters the four-way valve (three-way in the case of a single-acting cylinder), it must be filtered to protect the components from contaminants. A regulating valve and a lubricator are also necessary (the lubricator supplies an oil mist to the components in the system). Air is returned to the atmosphere instead of a reservoir. A muffler is sometimes attached to the discharge line to reduce the noise caused by the discharged air.

Air motors are available to the designer. These are often used in power tools for mass production. Air motors can be of the vane type or the piston type. For mass assembly applications, they can be used for driving (tightening) several nuts simultaneously, or for such operations as polishing, grinding, and drilling. This type of motor is much lighter in weight than its electrical counterpart. Air tools are reliable, safe, and easy to operate. On the other hand, they are less efficient than electrically operated tools.

A pneumatic or hydraulic cylinder is reasonably easy to select. The "push" and "pull" values of the piston rod depends on the pressure of the fluid. Air cylinders are commonly available for pressures up to 150 psi; many manufacturers produce hydraulic cylinders for pressures up to 3000 psi. The force of the piston is found by multiplying the pressure by the piston area:

$$P = pA,$$

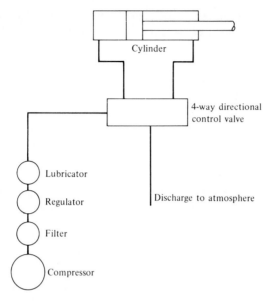

Figure 17-7
Simple pneumatic circuit.

where

P = force exerted by rod (lb),

p = pressure (psi),

A = piston area (sq. in.).

The following problem shows how to calculate "push" and "pull" values.

◆ *Example* _____

A double-acting hydraulic cylinder is operated at 1000 psi pressure. The cylinder has a 6-in. bore and a 12-in. stroke. The piston rod is 1.5 in. in diameter. Find the "push" and "pull" exerted by the piston rod.

Solution. The "push" value is

$$P = pA = 1000(\pi/4)(6)^2$$

$$= 28{,}274 \text{ lb}.$$

The "pull" value is

$$P = pA = 1000(\pi/4)\left[(6)^2 - (1.5)^2\right]$$
$$= 26{,}507 \text{ lb.}$$

Note that the length of the stroke has no bearing on the operating force.

In specifying a single- or double-acting cylinder, it is essential that both bore and stroke be given. Certain stock bores are available, but the stroke must be specified for each application. Some manufacturers specify strokes in even-inch increments only. A basic price is often quoted with no stroke; then a charge per inch of stroke is made. Each inch of stroke adds an inch to the length of the cylinder as well as to the length of the rod. Mounting dimensions and clearances should be carefully explored from a manufacturer's catalog before completing a project using cylinders.

Cushions can be specified to allow deceleration at the ends of strokes. A cushion is an elaborate porting arrangement to direct fluid flow (and speed). On exceptionally large strokes, the piston rod must be considered a slender column and checked accordingly. Sometimes oversized rods are used (and predesigned) so that the "push" value is double the "pull" value. Oversizing also strengthens the rod as a column.

The choice between a single- or double-acting cylinder often depends on the particular application. A double-acting cylinder usually requires the use of a four-way valve. With such a system, the "push" and "pull" forces do not differ greatly in numerical value. The four-way valve, however, is expensive. If the force for the return stroke need not be high, a circuit can be designed using a three-way valve and a control valve to use a lower back pressure for the return stroke. The use of a single-acting cylinder and a three-way valve simplifies the circuit and reduces the cost of components. In this case, the force varies along the stroke because the pressure against the piston must compress the spring. A two-way valve can sometimes be used with a single-acting cylinder, if the operating force need not be large and the cylinder rod is extended for short periods only. This inexpensive arrangement is often used for air clamps. Small-bore (and stroke) cylinders are commercially available for clamping uses. Since most modern factories are equipped with a system of air lines as complete and convenient as electric power outlets, air-operated clamps and tools are quite popular.

Figure 17-8 shows a simplified sketch of a fluid cylinder. Not shown are the tie rods (often used to secure the barrel to the heads) and the rod wiper and sealing assembly at the rod end of the cylinder.

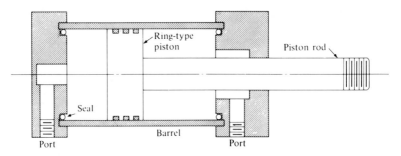

Figure 17-8
Fluid cylinder.

Friction Losses in Hydraulic Circuits

When water, oil, or any other fluid flows through pipes or tubing, friction and the resulting losses must be considered. The total head (pressure, potential, and velocity) at the beginning of a circuit is equal to the total head at the end of the circuit plus the head lost to friction. Frictional losses include pipe or tubing losses (particularly in long lengths), bend losses, expansion or contraction losses, and losses caused by fittings. All these losses generate heat. Wherever possible, the system should be designed for laminar flow—i.e., flow in which the fluid moves in layers. Laminar flow, sometimes called streamline flow, is the direct opposite of turbulent flow. Friction factors for various types of piping and for fittings, bends, and other troublesome circuit points can be found in many texts on fluid mechanics.

In general, connecting pipes or hoses should be kept as short as possible. Unnecessary bends should be avoided; components should be arranged so that excessively sharp bends are not necessary. Careful planning can greatly reduce friction losses and contribute to the compactness of a design.

Lines and Fittings

In electrical circuits, wires and cables are used to connect the various components. The type of conductor used depends on the current to be handled and on surrounding conditions, such as temperature and moisture. Connector plugs and terminal blocks are used for fast connecting and disconnecting. Similar components are available in the fluid-power field. Several types of hoses and pipes are available for conducting fluid (air or oil) from one component to another. The size and type of hose or pipe depends on the fluid used and on the amount of pressure to be handled; sometimes resistance to certain chemicals must also be considered. Thermoplastic tubing is often used for moderate pressures. Polyethylene can be effectively used for temperatures up to 175 °F and working pressures up to 150 psi. Nylon tubing can be used if the temperature does not exceed 250 °F and the working pressure is kept below 375 psi. Plastic tubing can be used with metal or plastic fittings. Compression-type tube fittings are available for the usual piping configurations, such as tees, elbows, and so on.

Regardless of the type of line employed, it is desirable to use manifolds as much as possible. These are merely blocks with interconnected passageways and connections for other fittings. Unused connections can easily be plugged. The use of manifolds compensates for the fact that tubing or piping is not as flexible as the wires used in electrical circuits.

If higher pressures (and temperatures) are to be encountered, metal tubing should be used. Flared or compression fittings are used to make the connection leak-proof; many variations of these fittings are commercially available. The use of metal tubing, however, prevents movement; thus, equipment must be kept stationary.

Synthetic rubber hosing is flexible and can be used at pressures as high as 12,000 psi. Teflon is sometimes used in place of rubber. The hose sometimes consists of three main parts: an inner tube of rubber, a braided wire sheath, and a tough rubber outer covering. The outer covering is designed to resist abrasion and deterioration from ambient conditions. Flexible hoses are usually equipped with pipe threads to ensure that leakage does not occur. Table 17-1 lists some of the common pipe sizes. Numerous hydraulic and pneumatic components are equipped with the $\frac{1}{8}$-in. and $\frac{1}{4}$-in. pipe threads.

Self-sealing couplings are used if hoses must be frequently connected or disconnected. These couplings prevent the loss of fluids from either line. One type for air lines uses a three-way valve to prevent any "kick" as the lines are disconnected; this ensures the safety of the operator.

Circuitry

Figures 17-6 and 17-7 show hydraulic and pneumatic circuits with plain block diagrams in which each component must be labeled. To simplify the drawing and understanding of a circuit, the Joint Industry Committee (consisting of several industrial associations) formulated the JIC standards for hydraulics and pneumatics. This material is beyond the scope of this text; anyone associated with fluid-power applications should obtain copies of these standards and become versed in their usage. One symbol, however, will be discussed to indicate the simplicity of the standards. Figure 17-9

Table 17-1
Pipe Threads

National Pipe Threads

Nominal size, in.	Thread/in.	Pipe outer dia., in.
$\frac{1}{8}$	27	0.405
$\frac{1}{4}$	18	0.540
$\frac{3}{8}$	18	0.675
$\frac{1}{2}$	14	0.840
$\frac{3}{4}$	14	1.050
1	$11\frac{1}{2}$	1.315
$1\frac{1}{4}$	$11\frac{1}{2}$	1.660
$1\frac{1}{2}$	$11\frac{1}{2}$	1.900

ISO Metric Pipe Threads

Nominal size, mm	Pipe outer dia., mm
6	10.2
8	13.6
10	17.1
15	21.4
20	26.9
25	33.8
32	42.4
40	48.4

Figure 17-9
JIC symbol for a three-position four-way valve (manual operation).

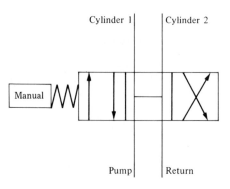

Figure 17-10
Air-pilot valve.
(Courtesy of Mead Fluid Dynamics
Division, Abex Corp.)

shows a three-position four-way valve similar to those shown in Figs. 17-6 and 17-7. In this case, the valve is manually operated. A slight change in the symbol can indicate other means of actuation.

Note that these circuit symbols produce a circuit drawing similar in nature to the electrical or electronic schematic drawing. It merely indicates the function of the circuit and the flow of fluid. Regular production drawings, either orthographic or isometric, must be used to show the position of the actual parts.

In air circuits, miniature three-way valves can be used for remote control of a master valve; these three-way valves are used as air-pilot valves. They are comparable in size to small electric limit switches; similar types of actuators are available so that these valves can be operated by other mechanical parts, such as cams and levers. Figure 17-10 illustrates this type of air-pilot valve. A designer should carefully study mounting conditions when incorporating such units with other components. The interior of a three-way air-pilot valve is shown in Fig. 17-11.

Figure 17-11
Cutaway of air-pilot valve.
(Courtesy of Mead Fluid Dynamics
Division, Abex Corp.)

In evaluating the possible use of pneumatic circuits or electrical circuits, consideration must be given to surrounding conditions. Air circuits present no sparking hazards; the ever-present danger of electrical shock is often completely eliminated with the proper use of pneumatics. However, if pressure is not carefully controlled in an air circuit, there is the danger of a hose rupturing.

Mechanical Advantage Through Hydraulics

Assuming a liquid to be a noncompressible fluid, a designer can gain a great advantage in force by sacrificing distance (see Fig. 17-12). If piston A has an area of 1 sq in. and piston B has an area of 10 sq in., a given force pushing downward on A will place the liquid under a pressure (psi) which will be transmitted to all the enclosed parts of the system. This constant pressure will push upward on piston B. Since B has a larger area (10 sq in.), it will exert an upward force 10 times that at A. Piston B, however, will travel only one-tenth the distance that piston A is moved. This idea is sometimes used for force advantage, and more often for displacing a load (in, for instance, the automobile hoists found in repair shops). This hydrostatic principle is often used in the field of mechanical design. An example is hydraulically operated automotive brakes.

Pneumatic-Hydraulic Boosters

Figure 17-13 is a sketch of a booster used to transform low-pressure air into useful hydraulic pressure. The differences in piston diameters are usually much greater than those shown in the sketch. The basic principle is simple and is similar to the hydraulic mechanical advantage previously discussed. When dual pressure operation is used, however, the circuitry can become somewhat complicated. In the simple application shown, air under low pressure pushes against the ram. The larger of the two areas contacts the air; the smaller area applies pressure to the oil. The ratio of air pressure

Figure 17-12
Mechanical advantage through hydraulics.

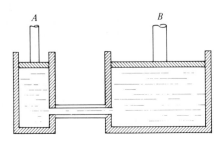

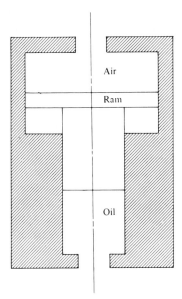

Air

Ram

Oil

Figure 17-13
Pneumatic-hydraulic booster.

to oil pressure is approximately inversely proportional to the two ram areas. (It must be remembered that air is more compressible than oil; thus, there is a slight error in the theoretical calculation.) Thus, ordinary shop air at a pressure of 80 to 100 psi can be used to put oil under the higher pressure needed for operating hydraulic equipment without the use of expensive pumps. Such a system is most useful for short hydraulic strokes in which the volume of oil is small.

Boosters also have an advantage over plain air cylinders. Excessively large air cylinders are needed if a large pressure is to be obtained from air alone. Boosters can save both weight and money. This equipment can be used on such portable devices as spot welders, clamps, and riveters.

Oil tanks must also be provided when designing booster-operated equipment. The capacity of the oil tank can be easily found by multiplying the piston area by the stroke and allowing for oil compressibility. (The compressibility can be approximated by adding 1% of the total volume to the computed volume.) Boosters are commercially available with built-in tanks; the use of such equipment provides simplicity and compactness of design.

Still higher hydraulic pressures can be gained by using a setup similar to Fig. 17-13, except that oil is used in place of air; this oil can be put under higher pressure than would be obtainable with air. The effectiveness of seals must be carefully investigated in all high-pressure applications. Extremely high pressures are obtainable with hydraulic-hydraulic boosters. Thus, safety precautions must be carefully considered.

Figures 17-12 and 17-13 should be studied carefully for similarities. In Fig. 17-12, mechanical advantage is gained by using two pistons and one fluid between the two. In Fig. 17-13, two fluids are used with one ram or piston. Pressures are changed by providing different end areas on a single ram placed between the two fluids. In the first case, the device regulates force; in the second, the device regulates pressure.

In any booster application, ports are usually provided with pipe threads in common sizes so that ordinary fluid lines can be easily connected.

Air Supply

A source of clean air is necessary when any pneumatic equipment is used. All air entering a compressor is filtered; additional filtering is necessary before the air is used in air tools or air cylinders. Equipment such as cylinders and directional control valves is made with great precision. Moisture, rust, grit, or scale could cause costly internal damage. Thus, it is important to install filtering equipment just ahead of the working components of a pneumatic circuit. A preassembled unit consisting of a filter, regulator, and lubricator is often used; such a unit is illustrated in Fig. 17-14. The filter part is a separator and filter combination; its function is to

Figure 17-14
Combination filter, regulator, and lubricator.
(Courtesy of C. A. Norgren Co.)

remove moisture and contaminants. It often contains a sintered bronze (porous) element. After the air passes through the filter, it passes through a regulator; this is a valve (with gage) for adjusting the air to an exact operating pressure. After passing through the regulator, the air stream runs through a lubricator, a device that supplies an oil mist to the working air. In this way, mechanical parts of the system are continuously lubricated.

Air Muffling

Air is exhausted into the atmosphere from pneumatic circuits; this causes noise that can be obnoxious, particularly when many pneumatic devices are used in a rather small area of a plant. The noise level can be substantially reduced by installing a silencer or muffler (made of sintered bronze) in the discharge port of the directional valve. This silencer can also include a valve for regulating the flow. Muffling of noise is usually needed to conform to noise abatement regulations.

Air Tools

Other types of air tools are available in addition to the multiple-use cylinder, clamp, or actuator. Air tools are somewhat less efficient than their electrical counterparts. However, the lighter weight of the pneumatic tools can sometimes effect labor savings because these tools require one operator rather than two. With air tools, there is no danger of electrical shock; gasoline-operated portable compressors make air available in places where electricity is inconvenient. Air hammers and paving breakers are commonly used in street-maintenance projects. Factory use of portable air tools is commonplace. The torque of power screwdrivers and wrenches is easily controlled by adjusting a valve. The direction of rotation and the speed of an air motor are easy to control. The danger of overheating is nonexistent. This type of motor can be used where electric sparks could cause explosions; it is popular in chemical plants. In deciding between air and electric operation, one must consider the initial equipment cost, the cost of operation (including labor), and the safety of the operators.

Pump Applications

Pumps have other uses in addition to the closed hydraulic circuit. The main purpose of pumping is to move a liquid from one location to another. Sometimes this movement is upward, sometimes it is horizontal; often it is a combination of the two. Water is a common pumped fluid; however, pumps are also used to move liquids with a specific gravity lower or higher than that of water. The service requirements of the pump must be known; some

Figure 17-15
Cutaway of rotary pump.
(Courtesy of Viking Pump Div., Houdaille Industries, Inc.)

pumps operate continuously, whereas others run for short periods only. Fluids that are pumped include oils, syrups, paints, beer, and thousands of other liquids with varying viscosities.

Pump Selection

The three general classes of pumps are *rotary*, *centrifugal*, and *reciprocating*. In selecting a suitable pump for a specific job, one should first determine the available space by making a rough layout of the general design. The total head must be determined and the capacity requirements found. The fluid to be handled should be studied in terms of viscosity, temperature, and contaminants. Thought must be given to the suction and discharge piping and fittings. Finally, the input horsepower requirements must be calculated so that total head and friction losses can be overcome and the fluid moved as required by the job.

Figure 17-15 shows a cutaway of a rotary pump; this particular model is a heavy-duty type. Note that rotary pumps can be of the lobe or gear type. In the cutaway, note the methods used for mounting and sealing the shaft. A heavy-duty centrifugal pump is illustrated in Fig. 17-16. The suction or intake is in the "eye" on the right side; the discharge port is at the top. Note the sealing method. Note also the valve and port at the top of the volute; this is necessary for priming in certain applications. Figure 17-17 illustrates a single-acting reciprocating pump, also of the heavy-duty type. Arrows indicate fluid flow. In the previously mentioned hydraulic circuits, pumps can range from light-duty to heavy-duty, depending on the use to which the

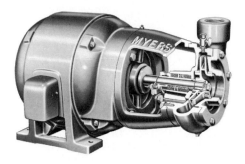

Figure 17-16
Centrifugal pump.
(Courtesy of F. E. Myers Co.)

hydraulic circuit is applied. In chemical plants and in other process industries where pumping is done primarily to move fluids long distances, the
pumps can be operated by means of steam or electric motors.

Head

Two fluid connections are made to any pump. The fluid enters the
pump in the *suction* line and leaves the pump through the *discharge* line. The
word suction is somewhat misleading here. The fluid is not "sucked" into
the suction line; rather, the pump creates a lower pressure in the suction
line, and the higher pressure (often atmospheric) at the supply level forces
the liquid up into the suction line (refer to Fig. 17-18). A designer must have
a good understanding of the following lift and head definitions.

Static suction lift: the distance (feet) from the surface of the supply
reservoir to the centerline of the pump, the pump being located above the
level of the supply.

Static suction head: the same distance as the static suction lift, except
that the pump is located below the surface of the fluid supply.

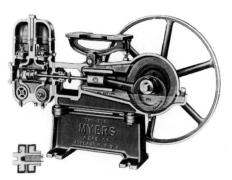

Figure 17-17
**Single-acting reciprocating
pump.**
(Courtesy of F. E. Myers Co.)

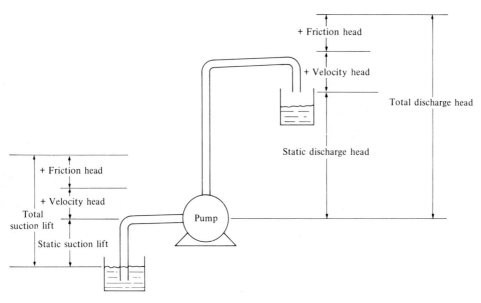

Figure 17-18
Head definitions.

Static discharge head: the vertical distance from the centerline of the pump to the point where the fluid is being delivered.

Total static head: the total distance from the level of the supply reservoir to the level of free discharge.

Friction head: the total pressure required to overcome the resistance of the pipes and fittings.

Velocity head: the kinetic energy of the liquid as determined by its rate of flow through the piping system. It can be computed from the following equation:

$$H = \frac{V^2}{2g}, \tag{17-1}$$

where

H = head (ft),

V = velocity (ft/sec).

Velocity head is often neglected in calculations because of its low value in comparison with the other heads.

Suction lift is equal to the static suction head plus the sum of the friction and velocity heads.

Suction head is equal to the static suction head minus the sum of the friction and velocity heads.

Discharge head is the sum of the static discharge head, the discharge friction head, and the velocity head.

Total head is equal to the sum of the suction lift and the discharge head. In cases where there is suction head, the total head is the difference between the discharge and suction heads. In general, the total dynamic head is the number of feet between the supply level and the free discharge level plus the frictional losses in all pipes and fittings.

The head can be expressed in feet or in pounds per square inch of pressure. In rotary and reciprocating pumps, the total head is expressed in pounds per square inch; in centrifugal pumps, it is expressed in feet. The following equation can be used to convert from one set of units to the other:

$$H = \frac{2.31p}{\text{specific gravity}},$$

where

H = head (ft),

p = pressure (psi).

Efficiency and Horsepower

The following equations are useful in computing efficiency and brake horsepower:

$$e = \frac{\text{hp}_{\text{out}}}{\text{hp}_{\text{in}}}(100) = \frac{\text{water hp}}{\text{brake hp}}(100), \qquad \text{(17-2)}$$

where

e = efficiency(%);

$$\text{brake hp} = \frac{Qp(100)}{1715e}, \qquad \text{(17-3)}$$

where

Q = discharge (gpm);

and brake horsepower is the shaft horsepower of the driving unit.

$$\text{Brake hp} = \frac{QH(\text{specific gravity})(100)}{3960e}; \qquad \text{(17-4)}$$

$$\text{brake hp} = \frac{(\text{watts input})(\text{efficiency of motor})}{746(100)} \qquad \text{(17-5)}$$

◆ *Example* _____

Water is to be pumped from an open tank on one floor of a chemical plant to an open tank 30 ft above (water level to water level) at the rate of 50 gal/min. Frictional losses in the system amount to 1.8 psi. The pump efficiency is 70% and the centerline of the pump is 4 ft above the surface of the lower tank. Assuming a rotary pump and no velocity head, find the brake horsepower.

Solution. Static suction lift = 4 ft. Static discharge head = 26 ft. Total static head = 30 ft. Friction head = 1.8 psi. Total head = [(4 + 26)/2.31] + 1.8 = 14.8 psi. Thus, the brake horsepower is

$$\frac{Qp(100)}{1715e} = \frac{50(14.8)(100)}{1715(70)}$$

$$= 0.616 \text{ hp.}$$

_____

Pump Analogies

Pumps have similarities to other mechanical assemblies. A rotary pump is similar to a simple gear train; in the pump, fluid is carried between the teeth (or lobes) and the housing. The piston type of pump is similar to the hydraulic cylinder; however, in the pump the fluid is being moved instead of providing the actuating force. The centrifugal pump is similar to a turbine, but again, the fluid is being moved instead of providing the driving force. For the centrifugal pump, laws similar to the fan laws can be stated:

1. The capacity varies directly as the speed.
2. The head varies as the square of the speed.
3. The horsepower varies as the cube of the speed.

Pumping Systems

Pumps form the nucleus of many mechanical systems that involve such components as valves, pipes, tanks, and special equipment. Chemical and food processing industries move fluids from one piece of process equipment to another, as well as to and from storage tanks. Thus, pumping systems should be designed around the proper pump; pump characteristics must be known and carefully evaluated.

Fluid Power Metrication

Hydraulic and pneumatic equipment, such as valves, cylinders, and similar components, must have mountings compatible with metric fasteners if the circuitry is designed within metric parameters. Thus, mounting and clevis holes must accommodate metric bolts and clevis pins.

The pressure of fluids, such as liquids and gases, has traditionally been expressed in pounds per square inch (psi) in the United States for pressures above atmospheric and in inches of mercury for pressures below atmospheric. There are several metric methods of specifying pressure. The most basic is the newton per square meter (N/m^2). However, it is convenient to use the term *pascal* (Pa), which represents one newton per square meter. In this way, the pascal is associated with pressure and not with stress. Segments of the fluid power industry prefer the term *bar*, which is equal to 100 000 pascals. Following are relationships that can be used for converting to metric units:

$$1 \text{ bar} = 100\,000 \text{ Pa} = 100\,000 \text{ N}/m^2 = 14.5 \text{ psi},$$

$$1'' \text{ mercury (at } 60\,°F) = 0.034 \text{ bars},$$

$$1 \text{ atmosphere} = 101\,325 \text{ Pa}.$$

Some U.S. manufacturers of fluid power equipment prefer to express gauge pressure in units of kg/cm^2. For comparison,

$$1 \text{ psi} = 0.07 \text{ kg}/cm^2.$$

Fluid flow in the United States has traditionally been expressed as gallons per minute for liquids and cubic feet per minute for gases. Cubic meters per minute or liters per minute are usable metric quantities. The following represent relative magnitudes:

$$1 \text{ gpm} = 3.79 \text{ l/min} = 0.003\,79 \text{ m}^3/\text{min}.$$

Gaseous flow can be expressed in cubic decimeters per second. Following is a comparison between the U.S. and metric units for gaseous fluid flow:

$$1 \text{ cfm} = 0.472 \text{ dm}^3/\text{s}.$$

Pressure gauges calibrated in both pounds per square inch and bars are commercially available.

Metric pipe and fittings are readily available; their linear dimensions are expressed in millimeters and they incorporate metric threads. Typical pipe couplings include:

1. straight male stud coupling,
2. straight coupling,
3. cone reduction,

4. male stud tee coupling,
5. male stud elbow coupling,
6. equal tee coupling,
7. male stud "L" coupling,
8. standard tube connector with nut, and
9. equal cross.

Conical (tapered) screw-in threads are available in the very light series. In general, the *very light* series is recommended for hydraulic circuits, refrigeration equipment, heating systems, and similar applications. The *light* series is recommended for pressure piping and where higher mechanical stresses may be present. The *heavy* series is recommended for heavy applications in the chemical, mining, and shipbuilding industries where extremely high pressures are encountered.

Tube sizes range from 4 mm to 80 mm in outside diameter. Inside diameters range from 3 mm to 60 mm. Thus, wall thicknesses vary from 0.5 mm in the smaller sizes to 10 mm in the larger sizes. Tube materials include steel (SAE 1010), stainless steel, annealed copper, and half-hard copper.

Thread sizes usually start at M8 × 1 for the smaller sizes and extend to M52 × 2 for outside tube diameters of 38 mm. Special fittings are generally required for extremely large sizes.

Manufacturers of tubes and fittings generally specify the working pressure in bars and the tensile strength of the tube material in daN/mm². A factor of safety is usually included; a popular value is 4.

 **Governors**

A governor is defined as an automatic speed-regulating device. It is needed to compensate for changes in the load on a machine that would otherwise increase or decrease the speed. Governors are used to regulate motors, engines, turbines, and other rotating machines. The usual method is to adjust the amount of steam, water, or fuel to maintain the desired speed. Sensitivity is one of the prime considerations. Some speed regulating devices are extremely complex and expensive; others are simple in nature. One of the simplest is an air-vane governor that is sometimes used in self-propelled power lawn mowers. This consists of a simple spring-loaded pivoted plate. An airstream from a vaned flywheel provides a force against the plate. When the speed of the engine is reduced, the force from the air stream is also reduced; this actuates the throttle to increase the engine speed. Thus, when the mower is climbing a terrace, it would naturally decrease in speed until the air-vane governor takes effect.

Principal Types of Governors

Mechanical governors can be divided into two general categories. One type responds to a change of speed; centrifugal governors are of this type. Another type responds to the rate of change of rotational speed; inertia governors are of this type.

Figure 17-19 shows the general configuration of a centrifugal shaft-type governor. The collar on the left is rigidly attached to the shaft; the one on the right is free to slide along the shaft. The actuating collar can engage with another load shaft and regulate fuel supply or regulate speed by actuating a friction brake. In Fig. 17-19, the position of the actuating collar is controlled by the size of the compression spring. When the shaft speed becomes excessive, the flyweights move outward by centrifugal action, and connecting links move the actuating collar to the left. The movement of the collar then regulates the speed.

In the inertia type of shaft governor, a weight is pivoted at a distance from the center of the shaft. As the shaft speed changes the pivoted weight turns about its center; this controls a regulating valve. Note that the inertia type acts immediately when there is a change of speed.

There are several types of speed controls for electric motors, some tied in with a permanent-magnet dc tachometer. If any change in rotational speed begins to occur, the speed of the driven machine is regulated. Some electrical control units are built directly into the equipment and provide regulation within 1% or less.

Figure 17-19
Centrifugal shaft-type governor.

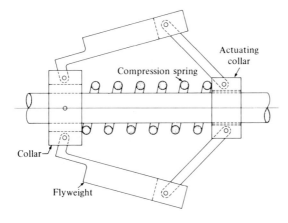

One must not confuse the manual regulation of speed with the governor type of action. The governor regulates the speed automatically. Equipment sometimes indicates when a speed change occurs and the setting is adjusted manually; in such cases, the indicating equipment is not considered a governor.

In selecting governors or speed-regulating equipment, one should consider the cost of the equipment and the amount of regulation desired. Sensitivity usually requires expensive and sophisticated equipment. If the design does not warrant such equipment, cheaper means of regulation should be used.

 Fluid Torque Converters

The term *torque converter* is not necessarily associated with the word *fluid*. In fact, any type of transmission or speed reducer is a torque converter, as well as a coupling. When a device reduces the speed that it receives from its input shaft without frictional braking, the torque is increased. One of the biggest problems in torque converters is to maintain efficiency.

A fluid torque converter is simple to construct from the theoretical standpoint, but complicated from the standpoint of precision. There are three essential parts in the simplest type of fluid torque converter. A *pump* is connected to the driving shaft and a *turbine* is connected to the output shaft. The third principal component is a *stator*, a stationary part. The fluid is accelerated by the pump and directed at high velocity to the blades of the turbine. After delivering energy to the turbine blades, the fluid returns to the pump through the stator. The entire operation is based on kinetic energy and hydrodynamic theory.

Turbine blades and the entire internal construction is critical in the following areas:

1. dimensional accuracy of all parts and proper positioning,
2. proper balancing (static and dynamic),
3. correct contours for the three principal parts,
4. suitable materials and sufficient strength.

Fluid torque converters are used in the heavy-automotive and construction fields. They are particularly useful when large engines are subjected to frequently changing speed and load requirements. It is essential that the proper oil be used in a fluid torque converter.

 **Summary**

Selecting a power unit and appropriate controls is an important decision in most mechanical-design projects. Typical factors to be considered are:

1. *Cost:* This includes initial cost, cost of operation, and cost of maintenance.
2. *Environmental conditions:* This is primarily a matter of temperature and humidity. Some equipment is operated completely indoors under ambient temperature conditions that are nearly the same in all seasons of the year. Other equipment is operated outdoors under extremes of temperature and exposure to the elements. Marine equipment meets varying conditions.
3. *Power availability:* If one specifies electric power, one should be certain that the proper voltage and phase is available at the location of the machine. If the equipment is to be air-operated, one must make sure that shop air is available at the needed pressure. If the equipment is to be air-operated under field conditions, one must make sure that a suitable portable air compressor is available that provides the proper air pressure.
4. *Vibrations:* Allowable shock and vibration conditions must be checked before selecting the power unit. Internal-combustion engines produce more vibrations than electric motors; the relative merits of each should be checked.
5. *Portability:* This is closely associated with power requirements and environmental conditions. One must know where and how a given piece of equipment is to be used.
6. *Safety hazards:* Potential safety hazards play an important role in selecting power units. For example, electric motors are not very useful in outdoor locations subjected to the elements. On the other hand, if internal-combustion engines are used indoors, proper provision must be made for venting exhaust gases to avoid the fatal effects of carbon monoxide.

The designer must know the type of motion that the power unit produces. Rotary motion can be converted to reciprocating motion and vice versa, but this involves other mechanical equipment, which adds to costs. One should therefore be well versed in the type of power directly produced by the power unit. Rotary motion is delivered by electric motors, internal-combustion engines, and air motors. Air or hydraulic cylinders produce linear "muscle power." This reciprocating motion can be converted to other types, but this, too, involves extra equipment.

In selecting speed reducers and related equipment, one should consider input and output speeds, torque or horsepower ratings, shaft sizes and locations, the type of drive, and mounting conditions. Speed reducers are commercially available both with and without integral motors.

Pumps are used to move fluids from one location to another. One must consider the type of fluid being moved and the type of service needed. The location of the pump with respect to the suction and discharge positions is extremely important. One should thoroughly understand all the possible heads involved, including the friction head, the velocity head, and the static discharge head.

Selecting controls is essential in both the electrical and fluid fields. Problems of cost and compactness often test the ingenuity of the designer.

Connecting lines such as wires, pipes, tubes, and hoses can complicate an assembly. Printed circuits and terminal blocks can be very helpful in the electrical field. In the mechanical field, manifolds can be used to save space and thus make a product more compact.

The failure of the power unit stops the operation of any machine. So-called "down time" must be avoided whenever possible. Thus, the reliability of any power unit must be carefully studied.

Questions for Review

1. How does a two-stroke gasoline engine differ from a four-stroke engine? List the complete cycle for each.

2. Why would a diesel engine be undesirable for a power lawn mower? List three diesel-engine applications.

3. What controls are needed for operating an electric motor? Explain why each type is used.

4. What advantages do electric motors have over gasoline engines? What disadvantages?

5. Check the nameplate of a large electric motor. What information is given?

6. How does single-phase electricity differ from three-phase electricity? What advantages do multiphase electric motors have over single-phase motors? What disadvantages?

7. List some applications of photoelectric controls in the mechanical-design field.

8. What items other than input and output speed should be checked in selecting a speed reducer? Why?

9. How does a double-acting hydraulic cylinder differ from a single-acting cylinder?

10. How does a four-way valve differ from a three-way valve in a fluid circuit? Sketch an application for each type.

11. Why are "push" and "pull" values different for an air and a hydraulic cylinder?

12. What is the metric counterpart of the horsepower rating of an electric motor?

13. What is the purpose of a *cushion* in a hydraulic cylinder? Why is it needed?

14. What frictional losses occur in fluid circuits?

15. What is the purpose of the following in a pneumatic circuit?
    a) filter
    b) regulator
    c) lubricator

16. Why is a muffler used in air circuitry?

17. What advantages do self-sealing couplings have over other types?

18. What is the purpose of a pneumatic-hydraulic booster? Give some applications.

19. Differentiate between the terms *bar* and *pascal* as applied to fluid pressure.

20. What metric units are appropriate for specifying gaseous flow?

21. What metric units are appropriate for specifying liquid flow?

Problems

1. Convert the following pressures to bars:
   a) 80 psi,
   b) 1000 psi,
   c) 30 inches Hg,
   d) 60 inches H_2O at 60 °F.

2. Convert the pressures in Problem 1 to pascals.

3. Convert 12 gpm to liters per minute.

4. Convert 30 cfm to cubic decimeters per second.

5. A pneumatic cylinder with a bore of 4 in., a stroke of 6 in., and a rod diameter of $\frac{3}{4}$ in. operates on 80-psi shop air. It is controlled by means of a four-way valve. Find the "push" and "pull" of the piston rod.

6. An air cylinder with a bore of 25 mm is operated with shop air at a pressure of 6.3 bars (approximately 90 psi). Find the "push" exerted by the piston rod, in newtons.

7. A hydraulic cylinder with a bore of 200 mm is operated by oil at a pressure of 200 bars. Find the "push" exerted by the piston, in newtons.

8. An air cylinder with a bore of 150 mm, a stroke of 75 mm, and a piston-rod diameter of 25 mm is operated at a pressure of 0.53 MPa. Find its "push" and "pull" in meganewtons.

9. In an arrangement similar to that in Fig. 17-13, oil at a pressure of 5 MPa operates the ram from the top. If the two cylinders are 25 mm and 75 mm in diameter, respectively, find the pressure developed at the bottom.

PROFESSIONAL PRACTICE

18

Comprehensive design problems provide an avenue for understanding the complete design process. Most are lengthy projects. Certain assumptions are made by the student. Materials are selected and sizes calculated; then the appropriate layout, detail, and assembly drawings are made to complete the project. Projects of this type force the student to put together (and review) previous knowledge of materials, processes, mechanics, strength of materials, and other subjects along with the theoretical applications in machine design so as to develop a complete design. This type of problem can be handled in the drafting room, since each is actually a design project. A rather complete drafting-room reference file is essential, in addition to the regular library containing reference books and handbooks. Manufacturers' catalogs form the backbone of such a reference file. A complete study of similar existing designs is also helpful to a student. Design layouts must often be made; force-analysis drawings and kinematic schemes are important first steps that a prospective designer often has to cope with before proceeding to any real calculations. Conforming to MIL standards (or appropriate company standards) is an important experience that can do much to ease the transition from school to industry.

Completeness is most important in all problems. Final drawings should be complete enough that the product could be easily manufactured and made ready for shipment to the user.

 ## Organization of the Design Function

Employment titles in the engineering field are not standardized; this results in much confusion. Engineering is often divided into two categories. One is product-design engineering; the other is plant or facilities engineering.

Plant engineering generally involves such areas as plant layout, materials handling, and maintenance. Design in this category deals with the equipment needed for performing these functions.

Product engineering covers the design of items manufactured by the plant. If a manufacturer produces household appliances, the product-design group will handle all the engineering of the appliances. Necessary tooling and materials-handling equipment, together with an efficient plant layout, are the responsibility of the plant-engineering group.

A *product-design group* consists of many individuals with various talents. The personnel involved with design work include such job classifications as design engineer, designer (often junior and senior), and design drafters. These individuals are supported by layout and detail drafters with various backgrounds—mechanical, electrical, and structural. The background required depends on the product being designed.

The key person in an engineering function is often called the *project engineer*. The project engineer is responsible for supervising the design group and ensuring the proper functioning of the finished product. The project engineer is involved in liaison with various groups both inside and outside the organization, and is in constant contact with management concerning the progress of the design. Contacts with the sales department can affect some aspects of the finished product. The project engineer is responsible for customer relations in the case of custom-built equipment. During the initial stages of a new design, close coordination is necessary with one's purchasing department and with outside sales representatives regarding prices and the availability of materials and stock items. After drawings are released for production, the project engineer must work with the various production departments to "debug" the finished product. Initial contacts are made with the production departments during the early stages of design in order to incorporate features that will facilitate production. The project engineer carries out these functions in addition to supervising the design group.

Since the project engineer is responsible for the design, decisions of great importance must be made—usually after consultation with senior staff members. For instance, the project engineer must select suitable factors of safety unless codes specify this value.

The need for a product and the general guidelines to be followed are usually supplied by the management. These are usually determined by sales contacts or by surveys conducted by either a special group within the organization or by an outside management-services firm. Management also makes decisions concerning the use of consulting firms or so-called "jobbing shops." These organizations are generally used when the engineering deadlines are too tight to handle within the organization or when necessary specialists are not available.

Ethics

Ethical conduct is important in any area of work, but particularly in design engineering. A designer is providing a service that eventually leads to a product. This product must be safe and functional. Relations with individuals within the firm or agency must be handled in a professional manner. Honesty and integrity are important attributes that one must possess.

In an effort to improve the status of engineers, most professional societies and the Accreditation Board for Engineering and Technology (formerly known as Engineers' Council for Professional Development) have established canons of ethics. These set forth some principles of engineering ethics as they relate to the general public, clients, employers, and other engineers.

Professional Registration

Professional engineers are licensed by boards of registration established by each state, commonwealth, territory, and the District of Columbia. The registration procedure was established to protect the public from unqualified practitioners. Registration is important from an individual's standpoint because courts will usually not recognize one's status as an engineer unless one is registered in the district hearing the case. Many engineers hold multiple registrations and thus can testify as expert witnesses in any state where they hold a valid registration. Professional registration is a condition of employment in many consulting offices, in government agencies, and in senior positions in industry.

Each registration board sets its own criteria for licensing, although many elements are common to all. The three chief elements are educational background, responsible engineering experience, and competence as demonstrated by a written examination. In recent years there has been a movement toward a standardization of registration laws. Almost all boards now use a national examination administered by the National Council of Engineering Examiners. NCEE's standardized examinations are each eight hours in length. The first, entitled "Fundamentals of Engineering," is generally taken by seniors in an ABET-accredited engineering program (or in an ABET-accredited engineering technology program, in some states) and by recent graduates in the process of gaining some practical experience. Each examination is different from the preceding ones, but the main components covered include mathematics, physics, chemistry, statics, dynamics, mechanics of materials, fluid mechanics, thermodynamics, electrical theory, nucleonics, wave phenomena, and computer applications. This examination was known as the "engineer-in-training" or EIT examination before the advent of the national examination. Some positions require that the candidate hold an "EIT certificate," which means passing the fundamentals examination.

The second examination is usually taken after one acquires the necessary practical experience. It is popularly known as the "Professional" or "Principles and Practices" examination. This eight-hour examination directly relates to a particular field of engineering.

Both are open-book examinations. Therefore, one must be familiar with the textbooks and handbooks that cover the specified areas.

Patent Agreements

Design engineering work can be creative, a matter of developing a new product; or it can be routine, a matter of ordinary calculations using common engineering and scientific principles. Patents are issued to an individual, not a corporation. Therefore, the firm must protect its rights if one uses its research facilities to develop products and is being paid for that service. Most employers require as a stipulation of employment that a new employee sign some type of patent agreement. Such an agreement states that the firm will handle the legal and financial aspects of the patent process, and that the employee will be rewarded in accordance with the terms of the agreement. Such agreements often extend for a period after an employee's termination of employment.

Note that such agreements are designed to prevent ruthless employees from using a firm's facilities and expertise to develop new products and then leaving the firm when a new breakthrough occurs. The firm is paying them for developing the product. Most organizations have an award system based on major profitable developments; such awards take the forms of bonuses, promotions, royalties, or a combination of these. Employers wish to encourage creativity; they have the patent agreement only for their protection.

In new development work, employers insist on accurate record keeping to protect any evidence relating to patents. Therefore, the following steps must be taken to assure indisputable facts:

1. maintain a ledger-type notebook (not a loose-leaf type);
2. date every entry;
3. record in ink;
4. make no erasures (cross out any changes instead);
5. update daily.

 **Cost Figures**

The cost of producing a manufactured product is an important consideration in its design. Labor and materials costs vary from day to day. A practicing designer has a decided advantage over a student when it comes to estimating costs. Figures for plant labor costs are readily available, and the costs of raw materials and stock components are easily found by consulting the firm's purchasing agent or the supplier. Manufacturers frequently give discounts for quantity purchases; these can greatly lower unit costs. Thus, the quantity of a production run greatly affects costs. Mass-produced products can be made for a lower unit cost; both labor and materials costs are lower. The labor cost is lower because unskilled or semiskilled employees can be trained to perform one simple task with great efficiency, and there is enough work available to justify breaking down each operation into simple tasks. Automobile and appliance assembly plants are good examples of this type of operation. In addition to labor cost savings, the quantity discount for large amounts of raw materials and stock parts (nuts, bolts, hardware, and other "off-the-shelf" items) is a substantial item.

A student cannot pinpoint specific costs; however, he or she can use catalog guidelines to apply judgment to a design. Any machine can be built with precision and designed for long life. The type of equipment and the type of use or abuse it gets will often dictate the amount of precision that must be built into it. Testing and laboratory equipment that is operated by skilled technicians will certainly differ in quality from such appliances as washing machines and vacuum cleaners.

The following figures are given in round numbers to aid the designer in selecting the appropriate material for a specific application.

The approximate costs per pound of the following metals as compared to carbon steel are:

Aluminum	6.5 times as much
Aluminum alloys	7 times as much
Titanium	35 times as much
Phosphor bronze	20 times as much
Bronze	14.5 times as much
Naval brass	14 times as much
Copper	9.5 times as much

Costs cannot be determined from such results. One can, however, use such comparisons to determine the optimal metal for a given set of necessary or sufficient physical and mechanical properties. Overspecifying of materials can cause high production costs.

 Format for Drawings

Most firms use printed drawing formats cut to the proper sheet size. Finished sizes are fairly well standardized at $8\frac{1}{2} \times 11$, 11×17, 17×22, 22×34, and 34×44. For larger drawings, 36-in. rolls of vellum, linen, or drafting film are used. Thus, the drawing is 36 in. wide and as long as convenient—often, several feet. The cut sheet sizes were chosen to allow easy filing. An $8\frac{1}{2} \times 11$ sheet is the same size as a standard letter form and hence can be attached or filed with letters. The next size (11×17) becomes

$8\frac{1}{2} \times 11$ when folded once, the next (17×22) when folded twice, and so on until one reaches the roll size. Some firms prefer to use slightly larger sheets but to include so-called "trim lines" at the usual sheet size. This gives the drafter some room for scrap while working. After the drawing is finished, it is trimmed to size.

Typical metric drawing sizes are 210×297 mm, 297×420 mm, 420×594 mm, 594×841 mm, and 841×1189 mm. Note that each larger size doubles the shorter dimension of the preceding size much as in the traditional U.S. system; thus, filing and mailing are a simple matter of folding. The metric sizes have one advantage over the U.S. sizes. Regardless of the drawing size, the ratio of the long side to the short side is always $\sqrt{2}$ to 1 or 1.414 to 1, as shown in Fig. 18-1. This proportion simplifies microfilming since only the camera distance needs to be changed when filming different sizes. The full area of the film is always used, and there is no wasted space.

Many information retrieval systems use microfiche, a system in which many drawings are placed on a film sheet about the size of a file card. A microfiche reading device is then used to move the "card" horizontally or vertically so that one can view a particular image. The standard microfiche image is 105 mm by 148 mm, which conforms to the $\sqrt{2}$ to 1 ratio common to metric drawings.

The so-called golden section (or golden rectangle) is a geometric proportion that has esthetic appeal. The golden rectangle is a rectangle in which the ratio of length to width is 1.618 to 1. Metric drawing formats, with their 1.414 to 1 ratio, nearly meet this value. The golden rectangle has been used

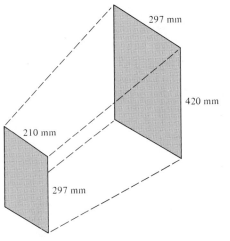

297 mm

420 mm

210 mm

297 mm

Figure 18-1
Proportionality of metric paper sizes.

by artists, architects, and designers since the days of the ancient Greeks. Windows, paintings, and room dimensions are often designed using this ratio.

Whether the dimensions are U.S. or metric, the printed format simplifies the work of the drafter. The drafter begins by completing such blanks as

1. drawing number,
2. scale,
3. name,
4. date,
5. drawing title.

Blank spaces are provided on all drawing sheets for checking and approval. In many organizations, one person approves the drawing from the structural standpoint, another from the production standpoint, and still another from the standpoint of standardization; in addition to all these, a special checker often verifies the correctness of views and dimensions. Some firms issue check prints to speed the checking process. In this way, several approvals can be made simultaneously. Printed formats also provide for engineering change records. Usually, detail sheets and subassembly sheets have a "used on" blank; this is done so that one immediately knows to what assembly this particular part belongs.

In choosing a particular size of sheet, the drafter should allow adequate room for the drawing without crowding dimensions and should allow space for general notes. The scale should be compatible with the configuration of the object. Some small parts must be greatly enlarged to make the drawing comprehensible. The drawing number is usually specially assigned and recorded so that two drawings will never be issued with the same number. Such mistakes could be costly.

Some type of general tolerance block is usually provided on a drawing so that the drafter need not place a tolerance on each individual dimension. This arrangement applies to dimensions that do not have a specific standard tolerance. Table 18-1 shows a commonly used general tolerance block, often printed directly on the drawing format. In companies where a wide variety of products is made (particularly with respect to size and precision), printed blocks are not too effective. One useful alternative is to have decals or rubber stamps made with varying tolerance limits. The drafter then selects the appropriate stamp or decal to fit the project. Tolerancing is important; properly handled, it can produce the desired quality of product at a competitive cost.

A general type of tolerance block such as that shown in Table 18-1 partially compensates for short and long dimensions, so that overlapping

Table 18-1
Typical Tolerance Block
(Variations on Finished Dimensions Unless Otherwise Marked)

Basic Dimensions	Fractional Dimensions	Decimal Dimensions
Up to $\frac{1}{4}$	$\pm \frac{1}{128}$	± 0.005
Above $\frac{1}{4}$ to 6	$\pm \frac{1}{64}$	± 0.005
Above 6 to 24	$\pm \frac{1}{32}$	± 0.010
Above 24	$\pm \frac{1}{16}$	± 0.015

All angles $\pm \frac{1}{2}$ deg.

tolerances can be somewhat minimized. It is always advisable to eliminate the least important dimension when an overall dimension is used with many intermediate distances. If *all* dimensions must be included, one should mark "REF" over the least important dimension. This indicates that no tolerance is applied to this particular dimension. In using the table, merely expressing a dimension in fractional or decimal form on the drawing affects the tolerance for that particular dimension.

Angular tolerance should be used with discretion. One should be particularly careful in specifying any angle in degrees, because it is usually difficult to lay out angles accurately when they are expressed in degrees. This is true whether the degrees are broken down into minutes and seconds or decimals. It is better to establish angles by using Cartesian coordinates to set an appropriate endpoint. Such angles as 90, 60, 45, or 30 degrees can be accurately established by either method. Remember that protractors are not accurate instruments; angle accuracy is best checked with a sine bar used in conjunction with gage blocks or by optical projection. Linear dimensions can be checked in many ways using various types of measuring equipment.

The titles given to assembly and detail drawings should be as descriptive as possible. The principal noun should be given first, followed by descriptive adjectives and other pertinent information. For example, an overrunning clutch would be labeled "Clutch, overrunning." Thus, anyone within the plant will understand that the subject matter is classified under *clutches*.

On some formats, two spaces are provided for dates—one for the starting date, the other for the completion date. When only one blank is provided, it is usually intended for the completion date.

The material from which the object is to be fabricated must be shown on detail drawings. This can take the form of a general note, or it may be written into a space provided in the printed format. The listed material should also be shown on the assembly-drawing bill of material.

In any design project, the drawing is the connecting link between the engineering department and the manufacturing facility. Thus, accuracy is important.

 Surface Finishes

The specifying of surface finishes is becoming standard practice in many industries. Formerly, a drawing merely stated that a surface was to be "ground smooth." Such a specification leaves much to be desired; the smoothness of the surface becomes a negotiable item between the operator, inspector, assembler, and user. Numerical values for surface roughness are more desirable. These values can be checked by electromechanical instruments, surface comparators, or surface blocks (surface blocks should not be confused with gage blocks). MIL standards and many industries use the microinch root-mean-square value as a measure of the roughness (or smoothness) of a surface. Figure 18-2 shows an electromechanical device for accurately checking surface finishes.

Root Mean Square

No surface is ever perfectly smooth. The root-mean-square average represents an adequate measure of the surface condition because it takes into account the deviations (peaks and valleys) above and below the mean

Figure 18-2
Surf-Indicator® Model AD-22 Surface Measurement Instrument for R_a.
(Courtesy of Federal Products Corporation, Providence, RI.)

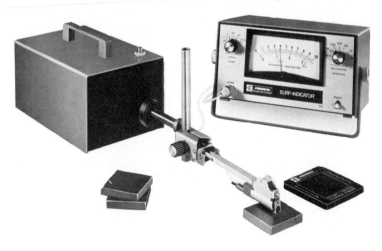

surface. The RMS value is expressed in microinches (millionths of an inch). Mathematically, the RMS value is found from the following relationship:

$$\text{RMS} = \sqrt{\frac{a^2 + b^2 + c^2 + \cdots}{n}}\;,$$

where a, b, c, etc., represent deviations from the mean surface in microinches and n is the number of measurements. The heights and depths of variations must be taken at equal intervals to be representative. The major advantage of RMS values from the designer's standpoint is that they represent specific surface conditions. Microinch finishes have been standardized at the following values: 2, 4, 8, 16, 32, 63, 125, 250, 500. Surfaces rougher than 500 are usually not specified. Values smaller than 2 microinches RMS are sometimes used in precise work. The microinch value is used in conjunction with the surface-finish symbol. A combined symbol is sometimes used that incorporates the following additional values:

1. *surface roughness width:* the distance between peaks and valleys;
2. *waviness:* the height of waves caused by warping, machine tool vibrations, deflections, etc.;
3. *lay:* the surface patterns of abrasions, such as parallel, perpendicular, circular, multidirectional, etc.

Limitations of Surface Finish

Practical values should be considered when specifying surface finishes. Overfinishing can be very expensive. For example, a finish of 8 microinches could cost four times as much as one of 125 microinches. The designer should therefore specify the roughest finish possible that will not interfere with the function of the part. Some designers feel that the simple arithmetical average is just as reliable as the root-mean-square average; the arithmetical average produces slightly lower values than the RMS value. The designer should understand what RMS values are obtainable by typical manufacturing processes. The following list should serve as a guide. Note carefully the methods needed to attain some of the lower values of surface roughness.

Casting	500
Forging	500, 250, 125
Die casting	125, 63, 32, 16
Sawing	500
Turning, Shaping	250, 125, 63, 32
Drilling	125, 63
Reaming	125, 63, 32
Grinding	125, 63, 32, 16, 8
Lapping	16, 8, 4, 2

Judgment must be used in specifying surface finishes that will allow proper performance. Accurate finishes naturally reduce friction. In the case of parts moving relative to each other, such as a shaft rotating in a sleeve bearing, proper finish is important. If the surface is too rough, the peaks will penetrate the lubricant and cause metal-to-metal contact with resultant wear; if the surface is too smooth, it will not maintain an oil film. Dimensional tolerances have a direct bearing on surface finishes. Loose dimensional tolerances must be used if dimensions are taken from a rough surface. It is better practice to specify dimensions from finished surfaces.

 ## Tolerances

Dimensional variations must be considered carefully as an important phase of design. If dimensional tolerances are too tight, the cost of the finished product may become excessive; if they are too loose, the equipment could malfunction. Realistic values must be chosen to fit the situation. Mating parts and cylindrical fits are places where tolerances must be carefully considered. Other tolerances are sometimes important as well. For instance, concentricity is vital where many concentric circular surfaces must align with mating concentric circles. A rifle cartridge is a product in which concentricity is of utmost importance. Positional dimensioning to ensure roundness, parallelism of surfaces, perpendicularity of adjoining surfaces, and flatness must be specified for some products.

 ## Protective and Decorative Coatings

Corrosion problems arise when any ferrous metals are used. Protective coatings are needed to prevent oxidation or chemical reaction. Paints, greases, and metallic coatings are extensively used. Metallic coatings serve many purposes besides rust protection. They can be decorative, they can prevent wear, and they may be used to increase the dimensions of worn parts. Properly coated sheet steel can often be used instead of more expensive metals, such as aluminum and stainless steel. Several methods are used to produce metallic coatings, of which electroplating is by far the most common. In this process, the part to be plated is placed in an electrolytic solution. The anode is the metal being used for plating, the cathode is the part receiving the plating, and the bath is a solution containing a compound of the plating metal in liquid form. An electric current activates the plating process. Plating is a slow and expensive process; therefore, the coatings are relatively thin. The commonly used plating metals are chromium, cadmium, zinc, nickel, copper, silver, gold, tin, lead, and some alloys such as brass and bronze. Chromium is one of the more expensive metals, but it imparts the greatest number of desirable qualities. Chrome plating is highly decorative and resists wear, heat, and corrosion. When used primarily for corrosion resistance, the part is usually plated with either copper or nickel before adding the coat of chromium.

Several other methods of applying metallic coatings are also used. The simplest is hot dipping; the part being coated is dipped into a molten bath of a material such as solder, tin, zinc, or lead. A similar method is spraying molten metal. In this technique, such nonferrous metals as tin, lead, copper, Monel, cadmium, aluminum, or brass are melted, atomized, and sprayed directly on the part. This process produces a somewhat porous coating (compared to plating), but it does have an advantage in that the coating metal can be brought to the part being coated. This method is advantageous when coatings are applied to large parts. Metal cladding is sometimes used; such metals as copper, tin, and aluminum are "clad" by pressing or rolling them on the surface of the heated base metal.

Cementation is still another way of protecting metals against corrosion. In this procedure, the surface of the base metal is alloyed by packing a nonferrous metal powder around the part and heating the aggregate to a temperature slightly below the critical temperature of the more fusible metal. Several metals can be used in this process. If zinc is used, the process is known as sheradizing; if aluminum is used, it is called calorizing; if chromium is used, it is called chromizing.

Plastic coatings are frequently used as a combination of temporary protection and packaging. When the product is put in use, the plastic coating is peeled away.

 **Numerical Control**

Over the years, manufacturing changes have progressed from hand operations to mechanized processes. A newer trend has been toward automation. Mechanization first led to mass production methods; this resulted in lower costs for products that could be mass-produced. Typical examples of mass-produced items are automobiles and appliances. Automation is merely continuous automatic production; practically all functions of manufacturing are handled automatically. Closely related to automation is the concept of numerical control. This idea is not entirely new—the player piano, for instance, was numerically controlled. Several inputs have been used successfully in the evolution of numerical control. Punched cards were the first device used in the automatic factory. A (somewhat) standard tape has been developed; this is a paper tape 1 in. wide with 8 "tracks." One versed in numerical control can read the tape prior to starting an operation; thus, rejections during a run can be minimized, if not completely eliminated. Magnetic tape used with a computer is another possibility.

Two basic systems can be employed. The *positioning* system is useful for such operations as drilling. The *continuous-path* system is suited for such functions as contour milling. If a continuous-path system is available, it can be used for positioning operations; the reverse does not apply. Certain shop processes and related functions of plant operations can be carried out under numerical control. Some of these are drilling, boring, milling, turning, plotting, drafting, measuring, inspecting, riveting, bending, welding, and flame cutting.

Note that changing production methods have had an effect on drafting and design procedures. Mass-production methods led to greater emphasis on such areas as tolerancing, for interchangeability of parts. Positional dimensioning has come to the forefront in many industries; baseline dimensioning has become more prominent over the years. Now, with numerical control, coordinate dimensioning has taken the spotlight. Coordinate dimensioning assists the programmer in setting up the numcrical-control system. If regular shop drawings are the only ones available, it is usually necessary to set up a coordinate system before programming.

Numerical control has many advantages, although the initial costs are high. Parts manufactured under numerical control are identical. After tapes have been prepared, they can be stored for future use in various parts of a factory or in other plants. In the machine-tool industry, setup time is greatly reduced when numerically controlled equipment is used.

 Human Engineering*

Psychologists and industrial engineers constantly strive to fit humans to machines. The designer can do much to fit machines to humans. While it is difficult to define the average human being, certain dimensions can be used when designing a piece of equipment. Much information has recently been collected in the area of human engineering. Important items include the diameter of a handle, the diameter of a handwheel, the finger room needed on a knob, the size of dial calibrations, signal-light intensity and color, the reach of a lever, the height of a working area, and human limitations on lifting, pulling, pushing, and reading. This is an area that will receive greater attention in future years; space exploration has focused attention on it.

*For information in this area, consult Dreyfuss, *The Measure of Man*, Watson-Guptill, 1967.

 Standard Parts

Common hardware and catalog parts are frequently used in design work, since it is convenient (and cheaper) to use such items rather than to produce special parts. Most moderately large companies have a standards division that sets engineering standards. These list the preferred sizes to be used when possible. *This is important!* If one considers all the materials, bolt-head types, screw series, classes of fits, lengths, and diameters that are available for even a common bolt, huge inventories would be necessary if a designer exercised unrestricted choice. The engineering standards show only the ones that are used in great quantities. One should therefore select one of the recommended sizes to avoid stocking seldom-used types.

Future replacement difficulties should be considered when selecting such catalog items as electric motors and ball bearings. A popular type and size that can be obtained "off the shelf" in all localities should be used if possible.

 Engineering Changes

In spite of all the careful planning that goes into the design of a product, changes often have to be made in the design stage and after the product reaches production. Dimensions, tolerances, or views should not be changed on a drawing without carefully recording the revisions. This is necessary because personnel from various departments can become involved. Most firms have a regular form for instituting engineering changes. After the form is completed and approved, drawing changes are made and new prints are issued.

The EC form varies in different companies, but it usually includes most of the following items. First, the identity of the part or assembly must be recorded, by title and by drawing number. The effective date of the change must be given. This can be a specific date, or the change can be effective with a certain lot number. An important point is to specify the disposition of the material on hand. Parts sometimes need to be reworked or matched with parts made prior to the change. Often a change involves new tooling or equipment. Previously made parts should not be scrapped if at all possible. One of the most important items on the EC form is the actual description of the change. This should be carefully worded to avoid misunderstanding. Usually, a short, terse statement in the past tense is preferred. An example of wording might be "$2\frac{1}{2}$-in. dim was $2\frac{3}{4}$ in." This informs all concerned that the $2\frac{3}{4}$-in. dimension is being changed to $2\frac{1}{2}$ in. Abbreviations are encouraged in most industries. Another important item on the EC form is the reason for the change. Any number of different reasons can be specified; however, the costs of the change are often charged to specific departments. Thus, careful wording is important. Typical reasons for a change might be *design change*, *engineering error*, *production request* (usually to cut costs or use available equipment), or *to facilitate purchasing*.

Engineering-change numbers, like drawing numbers, must be carefully assigned to avoid duplication. A revision letter usually accompanies the change; this appears on the drawing as revision A, B, C, etc. Sometimes, the revision letter is placed in a circle over the changed dimension to indicate what was altered; in addition, most drawings contain an EC column for a record of the changes. In complex products, changes can exhaust the alphabet; the revisions are then titled AA, BB, etc. When the number of revisions becomes excessive, a new drawing is made to supersede the old one.

Since many ideas become obsolete before an end product is produced, the designer must keep an open mind with respect to changes and lost effort.

 Computer-Aided Design

The computer is a tool helpful to mechanical designers for lightening tedious calculations and providing extended analysis of available data. Computer-based *interactive systems* have made possible the concepts of computer-aided design (CAD) and computer-aided manufacturing (CAM). Such systems allow one to transmit ideas to punched tapes for numerical machine control without producing formal working drawings. This is an extreme situation that is useful for producing only certain components. However, computer-aided design is a widely used technique that larger firms are finding useful in developing drawings at minimal expense.

Figure 18-3 shows a system that has made CAD and CAM a reality. This three-axis interactive-design system provides all the advantages of interactive graphics to the mechanical designer and the N/C programmer. The system combines a full three-dimensional design and drafting capability with a set of algorithms that use the drawing data to automatically generate the tool paths for N/C machining.

The software is constructed as three sets of applications functions, all working on the same data base. The geometric construction subsystem allows the mechanical designer to describe the geometry of a part by using interactive graphics instead of the conventional drafting tools. The designer

Figure 18-3
Three-axis interactive design system.
(Courtesy of Gerber Scientific, Inc.)

can then call upon the drafting subsystem, which aids him or her in dimensioning, labeling, and adding notes to the drawing.

To complete the transition from design concept to manufacturing, a programmer uses the N/C portion of the system. Working with the geometric description created during the design process, the programmer selects a sequence of machining operations. The system automatically generates the necessary N/C tool paths to machine the part. Included in the system are algorithms for drilling, pocket and profile cutting, and three- to five-axis surface contouring. During the N/C programming process, the programmer receives immediate visual verification of each tool path. In this way, errors can be eliminated without test cutting.

The system is built around an interactive-design-system central processor that includes a 24K minicomputer with floating-point hardware and a 2.4 million word disc. Up to six remote terminals can be used with this system. Peripheral terminal equipment can include such items as a cathode ray tube with cursor and a digitizer/plotter. Coupled with the hardware components of the system is a software system that graphically couples designer and machine.

Questions for Review

1. Check current periodicals for three products for which CAD and CAM were used.

2. What advantages do metric-size drawing sheets have over the traditional U.S. sizes?

3. Give three uses of the golden rectangle in addition to those given in this chapter.

4. What advantage does an employee gain by signing a patent agreement with his or her employer?

5. Explain what is meant by a *canon of ethics* for professional engineers. Indicate the importance of such a document to the design field.

6. Explain why engineering changes must be documented with special forms as well as being noted on the drawings.

7. Explain how the cost of production relates to proper tolerances and surface finishes.

8. You are to design a gasoline-powered lawnmower. What items should you check in addition to the design of the moving components?

APPENDIXES

A. Tables of properties of common engineering materials
B. Unified screw threads
C. Selected reference books
D. Patent facts
E. Technical societies and trade associations
F. Commonly used mathematical relationships and constants
G. Viscosity conversion factors
H. Selected general conversion factors
 I. Wire and sheet-metal gages
 J. Selected U.S./metric conversions
K. Listing of principal formulas
L. Computer programs

Appendix A   Tables of Properties of Common Engineering Materials

Table 1
Average Tensile Properties of Materials

Material	Condition	Tensile Strength, psi	Yield stress, psi	Percent Elongation in 2 in.	Remarks and Typical Applications
Wrought iron	hot rolled	48,000	30,000	—	
Ductile cast iron		62,000	34,000	14	
Gray cast iron		25,000	6,000	—	
Structural steel		66,000	33,000	21	
18-8 stainless steel	annealed	80,000	30,000	50	
AISI 1015*	hot rolled	61,000	45,500	39	sheet-metal applications
AISI 1018*	hot rolled	69,000	48,000	38	sheet-metal applications
	cold drawn	82,000	70,000	20	
AISI 1020*	hot rolled	65,000	48,000	36	fan blades, camshafts
	cold drawn	78,000	66,000	20	
AISI 1025*	hot rolled	67,000	45,000	36	forgings
	cold drawn	80,000	68,000	20	
AISI 1030†	hot rolled	80,000	50,000	31	keys, links, levers
AISI 1040†	hot rolled	90,000	60,000	24	bolts, camshafts, links, levers
AISI 1045†	hot rolled	98,000	59,000	24	bolts, camshafts, links, levers
AISI 1050†	hot rolled	103,000	60,000	20	forgings, shafts, gears
AISI 1060†	hot rolled	118,000	70,000	16	lock washers
AISI 1080†	hot rolled	141,000	83,000	12	leaf springs
AISI 1095†	hot rolled	141,000	83,000	9	springs, tools
AISI 1117*	hot rolled	70,600	44,300	33	free-cutting applications
AISI 1118*	hot rolled	75,600	45,900	32	free-cutting applications

Material	Condition				Applications
AISI 1137†	hot rolled	91,000	54,000	28	free-cutting applications
AISI 1141†	hot rolled	98,000	52,000	22	free-cutting applications
AISI 1144†	hot rolled	102,000	61,000	21	free-cutting applications
AISI 4130‡	hot rolled (anneal)	86,000	56,000	29	axles, links
	cold drawn (anneal)	98,000	87,000	21	
AISI 4320§	hot rolled (anneal)	86,000	61,600	29	gears, pins, shafts
AISI 4340‖	hot rolled (anneal)	101,000	69,000	21	
	cold drawn (anneal)	111,000	99,000	16	
AISI 6150‖	hot rolled (anneal)	91,000	58,000	22	springs, shafts, gears, forgings
AISI 8740‖	hot rolled (anneal)	95,000	64,000	25	heavy-duty machine parts
	cold drawn (anneal)	107,000	96,000	17	
Aluminum alloy					
2014-O	annealed	27,000	14,000	18	
Monel	cold drawn	100,000	80,000	25	
Monel	cast	80,000	35,000	35	
Phosphor bronze	cold rolled	100,000	75,000	3	
Beryllium copper	cold rolled	120,000	90,000	4	springs
Titanium	cold rolled	145,000	125,000	10	
Nylon	molded compound	7000 to 12,000	—	25 to 320	
Polyethylene		1200 to 3500	—	50 to 500	

Note: The properties presented in these tables are for use in solving problems in this text; only a few average values are given to serve as a guide. Specific information for design applications should be obtained from the producers of the materials. For steels, the shearing stress is approximately three-quarters ultimate stress; the endurance limit is approximately one-half the ultimate value.

*Low-carbon type, case hardening.
† Water and oil hardening grades.
‡ Alloy steel, water hardening grade.
§Alloy steel, carburizing grade.
‖Alloy steel, oil hardening grade.

Table 2
Average Modulus-of-Elasticity Values
(All values should be multiplied by 10^6)

Material	Tension (E), psi	Shear (G), psi
Wrought iron	27	10
Steel	30	12
ASTM gray cast iron	15	6
Malleable cast iron	23	9.2
18-8 stainless steel	27.6	10.6
Monel	26	10
Bronze	15	6
Titanium	16.5	6.5
Aluminum alloys	10	4
Nylon molding compound	4	____
Polyethylene	0.4	____

Table 3
SAE Numbering System for Steels

1*	carbon steel[†]
11	carbon steel (free-cutting, screw stock)
2	nickel
3	nickel chromium
4	molybdenum
5	chromium
6	chromium vanadium
7	tungsten
9	silicon manganese

*A coded system for designating steels is commonly used by industry, by the Society of Automotive Engineers, and by the American Iron and Steel Institute. A four- or five-digit number indicates much valuable information. The first digit designates the type of steel.

[†]The second digit gives the percentage of the principal alloying element. The remaining digits indicate the points of carbon, or the percentage of carbon content in hundredths of a percent. For example, SAE 1060 steel is a plain carbon steel with a carbon content averaging 0.60%; SAE 5150 contains approximately 1% chromium and approximately 0.50% carbon.

Table 4
Effects of Alloying Elements on Steel

Alloying element	Effects
Aluminum	deoxidizer and degasifier
Boron	increases hardenability
Chromium	increases hardenability
	promotes corrosion resistance
Manganese	increases hardenability
	increases strength and hardness
Molybdenum	promotes abrasion resistance
	increases tensile strength and creep strength at elevated temperatures
	increases hardenability
Nickel	increases corrosion resistance
	increases toughness
Silicon	serves as deoxidizer
	increases strength
Tungsten	adds abrasion resistance
	promotes strength and hardness at elevated temperatures
Vanadium	deoxidizer
	increases fatigue-life limit

Appendix B   Unified Screw Threads

Size	Threads / in. (coarse)	Threads / in. (fine)
0 (0.060)	——	80
1 (0.073)	64	72
2 (0.086)	56	64
3 (0.099)	48	56
4 (0.112)	40	48
5 (0.125)	40	44
6 (0.138)	32	40
8 (0.164)	32	36
10 (0.190)	24	32
12 (0.216)	24	28
$\frac{1}{4}$	20	28
$\frac{5}{16}$	18	24
$\frac{3}{8}$	16	24
$\frac{7}{16}$	14	20
$\frac{1}{2}$	13	20
$\frac{9}{16}$	12	18
$\frac{5}{8}$	11	18
$\frac{3}{4}$	10	16
$\frac{7}{8}$	9	14
1	8	12
$1\frac{1}{8}$	7	12
$1\frac{1}{4}$	7	12
$1\frac{3}{8}$	6	12
$1\frac{1}{2}$	6	12
$1\frac{3}{4}$	5	12
2	$4\frac{1}{2}$	12
$2\frac{1}{4}$	$4\frac{1}{2}$	12
$2\frac{1}{2}$	4	12
$2\frac{3}{4}$	4	12
3	4	12
$3\frac{1}{4}$	4	12
$3\frac{1}{2}$	4	12
$3\frac{3}{4}$	4	12
4	4	12

ASA B1.1 (1960)

Appendix C   Selected Reference Books

Handbooks

AISC, *Manual of Steel Construction*, American Institute of Steel Construction

ASTME, *Tool Engineers Handbook*, McGraw-Hill Book Co., Inc.

Bolz and Hagermann, *Materials Handling Handbook*, Ronald Press Co.

Chains for Power Transmission and Material Handling—Design and Applications Handbook, American Chain Association

Dudley, *Gear Handbook*, McGraw-Hill Book Co., Inc.

Eshbach, *Handbook of Engineering Fundamentals*, John Wiley & Sons, Inc.

Greenwood, *Product Engineering Design Manual*, McGraw-Hill Book Co., Inc.

Hudson, *The Engineers' Manual*, John Wiley & Sons, Inc.

Jones, *Ingenious Mechanisms for Designers and Inventors*, The Industrial Press

Kent, *Mechanical Engineers' Handbook*, John Wiley & Sons, Inc.

McNeese and Hoag, *Engineering and Technical Handbook*, Prentice-Hall, Inc.

Mantell, *Engineering Materials Handbook*, McGraw-Hill Book Co., Inc.

Marks and Baumeister, *Standard Handbook for Mechanical Engineers*, McGraw-Hill Book Co., Inc.

Oberg and Jones, *Machinery's Handbook*, The Industrial Press

Stanier, *Plant Engineering Handbook*, McGraw-Hill Book Co., Inc.

Fairweather and Sliwa, *VNR Metric Handbook*, Van Nostrand Reinhold Co.

Textbooks

Bassin, Brodsky, and Wolkoff, *Statics and Strength of Materials*, McGraw-Hill Book Co., Inc.

Begeman and Amstead, *Manufacturing Processes*, John Wiley & Sons, Inc.

Blodgett, *Design of Weldments*, The James F. Lincoln Arc Welding Foundation

Brown, *Package Design Engineering*, John Wiley & Sons, Inc.

Buckingham, *Manual of Gear Design* (Sections 1, 2, 3), The Industrial Press

Den Hartog, *Mechanical Vibrations*, McGraw-Hill Book Co., Inc.

Doughtie, Vallance, and Kreisle, *Design of Machine Members*, McGraw-Hill Book Co., Inc.

Earle, *Engineering Design Graphics*, Addison-Wesley Publishing Co., Inc.

Edgar, *Fundamentals of Manufacturing Processes*, Addison-Wesley Publishing Co., Inc.

Fitzgerald, *Strength of Materials*, Addison-Wesley Publishing Co., Inc.

French and Vierck, *Fundamentals of Engineering Drawing and Graphic Technology*, McGraw-Hill Book Co., Inc.

Friedman and Kipness, *Industrial Packaging*, John Wiley & Sons, Inc.

Greenwood, *Mechanical Power Transmission*, McGraw-Hill Book Co., Inc.

Henke, *Introduction to Fluid Mechanics*, Addison-Wesley Publishing Co., Inc.

Hicks, *Pump Selection and Application*, McGraw-Hill Book Co., Inc.

Higdon and Stiles, *Engineering Mechanics* (Volume 1—*Statics;* Volume 2—*Dynamics*), Prentice-Hall, Inc.

Hinkle, *Kinematics of Machines*, Prentice-Hall, Inc.

Jensen and Chenoweth, *Applied Engineering Mechanics*, McGraw-Hill Book Co., Inc.

Johnson, *Optimum Design of Mechanical Elements*, John Wiley & Sons, Inc.

Olsen, *Strength of Materials*, Prentice-Hall, Inc.

Pippenger and Hicks, *Industrial Hydraulics*, McGraw-Hill Book Co., Inc.

Roark and Young, *Formulas for Stress and Strain*, McGraw-Hill Book Co., Inc.

Rothbart, *Cams: Design, Dynamics and Accuracy*, John Wiley & Sons, Inc.

Seely and Smith, *Advanced Mechanics of Materials*, John Wiley & Sons, Inc.

Shigley, *Mechanical Engineering Design*, McGraw-Hill Book Co., Inc.

Tao, *Fundamentals of Applied Kinematics*, Addison-Wesley Publishing Co., Inc.

Timoshenko and Young, *Elements of Strength of Materials*, Van Nostrand Reinhold Co.

Young, *Materials and Processes*, John Wiley & Sons, Inc.

Periodicals

Machine Design, Penton/IPC
Materials Engineering, Penton/IPC
Production Engineering, Penton/IPC

Standards

Numerous standards are available in the area of mechanical design. A few of the organizations and societies having appropriate material are the following:

American Gear Manufacturers Association
American Society of Mechanical Engineers
American Society for Testing and Materials
Anti-Friction Bearing Manufacturers Association
MIL standards
Society of Automotive Engineers
United States of America Standards Institute

Appendix D   Patent Facts

As a stipulation for employment, most technical personnel are required to sign a patent agreement with their company. This is done primarily so that the company's research and development projects will not fall into the hands of competitors. Patent law is very complex; patent attorneys should be consulted on most inventions. The Patent Office of the U. S. Department of Commerce publishes much information for the use of the public.

Sketches and calculations *must* be signed by those involved in the design. This practice cannot be overemphasized.

The following statements are a very few of the many facts about patents:

1. Patents are granted by the United States Government for a period of 17 years.
2. Patents are granted to individuals, not to organizations.
3. Sex, age, or citizenship have no bearing on one's eligibility for obtaining a patent.
4. Subject matter for a patent includes any new process, machine, manufacture, or composition of matter.
5. Useless devices are excluded from patenting.

6. A design patent (ornamental design) is issued for $3\frac{1}{2}$, 7, or 14 years; this is the option of the applicant.
7. Models are very seldom required in obtaining a patent.
8. Printed copies of any patent can be obtained for a nominal charge.
9. The Patent Office maintains copies of every patent in numerical order and by subject matter.
10. Libraries in 22 cities maintain copies in numerical order.
11. Drawings must be done in India ink on 10×15 sheets of specified white paper.
12. Hatching and shading specifications on drawings must be followed.
13. A patent grants the right to exclude others from making, using, or selling a patentee's invention.
14. A patent is considered a personal property; it can be sold, mortgaged, or bequeathed.
15. The term "patent pending" merely means that an application has been filed; it does not guarantee issuance.
16. A patent cannot be issued if the invention was described in any printed publication (worldwide) more than a year before the patent application was filed.
17. Legal suits for patent infringements are instituted by the patentee. A patent, even though issued, is strengthened by being declared valid by a court decision.

Appendix E Technical Societies and Trade Associations

Since mechanical design cuts across all types of industries, the designer should be familiar with organizations that disseminate authoritative information in particular areas of interest. Means of dissemination include technical meetings, seminars, technical papers, periodicals, expositions, standards, and handbooks. The following list includes *some* of the many organizations that provide valuable design information:

American Chain Association, 1133 Fifteenth St., N.W., Washington, D.C. 20005

American Foundryman's Association, Golf and Wolf Roads, Des Plaines, Illinois 60016

American Gear Manufacturers Association, 1901 North Myer Drive, Arlington, Virginia 22209

American Institute of Steel Construction, 400 North Michigan Avenue, 8th Floor, Chicago, Illinois 60611

American Iron and Steel Institute, 1000 16th Street, N.W., Washington, D.C. 20036

American National Standards Institute, 1430 Broadway, New York, New York 10018

American Nuclear Society, 555 North Kensington Avenue, La Grange Park, Illinois 60525

American Petroleum Institute, 2101 L Street, N.W., Washington, D.C. 20037

American Society for Engineering Education, 11 Dupont Circle, Washington, D.C. 20036

American Society of Lubrication Engineers, 838 Busse Highway, Park Ridge, Illinois 60068

American Society for Metals, Metals Park, Ohio 44073

American Society for Non-Destructive Testing, 3200 Riverside Drive, Columbus, Ohio 43221

American Society of Safety Engineers, 850 Busse Highway, Park Ridge, Illinois 60068

American Society for Testing and Materials, 1916 Race Street, Philadelphia, Pennsylvania 19103

American Society of Heating, Refrigerating and Air Conditioning Engineers, 1791 Tullie Circle, N.E., Atlanta, Georgia 30329

American Society of Mechanical Engineers, 345 East 47th Street, New York, New York 10017

American Welding Society, 2501 N.W. 7th Street, Miami, Florida 33125

Anti-Friction Bearing Manufacturers Association, Century Building, Suite 1015, 2341 Jefferson Davis Highway, Arlington, Virginia 22202

Compressed Air and Gas Institute, 1230 Keith Building, Cleveland, Ohio 44115

Institute of Electrical and Electronics Engineers, 345 East 47th Street, New York, New York 10017

Instrument Society of America, 400 Stanwix Street, Pittsburgh, Pennsylvania 15222

Mechanical Power Transmission Association, 1717 West Howard Street, Evanston, Illinois 60202

National Electrical Manufacturers Association, 2101 L Street, N.W., Washington, D.C. 20037

National Fluid Power Association, 3333 North Mayfair Road, Suite 311, Milwaukee, Wisconsin 53222

The Rubber Manufacturers Association, 1901 Pennsylvania Avenue, Washington, D.C. 20006

Society of Automotive Engineers, 400 Commonwealth Drive, Warrendale, Pennsylvania 15096

Society of Manufacturing Engineers, One SME Drive, Box 930, Dearborn, Michigan 48128

Society of Plastics Engineers, 656 West Putnam Avenue, Greenwich, Connecticut 06830

Appendix F Commonly used Mathematical Relationships and Constants

Trigonometric Functions

Angle	sin	cos	tan
0	0.0000	1.0000	0.0000
10	0.1736	0.9848	0.1763
14.5	0.2504	0.9681	0.2586
15	0.2618	0.9659	0.2679
23	0.3907	0.9205	0.4245
30	0.5000	0.8660	0.5774
45	0.7071	0.7071	1.0000
60	0.8660	0.5000	1.7321
90	1.0000	0.0000	∞

$e = 2.718$
$\pi/4 = 0.7854$
$\log_{10} e = 0.4343$
$\log_e N = 2.3026 \log_{10} N$
1 degree $= 0.01745$ radians

57.29 degrees $= 1$ radian
180 degrees $= \pi$ radians
Arc length $= r\theta$,
  where θ is expressed
  in radians

Appendix G Viscosity Conversion Factors

Multiply	By	To obtain
Pa $\cdot$ s	1000	cP
P (poise)	100	cP
St (stoke)	100	cSt
m^2/s	10^6	cSt
ft^2/s	9.29×10^4	cSt
in^2/s	6.45×10^2	cSt

Appendix H   Selected General Conversion Factors

Multiply	By	To obtain
Btu	778	ft-lb
Btu/min	13	ft-lb/sec
Btu/min	0.0236	hp
cu ft	7.48	gal
feet of water	0.434	psi
fpm	0.0114	mph
fps	0.682	mph
fps^2	0.682	mph/sec
ft-lb	1.29×10^{-3}	Btu
ft-lb/min	3.03×10^{-5}	hp
gal (water)	8.35	lb (water)
gpm	2.23×10^{-3}	cfs
hp	42.4	Btu/min
hp	33,000	ft-lb/min
mi/hr/sec	1.47	fps^2
mph	1.47	fps
rad/sec	9.55	rpm
rpm	0.105	rad/sec
watts	3.41	Btu/hr
watts	44.3	ft-lb/min
watts	1.34×10^{-3}	hp

Appendix I   Wire and Sheet-metal Gages

Gage No.	American or B & S (nonferrous sheet and wire)	Washburn and Moen (ferrous wire)	Piano wire (ferrous wire)	U.S. Standard (ferrous sheet and plate)
0000000	0.6513	0.4900	————	0.5000
000000	0.5800	0.4615	0.004	0.4688
00000	0.5165	0.4305	0.005	0.4375
0000	0.4600	0.3938	0.006	0.4063
000	0.4096	0.3625	0.007	0.3750
00	0.3648	0.3310	0.008	0.3438
0	0.3249	0.3065	0.009	0.3125
1	0.2893	0.2830	0.010	0.2813
2	0.2576	0.2625	0.011	0.2656
3	0.2294	0.2437	0.012	0.2500
4	0.2043	0.2253	0.013	0.2344
5	0.1819	0.2070	0.014	0.2188
6	0.1620	0.1920	0.016	0.2031
7	0.1443	0.1770	0.018	0.1875
8	0.1285	0.1620	0.020	0.1719
9	0.1144	0.1483	0.022	0.1563
10	0.1019	0.1350	0.024	0.1406
11	0.0907	0.1205	0.026	0.1250
12	0.0808	0.1055	0.029	0.1094
13	0.0720	0.0915	0.031	0.0938
14	0.0641	0.0800	0.033	0.0781
15	0.0571	0.0720	0.035	0.0703
16	0.0508	0.0625	0.037	0.0625
17	0.0453	0.0540	0.039	0.0563
18	0.0403	0.0475	0.041	0.0500
19	0.0359	0.0410	0.043	0.0438
20	0.0320	0.0348	0.045	0.0375
21	0.0285	0.0317	0.047	0.0344
22	0.0253	0.0286	0.049	0.0313
23	0.0226	0.0258	0.051	0.0281
24	0.0201	0.0230	0.055	0.0250
25	0.0179	0.0204	0.059	0.0219
26	0.0159	0.0181	0.063	0.0188
27	0.0142	0.0173	0.067	0.0172
28	0.0126	0.0162	0.071	0.0156
29	0.0113	0.0150	0.075	0.0141
30	0.0100	0.0140	0.080	0.0125
31	0.0089	0.0132	0.085	0.0109
32	0.0080	0.0128	0.090	0.0102
33	0.0071	0.0118	0.095	0.0094

Gage No.	American or B & S (nonferrous sheet and wire)	Washburn and Moen (ferrous wire)	Piano wire (ferrous wire)	U.S. Standard (ferrous sheet and plate)
34	0.0063	0.0104	0.100	0.0086
35	0.0056	0.0095	0.106	0.0078
36	0.0050	0.0090	0.112	0.0070
37	0.0045	0.0085	0.118	0.0066
38	0.0040	0.0080	0.124	0.0063
39	0.0035	0.0075	0.130	——
40	0.0031	0.0070	0.138	——

All dimensions in decimal parts of an inch.

Appendix J Selected U.S./Metric Conversions

The following list gives approximate conversions from U.S. to metric units. A few of these are exact. The listing is classified by the physical quantities that are most useful in mechanical design work.

Multiply	By	To obtain
Acceleration		
foot/second^2	3.048×10^{-1}	meter/second^2
Area		
foot^2	9.29×10^{-2}	meter^2
inch^2	6.45×10^{-4}	meter^2
mile^2 (U.S. statute)	2.59×10^6	meter^2
Bending moment or torque		
kilogram (force)-meter	9.81	newton-meter
pound (force)-inch	1.13×10^{-1}	newton-meter
pound (force)-foot	1.36	newton-meter
Energy-work		
British thermal unit	1.06×10^3	joule
pound (force)-foot	1.36	joule
kilowatt-hour	3.60×10^6	joule
Flow		
gallons per minute	3.79	liters per minute
cubic feet per minute	4.72×10^{-1}	cubic decimeters per second
Force		
kilogram (force)	9.81	newton
kip (force)	4.45×10^3	newton

Multiply	By	To obtain
Force		
pound (force)	4.45	newton
Length		
foot	3.048×10^{-1}	meter
inch	2.54×10^{-2}	meter
microinch	2.54×10^{-8}	meter
mile (statute)	1.609×10^{3}	meter
Mass		
pound (mass)	4.54×10^{-1}	kilogram
tonne	1.00×10^{3}	kilogram
Moment of inertia		
inches^4	4.16×10^{5}	10^6 millimeter^4
Power		
horsepower	7.46×10^{2}	watt
Pressure		
bar	1.00×10^{5}	pascal
inch Hg (60 °F)	3.38×10^{3}	pascal
inch H_2O (60 °F)	2.49×10^{2}	pascal
psi	6.89×10^{3}	pascal
pascal	1.00	newton/meter^2
Stress		
kip/inch^2	6.89×10^{6}	newton/meter^2
psi (lbf/inch^2)	6.89×10^{3}	newton/meter^2
psi (lbf/inch^2)	6.89×10^{3}	pascal
Velocity		
foot/minute	5.08×10^{-3}	meter/second
foot/second	3.048×10^{-1}	meter/second
mile/hour	1.609	kilometer/hour
Viscosity		
centistoke	1.00×10^{-6}	meter^2/second
Saybolt universal second	2.157×10^{-1}	centistoke (at 38 °C)
Volume		
foot^3	2.83×10^{-2}	meter^3
gallon (U.S. liquid)	3.785×10^{-3}	meter^3
inch^3	1.64×10^{-5}	meter^3
liter	1.00×10^{-3}	meter^3

Appendix K   Listing of Principal Formulas

Static equations of equilibrium:

$$\sum F_x = 0, \qquad \sum F_y = 0, \qquad \sum F_z = 0, \qquad \sum M = 0 \qquad \text{(2-1)}$$

Angular velocity:

$$\omega = \frac{\theta}{t} \qquad \text{(2-2)}$$

Linear-angular velocity:

$$v = r\omega \qquad \text{(2-3a)}$$
$$\omega = 2\pi N \qquad \text{(2-3b)}$$
$$v = \pi DN \qquad \text{(2-3c)}$$

Displacement-velocity-acceleration:

$$s = v_1 t + \frac{at^2}{2}, \qquad \theta = \omega_1 t + \frac{\alpha t^2}{2} \qquad \text{(2-4a)}$$
$$v_2 = v_1 + at, \qquad \omega_2 = \omega_1 + \alpha t \qquad \text{(2-4b)}$$
$$v_2^2 = v_1^2 + 2as, \qquad \omega_2^2 = \omega_1^2 + 2\alpha\theta \qquad \text{(2-4c)}$$

Normal acceleration:

$$a_n = v\omega = \frac{v^2}{r} = \omega^2 r \qquad \text{(2-5)}$$

Tangential acceleration:

$$a_t = \alpha r \qquad \text{(2-6)}$$

Torque:

$$T = Fr \qquad \text{(2-7)}$$

Horsepower:

$$\text{hp} = \frac{NT}{63{,}000} \qquad \text{(2-8)}$$

Power (metric):

$$\text{kW} = \frac{2\pi NT}{60\,000} \qquad \text{(2-8a)}$$

Newton's second law:

$$F = ma = \frac{Wa}{g} \qquad \text{(2-9)}$$

Kinetic energy:

$$K = \frac{mv^2}{2}$$

(2-10)

Efficiency:

$$e = \frac{\text{output } (100)}{\text{input}}$$

(2-11)

Centroid of composite figure:

$$A\bar{x} = A_1 x_1 + A_2 x_2 + \cdots$$

(2-12)

Moment of inertia:

$$I_x = \int y^2 \, dA$$

$$I_y = \int x^2 \, dA$$

(2-13)

Polar moment of inertia:

$$J_z = I_y + I_x$$

(2-14)

Parallel-axis theorem:

$$I = \bar{I} + Ad^2$$

(2-15a)

$$J = \bar{J} + Ad^2$$

(2-15b)

Radius of gyration:

$$k = \sqrt{I/A}$$

(2-16)

Simple stress (tension or compression):

$$S = \frac{P}{A}$$

(2-17)

Simple stress (shear):

$$S_s = \frac{P}{A}$$

(2-18)

Strain:

$$\delta = \frac{\Delta L}{L}$$

(2-19)

Modulus of elasticity:

$$E = \frac{PL}{A(\Delta L)}$$

(2-20)

Shear modulus of elasticity:

$$G = \frac{E}{2(1 + \mu)}$$

(2-21)

Thermal-expansion stress:

$$S = EC(t_2 - t_1)$$

(2-22)

Flexure:

$$S = \frac{Mc}{I} = \frac{M}{Z}$$

(2-23)

Horizontal shear (beams), for rectangular sections:

$$S_H = \frac{3V}{2A}$$

(2-24a)

Horizontal shear (beams), for circular sections:

$$S_H = \frac{4V}{3A}$$

(2-24b)

Vertical shear (beams):

$$S_V = \frac{V}{td}$$

(2-25)

Torsional shear stress:

$$S_s = \frac{Tc}{J}$$

(2-26)

Direct stress combined with flexure:

$$S = \pm \frac{P}{A} \pm \frac{Mc}{I}$$

(2-27)

Maximum resultant shearing stress:

$$S_{s_{max}} = \frac{\sqrt{S^2 + (2S_s)^2}}{2}$$

(2-28)

Maximum resultant tensile stress:

$$S_{max} = \frac{S}{2} + S_{s_{max}}$$

(2-29)

Sliding friction:

$$f = \frac{F}{N}$$

(3-1)

Rolling resistance:

$$a = \frac{Fr}{W} \tag{3-2}$$

Pivot frictional torque (conical pivot):

$$T_f = \frac{2}{3}\left(\frac{r}{\sin\theta}\right)fW \tag{3-3a}$$

Pivot frictional torque (flat pivot):

$$T_f = \tfrac{2}{3}rfW \tag{3-3b}$$

Frictional radius (collar):

$$r_f = \frac{2}{3}\left(\frac{r_o^3 - r_i^3}{r_o^2 - r_i^2}\right) \tag{3-4}$$

Collar-friction torque:

$$T_f = \frac{2}{3}\left(\frac{r_o^3 - r_i^3}{r_o^2 - r_i^2}\right)fW \tag{3-5}$$

Gravity, API:

$$\text{gravity, API} = \frac{141.5}{\text{specific gravity}} - 131.5 \tag{3-6}$$

Absolute viscosity:

$$Z = Z_k\rho \tag{3-7}$$

Sleeve bearing pressure:

$$p = \frac{F}{LD} \tag{4-1}$$

Heat-radiating capacity (for sliding bearings):

$$Q = \frac{(t_o - t + 33)^2}{k} \tag{4-2a}$$

$$Q = \frac{FV}{LD} \tag{4-2b}$$

Torsional strength (solid shaft):

$$T = \frac{S_s\pi D^3}{16}$$

$$= \frac{63,000(\text{hp})}{N} \tag{5-1}$$

Torsional strength (hollow shaft):

$$T = \frac{S_s \pi \left(D_o^4 - D_i^4 \right)}{16 D_o}$$

$$= \frac{63,000(\text{hp})}{N} \tag{5-2}$$

Solid shaft diameter (stiffness):

$$D = \sqrt[4]{\frac{32 T L}{\pi G \theta}} \tag{5-3}$$

Hollow shaft diameters to replace a solid shaft with equal strength but one-half the weight:

$$D_o = \frac{D}{2} + \frac{D\sqrt{2}}{2} \tag{5-4a}$$

$$D_i = \sqrt{D_o^2 - \frac{D^2}{2}} \tag{5-4b}$$

Equivalent twisting moment (solid shaft):

$$T_E = \sqrt{M^2 + T^2} \tag{5-5}$$

Equivalent bending moment (solid shaft):

$$M_E = \frac{M + T_E}{2} \tag{5-6}$$

Combined torsion and bending (solid shaft):

$$D = \sqrt[3]{\frac{16 T_E}{\pi S_s}} \tag{5-7}$$

$$D = \sqrt[3]{\frac{32 M_E}{\pi S}} \tag{5-8}$$

Critical speed, solid shaft (uniformly distributed weight):

$$\omega_c = \sqrt{\frac{384 E I}{5 L^3} \frac{12 g}{W}} \tag{5-9}$$

$$N = 4,270,000 \frac{D}{L^2} \tag{5.10}$$

Critical speed, shaft with disc mounted midway:

$$\omega_c = \sqrt{\frac{576EIg}{WL^3}}$$

(5-11)

Critical speed, shaft with disc mounted off middle:

$$\omega_c = \sqrt{\frac{36EILg}{a^2b^2W}}$$

(5-12)

Critical speed for system:

$$\frac{1}{\omega_{c_s}^2} = \frac{1}{\omega_{c_1}^2} + \frac{1}{\omega_{c_2}^2} + \frac{1}{\omega_{c_3}^2} + \cdots$$

(5-13)

Shear in square shaft (approximate):

$$S_s = \frac{4.8T}{B^3}$$

(5-14)

Stress concentration factor:

$$K_t = \frac{S_{max}}{S}$$

(5-15)

Bolt size (rigid couplings):

$$D = \sqrt{\frac{4T}{\pi S_s r_1 n}}$$

(6-1)

Drive pin in double shear:

$$D = \sqrt{\frac{2T}{\pi r S_s}}$$

(6-2)

Key size:

$$A = Lw = \frac{P}{S_s}$$

(6-3)

Torque capacity (spline):

$$T = 1000hLNr_m$$

(6-4)

Weld treated as line:

$$F = \frac{V}{L}$$

(6-5)

$$F = \frac{M}{Z}$$

(6-6)

Pulley speeds:

$$\frac{D_1}{D_2} = \frac{N_2}{N_1} \tag{7-1}$$

Belt contact:

$$\text{contact} = 180° \pm 2 \text{ arc sin} \left[\frac{(D_1/2) - (D_2/2)}{C} \right] \tag{7-2}$$

Length of belt:

$$L = 2C + 1.57(D_1 + D_2) + \frac{(D_1 - D_2)^2}{4C} \tag{7-3}$$

Belt pull ratios:

$$\frac{F_1}{F_2} = e^{f\theta} \tag{7-4}$$

Centrifugal force (belts):

$$F_c = \frac{wbtv^2}{2.68} \tag{7-5}$$

Net belt tension:

$$F_1 - F_2 = bt\left(S - \frac{wv^2}{2.68}\right)\left(\frac{e^{f\theta} - 1}{e^{f\theta}}\right) \tag{7-6}$$

Belt power equation:

$$\frac{550(\text{hp})}{v} = bt\left(S - \frac{wv^2}{2.68}\right)\left(\frac{e^{f\theta} - 1}{e^{f\theta}}\right) \tag{7-7}$$

V-flat drive:

$$\frac{F_1 - F_c}{F_2 - F_c} = e^{f\theta/\sin(\alpha/2)} \tag{7-8}$$

Roller chain length (approximate):

$$L = 2C + \frac{N + n}{2} + \frac{(N - n)^2}{4\pi^2 C} \tag{8-1}$$

Roller chain center distance (approximate):

$$C = \frac{L - \dfrac{N + n}{2} + \sqrt{\left(L - \dfrac{N + n}{2}\right)^2 - 8\dfrac{(N - n)^2}{4\pi^2}}}{4} \tag{8-2}$$

Roller chain center distance:

$$C = \frac{L - n\left(\dfrac{90 - \alpha}{180}\right) - N\left(\dfrac{90 + \alpha}{180}\right)}{2 \cos \alpha} \tag{8-3}$$

Radial drum pressure:

$$P = \frac{L}{RD} \tag{8-4}$$

Cable-reel capacity:

$$L = \frac{0.2618Wh(B + h)}{d^2} \tag{8-5}$$

Block brake:

$$F = \frac{T}{r} \quad \text{and} \quad N = \frac{F}{f} \tag{9-1}$$

Band brake:

$$P = \frac{T}{cr}\left(\frac{a - e^{f\theta}b}{e^{f\theta} - 1}\right) \tag{9-2}$$

Caliper disc brake:

$$T = 2Prf \tag{9-3}$$

Braking time:

$$T = \frac{(WR^2)N}{308t} \tag{9-4}$$

Cone clutch:

$$T = \frac{Pfr_m}{\sin \alpha} \quad \textit{Uniform wear} \tag{10-1}$$

$$T = pb\pi D_m f\left[\frac{2}{3}\left(\frac{r_o^3 - r_i^3}{r_o^2 - r_i^2}\right)\right] \quad \textit{Uniform pressure} \tag{10-2}$$

Disc clutch:

$$T = fP\left[\frac{2}{3}\left(\frac{r_o^3 - r_i^3}{r_o^2 - r_i^2}\right)\right] \quad \textit{Uniform pressure} \tag{10-3}$$

$$T = fP\left(\frac{r_o + r_i}{2}\right) \quad \textit{Uniform wear} \tag{10-4}$$

Multiple-disc clutch:

$$T = fPn\left[\frac{2}{3}\left(\frac{r_o^3 - r_i^3}{r_o^2 - r_i^2}\right)\right] \qquad \textit{Uniform pressure} \qquad \textbf{(10-5)}$$

$$T = fPn\left(\frac{r_o + r_i}{2}\right) \qquad \textit{Uniform wear} \qquad \textbf{(10-6)}$$

Band clutch:

$$\text{hp} = \frac{Nr}{63,000}\left(F_1 - \frac{F_1}{e^{f\theta}}\right) \qquad \textbf{(10-7)}$$

Square thread torque (raising):

$$T_f = \frac{WD_m}{2}\left(\frac{L + \pi fD_m}{\pi D_m - fL}\right) \qquad \textbf{(11-1)}$$

Square thread torque (lowering):

$$T_f = \frac{WD_m}{2}\left(\frac{L - \pi fD_m}{\pi D_m + fL}\right) \qquad \textbf{(11-2)}$$

Acme thread torque (raising):

$$T_f = \frac{WD_m}{2}\left(\frac{0.968L + \pi fD_m}{0.968\pi D_m - fL}\right) \qquad \textbf{(11-3)}$$

Acme thread torque (lowering):

$$T_f = \frac{WD_m}{2}\left(\frac{0.968L - \pi fD_m}{0.968\pi D_m + fL}\right) \qquad \textbf{(11-4)}$$

Trapezoidal metric thread (raising):

$$T_f = \frac{WD_m}{2}\left(\frac{0.966L + \pi fD_m}{0.966\pi D_m - fL}\right)(10^{-3}) \qquad \textbf{(11-5)}$$

Trapezoidal metric thread (lowering):

$$T_f = \frac{WD_m}{2}\left(\frac{0.966L - \pi fD_m}{0.966\pi D_m + fL}\right)(10^{-3}) \qquad \textbf{(11-6)}$$

American Standard thread torque (raising):

$$T_f = \frac{WD_m}{2}\left(\frac{0.866L + \pi fD_m}{0.866\pi D_m - fL}\right) \qquad \textbf{(11-7)}$$

American Standard thread torque (lowering):

$$T_f = \frac{WD_m}{2}\left(\frac{0.866L - \pi fD_m}{0.866\pi D_m + fL}\right)$$

(11-8)

Collar torque:

$$T = Wr_f f$$

(11-9)

Translation-screw efficiency:

$$e = \frac{WL(100)}{Pr(2\pi)}$$

(11-10)

Diametral pitch:

$$P = \frac{t}{D}$$

(12-1)

Circular pitch:

$$P_c = \frac{\pi}{P}$$

(12-2)

Addendum:

$$a = \frac{1}{P}$$

(12-3)

Dedendum:

$$d = \frac{1.157}{P}$$

(12-4)

Base-circle diameter:

$$D_B = D\cos\phi$$

(12-5)

Module (metric):

$$m = \frac{D}{t}$$

(12-6)

Addendum (metric):

$$a = m$$

(12-7)

Dedendum (metric):

$$b = 1.25\,m$$

(12-8)

Circular pitch (metric):

$$P_c = m\pi$$

(12-9)

Center distance for spur gears (metric):

$$C = \frac{m(t_1 + t_2)}{2}$$
(12-10)

Outside diameter (metric):

$$D_o = D + 2m$$
(12-11)

Contact ratio:

$$CR = \frac{\sqrt{R_o^2 - R_B^2} + \sqrt{r_o^2 - r_B^2} - C\sin\phi}{P_c \cos\phi}$$
(12-12)

Gear ratios:

$$R = \frac{D_f}{D_d} = \frac{t_f}{t_d} = \frac{N_d}{N_f}$$
(12-13)

Gear-train values:

$$TV = \frac{t_d}{t_f} = \frac{D_d}{D_f} = \frac{N_f}{N_d}$$
(12-14)

Planetary gear train:

$$N_s = (TV)N_r + N_c[1 - (TV)]$$
(12-15)

Tangential force (gears):

$$F = \frac{63,000(\text{hp})}{Nr}$$
(12-16)

Separating force (gears):

$$Q = F\tan\phi$$
(12-17)

Total load:

$$P = \frac{F}{\cos\phi}$$
(12-18)

End thrust (helical gear):

$$Q_e = F(\pi D / L)$$
(12-19)

Bevel-gear ratio (external):

$$A = \arctan\left(\frac{\sin C}{R + \cos C}\right)$$
(12-20)

Bevel-gear ratio (internal):

$$A = \arctan\left(\frac{\sin C}{R - \cos C}\right) \tag{12-21}$$

Beam strength (spur gears):

$$W_b = SP_c by \tag{12-22}$$

Power (spur gear):

$$\text{hp} = \frac{SbYkV}{33,000P} \tag{12-23}$$

AGMA strength horsepower:

$$\text{hp} = \frac{N_p D_p S_a bJK_v K_L}{126,000 PK_m K_s K_o K_R K_T} \tag{12-24}$$

AGMA durability horsepower:

$$\text{hp} = \frac{N_p bC_v I}{126,000 C_s C_m C_f C_o} \left(\frac{S_a D_p C_L C_H}{C_p C_T C_R}\right)^2 \tag{12-25}$$

Circular pitch (helical gear):

$$P_{cn} = P_c \cos \psi \tag{12-26}$$

Diametral pitch (helical gear):

$$P_n = \frac{P}{\cos \psi} \tag{12-27}$$

Jerk:

$$J = \frac{da}{dt} \tag{13-1}$$

Helical spring load:

$$P = \frac{S_s \pi d^3}{8KD_m} \tag{14-1}$$

Helical spring deflection:

$$\Delta = \frac{8PD_m^3 n}{Gd^4} \tag{14-2}$$

Helical spring constant or gradient:

$$k = \frac{P}{\Delta} = \frac{Gd}{8C^3 n} \tag{14-3}$$

Equivalent spring (series):

$$k_e = \frac{1}{(1/k_1) + (1/k_2) + (1/k_3) + \cdots}$$

(14-4)

Equivalent spring (parallel):

$$k_e = k_1 + k_2 + k_3 + \cdots$$

(14-5)

Impingement load on springs:

$$W(d + \Delta) = \frac{P\Delta}{2}$$

(14-6)

Graduated-leaf cantilever spring:

$$S = \frac{6PL}{nbt^2}$$

(14-7)

$$\Delta = \frac{6PL^3}{nbt^3E}$$

(14-8)

Semielliptical leaf spring:

$$P = \frac{Snbt^2}{3L}$$

(14-9)

$$\Delta = \frac{2L^2SK}{tE}$$

(14-10)

Torsion bar:

$$k = \frac{T}{\theta} = \frac{\pi D^4 G}{32L}$$

(14-11)

Belleville springs:

$$P = \frac{4E\Delta}{M(1 - \mu^2)(D)^2}\left[\left(h - \frac{\Delta}{2}\right)(h - \Delta)t + t^3\right]$$

(14-12)

$$S = \frac{4E\Delta}{M(1 - \mu^2)(D)^2}\left[C_1\left(h - \frac{\Delta}{2}\right) + C_2 t\right]$$

(14-13)

Coefficient of fluctuation:

$$C_f = \frac{N_1 - N_2}{N} = \frac{v_1 - v_2}{v}$$

(15-1)

Flywheel energy:

$$\Delta E = \frac{W}{2g}\left(v_1^2 - v_2^2\right)$$

(15-2)

$$\Delta E = \frac{W}{g} C_f v^2 \tag{15-3}$$

Flywheel effect:

$$E = 0.000171 N^2 (WR^2) \tag{15-4}$$

Flywheel torque:

$$T = \frac{0.0391(N_2 - N_1)(WR^2)}{t} \tag{15-5}$$

Flywheel hoop stress:

$$S = \frac{12wv^2}{g} \tag{15-6}$$

Flywheel bending stress:

$$S = \frac{12wv^2 \pi^2 D}{gn^2 t} \tag{15-7}$$

Total flywheel stress:

$$S = \frac{3wv^2}{g} \left(3 + \frac{\pi^2 D}{n^2 t} \right) \tag{15-8}$$

Elliptical-arm design:

$$a = D \sqrt[3]{\frac{3S_s}{8Sn}} \tag{15-9}$$

Solid-disc flywheel:

$$W = \frac{4g(\Delta E)}{v_1^2 - v_2^2} \tag{15-10}$$

Shear-pin torque:

$$T = \frac{S_s \pi D^2 r}{4} \tag{16-1}$$

Thin-walled cylinders (based on circumferential stress):

$$t = \frac{pD}{2S} \tag{16-2}$$

Thin-walled cylinders (based on longitudinal stress):

$$t = \frac{pD}{4S} \tag{16-3}$$

Thick-walled cylinders:

$$S_1 = p\left(\frac{a^2}{b^2 - a^2}\right) \tag{16-4}$$

$$S_2 = p\left[\frac{a^2(b^2 + r^2)}{r^2(b^2 - a^2)}\right] \tag{16-5}$$

$$S_3 = p\left[\frac{a^2(b^2 - r^2)}{r^2(b^2 - a^2)}\right] \tag{16-6}$$

Rectangular flat plates simply supported at four edges:

$$t = ab\sqrt{\frac{p}{2S(a^2 + b^2)}} \tag{16-7}$$

Rectangular flat plates fixed at edges:

$$t = ab\sqrt{\frac{p}{3S(a^2 + b^2)}} \tag{16-8}$$

Circular flat plates:

$$t = \frac{d}{2}\sqrt{\frac{p}{S}} \tag{16-9}$$

Dished heads:

$$t = \frac{5pr}{6S} \tag{16-10}$$

Euler formula:

$$\frac{P}{A} = \frac{\pi E}{(L/k)^2} \tag{16-11}$$

J. B. Johnson formula:

$$P = AS_y\left[1 - \frac{S_y(L/k)^2}{4c\pi^2 E}\right] \tag{16-12}$$

Velocity head:

$$H = \frac{V^2}{2g} \tag{17-1}$$

Pump efficiency:

$$e = \frac{\text{water hp}}{\text{brake hp}}(100) \tag{17-2}$$

Pump horsepower:

$$\text{brake hp} = \frac{Qp(100)}{1715e} \tag{17-3}$$

$$\text{brake hp} = \frac{QH(\text{specific gravity})(100)}{3960e} \tag{17-4}$$

$$\text{brake hp} = \frac{(\text{watts input})e}{746(100)} \tag{17-5}$$

Appendix L

The following design-related computer programs are written in the FORTRAN language with DO loops.

Collar Friction Radius and Frictional Torque

$$T_f = \frac{2}{3}\left(\frac{r_o^3 - r_i^3}{r_o^2 - r_i^2}\right) fW$$

LIST

```
100      PROGRAM FRICRAD(INPUT,OUTPUT,TAPE5=INPUT,TAPE6=OUTPUT)
110      DIMENSION TORFIC(10),RADFIC(10)
120      DATA W,COEF,TWO/500.,.15,.66/
130  C PROGRAM CALCULATES THE FRICTIONAL RADIUS AND FRICTIONAL
140  C TORQUE,INNER AND OUTER RADII ARE INCREASED BY AN
150  C INCREMENT OF HALF AN INCH. STARTING AT 1 AND 3
160  C INCHES RESPECTIVELY.
170  C ************************************************************
180  C PROGRAM FOR MACHINE DESIGN COURSE
190  C PROGRAMMER
200      RO=3.0
210      RI=1.0
220      DO 15, K=1,10
230      RF=TWO*((RO**3-RI**3))/(RO**2-RI**2)
240      TORFIC(K)=RF*COEF*W
250      RADFIC(K)=RF
260      RO=RO+.5
270      RI=RI+.5
280   15 CONTINUE
290      WRITE(6,18)
300   18 FORMAT(10X,'FRICTIONAL RADIUS AND FRICTIONAL TORQUE')
310      WRITE(6,19)
320   19 FORMAT(10X,'OF COLLARS INNER AND OUTER RADIUS INCREASED',
330      + 2X,'AT HALF IN. INCREMENTS'//)
340      WRITE(6,20)
350   20 FORMAT(5X,'FRICTIONAL RADIUS',10X,'FRICTIONAL TORQUE'/)
360      DO 35 K=1,10
370      WRITE(6,22)RADFIC(K),TORFIC(K)
380   22 FORMAT(10X,F12.2,18X,F12.2/)
390   35 CONTINUE
400      STOP
410      END
```

```
PUNCH
READY.

PRINT
READY.

PW=132
PW ACCEPTED..
RUN


   FRICRAD    14:59    FTN5

            FRICTIONAL RADIUS AND FRICTIONAL TORQUE
            OF COLLARS INNER AND OUTER RADIUS INCREASED  AT HALF IN. INCREMENTS


        FRICTIONAL RADIUS         FRICTIONAL TORQUE

                2.15                    160.88

                2.61                    195.53

                3.08                    231.00

                3.56                    266.95

                4.04                    303.19

                4.53                    339.63

                5.02                    376.20

                5.51                    412.88

                6.00                    449.63

                6.49                    486.43

    READY.
```

Horsepower (Based on Torque and Rotational Speed)

```
LIST

100        PROGRAM CHEVAL(INPUT,OUTPUT,TAPE5=INPUT,TAPE6=OUTPUT)
110        DIMENSION HPOWER(10),TORID(10)
120        DATA ANGVEL,CONST/1000.,63000./
130        TORQUE=1000.
140 C ***********************************************
150 C PROGRAM FOR MACHINE DESIGN COURSE PRINTS A TABLE
160 C OF HORSEPOWER AND TORQUE AS THE TORQUE IS
170 C INCREASED IN INCREMENTS OF 1000.,FROM
180 C 1000 TO 10000 IN-LB OF TORQUE ANGULAR
190 C VELOCITY IS CONSTANT AT 1000 RPM.
200 C ***********************************
210        DO 15 K=1,10
220 C LOOP INCREMENTS TORQUE
230        HP=ANGVEL*TORQUE/CONST
240        TORID(K)=TORQUE
250        HPOWER(K)=HP
260        TORQUE=TORID(K)+1000.
270    15 CONTINUE
280 C TABLE AND RESULTS ARE OUTPUT
290        WRITE(6,18)
300    18 FORMAT(2X,//////)
310        WRITE(6,20)
320    20 FORMAT(10X,'TABLE OF INCREASE IN HORSEPOWER'/
330        + 7X,'TORQUE BEING INCREASED IN INCREMENTS OF 1000.'//)
340        WRITE(6,25)
350    25 FORMAT(15X,'TORQUE',15X,'HORSEPOWER'/)
360        DO 30 K=1,10
370        WRITE(6,27)TORID(K),HPOWER(K)
380    27 FORMAT(16X,F6.0,10X,F12.2/)
390    30 CONTINUE
400        STOP
410        END

RUN


CHEVAL    10:43   FTN5


          TABLE OF INCREASE IN HORSEPOWER
       TORQUE BEING INCREASED IN INCREMENTS OF 1000.
```

TORQUE	HORSEPOWER
1000.	15.87
2000.	31.75
3000.	47.62
4000.	63.49
5000.	79.37
6000.	95.24
7000.	111.11
8000.	126.98
9000.	142.86
10000.	158.73

```
READY.
```

Diameter of Solid Steel Torsion Shaft

$$D = \sqrt[3]{\frac{16T}{\pi S_s}}$$

LIST

```
100          PROGRAM SHAFT(INPUT,OUTPUT,TAPE5=INPUT,TAPE6=OUTPUT)
110          DIMENSION DIAM(10),TORID(10),HORSE(10)
120  C  ***************************************************
130  C  MACHINE DESIGN PROGRAM FOR CALCULATING THE DIAMETER
140  C  OF A SOLID STEEL SHAFT AISI 1050 PSI IS ASSUMMED.
150  C  THE USER INPUTS TEN DIFFERENT RPM AND HORSE POWERS
160          DO 25 K=1,10
170          WRITE(6,7)
180        7 FORMAT(5X,'PLEASE ENTER THE HORSEPOWER AND THE RPM',
190       +  2X,'OF THE SHAFT AS REAL NOS.'/)
200          READ(5,*)HP,RPM
210          TORQUE=63000.*HP/RPM
220          D1=(16.*TORQUE)/(3.1416*6000.)
230          D2=ALOG10(D1)
240          CUBLOG=D2/3.0
250          D3=10.0**CUBLOG
260          DIAM(K)=D3
270          TORID(K)=TORQUE
280          HORSE(K)=HP
290       25 CONTINUE
300          WRITE(6,26)
310       26 FORMAT(1X,/////)
320          WRITE(6,28)
330       28 FORMAT(10X,'TABLE OF TORQUE HORSEPOWER AND',
340       +  2X,'SHAFT DIAMETER IN INCHES'/)
350          WRITE(6,30)
360       30 FORMAT(10X,'TORQUE',11X,'HORSEPOWER',5X,'SHAFT DIAMETER'/)
370          DO 40 K=1,10
380          WRITE(6,35)TORID(K),HORSE(K),DIAM(K)
390       35 FORMAT(10X,F6.0,11X,F9.2,8X,F12.3/)
400       40 CONTINUE
410          STOP
420          END
```

```
RUN

SHAFT     11:14    FTN5

     PLEASE ENTER THE HORSEPOWER AND THE RPM. OF THE SHAFT AS REAL NOS.
? 2.,200.
     PLEASE ENTER THE HORSEPOWER AND THE RPM  OF THE SHAFT AS REAL NOS.
? 3.,300.
     PLEASE ENTER THE HORSEPOWER AND THE RPM  OF THE SHAFT AS REAL NOS.
? 4.,400.
     PLEASE ENTER THE HORSEPOWER AND THE RPM  OF THE SHAFT AS REAL NOS.
? 5.,500.
     PLEASE ENTER THE HORSEPOWER AND THE RPM  OF THE SHAFT AS REAL NOS.
? 6.,800.
     PLEASE ENTER THE HORSEPOWER AND THE RPM  OF THE SHAFT AS REAL NOS.
? 7.,1000.
     PLEASE ENTER THE HORSEPOWER AND THE RPM  OF THE SHAFT AS REAL NOS.
? 8.,1200.
     PLEASE ENTER THE HORSEPOWER AND THE RPM  OF THE SHAFT AS REAL NOS.
? 9.,1300.
     PLEASE ENTER THE HORSEPOWER AND THE RPM  OF THE SHAFT AS REAL NOS.
? 10.,1500.
     PLEASE ENTER THE HORSEPOWER AND THE RPM  OF THE SHAFT AS REAL NOS.
? 12.,1780.
```

TABLE OF TORQUE HORSEPOWER AND SHAFT DIAMETER IN INCHES

TORQUE	HORSEPOWER	SHAFT DIAMETER
630.	2.00	.812
630.	3.00	.812
630.	4.00	.812
630.	5.00	.812
473.	6.00	.737
441.	7.00	.721
420.	8.00	.709
436.	9.00	.718
420.	10.00	.709
425.	12.00	.712

```
READY.
```

```
RUN

SHAFT      11:17    FTN5
        PLEASE ENTER THE HORSEPOWER AND THE RPM  OF THE SHAFT AS REAL NOS.
? 5.,200.
        PLEASE ENTER THE HORSEPOWER AND THE RPM  OF THE SHAFT AS REAL NOS.
? 10.,500.
        PLEASE ENTER THE HORSEPOWER AND THE RPM  OF THE SHAFT AS REAL NOS.
? 15.,800.
        PLEASE ENTER THE HORSEPOWER AND THE RPM  OF THE SHAFT AS REAL NOS.
? 20.,1100.
        PLEASE ENTER THE HORSEPOWER AND THE RPM  OF THE SHAFT AS REAL NOS.
? 25.,1400.
        PLEASE ENTER THE HORSEPOWER AND THE RPM  OF THE SHAFT AS REAL NOS.
? 30.,1700.
        PLEASE ENTER THE HORSEPOWER AND THE RPM  OF THE SHAFT AS REAL NOS.
? 35.,2000.
        PLEASE ENTER THE HORSEPOWER AND THE RPM  OF THE SHAFT AS REAL NOS.
? 40.,2300.
        PLEASE ENTER THE HORSEPOWER AND THE RPM  OF THE SHAFT AS REAL NOS.
? 45.,2600.
        PLEASE ENTER THE HORSEPOWER AND THE RPM  OF THE SHAFT AS REAL NOS.
? 50.,2900.
```

```
        TABLE OF TORQUE HORSEPOWER AND  SHAFT DIAMETER IN INCHES
        TORQUE          HORSEPOWER    SHAFT DIAMETER

          1575.            5.00            1.102

          1260.           10.00            1.023

          1181.           15.00            1.001

          1145.           20.00             .991

          1125.           25.00             .985

          1112.           30.00             .981

          1103.           35.00             .978

          1096.           40.00             .976

          1090.           45.00             .975

          1086.           50.00             .973
READY.
```

```
RUN

SHAFT     11:10    FTN5

     PLEASE ENTER THE HORSEPOWER AND THE RPM  OF THE SHAFT AS REAL NOS.
? 5.,500.
     PLEASE ENTER THE HORSEPOWER AND THE RPM  OF THE SHAFT AS REAL NOS.
? 10.,1000.
     PLEASE ENTER THE HORSEPOWER AND THE RPM  OF THE SHAFT AS REAL NOS.
? 15.,1200.
     PLEASE ENTER THE HORSEPOWER AND THE RPM  OF THE SHAFT AS REAL NOS.
? 18.,1300.
     PLEASE ENTER THE HORSEPOWER AND THE RPM  OF THE SHAFT AS REAL NOS.
? 19.,1500.
     PLEASE ENTER THE HORSEPOWER AND THE RPM  OF THE SHAFT AS REAL NOS.
? 20.,1600.
     PLEASE ENTER THE HORSEPOWER AND THE RPM  OF THE SHAFT AS REAL NOS.
? 21.,1900.
     PLEASE ENTER THE HORSEPOWER AND THE RPM  OF THE SHAFT AS REAL NOS.
? 22.,2000.
     PLEASE ENTER THE HORSEPOWER AND THE RPM  OF THE SHAFT AS REAL NOS.
? 24.,2500.
     PLEASE ENTER THE HORSEPOWER AND THE RPM  OF THE SHAFT AS REAL NOS.
? 26.,3000.
```

```
         TABLE OF TORQUE HORSEPOWER AND  SHAFT DIAMETER IN INCHES
```

TORQUE	HORSEPOWER	SHAFT DIAMETER
630.	5.00	.812
630.	10.00	.812
788.	15.00	.874
872.	18.00	.905
798.	19.00	.878
788.	20.00	.874
696.	21.00	.839
693.	22.00	.838
605.	24.00	.801
546.	26.00	.774

```
READY.
```

ANSWERS TO SELECTED PROBLEMS

Chapter 2

1. 15,000 psi; next to wall
2. 5" from top; neutral axis
3. 0.25
4. 16.6 N · m; 27.1 m/s
5. 16 590 N
6. 23,600 lb
7. 5.00 in.
8. 50 fpm; 78.5 fpm; 50 rpm
9. 245 lb
10. midpoint; rt. end; left end; decreased
11. 0.201 in.
12. 144 MN/m² at base
13. 190 MN/m² at midpoint
14. 830 lb
15. 15,300 psi
16. 6370 psi
17. 21.2 in.
18. 7515 ft-lb
19. 3751 lb; 4249 lb
20. 3150 lb
21. 40 km/h
22. AE = 15 450 N; AB = 14 925 N; DE = 15 450 N; BE = 0; BD = 8000 N; BC = 21 853 N
23. 1.67 m/s²

24. 18 850 N
25. 0.25
26. 10 053 N
27. 205 GN/m²
28. 6 kg
29. 38.4 mm
30. 2.04 mm
31. 176.8 rpm; 707 rpm
32. 2767 N
33. 59 800 N
34. 15.3 N/mm²
35. 107 MN/m²
36. 127 MN/m²; 13.8 mm
37. 812 fpm
38. 1730 lb; 1470 lb
39. $\bar{x}$ = 13.6 mm; $\bar{y}$ = 26.1 mm
40. 20.62 lb; 17.31 lb; 11.13 lb

Chapter 3

1. 36 lb
2. Yes; 441 N > 350 N
3. 22.1 N
4. 0.00187 hp
5. 7.5 in-lb
6. 0.0925 hp

7. 0.476 hp
8. 50 lb
9. 41° API
10. 54.9 lb; 3.1 lb
11. 4.08 in.
12. 1309 lb
13. 36 lb
14. 7482 N
15. approx. 650 lb; 350 lb
16. 93.2 watts
17. 5.85 N
18. 60 lb
19. approx. 95 lb
20. 112 lb
21. approx. 47 lb
22. 0.523 kW
23. 29.7 N
24. 0.44 hp
25. 0.925 hp
26. 12.9×10^{-6} m²/s
27. 3.02 N
28. 140 lb
29. approx. 860 N

Chapter 4

1. 85 mm; 2.36
2. 2 MN/m²
3. 1.82 MN/m²
4. 71.4 mm
5. 200 psi; 1.78
6. 2583 N
7. 373 psi; *B*
8. 4.53 in.; 37.4 psi
9. 172 °F
10. 1.21 Btu/min.
11. 5630 lb; MR-14
12. 1440 lb
13. 0.55×10^{6} Pa
14. 43.3 N
15. 1442 N; 80 N · m
16. 88,194 miles

17. 4938 hours; approx. 5000 hours
18. 0.727 MPa; 0.91

Chapter 5

1. 0.27 in.
2. 77.86 MN/m²
3. 6.62 mm
4. 41.5 mm
5. 1.32 in.
6. 0.342 in.
7. 37.4 mm
8. 126 mm
9. 110 mm
10. 104 mm
11. 0.567°
12. 0.584 hp
13. 3.37 in.
14. 3.21°
15. 1010 psi
16. 11,460 psi
17. 2370 rpm
18. 2370 rpm
19. 91.1 kW; 70.7 kW
20. 6406 psi
21. 2040 rpm
22. 2070 rpm
23. 283 psi
24. 23,000 psi
25. 0.0975 hp
26. 8150 psi
27. 3.84 in.; 5.79 in.
28. 3.52 in.
29. 0.525 hp
30. 4.5°
31. 2.32 in.
32. 2360 psi
33. 1.34 in.; 1.09 in.
34. 6.57 hp
35. 0.0219°
36. 111%; 7.41%; 11.1%

37. 2.20 in.; 2.25 in.; 0.716°;
    5630 psi; 12
38. 24.1 mm
39. 13.6 N · m
40. 46.0 MN/m^2
41. 58.9 MN/m^2
42. 34.1 mm; 0.104 radians
43. 11.1 mm; 11.6 mm
44. 2.85 in.
45. 172.5 MN/m^2; 102.3 MN/m^2
47. 29.4 mm; 26.7 mm

Chapter 6

1. 7630 psi; 2 and 4
2. 1810 psi; 707 lb; 0.128 in.
3. 3.35 hp
4. 123 lb
5. 0.200 in.; 5/16 in. setscrew
6. 157.6 lb; 1/4 in. setscrew
7. 0.629 hp
8. 889 watts
9. 131 N; 524 N; reduce
10. 120 lb; 40 lb
11. No shear on bolt B
12. 76 lb
13. 7.33 in.; 3.30 in.
14. 11.25 in.; 23,000 lb
15. 13,800 lb
16. 0.177 in.; or 3/16 in.
17. 2830 lb
18. 144×10^{-3} kN · m
19. 60 kgf/mm^2; 48 kgf/mm^2;
    589 MPa; 471 MPa
20. 10.9 mm
21. 4.68 or 5
22. 50.7 MN/m^2
23. 24 N · m
24. 7.86 or 8
25. 0.0219 MN

Chapter 7

1. 175.2°
2. 172.4 in.
3. 281.4 in.
4. 288 rpm
5. 15.7 hp
6. 4.44 in. or 5 in.; 5.33 in. or 6 in.
7. 20.2 hp
8. 6.24 hp
9. 8.20 in.
10. 4.56 or 5 belts
11. 0.00904 hp
12. 69.4 hp
13. 2.47 in.
14. 147.9 in.; 40.2 hp
15. 3252 fpm; 1.36 or 2 belts
16. 2126 fpm; 2.69 or 3 belts
17. 3710 mm; 3707 mm
18. 0.00283 kW

Chapter 8

1. 26 teeth; 50 teeth, type B
2. 20 teeth, type B
3. 15.9 hp
4. 187 N
5. 20.9 lb
6. 2120 lb
7. 3300 lb; 4670 lb
8. 597 lb
9. 12°
10. 4040 lb
11. 751 N
12. 1300 N
13. 5380 N
14. 1710 lb
15. 1.21 tons; $FS = 9.5$
16. 28.6 ft
17. 2.15°; no
18. 1.09 tons
19. 15.9 in.

20. $A = B = C = 2943$ N
21. $AB = 1666$ lb; $BC = 1116$ lb;
    $BD = 2157$ lb
22. 5.61

Chapter 9

1. 189 lb
2. 214 lb
3. 0.329
4. 40.5 N $\cdot$ m
5. 13.4 N
6. 492 in-lb
7. 5.2 lb
8. 1.28 in.
9. 57,500 psi; increase thickness
10. 0.28
11. 173 ft-lb
12. 12.9 mm
13. 0.322
14. 25 N
15. 96.3 lb-ft^2
16. 0.35; 1.34 kW

Chapter 10

1. 3068 in-lb
2. 370 N
3. 142 N $\cdot$ m
4. 498 in-lb
5. 2.29 hp
6. 91.4 lb
7. 0.318
8. 1.01 in.
9. 27.5 hp
10. 24 lb
11. 622 N $\cdot$ m
12. 46.7 N $\cdot$ m
13. 1.69 kW
14. 19.0 kW
15. 7.69 kW
16. 2.55 kW

Chapter 11

1. 0.75 in.; 5 thds/in.; 0.4 in.
2. 705 in-lb
3. 45.9 in-lb
4. 111 lb
5. 51%
6. 15.4 lb
7. 21.5%
8. 4.36 lb
9. 15.6 lb
10. 6550 lb
11. 182 in-lb
12. 5/6 fpm
13. 12.3 lb
14. 49.3 in-lb
15. 4.38 hp
16. 1382 in-lb
17. 2231 psi
18. 1020 psi
19. 6.50 N $\cdot$ m
20. 45.13 N $\cdot$ m
21. 10.62 in-lb
22. 33.26 N $\cdot$ m; 10.18 MN/m^2;
    7.85 MN/m^2; 39.1
23. 37.58 MN/m^2

Chapter 12

1. 0.785 in.; 0.250 in.; 0.289 in.;
   8.000 in.; 0.539 in.; 0.039 in.;
   7.745 in.
2. 2; 10 in.
3. 2.17
4. 2.500 in.; 3.750 in.
7. 3.30 lb
8. 8.07 hp
9. 6.73 hp
10. 47.6 hp; 10.4 hp
11. 13.3 hp
12. 5.91 hp
13. 88 mm; 22 teeth; 4 mm; 5 mm
14. 40 teeth; 60 teeth

15. 25.6 mm; 56.0 mm; 39.2 mm;
    2.513 mm

16. 1.63

17. 222 rpm

18. 48.2 lb

19. 15, 45, 15, 60 or other combinations

Chapter 14

1. 10 lb/in.; 0.116 in.; 0.928 in.;
   32.6 coils

2. 14.7 lb

3. 20 in.

4. 5 in.

5. 0.416 in.; 2.912 in.; 6.54 coils

6. 0.328 in.; 2.296 in.; 20.9 coils

7. 76,800 psi

8. 1.37 in.

9. 0.512 in.; 19,200 psi

10. 4.47 lb; 0.447 lb

11. 0.147 in.; 92,160 psi

12. 11,200 psi

13. $\dfrac{P}{\Delta} = \dfrac{nbt^3E}{6L^3K}$

14. $\dfrac{P}{\Delta} = \dfrac{nbt^3E}{12L^3K}$

15. 0.438; 3.57 in.

16. 0.0649 in.

17. 6.00 kg

18. 389 mm

19. 125 mm

20. 21.8 mm

21. 171.6 N; 98.7 N

22. 5 N/mm

23. 3.25 mm; 22.8 mm; 11.7 coils

24. 0.725 radians; 16.66 N · m/radian

Chapter 15

1. 334 lb; 1.07 in.

2. 5952 ft-lb

3. 932 lb

4. 74 lb

5. 0.793 in.; 3.18 in.

6. 382 psi; 394 psi

7. 1.87 in.^{2}

8. 2685 ft-lb

9. 97 in-lb

10. 192 rpm

11. 70.7 fps

12. 72.6 fps

13. 808 psi

14. 3650 psi

15. 7500 psi

16. 1.634 in.; 0.817 in.

17. 0.965 in.; 0.483 in.

18. 396 lb

19. 0.192

21. 1.75 lb at 2 in. radius; 3.50 lb at
    2 in. radius

22. 28.5 mm

23. 47.4 mm; 94.9 mm

24. 44.2 mm

Chapter 16

2. 9000 psi

3. 14,300 psi; 36,500 psi; 8000 psi

4. 15 teeth

5. 6.32 hp

6. 87.7 psi

7. 0.152 in.; increase it

8. 17,000 psi; 5000 psi; 2780 psi;
   2000 psi

9. 0.56 in.

10. 96; 8030 lb

11. 17.6 mm

12. 2.8 mm

13. 17 100 N

14. 2.56 mm

15. 16.9 mm

16. 2.68 MPa

17. 1.608 MPa

Chapter 17

1. 5.51 bars; 68.9 bars; 1.01 bars; 0.149 bars
2. 551 000 Pa; 6 890 000 Pa; 101 000 Pa; 14 900 Pa
3. 3.17 liters/m
4. 14.2 dm^3/s
5. 1005 lb
6. 309 N
7. 628 000 N
8. 0.009 37 MN; 0.009 11 MN
9. 45 MPa

INDEX